Carnivores in Ecosystems

Carnivores in Ecosystems: The Yellowstone Experience

Edited by   Tim W. Clark

A. Peyton Curlee

Steven C. Minta

Peter M. Kareiva

Yale University Press

New Haven and London

Set in Swift and Nobel type by The
Composing Room of Michigan, Inc.,
Grand Rapids, Michigan.
Printed in the United States of America
by BookCrafters, Inc., Chelsea, Michi-
gan.

Library of Congress Cataloging-in-
Publication Data

Carnivores in ecosystems : the
    Yellowstone experience / edited by
    Tim W. Clark . . . [et al.].
        p.   cm.
    Includes bibliographical references
    and index.
    ISBN 0-300-07816-1 (cloth : alk.
    paper)
    1. Carnivora—Yellowstone
National Park.   I. Clark, Tim W.
QL737.C2C345   1999
599.7′09787′52—dc21   99-28229

A catalogue record for this book is available from the British Library.

The paper in this book meets the guidelines for permanence and durability of
the Committee on Production Guidelines for Book Longevity of the Council
on Library Resources.

10 9 8 7 6 5 4 3 2 1

Contents

vi   **List of Figures**

viii   **List of Tables**

xi   **Preface**

1   **1 A Model Ecosystem for Carnivores in Greater Yellowstone**
Tim W. Clark, Steven C. Minta, A. Peyton Curlee, and Peter M. Kareiva

11   **2 Greater Yellowstone Carnivores: A History of Changing Attitudes**
Paul Schullery and Lee H. Whittlesey

51   **3 Yellowstone Bears**
Richard R. Knight, Bonnie M. Blanchard, and Paul Schullery

77   **4 The Ecology of Anthropogenic Influences on Cougars**
Kerry M. Murphy, P. Ian Ross, and Maurice G. Hornocker

103   **5 Wolves in the Greater Yellowstone Ecosystem: Restoration of a Top Carnivore in a Complex Management Environment**
Douglas W. Smith, Wayne G. Brewster, and Edward E. Bangs

127   **6 Coyotes and Canid Coexistence in Yellowstone**
Robert L. Crabtree and Jennifer W. Sheldon

165   **7 Mesocarnivores of Yellowstone**
Steven W. Buskirk

189   **8 Predicting the Effects of Wildfire and Carnivore Predation on Ungulates**
Francis J. Singer and John A. Mack

239   **9 Small Prey of Carnivores in the Greater Yellowstone Ecosystem**
Kurt A. Johnson and Robert L. Crabtree

265   **10 Evaluating the Role of Carnivores in the Greater Yellowstone Ecosystem**
Mark S. Boyce and Eric M. Anderson

285   **11 Genetic Considerations for Carnivore Conservation in the Greater Yellowstone Ecosystem**
F. Lance Craighead, Michael E. Gilpin, and Ernest R. Vyse

323   **12 Carnivore Research and Conservation: Learning from History and Theory**
Steven C. Minta, Peter M. Kareiva, and A. Peyton Curlee

405   **List of Contributors**

415   **Index**

Figures

2   **1.1 The Greater Yellowstone Ecosystem**

56   **3.1 Grizzly bear age structure**

59   **3.2 Grizzly bear litter size frequency**

60   **3.3 Correlation between litter size and age of female grizzly bears**

80   **4.1 Birth months of cougar litters**

113   **5.1 Wolf pack territories**

149   **6.1 Seasonal biomass intake by coyotes**

174   **7.1 Body-mass distributions of herbivorous and predatory mammals**

190   **8.1 Map of study area and elk summer and winter home ranges and migration routes**

206   **8.2 Increases in five species of ungulates following cessation of human controls**

208   **8.3 Changes in niche and diet overlaps among ungulates following population release from artificial controls**

211   **8.4 Projections of elk populations following the fires of 1988**

222   **8.5 Compensation between winter malnutrition losses of elk calves and wolf predation, one hundred wolves**

222   **8.6 Scenarios of compensation among sources of mortality and reduced harvests of elk, one hundred wolves**

223   **8.7 Compensation between winter malnutrition losses of elk calves and wolf predation, seventy-eight wolves**

224   **8.8 Scenarios of compensation among sources of mortality and reduced harvests of elk, seventy-eight wolves**

226   **8.9 Possible changes to ungulates and other predators following return of a keystone predator**

227   **8.10 An ecosystem macrohypothesis for changes in elk, vegetation, climate, and wolves**

267   **10.1 Functional responses showing the number of prey killed per predator per year as a function of prey abundance**

270   **10.2 Dynamics of a deterministic vegetation-elk interaction**

271   **10.3 Dynamics of a deterministic three-species system to anticipate dynamics of the vegetation-elk-wolf interaction**

272   **10.4 Dynamics of the three-species system plotted in figure 10.3, but with 20 percent stochastic variation in the carrying capacity for vegetation**

272   **10.5 Autocorrelation functions for vegetation, elk, and wolves from the time series plotted in figure 10.4**

273   **10.6 Dynamics of the three-species system plotted in figure 10.3, but with 20 percent stochastic variation in the functional response for predators**

274   **10.7 Autocorrelation functions for vegetation, elk, and wolves from the time series plotted in figure 10.6**

275   **10.8 Autocorrelation functions for elk and wolves from the time series generated by program WOLF5**

276   **10.9 Autocorrelation functions for moose and wolves**

277   **10.10 Autocorrelation functions for detrended counts of elk**

298   **11.1 Greater Yellowstone Ecosystem, showing national park and forest lands, the grizzly bear recovery zone, and bear dispersal distances**

301   **11.2 Grizzly bear recovery zones**

361   **12.1 Hypothesized strength of site selection by a common carnivore occupying a forested system at different spatial scales**

Tables

19   **2.1 Reports of herbivores in documentary record before 1882**

20   **2.2 Reports of carnivores in documentary record before 1882**

26   **2.3 Predator control in Yellowstone, 1904–35**

58   **3.1 Comparison of adult female grizzly bear weights and conditions**

58   **3.2 Comparison of weights and conditions of radio-marked research vs. management grizzly bears**

64   **3.3 Comparison of weights and conditions of three- and four-year-old female grizzly bears**

88   **4.1 Sources of death among adult and subadult radio-collared cougars**

115   **5.1 Predictions and appraisals of Yellowstone National Park wolf restoration**

130   **6.1 Comparison of coyote studies**

135   **6.2 Comparison of predation rates by large predators**

148   **6.3 Coyote food habits**

166   **7.1 Articles, monographs, and theses dealing with mesocarnivores**

173   **7.2 Observed or inferred interference competition involving mesocarnivores**

177   **7.3 Species richness, ranges of body weights, ratios of largest to smallest, and three measures of postcranial morphological diversity of some families of North American Carnivora**

191   **8.1 Numbers of ungulates available to carnivores**

192   **8.2 Biomass of ungulates available to predators before the fires of 1988**

199   **8.3 Sources of elk and caribou calf mortality during the first year of life**

220   **8.4 Predicted number of ungulates preyed upon by wolves**

242   **9.1 Scientific names and adult body weights of small mammals**

247   **9.2 Small mammals caught per minigrid in Yellowstone's Northern Range**

254   **9.3 Known and suspected small mammal–carnivore prey relations**

269   **10.1 Definitions and values of variables used during simulation trials**

278   **10.2 Regression analysis of herbivore per capita growth rate**

286   **11.1 Body size and estimated number of carnivores**

300   **11.2 Home range/territory size and dispersal distances of carnivores**

302   **11.3 Summary of carnivore genetic status**

Preface

The carnivores of the Greater Yellowstone Ecosystem are some of the most magnificent animals to grace our planet, and they inhabit an ecosystem renowned as the site of the world's first national park. There is much to be learned from an overview of these species in this region, and this book focuses on their current and past status, management, and conservation in one of the largest relatively intact northern temperate ecosystems. Some of these species have been continuously researched for almost forty years, and today many of them are under scrutiny by researchers using sophisticated ecological theory and field methods. All have been managed or affected by humans in various ways, and all remain under intense pressure from human influences, direct and indirect.

Carnivores in ecosystems: The Yellowstone experience is written by the researchers studying these animals in this singular ecosystem. The book is an outgrowth of a special symposium held in 1995, on the last day of Yellowstone National Park's third biennial scientific conference, which focused on Yellowstone's predators. The conference provided an ideal opportunity to carry out a major synthesis and analysis of the long-term research on the ecosystem's large carnivores. Yellowstone is remarkable among ecological research areas for the long-term data that have been collected on the resident large carnivores. It is unusual to have such long-term data for an entire carnivore guild. To use this information in the most effective and efficient manner for conservation and management, we need to synthesize it appropriately. This book is the beginning of such a synthesis.

We thank the contributors to this volume who shared their considerable talent and hard-earned experience. We are grateful to the many people who served as reviewers and made great improvements.

This book would not have been possible without the commitment and generous support of the Merrill G. and Emita E. Hastings Foundation, National Fish and Wildlife Foundation, the Nature Conservancy, Gilman Ordway, Catherine Patrick, J. Douglas Smith, the Wiancko Charitable Foundation, Wildlife Conservation Society, the Yellowstone Association, the Yellowstone Park Foundation, and an anonymous donor. Many thanks to John Varley and Paul Schullery of Yellowstone National Park, and the Northern Rockies Conservation Cooperative for their sponsorship and support of this project. The Fanwood Foundation, New-Land Foundation, and the Henry P. Kendall Foundation all provided the Northern Rockies Conservation Cooperative with general support funding that helped us carry out this project. Steve Minta was supported by the University of California, Santa Cruz, and by the National Council for Air and Stream Improvement, Peter Kareiva by the University of Washington, and Tim Clark by Yale University.

Several people were instrumental in helping us conceptualize the book, and their insights greatly increased the rigor of our writing and editing. They include Michael Soulé, William Settle, Marc Mangel, Kim Heinemeyer, and especially Richard Reading. Fred Allendorf and Mike Phillips helped us with the tough topics of conservation genetics and wolves. D. F. Lott, R. G. Schwab, and D. S. Zezulak were essential to our understanding of individual and social variation in carnivores. Denise Casey provided outstanding editing and helped us pull all the chapters together into a cohesive book.

Many thanks, too, to Jean Thomson Black, Dan Heaton, and the staff at Yale University Press for their considerable help with this project.

Carnivores in Ecosystems

Lynx and snowshoe hare (John Weaver)

Chapter One

A Model Ecosystem for Carnivores in Greater Yellowstone

Tim W. Clark, Steven C. Minta, A. Peyton Curlee, and Peter M. Kareiva

Carnivores of the Greater Yellowstone Ecosystem (GYE) include grizzly and black bears, wolves, mountain lions, and many smaller species, such as coyotes, red foxes, lynx, bobcats, wolverines, river otters, fisher, American martens, badgers, skunks, and weasels. The GYE, almost eight million hectares in the Rocky Mountains of Wyoming, Montana, and Idaho, is relatively wild and intact (figure 1.1). It is the southernmost point on the North American continent that still contains the full suite of carnivores present before Euro-Americans arrived. To the east and south, many of these species have been lost from their former ranges. This species assemblage and this ecosystem are globally unique today for this and other reasons. It is imperative to learn how these splendid animals interact with their world and how we can share an increasingly human-dominated landscape with them. Understanding and maintaining their ecology is a scientific, management, and conservation challenge of immense proportions. This book is about this cumulative challenge, what we know about these species, and what we might want to do in the future to conserve this unique biological heritage.

Carnivores in wildlands have always captured our imagination, and our fascination has been played out in oral stories, art, literature, and in private and government policies and programs. Historians have hypothesized that many English words have roots in ancient words relating to bears, which humans have long idolized and emulated, perhaps finding a better reflection of them-

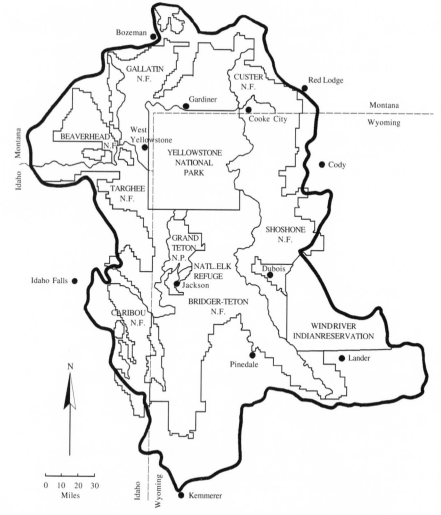

1.1. The Greater Yellowstone Ecosystem in the Rocky Mountains of the United States.

selves in bears than in any other animal (Shepard and Sanders 1985, Rockwell 1991). Bears and wolves are considered helpers, teachers, and brethren by many Native American peoples and are respected for their skill, cunning, and strength (Lopez 1979, Clark and Casey 1994).

The same qualities have made large carnivores formidable competitors with and even enemies of humans. Euro-American settlers in North America went to great lengths to eliminate large carnivores from their environment (Casey and Clark 1996). They were so effective in their eradication campaigns, which lasted until the middle decades of the twentieth century, that bears have been reduced to 2 percent of their historic range in the continental United States, and wolves, mountain lions, coyotes, lynx, wolverines, and other species have all been affected in dramatic ways. Many pressures remain, including direct killing, habitat usurpation, and competition for prey and space.

Some people still see a threat in these remnant carnivore populations and wildlands, but the image of these species is changing. Many people now see carnivores in positive and protective ways (Schmidt 1990, Kellert et al. 1996). Through the 1973 U.S. Endangered Species Act (ESA), society has made an explicit commitment to preserve even those threatened or endangered species that pose hazards to humans or may require considerable effort and accommodation to ensure their long-term survival. In 1995 wolves were reintroduced to Yellowstone National Park, a sign of our intention to reverse earlier attempts to destroy them.

Within the GYE, bears especially have taught us new ways of looking at our environment through their wide-ranging behavior. Based on years of fieldwork, pioneering grizzly bear researchers Frank and John Craighead were the first to relate bears' use of habitat in the Yellowstone region to a greater ecosystem model (Craighead 1979). Their notion of a "grizzly bear ecosystem" that includes Yellowstone and Grand Teton national parks and the surrounding national forests was the basis for our recognition in recent decades of a bioregion that we now call the Greater Yellowstone Ecosystem (Clark and Harvey 1988). This notion is growing and influencing national and international policy and management of carnivores and wild ecosystems (Golley 1993, Gunderson et al. 1995, Boyce and Haney 1997).

The GYE is characterized by large mammals such as grizzly bears, elk, and bison, extensive coniferous forests, large-scale natural processes such as wildfire, and the headwaters of three major watersheds (Clark and Minta 1994). The relatively high altitude and northern latitude of much of the ecosystem preclude it from having species richness comparable to the tropics, yet it is known for its great abundance of certain species. Seven floras converge at this spot in the Continental Divide and 1,700 plant species occur here. Ten fish, twenty-four amphibian and reptile, three hundred bird, and seventy mammal species reside here, as well as unknown thousands of invertebrates (Clark et al. 1989). Yellowstone has the largest elk herds in North America and one of the last free-ranging bison herds. It is also the site of the largest functional geothermal basin in the world.

The range of Yellowstone's grizzly bears links most of the habitats and associated species of the GYE. Bears taught us to look at the landscape in a way that is relevant for natural processes and native species rather than just for human land ownership and authority. Wolves, mountain lions, and other wide-ranging carnivore species have reinforced that lesson. A single wolf has been recorded traveling as far as 917 kilometers. Mountain lions have been known to travel 480 kilometers, bears 300 kilometers, wolverines 378 kilometers, and lynx as far as 1,100 kilometers (see Chapter 11, table 11.2). Reintroduced wolves and their offspring in the Yellowstone ecosystem are showing us once more that our national park and other geopolitical boundaries are meaningless in their search for habitat, and these boundaries are insufficient if we want these magnificent carnivores to have populations that are viable for the long term.

More than any other animals, carnivores have forced us to move from ecosystem theory to ecosystem management and conservation (see Keiter and Boyce 1991). They constitute one of the most concrete elements that we can use to "operationalize" in a practical fashion the past few decades of theoretical advances in ecology. Carnivores represent an element that many humans strongly identify with and cherish. Ecosystems are highly complex and multi-layered—perhaps too complex for us to truly understand or to model in a representative fashion. Thus carnivores and their interactions with the abiotic and biotic elements of their environment are useful proxies for understanding ecosystem function and signaling breaks in the integrity of the system (Gittleman 1996).

Perhaps it is because carnivore research has been driven largely by management objectives that ecosystem management and conservation biology will further develop and codify this approach. Legislation and regulation like the National Environmental Policy Act (NEPA) require us to attend to cumulative effects, though we still have difficulty modeling and analyzing multiple management actions and environmental changes. The ESA requirements for delisting populations are formidable because we must tend to habitat and landscape capability as part of a defendable population viability analysis (PVA). This will prove especially important for wide-ranging carnivores, and some of the best-known cumulative-effects modeling is focused on grizzly bears in the Yellowstone ecosystem (see Chapter 3).

Carnivores also underscore the limitations—only recently acknowledged—

in our abilities to control and predict change in our environment. The Clementsian view of a stable, distinct ecosystem has given way to a view of ecosystems as characterized by energy flows, material cycles, disturbance, and shifting mosaics (Borman and Likens 1979, Golley 1993). Many of our ecosystems may be more typified by episodic chaos than by stable equilibria. Confidence in our ability to control such systems has also been questioned. We no longer believe in our unlimited abilities to redirect rivers, stop recurrent natural wildfire, or restore critically endangered species. And we are realizing that by trying to stop natural processes, we may be causing many more long-term problems than we are alleviating (Keiter and Boyce 1991). Certain species and assemblages are adapted to such disturbance regimes, and without them biodiversity may become impoverished, and certain environmental processes, such as water purification, may become faulty.

The ecosystem concept began as a focal shift from species to systems and from science vs. management to science *and* management. Indeed, at large scales, science and management merge, as do experiment and art. We match management domains with study domains because every action is an experiment with both ecological and social consequences. In other words, researchers need to study those processes and features that are important to managers but that conform to experimental method and measurement. Although an operational definition of ecosystem management remains ambiguous and begs for prototypical management experiments, the overarching goal is sustainability of constituents and processes. We are increasingly able to offer a descriptive definition that emphasizes complexity, connectedness, scale, context, dynamism in space and time (disturbance, pattern, and probabilism), uncertainty, adaptability, and functional redundancy as a safety valve for evolutionary change.

Our goal for this book is to begin integration and synthesis of the remarkable scientific legacy that exists for GYE's carnivores and to evaluate how we can best position ourselves to meet the growing conservation challenge in this and other ecosystems around the world. Our traditional research and management focus on single species, specific sites, and short periods does not allow us to conceptualize or meet the task before us fully and practically. This is not to say that we have not had some impressive successes—note how quickly wolves seem to be reestablishing themselves in the GYE. However, given the current trends of human population growth, development, and habitat fragmentation, our task promises to get harder, not easier. Competition for land and natural resources is increasing.

Even in an area like the GYE that has a considerable amount of protected land, we will have a difficult time accommodating all the multiple, mushrooming demands for resources while preserving native biodiversity at a functional, not merely a residual, level. Taken together, the counties that make up the GYE have the highest rates of population growth in the United States. Some of the native species we are so proud of persist still in the GYE only at low numbers. For instance, 168 plant, thirty-two bird, eighteen mammal, twelve fish, and at least six amphibian and reptile species are at levels that may make them vulnerable to extinction (Clark et al. 1989). Every day we have evidence of the uncertain coexistence between wildlife and humans. As we drafted this introduction, six threatened grizzly bears were inadvertently killed by elk hunters and two wolves were hit by cars in the GYE within two weeks, and appeals were pending in a court ruling that the wolf reintroduction violated the intent of the ESA.

It is fitting that we do our best to figure out how to preserve intact ecosystems and diverse, functional wildlife assemblages in the GYE. Despite the region's history of preservation, we can no longer rest assured that if we fail here, people somewhere else—Canada or Alaska—will "do the job" to save these species and their ecosystems. It is becoming more and more evident that the fate of large carnivores in the United States depends on how well we manage the whole chain of ecosystems in the Rocky Mountains of North America (Clark et al. 1996). This lesson is also true for carnivore populations in other parts of the world, such as Asian tigers and South and Central American panthers. The GYE is increasingly becoming an island ecosystem, and many wildlife populations will not remain viable in the long term without some exchange or migration between populations in other ecosystems. Island populations are much more vulnerable to extinction from demographic, genetic, and environmental events, as the essays in this book well document.

Carnivore conservation will strongly test our abilities and commitment, but it also may act as a stimulus to improved human organization, policy, and management for better conservation outcomes. Although we cannot protect all species under the umbrella of large carnivores, we will protect many of them, along with many of the natural ecosystem processes that support native biodiversity and ecological integrity worldwide.

With those goals in mind, we invited prominent researchers from the GYE and farther afield to contribute their ideas about the status and future directions of our quest to understand carnivores and ecosystem function. This book provides an overview of the science and conservation of carnivores in the Greater Yellowstone Ecosystem. Despite the great variation in carnivore behavior, morphology, and ecology, and the great variation in study methods employed by contributing researchers, there are a number of themes that recur throughout this book.

Perhaps foremost is the difficulty of studying these often secretive, patchily distributed, and even dangerous animals. A few carnivores, such as wolves, are located readily and can be monitored reliably for demographic trends, but many cannot even be monitored reliably for a snapshot in time, much less with sufficient certainty to project population trends. After decades of research and many significant accomplishments, we still have large gaps in our knowledge about resource or habitat requirements, demographic trends, and tolerance of certain types of natural and human-caused disturbance. Hence we are left with the need for proxies and conceptual, analytic tools that can compensate for some of this uncertainty.

The chapters of this book also illustrate that we need a greatly improved understanding of the relation between carnivores and other species present in this and other ecosystems. Whereas we may have once sought a universal law for community or ecosystem function and structure, we now appreciate the variety of processes at play and are more interested in understanding what various perturbations may mean for ecosystem function and composition. In some cases it appears that if you remove or greatly reduce the ecosystem effects of a top predator, you will have a consequent loss of species diversity at other trophic levels. At the same time, ostensibly subtle changes at the foundation of the food web—vegetation and habitat—can greatly affect carnivores and other ecosystem consumers. The study of carnivores emphasizes that we need to do a better job of integrating various trophic levels in our research projects, and we need to improve our ability to model the complex reality of ecosystems.

A further weakness brought to light by this book is the need for new ways to integrate knowledge across disciplines and the importance of case history and comparative method approaches to informed synthesis and speculation, something anathema to many traditional scientists who only give credence to positivistic "purity" in research. That benchmark is not only often impossible to achieve in carnivore and ecosystem research, it may actually hinder further enlightenment because it is unsuited to clarifying the very sorts of nested, fluctuating relations we are seeking to understand. Instead of fairly competing experts being encouraged to learn each other's disciplines in order to communicate and promote synthetic and catalytic marriages of ideas, we compartmentalize knowledge, resources, and the education of professionals who will face real-world problems far more convoluted than their methods and disciplinary backgrounds can accommodate.

The pitfalls of a narrow disciplinary focus have been readily identified by many who have studied carnivores. Carnivore ecologists have had such trouble studying their subjects that they have embraced any technology or method that helps them understand the extraordinary variation exhibited by their subjects. The chapters in this book testify to the diversity of disciplines and methods embraced by reseachers. We encouraged synthesis and speculation from each contributor because we wanted to allow leading reseachers a free rein in getting across what they judge are the key components for appreciating ecosystems by way of understanding carnivores. The book itself uses a case-history approach, with each species account emphasizing an ecosystem perspective. In addition, synthetic chapters address such ecosystem themes as historical analysis, human attitudes and interactions, trophic connections, species restoration and coexistence, and evolutionary and genetic considerations.

The final overarching theme of the book, which is so ably summarized by contributors in their histories of carnivore eradication and recent restoration efforts, is the role of human attitudes and behavior as the ultimate arbiters of carnivore persistence. Our success or failure at conserving carnivores will result from how we conceptualize and organize our efforts, how society directs us in this values-driven endeavor, and how well we learn, individually and collectively. The social, economic, and policy aspects of carnivore conservation are covered elsewhere by other authors (e.g., Clark et al. 1996), though they deserve ongoing, serious, systematic attention. With this book we attempt the first step in integrating and synthesizing the reliable scientific knowledge base, which we hope will be used practically by managers, decision makers, and conservationists with great effect.

An overall challenge remains in practice and in policy—to conserve carnivores in viable populations and distributions in ways that enjoy broad, enduring public support. The biological sciences and study of carnivores and ecosystems are necessary to meet this challenge, but they are not sufficient. To continue progress toward effective carnivore conservation, we must significantly improve efforts and means to integrate what we already collectively know with new knowledge as it becomes available. This is an interdisciplinary challenge of integrating science, management, and policy. Although this book is largely focused on biological knowledge, what is needed extends far beyond that. It must include knowledge about the social sciences as it relates directly to carnivores and ecosystems—how people value them, how their economic behavior and social and political organization affect them. We must integrate knowledge about how management programs are devised and implemented, how professional systems are organized, and how organizations are structured and operated. We must learn how policies are developed, carried out, and evaluated—specifically, how problems are created and defined, how knowledge is used, and how policy learning and innovation take place. Integration across disciplines, at large scales, can guide future research and practical management and policy. But these objectives cannot be achieved on their own: people need skills as well as institutional backing to accomplish these ends. Both are currently in short supply in the GYE. As the contributions to this book testify, there have been efforts in recent years to organize and integrate knowledge about carnivores and ecosystems. These ongoing efforts have all been helpful to our current understanding of these subjects. We call for a continuation and acceleration of this trend and its extension to these difficult but less-well-studied needs. But in addition, we call for innovation to bring new practices to the forefront to advance our common goal of sustaining carnivores and ecosystems, as we are required by our own profession and by society. The carnivores of Greater Yellowstone are a unique system of global and timeless value. They must be protected.

Literature Cited: Borman, F. H., and G. E. Likens. 1979. Pattern and process in a forested ecosystem. Springer-Verlag, New York.

Boyce, M. S., and A. Haney. 1997. Ecosystem management: Applications for sustainable forest and wildlife resources. Yale University Press, New Haven.

Casey, D., and T. W. Clark. 1996. Tales of the wolf: Fifty-one stories of wolf encounters in the wild. Homestead Press, Moose, Wyoming.

Clark, T. W., and D. Casey. 1994. Tales of the grizzly: Thirty-nine stories of grizzly bear encounters in the wilderness. Homestead Press, Moose, Wyoming.

Clark, T. W., and A. H. Harvey. 1988. Management of the Greater Yellowstone Ecosystem: An annotated bibliography. Northern Rockies Conservation Cooperative, Jackson, Wyoming.

Clark, T. W., A. H. Harvey, R. D. Dorn, D. L. Genter, and C. Groves, eds. 1989. Rare, sensitive, and threatened species of the Greater Yellowstone Ecosystem. Northern Rockies Conservation Cooperative, Jackson, Wyoming.

Clark, T. W., and S. C. Minta. 1994. Greater Yellowstone's future: Prospects for ecosystem science, management, and policy. Homestead Press, Moose, Wyoming.

Clark, T. W., P. C. Paquet, and A. P. Curlee, eds. 1996. Large carnivore conservation in the Rocky Mountains of the United States and Canada. Conservation Biology 10:936–1058.

Craighead, F. C., Jr. 1979. Track of the grizzly. Sierra Club Books, San Francisco.

Gittleman, J. L., ed. 1996. Carnivore behavior, ecology, and evolution, vol. 2. Cornell University Press, Ithaca.

Golley, F. B. 1993. A history of the ecosystem concept in ecology. Yale University Press, New Haven.

Gunderson, L. H., C. S. Holling, and S. S. Light, eds. 1995. Barriers and bridges to the renewal of ecosystems and institutions. Columbia University Press, New York.

Keiter, R. B., and M. S. Boyce, eds. 1991. The Greater Yellowstone Ecosystem: Redefining America's wilderness heritage. Yale University Press, New Haven.

Kellert, S. R., M. Black, C. R. Rush, and A. J. Bath. 1996. Human culture and large carnivore conservation in North America. Conservation Biology 10:977–90.

Lopez, B. 1978. Of wolves and men. Charles Scribner's Sons, New York.

Rockwell, D. 1991. Giving voice to bear: North American Indian rituals, myths, and images of the bear. Roberts Rinehart Publishers, Niwot, Colorado.

Schmidt, K. J. 1990. Conserving Greater Yellowstone: A teacher's guide. Northern Rockies Conservation Cooperative, Jackson, Wyoming.

Shepard, P., and B. Sanders. 1985. The sacred paw: The bear in nature, myth, and literature. Viking Penguin, Toronto.

Feeding black bears in Yellowstone National Park in the early 1900s (Yellowstone National Park Archives, used by permission)

Chapter Two

Greater Yellowstone Carnivores: A History of Changing Attitudes

Paul Schullery and Lee H. Whittlesey

Yellowstone National Park (YNP) was established by an act of Congress on March 1, 1872 (Haines 1977). Created at the dawn of the modern wildlife conservation movement, the park attracted the attention and passions of each succeeding generation of conservationists and ecologists, both as a focus of the conservation struggle and as an unparalleled opportunity for learning about wildlands and their management. Yellowstone has repeatedly been the site of experiments with new policy directions on behalf of the national park system, and each new stage of its management has resulted in ever more public and professional attention to its activities.

Predator, or carnivore, management has passed through several stages in Yellowstone, each one reflecting national attitudes. Because of its unique combination of exclusive legal jurisdiction (which gives park administrators extraordinary freedom to act), its international constituency (which counterbalances mere local interests), and its largely intact native biota, Yellowstone became embroiled in a series of wildlife management controversies that eventually blended into one integrated controversy with numerous elements, most of which have some bearing on predator management.

We begin this essay by summarizing the known prehistory of Yellowstone predators and the early historical record of large mammals in the park area. We then review the halting and often contentious development of a wildlife management policy for YNP and the National Park Service (NPS), with special attention to case studies in the evolution of predator policy as a reflection of changing public attitudes. The rise of ecological studies as a significant factor in policy development, as exemplified in the work of the Murie brothers and the Craighead brothers, serves to exemplify NPS growing pains in predator management. Until late 1998 the agency had no formal mandate for science and yet could not achieve its mission of conservation without scientific research. Manager-scientist relations have been uneasy and often stormy. In the concluding sections of the essay, we suggest that managers' deliberations on predator management have largely been driven by ungulate management issues and, more recently, by changing concepts of nature. As a result, Yellowstone's predators today, though given a great deal more respect than in earlier times, are still perceived through a different set of perceptual filters than are herbivores and still rely upon a more overt and conscious tolerance by the public.

The Greater Yellowstone Landscape Before 1872

During the 12,000–14,000 years since the end of the last ice age, changing environmental conditions have caused well-documented variations in the natural community of Yellowstone (Hadly 1990, Engstrom et al. 1991, Meyer et al. 1992, Whitlock and Bartlein 1993, Barnosky 1994). Although it is now certain that Yellowstone's landscape has undergone dramatic changes over the past 12,000 years, central to the management debates for many years have been competing theories on the makeup of Yellowstone's faunal community before the park's establishment in 1872. Paleoecological research indicates that while members of the major ecological communities persist over thousands of years in Yellowstone, their relative abundance varies, primarily with climatic changes (Barnosky 1994, Whitlock and Bartlein 1993). Thus, although wildlife or vegetation communities may not be stable in the simplistic sense of all species maintaining consistent and unchanging abundance and distribution, they are stable in the sense of persistence in the ecosystem.

But there have long been disagreements about the presence and abundance of large mammals in Yellowstone before the park was established. Many theories have been advanced, but these four, with some variations, are probably the most frequently invoked:

1. Large mammals were not native to the Yellowstone Plateau before 1872.
2. Large mammals were rare on the Yellowstone Plateau before 1872.
3. Large mammals used the Yellowstone Plateau as summer range but not as winter range before 1872.
4. Large mammals used the Yellowstone Plateau as summer and winter range before 1872 as environmental conditions allowed.

Obviously, there is considerable overlap among these theories. More important is the long-standing, implicit presumption among many contestants in Yellowstone's management controversies that a single "prehistoric condition" or ecological state will serve us as a baseline for measuring not only modern conditions but also the success of modern management.

PALEONTOLOGICAL STUDIES

A variety of approaches have been used to determine wildlife status in Yellowstone before the park's establishment. Paleontology shows great promise for fleshing out the prehistory of the park area, but paleontological work on mammals in YNP began only in the past decade. (We use the term *prehistory* with growing unease because it is a European concept that does not do justice to the different historical self-perceptions of native cultures.) The first paleontological study of Yellowstone mammalian fauna (Hadly 1990, 1995, Barnosky 1994) was conducted in a cave site whose strata embraced the most recent three thousand years and yielded mammalian faunal remains from "6 orders, 16 families, 31 genera, and 40 species" (Hadly 1995:21). The study suggested that this Lamar Valley site has been occupied by a faunal assemblage similar to that found there today, but that climate changes over the past three thousand years have at times affected the presence and relative abundance of some small mammal species (Barnosky 1994). The site does not contain all species currently present in YNP, but it does contain twenty-eight of the thirty species associated with habitats within range of the site.

Predictably, the vast majority of the mammal bones found at Lamar Cave are those of smaller species, including small predators: pygmy shrew *(Sorex cf. hoyi)*, Merriam's shrew *(Sorex merriami)*, water shrew *(Sorex palustris)*, fox *(Vulpes vulpes)*, striped skunk *(Mephitis mephitis)*, American marten *(Martes americana)*, ermine *(Mustela erminea)*, long-tailed weasel *(Mustela frenata)*, and badger *(Taxidea taxus)*.

The larger predators are also represented. A total of thirty-seven coyote *(Canis latrans)* bones appear in seven of the cave's sixteen identified layers. Bones of the ever-controversial wolf *(Canis lupus)* are also present: twelve wolf bones appear in five layers. Nine grizzly bear *(Ursus arctos)* bones appear in four layers. Black bear *(Ursus americanus)* bones were not identified at Lamar Cave, though two ursid bones of indeterminate species were found. The absence of black bear bones at Lamar Cave probably results in part from its location in habitat types less frequented by black bears (Herrero 1985). Elk *(Cervus elaphus)*, deer *(Odocoileus hemionus)*, bison *(Bison bison)*, and bighorn sheep *(Ovis canadensis)* bones appear throughout the strata as well, indicating the presence of a prey base to support the predators.

ARCHAEOLOGICAL STUDIES

Archaeology has also been applied, but so far only sketchily, to the resolution of questions about Yellowstone wildlife prehistory. Wright (1982:11) described Yellowstone National Park as "one of the poorest known archaeological areas in North America." Cannon (1992:1–248) estimated that 0.19 percent of the park had been "intensively inventoried for archaeological sites." Despite the preliminary state of archaeological investigations in the park and nearby, some writers have emphasized the shortage of ungulate bones in park sites. However, investigations in and near the park in the past few years have revealed ungulate bones in many new sites. Cannon (1992) summarized investigations along the shore of Jackson Lake from 1986 to 1988 that identified bones of bison, elk, moose *(Alces alces),* and mule deer, as well as numerous predator species—canid, grizzly bear, wolverine *(Gulo gulo),* otter *(Lutra canadensis),* mink *(Mustela vison),* skunk, and bobcat *(Felis rufus).* Cannon (1992) also reported on an investigation near Corwin Springs, Montana, just north of the park, that identified elk, bison, deer, and an unknown canid. A bison kill site (human hunting) dating to A.D. 1150 was recently excavated along the shore of Yellowstone Lake (Cannon 1992). Cannon (1991) found a variety of ungulate bones, including elk and bighorn sheep, in park sites more than one thousand years old along the Yellowstone River near Gardiner, Montana. Allen (U.S. Forest Service, personal communication 1992) reported many ungulate bones at a Sphinx Creek site in Yankee Jim Canyon north of the park. Recent analysis of blood residue on prehistoric stone artifacts of many dates over the past nine thousand years from YNP have revealed blood from ursids, felids, and canids, as well as from bison, deer, elk, sheep, and rabbits *(Leporidae* spp.; Cannon and Newman 1994, Cannon 1995, Yellowstone Science 1995). Johnson (1997) reports bison bones in nine different archaeological sites in the park, three in Montana and six in Wyoming. Based on these recent findings, it seems probable that the earlier reports of a lack of ungulates in Yellowstone archaeological sites resulted largely from inadequate investigation.

After reviewing regional archaeological or paleontological research, Frison (1991), Hadly (1990), Cannon (1992), and Kay (1994) have all suggested possible changes in abundance or relative abundance of large ungulates in the mountain West, with possible increases in elk or other ungulates at some time over the past fifteen hundred years. If such changes occurred, they would have affected predators. These authors disagree on causes for the changes, and no doubt the debate will continue.

Archaeological and paleontological reports of YNP faunal remains only serve to demonstrate the presence of various species at times in the past; the studies have not yet been used successfully to compare abundance at some past time with abundance now. This may be regarded as unfortunate, because relative abundance of the large mammals, especially elk and bison, has become a central issue in modern scholarly dialogues over the status of Yellowstone's grazing lands (Houston 1982, Kay 1990, Schullery and Whittlesey 1992, Wagner et al. 1995a, Schullery 1997, Bishop et al. 1997), and of course has direct bearing on our understanding of modern predators in Greater Yellowstone. On the other hand, considering the variety of environmental variables that influence such abundance, even if research could establish precise abundance of wildlife species at some past time, those numbers might not be especially pertinent to the modern situation.

EURO-AMERICAN INFLUENCES BEFORE 1872

A further complication in clarifying the prehistoric status of the mammals of Greater Yellowstone is that the first white observers here did not necessarily see a landscape free of prior influence by their culture or technology. Potential influences included at least the following:

1. European livestock diseases may have reached the region in a variety of ways, including transmission from other wildlife populations that were infected by distant livestock. Such diseases could have affected relative abundance of prey species or abundance of all large prey species, thus affecting predator habits and abundance.

2. European cosmopolitan "crowd diseases," such as smallpox, may have affected regional Native American populations at any time after these diseases were established and spread with devastating effects in eastern North America in the sixteenth and seventeenth centuries, often far outrunning the Euro-Americans who introduced them (Dobyns 1983, Ramenofsky 1987, Verano and Ubelaker 1992, Stannard 1992, Williams 1989, Denevan 1992, Schlesier 1994). Changes in Native American populations would have had consequences of many kinds, including changes in fire regimes and wildlife harvests.

3. At various times after 1492, some eastern Native American groups began westward movements (Haines 1977, White 1978, Waldman 1985, Janetski 1987). These events could have led to significant changes from pre-1492 distribution of people in the Greater Yellowstone Ecosystem, thus affecting human use of the region.

4. The arrival of the horse in the mid-1700s (from Spanish stock, via Mexico) affected Native American groups profoundly (Roe 1955, Waldman 1985). Political balances shifted between groups, and the relations of groups to resources were often altered tremendously (White 1978). The Bannock Indians provide an excellent example, with significant implications for the

Yellowstone landscape. The later arrival of trade firearms (from French-Canadian traders) allowed them greater hunting range and effectiveness, as well as generally increased mobility.

5. Iron tools and weaponry reached many groups well in advance of European settlement (Waldman 1985). Unlike the horse, much of this material arrived from the north, through Canadian trade networks (Waldman 1985). Again, as firearms and other weapons arrived, political balances shifted (Janetski 1987), and in many cases, the relation with huntable wildlife changed.

6. Trade is known to have influenced most Native American groups living in regions with commercially valuable wildlife (Wood 1972). The surviving written record from the earliest known era of trapping activity in Greater Yellowstone, 1820–40 (Haines 1977, Schullery and Whittlesey 1992), contains several references to trading of furs between Native Americans and white trappers.

For now, our understanding of the magnitude or character of each of these six effects is quite limited, even conjectural, but theorizing continues on the influence of Native Americans on wildlife populations in prehistoric Greater Yellowstone. Kay (1994, 1995) hypothesizes that the Native American population of North America was much larger than previously believed, on the order of one hundred million, and that this high number of humans suppressed western wildlife populations to extremely low numbers. However, the citations Kay provides to support his estimate (Dobyns 1983, Ramenofsky 1987) do not propose this number, which is normally associated only with the largest estimates of the Native American population for the entire New World (80 percent of the number being in Central and South America). Dobyns (1983) gives the most generous estimate of the pre-Columbian North American human population that we have found in the literature—about eighteen million. But this number has been called "astronomically high" by Ramenofsky (1987:13), and typical estimates are around seven million (Kennedy 1994). In any event, based on his belief in a pre-Columbian population estimate of one hundred million, Kay hypothesizes that in the seventeenth century the population was decimated by European diseases, thus releasing ungulate populations from "aboriginal overkill" that had kept them extremely low. However, Kay (1990) also maintains that large mammal numbers were extremely low in the YNP area in the century before the creation of the park in 1872; he does not explain what kept the numbers low following his conjectural destruction of Native American populations.

Craighead et al. (1995), on the other hand, based on evidence from other areas, hypothesize that ungulates were abundant enough that Native Americans, by leaving excess meat at their kill sites at buffalo jumps, created "ecocenters" in prehistoric Greater Yellowstone that were heavily used by grizzly bears and other scavengers. Craighead et al. (1995:325) state that bison jump sites existed in counties around present YNP and in areas farther away in Wyoming, and they conclude that "it is probable" that such sites were used by a variety of scavengers, including grizzly bears. These authors attempt to establish the prehistoric existence of "ecocenters" on the same scale of influence as twentieth-century garbage dumps, but their evidence is not persuasive. For example, the only specific data they give on bison meat available to scavengers at a bison jump are for the Wardell site near Big Piney, Wyoming. Here, they emphasize that "more than a thousand bison were killed and rendered" (p. 325), but they also note that the site was active for five hundred years, which yields an average of two bison per year. Given the intermittent human use of bison jump sites, it seems improbable that they could provide the consistent nutrition of a modern garbage dump. But it seems equally probable that grizzly bears and other scavengers benefited occasionally from the inadvertent largesse of human hunters.

DOCUMENTARY RECORDS OF WILDLIFE IN THE NINETEENTH CENTURY

Perhaps the most commonly used means of determining wildlife conditions before the creation of YNP is analysis of early written accounts. Firsthand documentary accounts of the region date only from the early 1800s. By the time Euro-Americans saw what is now YNP, all six of the abovementioned influences could have affected the region. However, these written accounts remain our best source of specific information on wildlife just before the modern era, when YNP visitors and managers began aggressive manipulations of the park setting.

Skinner (1927), Murie (1940), Meagher (1973), Houston (1982), and Kay (1990) are among the many previous investigators to use some portion of the historical record to determine wildlife abundance in YNP in the early 1800s, but none used more than about twenty accounts of the area before 1880 in their analyses. Perhaps not surprisingly, their conclusions disagree strongly, ranging from scarcity to abundance of fauna. However, Schullery and Whittlesey (1992, 1995) have evaluated 168 accounts of the YNP area before 1882 and have reached several conclusions relating to predators:

1. Although the evidence is not detailed enough to make quantitative comparisons between animal numbers in the period 1800–1872 and today, the evidence is overwhelming that large mammals were common and widespread in what is now YNP and the surrounding area. More than 90 percent

of the observers who commented on the abundance of wildlife in the park expressed the belief that it was very abundant.

2. Ungulates were reported widely in present YNP and surrounding areas (table 2.1). Elk, bison, pronghorn *(Antilocapra americana),* and mule deer were common, with elk by far the most commonly observed species in large numbers. Bighorn sheep may have been more abundant then than now. Moose were common only in the southern part of what is now YNP. White-tailed deer were uncommon, and mountain goats *(Oreamnos americanus)* did not occur in what is now YNP.

3. Predators were likewise widely reported (table 2.2). Wolves, mountain lions *(Felis concolor),* grizzly bears, black bears, coyotes, wolverines, and fox were present.

4. Smaller mammals, herbivores and carnivores alike, were rarely mentioned by early observers, except for a few naturalists, but were probably common in appropriate habitats.

5. Portions of what is now YNP were used as winter range by large numbers of ungulates before 1882, presumably in accordance with environmental conditions.

We accept these conclusions for the purposes of this essay.

The Beginnings of Predator Policy

The creation of YNP coincided roughly with the final control and forced settlement of regional Native American groups, though sporadic use of the park area by these groups continued for some years after 1872 (Haines 1977, Janetski 1987). At times in the 1870s and 1880s, park managers complained that Indians were still using the park—ironically, they mentioned the setting of fires by Indians as one of the biggest problems—but after 1872 the only significant human effects on the YNP landscape would be those of Euro-Americans.

THE POLICY VACUUM

The park was created in something of a legal vacuum: there was little precedent to draw from beyond the general guidelines of some other early legislation, including that surrounding the setting aside of the Yosemite Valley, which would later become a national park (Bartlett 1974, Haines 1974, 1977). The Yellowstone Park Act provided only a framework for management, creating boundaries for the park and outlining authorities for its future protection and management. With respect to wildlife, the act required the secretary of the interior only to "provide against the wanton destruction of the fish and game found within said park, and against their capture or destruction for the

Table 2.1

Reports of herbivores in documentary record before 1882

	Sightings of individuals	Sightings of small groups	Sightings of herds	Reports of sounds	Reports of meat, skin, or other parts	Reports of tracks, trails, scat, or other sign	General statements of presence
Elk	43	53	35	9	25	31	84
Bison	6	8	17	—	8	9	26
Moose	4	2	—	2	—	2	23
Mule deer	38	29	3	1	10	18	70
White-tailed deer	1	—	—	—	—	—	10
Bighorn sheep	1	9	15	—	8	9	34
Pronghorn	18	24	24	—	7	1	45
Mountain goat	1	—	—	—	—	—	1
Beaver	—	—	—	1	4	5	21

Table 2.2

Reports of carnivores in documentary records before 1882

	Sightings of individuals	Sightings of small groups	Reports of sounds	Reports of meat, skin, or other parts	Reports of tracks, trails, scat, or other sign	General statements of presence
Wolf	2	2	10	1	2	17
Mountain lion	3	–	9	2	3	17
Coyote	1	–	1	–	–	8
Grizzly bear	14	7	–	–	3	9
Black bear	5	3	–	–	1	7
Bear, unspecified	24	20	–	5	16	47
Wolverine	8	–	–	1	–	7
Fox	4	1	–	–	–	6
Lynx	1	–	–	–	–	2
Bobcat	1	–	–	–	–	–

purposes of merchandise or profit" (Schullery 1997:74). Market hunting was thereby disallowed, but sport hunting, as well as hunting for provision of parties of visitors and residents, was regarded as an appropriate activity in YNP for eleven years. With little or no budget and little legal support, however, early park administrators were unable to stop market hunters from slaughtering park wildlife in great numbers, even if their policy framework had been more explicit or complete.

There may have been a policy vacuum, but there was certainly no attitude vacuum. Late-nineteenth-century America was the scene of profligate waste of natural resources. The infamous destruction of bison herds, known by every modern schoolchild, was mirrored by similar destruction of countless other wildlife populations (Dary 1974). In those places where game species were protected, their predators were persecuted with a religious fervor now hard to understand. It was only natural that YNP should experience the effects of these attitudes and that YNP managers should attempt to formulate their first policies in this social context.

Cahalane (1939:235) noted that "predator management in the parks followed the general trend and pattern of thought of the times." It cannot be overemphasized that YNP policies and programs have always been a reflection of contemporary trends. That YNP has often been at the lead in those trends is an important element of the Yellowstone story, but even at its most pioneering the park could not dramatically outrun its times.

MASSIVE DISTURBANCE OF THE WILDLIFE COMMUNITY, 1872–82

Numerous historians (Beal 1960, Hampton 1971, Reiger 1975, Haines 1977, Bartlett 1985) have described the market-hunting slaughter of YNP wildlife in the 1870s. They have reviewed firsthand accounts of extensive and repeated trips by market hunters that resulted in widespread destruction of wildlife, including thousands of elk and smaller numbers of other species. The occurrence of this several-year event, which was being duplicated throughout the West in that decade, has long seemed an incontrovertible fact among historical scholars. Kay (1990) recently attempted to prove that the widespread slaughter of thousands of park ungulates in the 1870s did not occur, but Schullery and Whittlesey (1992) published numerous local, firsthand, and even participatory reports that established with reasonable certainty that thousands of elk were killed by market hunters in the 1870s in Yellowstone.

It is, however, difficult to judge the effects of the slaughter on park wildlife, beyond generalizations and intriguing conjectures. It is impossible to verify the accuracy of any of these accounts, but it is likewise impossible to ignore the magnitude of the reported kills, especially because they only imitated similar slaughters elsewhere in the West and were therefore not exceptional. Accounts from the winter and spring of 1875 attest to as many as eight thousand elk killed in the northern part of the park (these reports did not include all of the Northern Range, much less all of the park), and various contemporary sources claim the wholesale slaughter of other species (Schullery and Whittlesey 1992). We are unable to authenticate such huge numbers, but harvests in the thousands, especially if sustained over a few years, would have affected predators in profound and complex ways—for one possible example, by providing a huge amount of carrion to some species and then a subsequent shortage of live animals to eat.

But large predators were without friends in late-nineteenth-century America, and their slaughter was incorporated into the process. Late in 1877, Superintendent Philetus Norris announced that the poisoning of wolves and wolverines was under way, and by late 1880, he reported that the destruction of the gray wolf, coyote, and mountain lion was far advanced, if not nearly complete:

> The large, ferocious gray or buffalo wolf, the sneaking, snarling coyote, and a species apparently between the two, of a dark-brown or black color, were once exceedingly numerous in all portions of the Park, but the value of their hides and their easy slaughter with strychnine-poisoned carcasses of animals have nearly led to their extermination.

> [The mountain lion's] tantalizing tendency to start false Indian alarms and stampede the animals has led to the persistent efforts of the mountaineers, with rifle, trap, and poison, to exterminate them, and so successful have their efforts proved that now the comparatively few survivors usually content themselves with slaughter of deer, antelope, and perhaps elk, at a respectful distance from camp. (Norris 1881:42)

Norris was one of very few early observers to leave any summary of the abundance of the smaller predators. He reported that foxes were "numerous and of various colors, the red, grey, black and the cross varieties (most valuable of all) predominating in the order named" (Norris 1881:42). Badgers were numerous, while otters *(Lutra canadensis)*, mink *(Mustela vison)*, martens, ermines, and sables (exactly what he meant by this name is unclear, but it was most likely a misidentified fisher or marten) were not. It may be important that Norris never wintered in the park and so lacked firsthand experience with track observations of some of these rarely seen species. Curiously, he reported that skunks (probably *Mephitis mephitis*) were extremely common, claiming that hundreds had to be slaughtered at Mammoth Hot Springs "before we could sleep peacefully" (Norris 1881:42).

Norris was an unschooled naturalist of unknown biases, so it is difficult to judge his credibility or knowledge in specific cases, but his accounts of the widespread destruction of the larger predators, especially wolves, mountain lions, and coyotes, are in keeping with similar destructive campaigns elsewhere in the West at the time (Curnow 1969, McIntyre 1995). We therefore assume that in the 1870s these three large predators were probably heavily poisoned and trapped.

Norris was also an aggressive and creative manager of wildlife populations, though his methods were as ill-informed as they were crude and imprecise. For example, he encouraged commercial trappers to destroy park beavers in the fear that if they were left unregulated they would flood large portions of the park. His report that "hundreds if not thousands" (Norris 1881:44) of beavers were taken from park waters annually during his administration (which began in 1877) may help answer the question of why the smaller furbearers were rare; a trapping campaign of this intensity was probably not aimed exclusively at beaver. On the other hand, it could be that his communications with trappers, who apparently had a large experience with park furbearers at this time, were the source of his information on the scarcity of these smaller furbearers.

THE NATIONAL PARK AS WILDLIFE RESERVE

Although the Yellowstone Park Act had expressly forbidden hunting for commercial gain, the uncontrolled market hunting of the 1870s apparently prompted Superintendent Norris to publish a regulation forbidding it. The regulation was reproduced in his 1878 annual report (Norris 1878).

The distinction between selling game inside and outside of the park allowed early park concessioners to provide guests with fish and game. Commercial fishing, to provide hotel restaurants with trout dinners, continued until well after the turn of the century (Varley and Schullery 1983). As we have seen, the no-hunting regulation was not backed by adequate law or even by sufficient police presence and was violated on an industrial scale.

Yellowstone National Park was created to protect its unusual geological and thermal features and to allow for their enjoyment by the public. The unstated but clearly appreciated recreational revenue derived from such activity was an underlying but driving force in the passage of the founding legislation (Bartlett 1974, Haines 1974, 1977, Schullery 1997). The eventual result of the destruction of park wildlife in the 1870s was a public campaign to improve park management and protection, a campaign that resulted in an abrupt and far-reaching broadening of the park's mission. Wildlife protection quickly became more significant as an element of YNP management. The way animals were to be "used" or "enjoyed" during the park experience changed, growing ever less consumptive.

In 1883, as Yellowstone's civilian administration struggled to meet the challenges of park management, pressures by conservationists (especially sportsmen) led to a more comprehensive prohibition of hunting in the park (Reiger 1975, Haines 1977). On January 15, 1883, Secretary of the Interior H. M. Teller wrote to the park superintendent as follows: "The regulations heretofore issued by the Secretary of the Interior in regard to killing game in the Yellowstone National Park are amended so as to prohibit absolutely the killing, wounding, or capturing at any time of any buffalo, bison, moose, elk, black tailed or white tailed deer, mountain sheep, Rocky Mountain goat, antelope, beaver, otter, martin, fisher, grouse, prairie chicken, pheasant, fool hen, partridge, quail, wild goose, duck, robin, meadow lark, thrush, goldfinch, flicker or yellow-hammer, blackbird, oriole, jay, snowbird, or any of the small birds commonly known as the singing birds" (Teller 1883:1).

Among the many interesting features of this list, aside from some species that did not appear in the park, is its selectivity. The only carnivores included are otters, martens, and fishers, all of those being commercially useful furbearers; they were obviously included for their commercial value rather than for any appreciation of their worth as predators. All other predators were excluded from protection and were still shot or otherwise killed when possible both by visitors and by park managers (Haines 1977, Schullery 1992).

The slaughter of large numbers of park animals appears to have tapered off by the early 1880s, for a variety of reasons (Norris 1881, Schullery and Whittlesey 1992, Schullery 1995b). Large-volume market hunting was replaced in the 1880s by a type of poaching more like the modern infraction—the taking of a smaller number of relatively more valuable animals, especially bison and the furbearers, although no doubt some level of local meat hunting also persisted. It cannot be assumed, however, that the wildlife species of the park simply resumed their preslaughter relations to one another or to their landscape (Schullery 1997).

THE BEGINNING OF FORMAL PREDATOR CONTROL

When the U.S. Cavalry assumed responsibility for protection of YNP in 1886, the first acting superintendent, Captain Moses Harris, decided to bring predators under the protection of park regulations, but not because of their inherent worth (Harris 1887). Harris had considered the matter because of concerns expressed by observers that the predators were causing harm to the herbivores, but he concluded that allowing the public free use of firearms and traps in the park would cause more harm to the preferred and protected game animals than the harm those animals might receive "from any ravages which might be feared from carnivorous animals" (Harris 1887:14). Harris, incidentally, shared

Norris's aversion to the skunk, singling it out as "chief among" the "noxious animals" not worthy of protection (p. 14).

Concerns over predators and their effects on other wildlife or on humans were expressed more or less continually through the time of the army administration (1886–1918). In 1897 Acting Superintendent S. B. M. Young said that though some people feared that the destruction of coyotes would release the "gopher" population to destroy the grasses, he would prefer to kill some coyotes because of their predation on elk calves, antelope fawns, and grouse. The next year, poisoning of coyotes was under way (Young 1897, Erwin 1898). By 1907 civilian scouts and noncommissioned officers at the soldier stations were "authorized and directed to kill mountain lions, coyotes, and timber wolves" (Secretary of the Interior 1907:21).

The preferred treatment of the smaller carnivores was not as clearly stated; the value of many of them as furbearers probably increased the likelihood of their protection by the army, because army officers were so determined to stop poaching of any kind that this duty would override their dislike of the animals. In some national parks, a variety of smaller carnivores were killed, and we assume that some (Harris's loathed skunks, for example) were killed in Yellowstone as well (Skinner 1927, Cahalane 1939).

Other predators were viewed with more ambivalence. The protection of bears in YNP after about 1886 resulted in a new tourist attraction, bear feeding. At first conducted primarily at garbage dumps near park hotels, but eventually also popular along park roads, bear feeding became institutionalized as one of the primary (and perhaps the most memorable) visitor activities in YNP (Schullery 1991, 1992). But its legality and appropriateness were both problematic for managers, who early on forbade visitors from feeding bears. In 1902 Acting Superintendent John Pitcher produced this circular:

The interference with, or molestation of bear or other wild game in the Yellowstone Park, by ANYONE is absolutely prohibited.

The custom of feeding the bear, on the part of the tourists, hotel employees, or enlisted men, must cease at once, as this practice will sooner or later, surely result in serious injury to some individual.

The bear of the park, while being absolutely wild, are perfectly harmless, but when rendered tame by being fed—in any other way except at the regular garbage piles at the various hotels—lose all fear of human beings and will enter kitchens and camps in search of food, thereby frequently doing considerable damage to property and provisions. (Pitcher 1902:1)

The ambivalence Pitcher expressed toward this charming and dangerous omnivore has never disappeared from bear management in North America. The bear was the first Yellowstone mammalian predator, indeed the first large

Table 2.3

Predator control in Yellowstone, 1904–35

Species	Years	Number taken
Mountain lions	1904–25	121
Wolves	1914–26	136
Coyotes	1907–35	4,352

Sources: Murie 1940, Weaver 1978.

North American mammalian predator, to develop an appreciative constituency of any size, and it seems likely that over time the popularity of the bear not only increased tolerance for its occasional predation on herbivores but also helped pave the way for eventual tolerance of other predators. But in 1902 managers still grappled with puzzlement over an animal that they could simultaneously describe as "absolutely wild," "perfectly harmless," and "rendered tame," all in the same sentence.

Murie (1940) and Weaver (1978) have documented the beginnings and development of systematic control programs for wolves, mountain lions, and coyotes. The decades after the turn of the century were the time of the famous federal predator purges on public lands, and Yellowstone was no exception to the rule (table 2.3). It has often been assumed that this was when Yellowstone lost its wolves and nearly lost its mountain lions, but we suspect that much of the mass destruction of wolves in the Yellowstone region was probably accomplished much earlier, in conjunction with the market hunting of elk and bison, and the poisoning of their carcasses. As we have earlier stated (Schullery and Whittlesey 1992, 1995), the final destruction of wolf pack activity in Yellowstone after the turn of the century may have been a "mopping up" operation of remnant wolves and perhaps some packs that represented a resurgence from the predator killing of the 1870s–90s.

An additional complication of turn-of-the-century wildlife management in Yellowstone was the emergence of a conviction that ungulates either were not native to the park area or occupied it only in summer. By the early 1900s it had become common knowledge among authorities on big game that elk were plains animals, pushed back into the mountains by human settlement. The impressions of dozens of earlier Yellowstone travelers notwithstanding (Schullery and Whittlesey 1992, as reviewed above), this belief was applied to the park. For the first six decades of the twentieth century, the large ungulate prey base of Yellowstone was often seen as a post-1872 addition to the local fauna.

It is perhaps unfair to think of these early YNP wildlife managers and their constituents as entirely primitive in their understanding of ecological processes. In their dialogues and records are occasional glimmerings of ecological thinking (the awareness that coyotes might help control small mammal numbers, for good example). Perhaps foremost among these was their realization that the park was not a complete, self-contained support system for its wildlife. When sportsmen pressured the secretary of the interior to outlaw all hunting in the park, their actions were not entirely altruistic. They already understood the concept of the game reservoir. Efforts beginning in the 1880s to expand park boundaries were largely motivated by the need to protect migratory wildlife (Grinnell 1882, 1883) and the forested headwaters of major river systems (Haines 1977). A primary justification for outlawing hunting in the park was to ensure the perpetual restocking, by migration from the park, of public hunting lands outside the park (Schullery 1997).

National Park Service Policy as a Dynamic Process

After several years of deliberation and growing public pressure, Congress created a National Park Service on August 25, 1916 (Haines 1977). The new agency was given management of a diverse set of reserves, including parks, monuments, and other sites. The transition from military protection of Yellowstone to civilian was hampered by political intrigue, however, as local businessmen who feared loss of revenue from the departing army used their political power to prevent funding of NPS staff in Yellowstone until 1918. During that two-year interim, troops continued to patrol and protect the park.

INHERITED PROGRAMS AND POLICIES

The creation of the NPS did not signal any immediate change in Yellowstone predator management, but the mandate of the new agency did contain the seeds of change. The National Park Service Act (64th Congress, 1st session, H.R. 15522) defined the purpose of the parks as follows: "to conserve the scenery and the natural and historic objects and the wild life therein and to provide for the enjoyment of the same in such manner and by such means as will leave them unimpaired for the enjoyment of future generations." The act left ample room for the removal of certain animals, however, declaring that the secretary of the interior "may also provide in his discretion for the destruction of such animals and such plant life as may be detrimental to the use of any said parks, monuments, or reservations."

A CAUTIOUS TOLERANCE

Curnow (1969) pointed out that the federal predator control program launched in the American West in 1915 occurred after the wolves and many other predators had already been dramatically reduced, but that antiwolf prejudice drove federal funding: "In their efforts to obtain bounty legislation,

stockmen generated a hatred toward the wolf which still remains. The publicity was deliberately exaggerated to obtain favorable bounty legislation to eradicate the wolf. This publicity aggravated the stockmen to such an extent that
their animosity toward the wolf became nearly pathological" (Curnow
1969:88).

This extreme characterization of the people who hated wolves in Montana
probably defined a much broader cross-section of American society than
merely stockmen; these attitudes were the product of many centuries of fear,
folklore, and actual experience with stock-killing wolves. Although public and
management attitudes may not have changed significantly by the 1920s and
probably still opposed protection of the most notorious predators (especially
mountain lions, wolves, and coyotes), a maturing community of ecological researchers and conservationists came to realize that predator control was problematic at best, and a danger to the parks at worst. Cahalane (1939), Wright
(1992), and Schullery (1992) have reviewed the growing opposition to predator
control and other forms of manipulation of wildlife in national parks. This opposition resulted in formal resolutions by the American Association for the Advancement of Science in 1921 and the Ecological Society of America in 1922
against additional introductions of exotic species and "all other unessential interference with natural conditions" (Moore 1925:353). In 1929 and 1930 specific
objections to predator control in the national parks came from the American
Society of Mammalogists, the Wilson Ornithological Club, the Cooper Ornithological Club, the New York Zoological Society, and the Boone and Crockett Club.

In response to this shift in attitude, the NPS developed a more tolerant policy. In 1931 Director Horace Albright issued a policy in the *Journal of Mammalogy* in which he stated that "predatory animals are to be considered an integral
part of the wild life protected in national parks, and no widespread campaigns
of destruction are to be countenanced. The only control practiced is that of
shooting coyotes and other predators when they are actually found making inroads upon herds of game or other animals needing special protection" (Albright 1931:186).

Although this statement did go on to ban poisons and severely restrict the
use of traps in national parks, it was transparently equivocal. The "only control
practiced" was the most objectionable kind—the ill-advised destruction of
large numbers of predators because of their imagined impact on grazers. (Albright was especially concerned at this time about pronghorns, whose numbers had been greatly reduced by human activities and whose final elimination he believed coyotes could accomplish.) It might not be cynical to say that
by 1931 Albright could comfortably order a cessation of killing wolves and

mountain lions in Yellowstone because they were already wiped out, or to point out that the year this policy was published Yellowstone staff killed 145 coyotes in the park. A more enlightened perspective was emerging, but it had not yet prevailed. An additional concern expressed repeatedly by these managers in the 1920s and 1930s was that parks must be "good neighbors"—that they must not harbor large populations of predators that would cause hardship to livestock growers on nearby public or private lands.

In 1927 Milton Skinner, a long-time Yellowstone resident and widely published naturalist, wrote an extended and impassioned summary of the many species of predators still being killed in national parks, and, perhaps more important, he recommended a heightened research effort to help direct future policy: "We do not know enough yet to lay out the details of future policy for the Yellowstone National Park. That would be for the expert staff to decide after making a thorough study of the situation" (Skinner 1927:268).

This brief statement contained two important concepts, neither original with Skinner and both requiring small revolutions against the implicit confidence of early wildlife managers. One was that certain types of information were absolutely required to manage the parks and that as of 1927 that information was lacking. The idea that research was essential to wildlife management had its shaky beginning in the 1920s (Wright 1992). Until that time, and for several decades afterward, Yellowstone's managers rarely lacked confidence that, if only their funding were adequate, they knew what to do.

The other concept in Skinner's statement was that only trained scientists, conducting disciplined, peer-reviewed field research, could provide the information needed. Until that time, the parks hired "naturalists" almost entirely to serve as public teachers, giving nature walks and campfire programs (Skinner 1927, Wright 1992). These people, known sarcastically as "Sunday supplement scientists" (R. Russell, NPS, personal communication 1977), sometimes engaged in thoughtful and even useful nature study, but few produced peer-reviewed publications that might influence managers. Perhaps their most important contribution would not be recognized for many years, that being the wildlife-sighting report system they initiated in the 1920s and maintained for several decades.

Several costly studies and reports published since the early 1960s have pointed out that the NPS legislative mandate, though it can be read to imply the need for research in order to do the job, does not adequately and explicitly require research (summarized in National Research Council 1992). There is striking evidence that Yellowstone administrators in the first decades of the NPS shared the confidence of their early forebears and still did not appreciate the extent of their own ecological ignorance (Schullery 1997).

Advances in predator policy were made, however, largely through the work of a few scientists. The NPS had very little in the way of a science program during the 1920s, but a beginning was made in 1930 with the establishment of the Branch of Research and Education (Sumner 1983). The philanthropist-naturalist George Wright spent two years on a self-funded survey of wildlife conditions in the parks, resulting not only in the first of a series of important scientific volumes (Wright et al. 1933) but also in a more strongly protective predator policy. Wright's position, which became policy in 1936, had an almost poetic rhetorical force, asserting—against all tradition—"that the rare predators shall be considered special charges of the national parks in proportion that they are persecuted everywhere else," and "that no native predator shall be destroyed on account of its normal utilization of any other park animal, excepting if that animal is in immediate danger of extermination, and then only if the predator is not itself a vanishing form" (Wright et al. 1933:147). Yellowstone ceased its slaughter of coyotes in 1935. NPS policy still left managers recourse if they judged a predator "harmful," but the opportunity for large-scale slaughter of predators was much smaller by 1936.

In the case of Yellowstone, Wright's desire to protect predators was specifically directed toward control of the ungulates, which then, as ever since, were widely perceived to be overpopulating their available range. During the 1936 superintendent's conference, Wright pointed out that "nobody knows much about" wildlife management, but that in the case of the "emergency" of the Yellowstone elk, "we must arrange for the predators to control the number of elk to the point where the devastation of the range will cease" (National Park Service 1939:n.p.). Wright's urgency was partly the result of his implicit acceptance of the common belief that these ungulate numbers were a departure from the park's pre-1872 conditions. Although it has proven difficult if not impossible to determine just how many ungulates occupied the park before 1872, and though that number would have changed from year to year depending upon environmental conditions anyway, recent historical research (Schullery and Whittlesey 1992) indicates that the common presumption that ungulates were rare in Yellowstone before 1872 was in error.

Simplistic ideas about predators and their relation to prey are deeply ingrained in our society. When YNP was established, it was universally accepted that predators must be destroyed to protect prey (and for other reasons, including human safety and protection of livestock). By the 1930s it was recognized that predators were sometimes important in controlling prey populations; this principle would eventually work its way into the consciousness of most of the public as the "balance of nature," by which predators were essential to control all prey populations. Although professionals recognized the complexities of predation in the 1940s (Errington 1946), the attractiveness of predation as nature's built-in prey management system has not faded with most people. Fifty years after Wright proclaimed the need for predators to control Yellowstone's elk, this same rationale—that wolves would be our salvation in solving the elk "problem"—was commonly expressed by advocacy groups to promote the restoration of wolves to the park. Even the most liberal models published in the 1990s (National Park Service 1990, Varley and Brewster 1992) did not suggest that ungulate numbers would be reduced enough to reduce their impact on aspen or willow, but science did not serve to disabuse these advocates of such a persuasive and rhetorically useful idea.

Ungulate management had been inextricably bound up with predator management in Yellowstone since the first administrators killed predators in the belief that they were protecting ungulates. Even the need for predators to control herbivores was not a new concept in Yellowstone (Young 1897, Roosevelt 1908). But by the 1930s NPS staff and other observers had firmed up their conviction that the elk were dangerously overabundant and were responsible for range deterioration. From the 1930s to the 1960s the largely unsuccessful attempts of managers and researchers to figure out what was "wrong" with the ungulates and their varying answers to that question influenced attitudes among those same people about predators.

The treatment of individual predator species had diverged as early as the late 1880s, when bears were recognized as popular tourist attractions. After its establishment in 1916, the NPS actively promoted this attraction, thus heightening the divergence by drawing more and more bears to the dumps. After 1915 and the arrival of the automobile, visitation increased rapidly (from 35,849 in 1916, to 227,901 in 1930, to 526,437 in 1940 [Haines 1977]). More visitors meant more garbage and justified the presence of many black bears along roadsides besides a growing number of grizzly bears at the dumps (Schullery 1992). Habituation of large numbers of bears resulted in hundreds of injuries to humans and hundreds of property damage cases as well as exacting a heavy toll on bears that were removed as too troublesome. For example, on average between 1930 and 1969 forty-six people were injured by black bears annually, resulting in an annual average of twenty-four black bears removed from the park (Schullery 1992).

Wright's first faunal survey objected to this sideshow approach to bear management, saying that although "the bear show has been one of the greatest assets of the national parks," it had now served its purpose: it made people familiar with the parks (Wright et al. 1933:84). Wright and his scientist colleagues concluded that "the sight of one bear under natural conditions is more stimulating than close association with dozens of bears." The eventual divorce of all bears from all human food sources would take forty more years and would be both controversial and unpopular with many people (McNamee 1986, Schullery 1992).

The feeding of park bears was institutionalized very early as an important element of the park experience (Skinner 1925, Schullery 1992) and would no doubt be welcomed by many visitors even today. But, as with predator control, the weight of scientific and esthetic appeals was such that the NPS took the opportunity presented by World War II, when visitation dropped drastically, to close the last public-viewing area at a park dump. Black bears would be fed along roads for another thirty years, and both species would dine at the dumps as long, but public attitudes about bears would no longer be shaped by the experience of watching them eat garbage in YNP dumps.

The ambivalence of treatment of the bear is all the more remarkable because the bear has been the only Yellowstone mammalian predator to injure humans in large numbers. Coyote-caused injuries are rare, and there is no record of a visitor injured by a mountain lion or a wolf in Yellowstone. On the other hand, Yellowstone Park bears killed people in 1916 and 1942 (Schullery 1992) and injured hundreds of others (Gunther 1994). The cultural underpinnings of an animal's public image are quite deep (Shepard and Sanders 1985) and complex enough that we will tolerate grievous personal harm from one species without demanding its destruction but will destroy another merely because of inherited hatred of it and its potential for causing us economic harm.

PIONEERING PREDATOR SCIENCE: MURIES AND CRAIGHEADS

With only a few notable and predictable exceptions among commercially important furbearers, carnivore research has suffered from a lag of several decades behind other game research in North America. So it is not surprising that the study of Yellowstone's predators was conducted fitfully and often amateurishly before World War II. A variety of park naturalists and the occasional visiting scientist published notes and observations on aspects of park wildlife, especially bears (reprinted in Schullery 1991, summarized in Schullery 1992). No comprehensive study of any predator species was undertaken until May 1937, when Adolph Murie began his two-year study that resulted in the fourth volume of the fauna series launched by Wright, *Ecology of the Coyote in the Yellowstone* (Murie 1940). Murie's study concluded that the coyote, long persecuted for its presumed effects on the ungulates, was in fact a minor factor in ungulate population sizes, far less significant than range availability and condition. Perhaps this finding had something to do with the eventual discarding of Wright's proposal that predators could be used to control ungulates. Murie's additional assertion that large mammals were abundant in Yellowstone before 1872 must have also been a surprise to many people.

Adolph Murie and his brother Olaus helped bring about an important step for Yellowstone predator science and management (indeed, for predators generally), into a more quantified and considered world. In a less well known but no less pathbreaking report completed in 1944, Olaus Murie examined the Yellowstone "bear problem," offering suggested guidelines for sanitation and management of habituated bears (Murie 1944). These guidelines were essentially in keeping with the bear management program that the NPS adopted decades later but were largely overlooked in the face of the postwar budget crisis and the overwhelming popularity of roadside bear feeding (Schullery 1992). As the increase in visitation accelerated (reaching one million in 1948), bears as a roadside attraction became harder to manage but more entrenched as an obligatory Yellowstone experience.

Wright (1992:24) has called the interval from 1940 to the early 1950s "one of the bleakest periods in the history of science in the NPS. Essentially no new information was produced, and the most lasting legacy of this era was the creation of an attitude toward research that persisted for many years." That attitude, according to Wright, involved neglecting wildlife "problems" (as they were simplistically called) "until they reached crisis proportions, and were hardly ever viewed in an ecosystem context" (p. 24). Park science budgets did revive slightly in the early 1960s as part of the promotional language associated with the Mission 66 program (Haines 1977, Wright 1992), but it would take a non-NPS-funded predator research project and the controversy that developed with it to move Yellowstone predator research into a new era.

The Murie brothers' studies in Yellowstone had been observational, based on extensive field observation of individual animals, their food sources, and their scat. Probably because of their background studying raptor populations, John and Frank Craighead's study of Yellowstone grizzly bears (1959–70) took a more population-oriented approach (J. Claar, U.S. Forest Service, personal communication 1993). Their eleven-year study was the first major ecological study of grizzly bears, resulting in the first quantification of the grizzly bear population's dynamics (Craighead et al. 1995). Through intensive capture and marking, as well as pioneering radio tracking, the Craigheads established population estimates based on known and identifiable animals, and likewise established a baseline that would influence all future studies and population estimates.

Following the 1967 publication of a preliminary Craighead report on the grizzly bears and the NPS decision to close the garbage dumps, a controversy developed that became one of the most bitter and famous in the history of North American wildlife management (Craighead and Craighead 1967, Craighead 1979, McNamee 1986, Schullery 1992, Craighead et al. 1995, Pritchard 1996). Whereas NPS managers and biologists insisted that the bears could be restored to a more natural condition and that the grizzly bear population could be divorced from garbage without harm, the Craigheads maintained that such a goal was unrealistic and that supplemental food in the form of garbage was now essential to the bear population's well-being. Subsequent studies demonstrated that native foods were abundant (Knight et al. 1983). The disputants have in some cases maintained their anger and respective positions for more than a quarter of a century (Craighead et al. 1995), and the arguments show no likelihood of subsiding.

The grizzly bear controversy's early years are revealing in several ways. The controversy revealed how complex and even tenuous the connection can be between management and science. *Biopolitics* became a buzzword that, depending upon one's viewpoint, symbolized either the fundamental reluctance (even inability) of management to incorporate scientific information into decision-making processes or the naïveté of the scientific community in presuming that the findings of one research project were the only factor to be considered when policies are developed. In the late 1960s, when Yellowstone Superintendent Jack Anderson refused to accept the Craigheads' management recommendations, he may have seen himself as having a philosophical difference with them over how the park should best be managed, or as taking the advice of other authorities who disagreed with the Craigheads, but he was usually portrayed as simply and willfully ignoring the world's premier scientific authorities on the subject. Implicit in that portrayal was the notion that the scientific authorities were qualified to guide management of a national park.

The grizzly bear controversy is in fact a showcase example of how science can function as a complicating, rather than a clarifying, element in management debates. As the controversy dragged on, it became clear that all sides could invoke respected scientific authorities on their behalf. This was especially true in the debates over the size of the population. As the decade in which the Craigheads actually studied the bears receded into the past, population modeling exercises by the Craigheads and others removed population analysis progressively further away from the "real" numbers recorded during the study. "The numbers game" was an appropriate but unfortunate nickname for these exercises in the 1970s and 1980s. It should have been no surprise that the public was left with the impression that the actual status of the population was anybody's guess.

Perhaps more important, the controversy heightened public awareness, albeit painfully. (The sad lesson here may be that controversy and pain are the only things that achieve such awareness.) It heightened awareness not only of the vulnerability of the grizzly bear population but also of the extent and unity of its habitat. In discussing the habitat of the grizzly bear in and around Yellowstone National Park, John Craighead is credited with first using the term Greater Yellowstone Ecosystem, which, despite continuing disagreement over its steadily enlarging definition (Schullery 1995a), has become a central concept in modern regional resource management.

VIGNETTES AND ILLUSIONS: THE LEOPOLD REPORT AND ITS CONTEXT

In the past century, managers of wild preserves have moved from intensive husbandry, apparently aimed at stabilizing all "important" elements of the setting (especially game animals and their primary foods), to a more open-ended approach, in which the continued functioning of the processes of that setting are the central management goal. Although a number of earlier observers discussed aspects of this variability and of the importance of maintaining the ecological processes of a park landscape (Grinnell and Storer 1916, Adams 1925, Moore 1925), the legendary Leopold Report (Leopold et al. 1963) offered the first popularly noticed codification of this newer view. Leopold and his colleagues affirmed the need for many forms of intervention (some of which we now might regard as ill-considered), but they hoped primarily to protect systems that mostly took care of themselves. Specifically, their statements about predators constituted another step in strengthening NPS predator protection: "The effort to protect large predators in and around the parks should be greatly intensified. At the same time it must be recognized that predation alone can seldom be relied upon to control ungulate numbers, particularly the larger species such as bison, moose, elk, and deer; additional artificial controls frequently are called for" (Leopold et al. 1963:38). This statement illustrates some important changes. NPS predator policy was no longer park-limited but reached to lands "in and around the parks." Since the Leopold Report was adopted by the NPS, park superintendents have had a policy obligation to pay attention to the behavior of their neighbors. On the other hand, the Leopold Report made it clear that its authors did not share Wright's earlier faith that predation was a simple answer for ungulate "problems," asserting that humans would have to augment the effects of predation with additional control actions. The decade of the Leopold Report was also the decade of a peak in the Yellowstone elk management controversy, when NPS elk-population control programs were stopped amid Congressional hearings and public outcry; almost all scientists agreed the elk required reduction, but the public would not allow it (Houston 1982).

It was in the spirit of the Leopold Report that NPS managers believed that they should not only close the garbage dumps but also initiate a variety of other significant resource management changes in the late 1960s and early 1970s. This was the beginning of the hotly debated "natural regulation" policy, sometimes described as an experiment, in which a broad attempt was made to honor, to a greater degree than ever before, the unhindered expression of the area's ecological processes.

In the minds of its proponents, natural regulation was nothing less than a redefinition of the park. Yellowstone, originally perceived as a collection of interesting things to be enjoyed for their individual value (whether a geyser or a deer), then redefined in the 1920s and 1930s as a collection of things that acted upon each other in ways we might be willing to tolerate up to a point (we could ration a certain number of deer to the coyotes, but no more), was now to be seen as a process whose dynamic nature was the source of our enjoyment and education. In principle, natural regulation anticipated Frankel and Soulé's definition (1981:98) of the purpose of a nature reserve—"to maintain, hopefully in perpetuity, a highly complex set of ecological, genetic, behavioral, evolutionary and physical processes and the coevolved, compatible populations which participate in these processes."

Whether right or wrong in any specifics of philosophy, science, or practical execution, Yellowstone under natural regulation again found itself in a leadership position similar to the one it experienced when it outlawed hunting, forbade the stocking of any more exotic fish species, or ceased predator control while other federal and state agencies were still enthusiastically setting traps and spreading poison. The National Park Service approached Yellowstone's centennial by championing natural fire programs, experimenting with self-regulating ungulate populations, enforcing less consumptive use of fishery resources, and launching other programs that seemed just as revolutionary.

Yellowstone Predators in an Environmental Age

Modern predator management in Yellowstone emerged from, and was fostered by, a rising environmental awareness in the public. Students of environmentalism are aware of the landmark events of this period, including the publication of *Silent spring,* the passage of the Wilderness Act, the first Earth Day, the elimination of 1080 poison stations, the banning of DDT, and other symbolic and influential developments. The public's access to environmental issues was heightened by greater media efficiency and interest; the media increasingly recognized these issues as news.

The process of attracting public interest was cumulative, as important new legislation, such as the National Environmental Policy Act (NEPA, 1970) and the Endangered Species Act (ESA, 1973, with earlier forms in 1966 and 1969), was applied and tested around the country (Bean 1977, Lund 1980). The legal machinery itself enforced a basic level of ecological education on participants and observers in environmental issues. Critical habitat, population viability, and other concepts that were evolving then (as they are now) became part of the language of land management debates.

YELLOWSTONE AS A MICROCOSM OF THE NATIONAL EXPERIENCE

Yellowstone's elk and grizzly bear management situations were widely reported in the popular media in the 1960s and 1970s, breaking ground for public exposure to other issues, and the fires of 1988 established Yellowstone as a consistently newsworthy topic. The Yellowstone management experience during the environmental era exemplified the national experience. Even in the sound-bite approach of most journalism, management topics that once were treated as individual, even isolated, issues were entangled with others, in an ecological tapestry that quickly draped across the region and beyond. The elk population's size may have been tough on the aspen, but it was critical to the survival of the grizzly bear (Mattson et al. 1986). The collapse, through overfishing, of the Yellowstone Lake cutthroat trout *(Oncorhynchus clarki bouvieri)* population in the 1950s was more than a disappointment to fishermen; it was a disaster for dozens of piscivorous wildlife species (Varley and Schullery 1995). The fate of Yellowstone Lake osprey *(Pandion haliaetus)* was tied to the use of DDT elsewhere in the United States (Varley and Schullery 1996). By the early 1980s management issues were increasingly seen from a global perspective, whether for the park's obligations to protect biodiversity or the park's vulnerability to global climate change (Schullery 1997). The most savvy observers realized that Yellowstone's plant and wildlife management issues were really all part of a single big issue of extraordinary complexity: management of the Greater Yellowstone Ecosystem.

Accompanying these changes, and often hastening them, were shifts in park and wildlife constituencies. For most of Yellowstone's first century, ungulate management (and therefore predator management) was primarily driven by the interests of sportsmen and their service agencies. George Bird Grinnell's early dream of Yellowstone as a game reservoir came true but was costly and complicated, as the elk did not migrate from the park with a predictability that could be readily accommodated by traditional hunting seasons, and elk were not uniformly welcome on all public and private lands outside the park. By the late 1960s, however, as hunting faced competition from a growing number of other uses of public lands and wildlife, as well as disapproval from a growing proportion of Americans, the park's wildlife constituency was evolving. Backpacking, wildlife photography, and other less directly consumptive activities were on the rise, and the attitudes and ethics of the hunting community itself continued to evolve.

Predators, so long "the enemy," were tolerated and even embraced as deserving of protection and celebration by these new constituencies, who considered predation on wild ungulates appropriate and had less sympathy than did earlier generations for the losses that regional stockgrowers might take from predation. The key event in this process of changing attitudes occurred on February 9, 1972, when President Nixon issued an executive order banning the poisoning of predators on federal lands or in federal programs (Bean 1977). This may not have signaled an end to the federal government's war on predators, but it was by almost any standards a cease-fire, the first in more than fifty years. The park had been ringed by such poisoning devices right up to the time of the ban.

But the 1960s and 1970s were also decades of confounding developments in wildlife management theory. Most important among these for Yellowstone were reconsiderations of overpopulation in wildlife populations in general (Caughley 1970, 1979) and in Yellowstone in particular (Barmore 1980, Houston 1971, 1982). In the 1980s and early 1990s, new studies of herbivory and other aspects of the ecology of Yellowstone's Northern Range would challenge traditional views (Barnosky 1994, Coughenour 1991, Coughenour and Singer 1991, 1996, Engstrom et al. 1991, Frank and McNaughton 1991, 1992, 1993, Houston 1982, Merrill and Boyce 1991, Merrill et al. 1994, Singer 1995, Singer and Norland 1994, Singer and Renkin 1995, Turner et al. 1994, Wallace and Macko 1993, Wallace et al. 1995, Whitlock et al. 1991). The long-standing "common knowledge" that Yellowstone's ungulates were significantly out of kilter with their range was challenged in all of its specifics, including traditional convictions that elk were terribly overpopulated, that ungulates were wildly accelerating erosion, that grasslands were experiencing declines in production and diversity, that aspen and willow declines were simply the result of too many elk, that elk and other ungulates were not native to the Yellowstone area, and that human control would be necessary to "restore" Yellowstone.

This new round of research findings has already generated much public interest and some scientific response (Kay 1990, Hamilton 1994, Engstrom et al. 1994, Wagner et al. 1995a, 1995b), and with luck the dialogue will advance our understanding of the issues. As has been observed, "One of the strengths of science is that it does not require that scientists be unbiased, only that different scientists have different biases" (Hull 1988:22).

RECOVERY AS A CONCEPT: THE COMPLICATIONS
OF ACCEPTING COMPLEXITY

Because other essays in this book will deal in detail with the many important new predator research projects of the 1980s and 1990s, ours concentrates on an overview of some important conceptual developments in this period.

The Endangered Species Act was a reflection of society's growing alarm over the loss of species; it has become one of the most powerful tools in the armament of conservationists. As with the park's organic act of 1872, the NPS Act of 1916, and the Leopold Report, it is difficult to read the ESA without doubting that its framers fully appreciated the extraordinary reach it might have in the management of Yellowstone. Although occasional commentators long expressed wishes that Yellowstone could be a wolf sanctuary (Heller 1925, Leopold 1944), the emergence of an actual wolf-recovery "movement" was entirely the product of a much later age. Under the impetus of the Leopold Report, NPS staff in Yellowstone had initiated what amounted to an ecological restoration project of broad and ambitious proportions in the late 1960s, including various programs we have mentioned, well before the concept of recovery was formalized in the ESA.

But the rather diffuse goals of natural regulation were complemented and modified, rather than clarified, by the ESA. As more was learned about the nature of wildland systems—for example, as those systems revealed themselves as more frequently nonequilibrial than earlier generations of observers ever imagined—and as concepts of island biogeography were applied with increasing sophistication to places like Greater Yellowstone, the supposed tidiness of earlier management ideas was lost. As our understanding of this landscape advanced, it became apparent that we may be just beginning to appreciate the episodic nature of change-forcing events in the ecosystem. Since 1872 ongoing changes, such as climate warming and a spectacular increase in human presence, have been punctuated by several unique shorter-term events that may have been of similarly great impact. These include the wildlife slaughter of the 1870s, the drought of the 1930s, the fires of 1988, and the dramatic increase in winter recreational use of the park. Each of these episodes was influenced by complex circumstances, and perhaps the most important point is this: the effects of each of these events are still being felt. The first and last were primarily the result of human activities, and all four resulted in a variety of significant human responses that complicate attempts to understand or measure the effects of the original events.

It seems certain that many other events on a similar scale of impact—an epidemic of wildlife or human disease, other abrupt changes in human use patterns, other extended wet or dry periods, and so on—have occurred in the more remote past and will continue to occur in the future. Yellowstone demonstrates that if an ecosystem might be said to have a trajectory over time, that trajectory is subject to countless influences (see also Chapter 12). We may be able to catalog potential influences, but we are far from able to predict their occurrence, much less the results of those occurrences. It should be a matter of great wonder to us that, even though we are unable to predict the weather more than about forty-eight hours in advance, and even though the weather is only one of the influences on Yellowstone's ecological processes, some people are still inclined to believe that we can manage this system so as to set these ecological processes on a given course and keep them there. Too stiff a devotion to a traditional, deterministic view of ecological process—that is, a "mechanistic" view in which the system is readily predictable and therefore easy to control— though comforting in the short run, will invariably give us fits in the long run. Furthermore, and perhaps more important in today's management dialogues, a system's trajectory, once subjected to any influence (whether by humans or other forces), may not be presumed to have the capacity or the tendency to recover its preinfluence trajectory. This may seem intuitive, but it is precisely this presumption that has driven most wildlife management policy in Yellowstone and elsewhere.

For example, many educated observers today consider Yellowstone's wildlife populations to be dysfunctional in some way, whether through a shortage of wolves, an overabundance of ungulates, a loss of some combination of ecological forces that existed in the park prehistorically, or some other combination of factors. There is a widespread assumption among almost all of these positions that Yellowstone's ecological system can be "gotten back on track" through some manipulation, implying that we can force the system back into its preinfluence trajectory. This assumption is especially audible in the public and in the advocacy community, where straightening out Yellowstone for one reason or another is common conversational currency, but it underlies much of the scientific dialogue on the park as well. It is no surprise, then, that in this age as in all previous ones in the park's history, there are powerful professional voices expressing confidence that surely we now know enough about how this system works to take an aggressive part in manipulating it for its own best interest (Wagner et al. 1995a).

The evidence from the first 125 years of Yellowstone National Park history suggests that we have not watched or studied this system long enough to grasp its potential for variation. For all our growing alertness to the dynamic nature of ecological processes, we continue to display great uneasiness in the face of such variation, and we tend to assume that such variation is simply our fault rather than the product of the system's complexity or the result of some complex combination of human influences and as-yet poorly understood ecological processes (Schullery 1997). Predators appear to be an especially instructive and challenging aspect of both human influences and ecological processes, precisely because they not only demonstrate the richness of system variation but also demand our tolerance of such variation despite the uneasiness it may cause us (Botkin 1990).

Wolf recovery abundantly illustrates the continuing change in our relation with predators as we struggle with our perception of this ecosystem. Although only twenty years ago park resource managers talked about restoration of a wolf population in a fairly straightforward manner ("bring them in and let them go"), now a host of subtle constraints are recognized. The papers published in two *Wolves for Yellowstone?* volumes (National Park Service 1990, Varley and Brewster 1992) suggested the breadth of these concerns: taxonomic issues surrounding "appropriate" subspecies for reintroduction, genetic complications and long-term viability of small isolated populations, potential competition among species that are endangered or of special concern, potential impacts on regional ungulates and livestock, potential economic impacts, human safety concerns, continuing questions of how nearly we can or should be trying to replicate pre-1872 Yellowstone, and the amazingly numerous logistical complications of simply moving far-ranging wild animals from one nation to another and releasing them in a way that encourages them to stay.

Many of the same concerns have emerged to enrich and muddle the grizzly bear recovery effort. The early successes of the 1995–96 wolf recovery program in Yellowstone are the result of (1) the comprehensive review by managers of an imposing array of traditional predator-related issues, (2) the anticipation of a multitude of consequences that may develop at any stage of the process, and (3) luck that so far the most important predictions (that is, the ones whose failure would cause the highest levels of public uneasiness) have come true.

What distinguishes the pre-reintroduction wolf research of the late 1980s and early 1990s from earlier Yellowstone predator research is its interdisciplinary breadth. To be sure, earlier generations of investigators did not have access to some of the most important tools that the authors of *Wolves for Yellowstone?* used, including various population modeling techniques, taxonomy based on increasingly sophisticated genetic analyses, and the Delphi process used to survey experts. Each generation of investigators has been better equipped than the last. But perhaps more significant was the extension of inquiry into the social sciences. Public-attitude surveys and economic studies often took center stage, testing suppositions about the public's attitudes toward wolves and the probable impacts of wolves. Quantification of the economic impacts of wolves on the region, suggesting a substantial economic gain (Duffield

1992), was probably the regional public's first exposure to natural resource economics, a field that has the power to play a key role in other future predator-related issues (Swanson et al. 1994). In the long run, it may be that the most important result of the many important new studies of predators conducted in Yellowstone since 1970 and highlighted in this volume will be cumulative and integrative.

Conclusions

People seeking a "baseline" historical or prehistoric date by which to judge current ecological conditions in Yellowstone National Park are doomed to disappointment. Used carefully, the historical and prehistoric evidence of past Yellowstone landscapes is helpful in understanding current ecological conditions and in developing management directions, but we are incapable of matching those conditions, and we have little if any reason to try.

The first decade of the park's existence was a critical and formerly underappreciated period in Yellowstone predator management. The wholesale ungulate slaughter and intensive trapping, poisoning, and shooting of carnivores from 1870 to 1882 may have significantly altered the ecological setting long before the much better documented and publicized predator control program of 1904–35. The consequences of those early alterations have not even been clearly imagined yet, but they obviously extended well beyond the end of the slaughter and should be suspected of being with us even today. The most important long-term lesson of these events may be that restoration itself is a very slippery concept. Given the enormous potential complexity and far-reaching ripple effects of past predator control programs, restoring the presence of a single eradicated species must not be presumed to be the same as restoring the entire ecological community to some "pristine" form.

Over the past century, especially the past seventy-five years, predators have made considerable progress in being respected as valid elements of wild settings such as Yellowstone National Park. Because of its peculiar place in our society, Yellowstone has often had a leadership (or "guinea pig") role in the development of new wildlife management policies and attitudes. The park, sometimes inadvertently through unwelcome controversies, has contributed enormously to public awareness of predator management issues. The most valuable lesson here may be that, because Yellowstone is such a powerful presence in world consciousness, it demands extraordinary protective attention. It is too valuable a tool to be neglected. A second lesson is that most of the times Yellowstone has served as a guinea pig, it was only retroactively that managers and scientists became fully aware of the value of whatever experiment was being conducted; sometimes it was not until much later that a given management action was recognized for its experimental value. For example, the removal of wolves had effects that can to some extent be measured retroactively as wolves are restored. The most alert and fortunate managers and researchers are those who, confronted with a vast and unruly wildland ecosystem, are able to identify such experimental opportunities (Varley and Schullery 1996).

On the other hand, despite growing management sympathy toward large predators, Yellowstone's management direction is still in good part the result of broader social trends. In many influential circles, large predators are still regarded somewhat as a luxury rather than as a necessity and must make their own way in a rapidly changing culture as well as in the modern political economy. Large predators are still afforded "first-class citizenship" in Yellowstone National Park only when:

1. they are successfully argued to have no "ill effects" on ungulates or other herbivores,
2. they are successfully justified as having "good effects" on ungulates or other herbivores,
3. their protection is required by additional laws beyond park-enabling legislation and NPS policy, or
4. all of the above.

The most important lesson here may be that even in a world-famous "nature reserve" of Yellowstone's caliber, predator conservation still requires intensive public education.

Predator management in Greater Yellowstone is like wildlife management elsewhere: our learning curve has steepened and shows no sign of reaching a plateau. Restoration of predator populations has become one of the greatest educational forces for modern park managers because it stretches their thinking and their resources and because it has a demonstrated popularity among a growing segment of the public and is thus defensible in a time of shrinking budgets.

Acknowledgments: We thank the following people for reading the manuscript: Norman Bishop (NPS, Yellowstone), Sarah Broadbent (NPS, Yellowstone), Kenneth Cannon (NPS, Lincoln, Nebraska), Douglas Houston (USGS, Olympic National Park), Ann Johnson (NPS, Yellowstone), Mary Meagher (USGS, Yellowstone), Richard Sellars (NPS, Santa Fe, New Mexico), and John Varley (NPS, Yellowstone). Additional conversations about Yellowstone predators with Bonnie Blanchard (USGS, Bozeman, Montana), Wayne Brewster (NPS, Yellowstone), Robert Crabtree (Yellowstone Ecosystem Studies, Bozeman, Montana), Marilynn French (Yellowstone Grizzly Foundation, Jackson, Wyoming), Steve French (Yellowstone Grizzly Foundation, Jackson, Wyoming), Mark Johnson (Wildlife Veterinary Resources, Gardiner, Montana), Richard Knight (USGS, Bozeman, Montana), Kerry Murphy (Yellowstone), Susan Rhoades Neel

(Montana State University), and Michael Phillips (NPS, Yellowstone) have influenced the interpretations in this essay.

We thank the editors of this volume for their many helpful comments and considerable guidance, not only on the manuscript but also throughout the process of organizing the special conference session upon which this book is based and for their vision in imagining a book of this breadth and worth.

We are likewise grateful to Yellowstone Center for Resources Director John Varley and Deputy Director Wayne Brewster for their interest in historical research in resource issues.

Last, we must as always express our gratitude to Aubrey Haines, long-time dean of Yellowstone historians, for his pioneering work in developing the Yellowstone archives and in laying such a firm scholarly foundation for all later students to build upon.

Literature Cited: Adams, C. C. 1925. The relation of wild life to the public in national and state parks. Roosevelt Wildlife Bulletin 2(4):371–402.

Albright, H. M. 1931. National parks predator policy. Journal of Mammalogy 12:185–86.

Barmore, W. J. 1980. Population characteristics, distribution, and habitat relationships of six ungulates in northern Yellowstone Park. Final report. Yellowstone Park files, Research Library, Yellowstone National Park.

Barnosky, E. H. 1994. Ecosystem dynamics through the past 2000 years as revealed by fossil mammals from Lamar Cave in Yellowstone National Park, USA. Historical Biology 8:71–90.

Bartlett, R. 1974. Nature's Yellowstone. University of Oklahoma Press, Norman.

———. 1985. Yellowstone: A wilderness besieged. University of Arizona Press, Tucson.

Beal, M. 1960. The story of man in Yellowstone. Yellowstone Library and Museum Association, Yellowstone National Park.

Bean, M. 1977. The evolution of national wildlife law. U.S. Government Printing Office, Washington.

Bishop, N., P. Schullery, F. J. Singer, and J. D. Varley. 1997. Yellowstone's Northern Range: Complexity and change in a wildland ecosystem. Yellowstone Center for Resources, Yellowstone National Park.

Botkin, D. B. 1990. Discordant harmonies, a new ecology for the twenty-first century. Oxford University Press, Oxford.

Cahalane, V. H. 1939. The evolution of predator control in the national parks. Journal of Wildlife Management 3:229–37.

Cannon, K. P. 1991. Site testing on the upper Yellowstone River, Yellowstone National Park. Paper presented at the Forty-ninth Plains Anthropological Conference, November 13–16, Lawrence, Kansas. Ms. provided by author.

———. 1992. A review of archaeological and paleontological evidence for the prehistoric presence of wolf and related prey species in the northern and central Rockies physiographic provinces. Pp. 1-175 to 1-265 *in* J. D. Varley and W. G. Brewster, eds., Wolves for Yellowstone? A report to the United States Congress, vol. 4, research and analysis. National Park Service, Yellowstone National Park.

———. 1995. Blood residue analysis of ancient stone tools reveal clues to prehistoric subsistence patterns in Yellowstone. Cultural Resource Management 18(2):14–16.

Cannon, K. P., and M. E. Newman. 1994. Results of blood residue analysis of a Late Paleoindian projectile point from Yellowstone National Park, Wyoming. Cultural Resource Preservation 11:18–21.

Caughley, G. 1970. Eruption of ungulate populations, with emphasis on Himalayan thar in New Zealand. Ecology 51:53–72.

———. 1979. What is this thing called carrying capacity? Pp. 2–8 *in* M. S. Boyce and L. D. Hayden-Wing, eds., North American elk: Ecology, behavior and management. University of Wyoming Press, Laramie.

Coughenour, M. 1991. Biomass and nitrogen responses to grazing of upland steppe on Yellowstone's northern winter range. Journal of Applied Ecology 28:71–82.

Coughenour, M., and F. J. Singer. 1991. The concept of overgrazing and its application to Yellowstone's northern winter range. Pp. 209–30 *in* R. B. Keiter and M. S. Boyce, eds., The

Greater Yellowstone Ecosystem: Redefining America's wilderness heritage. Yale University Press, New Haven.

———. 1996. Elk population processes in Yellowstone National Park under the policy of natural regulation. Ecological Applications 6:573–93.

Craighead, F. 1979. Track of the grizzly. Sierra Club Books, San Francisco.

Craighead, J., and F. Craighead. 1967. Management of bears in Yellowstone National Park. Unpublished report, Yellowstone Park files, Research Library, Yellowstone National Park.

Craighead, J., J. S. Sumner, and J. A. Mitchell. 1995. The grizzly bears of Yellowstone. Island Press, Washington, D.C.

Curnow, E. 1969. The history of the eradication of the wolf in Montana. M.S. thesis, University of Montana, Missoula.

Dary, D. 1974. The buffalo book. Swallow Press, Chicago.

Denevan, W. M., ed. 1992. The native population of the Americas in 1492. University of Wisconsin Press, Madison.

Dobyns, H. F. 1983. Their number become thinned: Native American population dynamics in eastern North America. University of Tennessee Press, Knoxville.

Duffield, J. W. 1992. An economic analysis of wolf recovery in Yellowstone: Park visitor attitudes and values. Pp. 2-31 to 2-87 *in* J. D. Varley and W. G. Brewster, eds., Wolves for Yellowstone? A report to the United States Congress, vol. 4, research and analysis. National Park Service, Yellowstone National Park.

Engstrom, D. R., C. Whitlock, S. C. Fritz, and H. E. Wright. 1991. Recent environmental changes inferred from the sediments of small lakes in Yellowstone's northern range. Journal of Paleolimnology 5:139–74.

———. 1994. Reinventing erosion in Yellowstone's northern range. Journal of Paleolimnology 10:159–61.

Errington, P. L. 1946. Predation and vertebrate populations. Quarterly Review of Biology 21:144–77, 221–45.

Erwin, J. B. 1898. Annual report of the superintendent, Yellowstone National Park. U.S. Government Printing Office, Washington.

Frank, D. A., and S. J. McNaughton. 1991. Stability increases with diversity in plant communities: Empirical evidence from the 1988 Yellowstone drought. Oikos 62:360–62.

———. 1992. The ecology of plants, large mammalian herbivores, and drought in Yellowstone National Park. Ecology 73:2043–58.

———. 1993. Evidence for the promotion of aboveground grassland production by native large herbivores in Yellowstone National Park. Oecologia 96:157–61.

Frankel, O. H., and M. E. Soulé. 1981. Conservation and evolution. Cambridge University Press, Cambridge.

Frison, G. C. 1991. Prehistoric hunters of the high plains. Academic Press, New York.

Grinnell, G. B. 1882. Their last refuge. Forest and Stream 19:382–83.

———. 1883. An important park order. Forest and Stream 19:481.

Grinnell, J., and T. Storer. 1916. Animal life as an asset of national parks. Science 44:375–80.

Gunther, K. 1994. Bear management in Yellowstone National Park. International Conference on Bear Research and Management 9:549–60.

Hadly, E. 1990. Late Holocene mammalian fauna of Lamar Cave and its implications for ecosystem dynamics in Yellowstone National Park, Wyoming. M.S. thesis, Northern Arizona University, Flagstaff.

———. 1995. Evolution, ecology, and taphonomy of Late-Holocene mammals from Lamar Cave, Yellowstone National Park, Wyoming. Ph.D. diss., University of California, Berkeley.

Haines, A. H. 1974. Yellowstone National Park, its exploration and establishment. U.S. Government Printing Office, Washington.

———. 1977. The Yellowstone story. Colorado Associated University Press and the Yellowstone Library and Museum Association, Boulder.

Hamilton, W. L. 1994. Comment: Recent environmental changes inferred from the sediments of small lakes in Yellowstone's northern range (Engstrom et al. 1991). Journal of Paleolimnology 10:156–59.

Hampton, H. D. 1971. How the U.S. Cavalry saved our national parks. Indiana University Press, Bloomington.

Harris, M. 1887. Annual report of the superintendent, Yellowstone National Park. U.S. Government Printing Office, Washington.

Heller, E. 1925. The big game animals of Yellowstone National Park. Roosevelt Wildlife Bulletin 2:404–67.

Herrero, S. 1985. Bear attacks, their causes and avoidance. Nick Lyons Books, New York.

Houston, D. B. 1971. Ecosystems of national parks. Science 172:648–51.

———. 1982. The northern Yellowstone elk: Ecology and management. Macmillan Company, New York.

Hull, D. L. 1988. Science as a process: An evolutionary account of the social and conceptual development of science. University of Chicago Press, Chicago.

Janetski, J. 1987. The Indians of Yellowstone Park. Bonneville Books, University of Utah Press, Salt Lake City.

Johnson, A. 1997. How long have bison been in the park? The Buffalo Chip, January–March.

Kay, C. 1990. Yellowstone's northern elk herd: A critical evaluation of the natural regulation paradigm. Ph.D. diss., Utah State University, Logan.

———. 1994. Aboriginal overkill: The role of native Americans in structuring western ecosystems. Human Nature 5:359–98.

———. 1995. Aboriginal overkill and native burning: Implications for modern ecosystem management. Pp. 107–18 *in* R. M. Linn, ed., Sustainable society and protected areas, contributed papers of the eighth Conference on Research and Resource Management in Parks and on Public Lands, April 17–21, Portland, Oregon. George Wright Society, Hancock, Michigan.

Kennedy, R. G. 1994. Hidden cities: The discovery and loss of ancient American civilization. Free Press, New York.

Knight, R., G. Brown, J. Craighead, M. Meagher, L. Roop, and C. Servheen. 1983. Final report, ad hoc committee to investigate the need and feasibility of the supplemental feeding of Yellowstone grizzly bears. Interagency Grizzly Bear Study Team, Bozeman.

Leopold, A. 1944. *Review of* The wolves of North America, by S. P. Young and E. H. Goldman. Journal of Forestry 42:928–29.

Leopold, A. S., S. A. Cain, C. M. Cottam, I. N. Gabrielson, and T. L. Kimball. 1963. Wildlife management in the national parks. Transactions of the North American Wildlife Conference 24:28–45.

Lund, T. A. 1980. American wildlife law. University of California Press, Berkeley.

Mattson, D., B. Blanchard, and R. Knight. 1986. Food habits of the Yellowstone grizzly bear. Paper presented at the seventh International Conference on Bear Research and Management, Williamsburg, Virginia.

McIntyre, R. 1995. War against the wolf. Voyageur Press, Stillwater, Minnesota.

McNamee, T. 1986. The grizzly bear. Alfred Knopf, New York.

Meagher, M. M. 1973. The bison of Yellowstone National Park. National Park Service Scientific Monograph Series, no. 1. U.S. Government Printing Office, Washington.

Merrill, E. H., and M. S. Boyce. 1991. Grassland phytomass, climatic variation and ungulate population dynamics in Yellowstone National Park. Pp. 263–74 *in* R. B. Keiter and M. S. Boyce, eds., The Greater Yellowstone Ecosystem: Redefining America's wilderness heritage. Yale University Press, New Haven.

Merrill, E. H., N. L. Stanton, and J. C. Hak. 1994. Responses of bluebunch wheatgrass, Idaho fescue, and nematodes to ungulate grazing in Yellowstone National Park. Oikos 69:231–40.

Meyer, G., S. Wells, R. Balling, and A. J. Jull. 1992. Response of alluvial systems to fire and climate change in Yellowstone National Park. Nature 357:147–49.

Moore, B. 1925. Importance of natural conditions in national parks. Pp. 340–55 *in* G. Grinnell and C. Sheldon, eds., Hunting and conservation: The book of the Boone and Crockett Club. Yale University Press, New Haven.

Murie, A. 1940. Ecology of the coyote in the Yellowstone. Fauna of the national parks, no. 4. U.S. Government Printing Office, Washington.

Murie, O. J. 1944. Progress report on the Yellowstone bear study. Unpublished administrative report, Research Library, Yellowstone National Park.

National Park Service. 1939. Exhibit "A" condensed chronology and discussion of National Park Service predatory policy, from Policy on predators and notes on predators, typescript, RG 79, Central Classified File, 715. Harpers Ferry National Park Service Library, Harpers Ferry, West Virginia.

——. 1990. Wolves for Yellowstone? A report to the United States Congress, vol. 2, research and analysis. National Park Service, Yellowstone National Park.

National Research Council. 1992. Science and the national parks. National Academy Press, Washington.

Norris, P. 1878. Report upon the Yellowstone National Park to the secretary of the interior by P. W. Norris, superintendent, for the year 1877. U.S. Government Printing Office, Washington.

——. 1881. Annual report of the superintendent of the Yellowstone National Park to the secretary of the interior for the year 1880. U.S. Government Printing Office, Washington.

Pitcher, J. 1902. Circular on wildlife feeding, August 6, 1902, Yellowstone National Park Archives, Army Section, Box 239, Yellowstone National Park.

Pritchard, J. A. 1996. Preserving natural conditions: Science and the perception of nature in Yellowstone National Park. Ph.D. diss., University of Kansas, Lawrence.

Ramenofsky, A. F. 1987. Vectors of death: The archaeology of European contact. University of New Mexico Press, Albuquerque.

Reiger, J. 1975. American sportsmen and the origins of conservation. Winchester Press, New York.

Roe, F. G. 1955. The Indian and the horse. University of Oklahoma Press, Norman.

Roosevelt, T. 1908. January 22, 1908, letter to Yellowstone Superintendent S. B. M Young, Research Library, Yellowstone National Park.

Schlesier, K. H., ed. 1994. Plains Indians, A.D. 500–1500. University of Oklahoma Press, Norman.

Schullery, P. 1991. Yellowstone bear tales. Roberts Rinehart, Niwot, Colorado.

——. 1992. The bears of Yellowstone. High Plains Publishing, Worland, Wyoming.

——. 1995a. The Greater Yellowstone Ecosystem. Pp. 312–14 *in* Our living resources. National Biological Service, Washington.

——. 1995b. Yellowstone ski pioneers. High Plains Publishing, Worland, Wyoming.

——. 1997. Searching for Yellowstone: Ecology and wonder in the last wilderness. Houghton Mifflin, Boston.

Schullery, P., and L. Whittlesey. 1992. The documentary record of wolves and related wildlife species in the Yellowstone National Park area prior to 1882. Pp. 1-3 to 1-173 *in* J. D. Varley and W. G. Brewster, eds., Wolves for Yellowstone? A report to the United States Congress, vol. 4, research and analysis. National Park Service, Yellowstone National Park.

——. 1995. A summary of the documentary record of wolves and other wildlife species in the Yellowstone National Park area prior to 1882. Pp. 63–76 *in* L. N. Carbyn, S. H. Fritts, and D. R. Seip, eds., Ecology and conservation of wolves in a changing world. Occasional publication no. 35. Canadian Circumpolar Institute, Edmonton.

Secretary of the Interior. 1907. Rules, regulations and instructions for the information and guidance of officers and enlisted men of the United States Army, and of the scouts doing duty in the Yellowstone National Park. U.S. Government Printing Office, Washington.

Shepard, P., and B. Sanders. 1985. The sacred paw: The bear in nature, myth, and literature. Viking, New York.

Singer, F. J. 1995. Grassland responses to elk and other ungulate grazing on the northern winter range of Yellowstone National Park. Northwest Science 69:191–203.

Singer, F. J., and J. E. Norland. 1994. Niche relationships within a guild of ungulate species in Yellowstone National Park, Wyoming, following release from artificial controls. Canadian Journal of Zoology 72:1383–94.

Singer, F. J., and R. A. Renkin. 1995. Effects of browsing by native ungulates on the shrubs in big sagebrush communities in Yellowstone National Park. Great Basin Naturalist 55:201–12.

Skinner, M. P. 1925. Bears in the Yellowstone. A. C. McClurg and Co., Chicago.

———. 1927. The predatory and fur bearing animals of the Yellowstone National Park. Roosevelt Wildlife Bulletin 4:163–281.

Stannard, D. E. 1992. American holocaust: Columbus and the conquest of the New World. Oxford University Press, New York.

Sumner, L. 1983. Biological research and management in the National Park Service: A history. George Wright Forum 3(4):3–27.

Swanson, C. S., D. W. McCollum, and M. Maj. 1994. Insights into the economic value of grizzly bears in the Yellowstone recovery zone. International Conference on Bear Research and Management 9:575–82.

Teller, H. M. 1883. Letter to superintendent of Yellowstone National Park, January 15. National Archives Record Group 48, No. 62, microfilm roll 2. Yellowstone Park Research Library, Yellowstone National Park.

Turner, M. G., Y. Wu, L. L. Wallace, W. H. Romme, and A. Brenkert. 1994. Simulating winter interactions among ungulates, vegetation, and fire in northern Yellowstone Park. Ecological Applications 4:472–96.

Varley, J., and W. G. Brewster, eds. 1992. Wolves for Yellowstone? A report to the United States Congress, vol. 4, research and analysis. National Park Service, Yellowstone National Park.

Varley, J., and P. Schullery. 1983. Freshwater wilderness: Yellowstone fishes and their world. Yellowstone Library and Museum Association, Yellowstone National Park.

———, eds. 1995. The Yellowstone Lake crisis: Confronting a lake trout invasion, a report to the director of the National Park Service. Yellowstone Center for Resources, Yellowstone National Park.

———. 1996. Yellowstone Lake and its cutthroat trout. Pp. 49–73 *in* W. Halvorson and G. E. Davis, eds., The evolution of ecosystem management in America's national parks. University of Arizona Press, Tucson.

Verano, J. W., and D. H. Ubelaker, eds. 1992. Disease and demography in the Americas. Smithsonian Institution Press, Washington.

Wagner, F. H., R. Foresta, R. B. Gill, D. R. McCullough, M. R. Pelton, W. F. Porter, and H. Salwasser. 1995a. Wildlife policies in the U.S. national parks. Island Press, Washington, D.C.

Wagner, F. H., R. B. Keigley, and C. L. Wamboldt. 1995b. *Comment on* Ungulate herbivory of willows on Yellowstone's northern winter range by Singer et al., 1994. Journal of Range Management 48:475–77.

Waldman, C. 1985. Atlas of the North American Indian. Facts on File Publications, New York.

Wallace, L. L., and S. A. Macko. 1993. Nutrient acquisition by clipped plants as a measure of competitive success: The effects of compensation. Functional Ecology 7:326–31.

Wallace, L. L., M. G. Turner, W. H. Romme, R. V. O'Neill, and Y. Wu. 1995. Scale and heterogeneity of forage production and winter foraging by elk and bison. Landscape Ecology 10(2):75–83.

Weaver, J. 1978. The wolves of Yellowstone. Natural Resources Report no. 14. National Park Service, Yellowstone National Park.

White, R. 1978. The winning of the West: The expansion of the western Sioux in the eighteenth and nineteenth centuries. Journal of American History 65:319–43.

Whitlock, C., and P. J. Bartlein. 1993. Spatial variations of Holocene climatic change in the Yellowstone region. Quaternary Research 39:231–38.

Whitlock, C., S. C. Fritz, and D. R. Engstrom. 1991. A prehistoric perspective on the northern range. Pp. 289–305 *in* R. B. Keiter and M. S. Boyce, eds., The Greater Yellowstone Ecosystem: Redefining America's wilderness heritage. Yale University Press, New Haven.

Williams, M. 1989. Americans and their forests. Cambridge University Press, New York.

Wood, W. R. 1972. Contrastive features of native North American trade systems. Pp. 153–69 *in* F. W. Voget and R. L. Stephenson, eds., For the chief: Essays in honor of Luther S. Cressman by some of his students. Anthropological Papers no. 4. University of Oregon, Eugene.

Wright, G. 1982. Archaeological research in Yellowstone National Park. Pp. 11–14 *in* the 33d annual field conference, Wyoming Geological Association Guidebook.

Wright, G., J. Dixon, and B. Thompson. 1933. Fauna of the national parks of the United States. U.S. Government Printing Office, Washington.

Wright, R. G. 1992. Wildlife research and management in the national parks. University of Illinois Press, Urbana.

Yellowstone Science. 1995. Blood residues on prehistoric stone artifacts reveal human hunting activities and diversity of local fauna. Yellowstone Science 3(2):19.

Young, S. B. M. 1897. Annual report of the superintendent, Yellowstone National Park. U.S. Government Printing Office, Washington.

Black bear feeding on an elk carcass in Hayden Valley (Frank Craighead)

Chapter Three

Yellowstone Bears

Richard R. Knight, Bonnie M. Blanchard, and Paul Schullery

The bears of Yellowstone are the most famous bears in the world. Black bears *(Ursus americanus)* were known internationally for their roadside begging antics from the late 1910s through the 1960s, after which this behavior was rigorously discouraged. The grizzly bear *(U. arctos)*, highly visible feeding at garbage dumps from the 1890s through the 1960s, became the subject of a much publicized controversy when the dumps were closed in the early 1970s.

More books and popular articles have been written about Yellowstone bears than about any other animal population. Grizzly bears in Yellowstone have been subjected to intensive research since 1959 and have been the focus of more scientific papers than any other bear population. Despite this knowledge, their current status remains one of the most controversial subjects in wildlife biology.

In this chapter we shall provide insight about Yellowstone's black and grizzly bear populations and their futures, based on information gathered by biologists who have studied the Yellowstone bears the longest and know them the best. We do not believe that this information will solve the impossible task of ending the controversies. Our major focus will be to discuss the complexities involved in estimating population size and determining carrying capacity of the Yellowstone ecosystem for grizzly bears.

History

Grizzly bears are known to have existed in the Greater Yellowstone Ecosystem since the Late Holocene (Cannon 1992, Hadly 1995; see Chapter 2 for more historical information). Influences of Native Americans and early settlers on bear populations remain speculative. With Euro-American settlement of southwestern Montana starting in the 1860s and the creation of Yellowstone National Park in 1872, human impacts became better documented. During the park's first decade, uncontrolled market hunting of ungulates and apparently widespread poisoning of carcasses seem to have significantly reduced some scavenger populations, including grizzly bears (Schullery and Whittlesey 1992). Grizzlies were also probably killed by sport hunters, as sport and limited subsistence hunting was legal in the park until 1883, and most carnivores were not clearly protected from public shooting until after 1886 (Haines 1977, Schullery 1995). Schullery (1992) cites accounts of Yellowstone National Park from the mid-1880s that indicate that bears were still numerous and being hunted (after the prohibition on hunting herbivores was enacted), so we are unable to determine the extent to which grizzly bear populations were affected during the park's least-protected decade.

Records of grizzly bear harvest are incomplete in the Yellowstone ecosystem outside of Yellowstone National Park. In the late 1800s and early 1900s unrestricted hunting, trapping, and poisoning occurred in all three surrounding states, along with kills resulting from livestock protection (Blair 1987). Poisoning of predators and scavengers was widespread, with unknown effects on grizzly bears, until the 1960s. The grizzly bear was reclassified from predator to game animal in Montana in 1923, restricting the kill to some extent, but Montana did not require a special hunting permit until 1967. Idaho prohibited grizzly bear hunting in 1946. Wyoming closed its grizzly bear hunting season after 1967, with limited hunting during some years after that. A moratorium was declared on grizzly bear hunting in the Yellowstone ecosystem in 1974, and the following year the grizzly bear was listed as threatened under the Endangered Species Act.

INFLUENCES OF HUMAN FOOD AT GARBAGE DUMPS

Visitation increased slowly over the park's first four decades. Annual visitation averaged 24,748 for the decade 1906–15, providing a relatively limited amount of garbage compared with later periods. As early as 1873 bears had been in trouble with humans over stored food (Schullery and Whittlesey 1992). By the late 1880s and early 1890s, shortly after sport hunting ceased and the army began to provide better protection, black and then grizzly bears began to appear at park dumps and soon became visitor attractions (Schullery 1991, 1992). After the mid-1890s superintendents frequently mentioned problems with bears in developments and occasionally discussed the need to remove a few of the more incorrigible animals.

Accounts suggest that before creation of the National Park Service in 1916, grizzly bear use of dumps was slight compared to later periods. Skinner (1925) stated that the most bears he had seen at one dump was twenty-nine in 1915. In fact, Skinner (1925) and Superintendent Horace Albright (1920) believed there were at least forty grizzly bears in the entire park during 1920.

Two interesting corollaries to these first population estimates emerge. The first is that in 1908 the hunter-naturalist-photographer William Wright, one of the best-informed early writers on grizzly bears, reported to the acting superintendent his certainty that U.S. Army soldiers responsible for protecting the park were killing many bears, to the point that the grizzly bear was threatened with extinction in the park (Wright 1908). Wright's observations were based mostly on what he saw at a few dump sites, but his argument is eloquent, and he claimed to quote unnamed soldiers who were actively involved in the clandestine killing of as many bears as possible. Perhaps the seemingly low estimate of forty was partly the result of this alleged killing. However, other reports contemporary with Wright's speak of the great abundance of bears during that period (Smith 1913, Franklin 1913).

The second curiosity about the grizzly bear population size at the time of the creation of the National Park Service is that Albright's estimate of forty bears in 1920 was replaced by progressively larger numbers, reaching 320 by 1940. Some writers have regarded this as a real increase of 700 percent in the Yellowstone National Park grizzly bear population (Craighead et al. 1995), but it appears that the growing availability of garbage would have been an important factor biasing observer opportunity. Visitation increased from 79,777 in 1920 to 526,437 in 1940 (560 percent); the corresponding increase in available garbage presumably attracted more bears into the dumps, where they were more visible. Army and National Park Service administrators made no formal effort to count bears beyond those seen at garbage dumps, apparently extrapolating from those seen to some imagined total (Schullery 1992). A more likely scenario is that in 1920, Yellowstone National Park and the surrounding wildlands contained a larger number of bears (certainly more than forty) that were eventually "recruited" into the dump-feeding group as more garbage became available.

Informal but officially published population estimates of grizzly bears in Yellowstone National Park ranged from 320 in 1940 to 180 in 1950. A change in the park's fiscal year and the date of the publication of the superintendent's annual report in the mid-1930s have caused several writers (Bray and Barnes 1967, Craighead and Craighead 1967, Craighead et al. 1995) to mistabulate grizzly and black bear population estimates, tabulating fifteen years of estimates incorrectly by one year. Through the 1960s the National Park Service's Annual Wildlife Census Reports listed a grizzly bear population in Yellowstone National Park of 200 or 250 animals. The Craighead study team's annual counts ranged from 154 in 1959 to 202 in 1966, and they believed that these represented almost all of the bears in the Yellowstone ecosystem, rather than only those in the park itself.

Craighead et al. (1974) reported that from 1959 to 1967, the grizzly bear population for the Yellowstone ecosystem grew from 172 to 195, a 13 percent increase that they believed was real. During the same period, however, visitation increased from 1,408,667 to 2,210,023, a 56 percent increase (Haines 1977). Garbage available to bears also increased. In an early report, the Craighead team stated that they believed that "almost all" of the bears in the ecosystem used the dumps and were counted by them (Craighead and Craighead 1967). However, by 1974 they estimated that 23 percent did not use the dumps (Craighead et al. 1974). It cannot be safely assumed that all the available grizzly bears were ever recruited to the dumps, and this confounded estimates of population size and trend during this period.

Changes in National Park Service management policy in the 1960s resulted in the controversial phasing out and closing of park dumps in the late 1960s and early 1970s for public safety reasons (Craighead et al. 1974, Schullery 1992). The Park Service supported abrupt closure of dumps, whereas the Craighead study team recommended a gradual closure over several years.

Effects of Closing Garbage Dumps

Exceptionally high mortalities of grizzly bears occurred after the dumps were abruptly closed, with 229 grizzly bears removed from the Yellowstone ecosystem between 1967 and 1972 (Craighead et al. 1988). Bears that had habitually used dumps sought similar foods in campgrounds and developments and were eliminated for public safety.

No grizzly bear research occurred between 1970 and 1973. As data collected by the Craighead research project were subjected to repeated analysis by various distinguished but naïve observers (NAS 1974, McCullough 1981), discussions about the status of the grizzly bear population became increasingly conjectural. The population had been affected not only by heavy mortality but by significant changes in available foods. The dumps were closed at a time when bison *(Bison bison)*, elk *(Cervus elaphus)*, and cutthroat trout *(Oncorhynchus clarki)* populations were low, so major alternate food sources were not plentiful. These populations rebounded in the years following dump closures, while the bear population was experiencing high mortality rates. The northern Yellowstone elk herd increased from about three thousand in 1968 to approximately 16,000 in 1986, bison numbers increased from about two hundred in 1966 to two thousand in 1986, and cutthroat trout increased from about 644 metric tons of spawners around Yellowstone Lake in 1966 to 834 metric tons by 1985. In response to controversy surrounding the status of the population, the Interagency Grizzly Bear Study Team (IGBST) was formed by the Department of the Interior in 1973. At that time, information was scarce and the status of the population was regarded as somewhere between passable and grave.

Data on population dynamics accumulated slowly. Radio marking individuals inside Yellowstone Park was not permitted until 1975, and then only on a limited basis—one to two animals per year. By 1980 only sixty-one grizzly bears had been radio-marked: thirty-nine males and twenty-two females, but eleven males and seven females had died during the interim. Only one of those captured bears carried an ear tag from the Craighead study. This was a twenty-year-old female that had been transported to Yellowstone Lake as a cub and apparently learned to capture spawning cutthroat trout.

There were gaps in the age structure (figure 3.1), and males outnumbered females almost two to one. Most older females were associated with garbage dumps still active on the periphery of the park. There was a paucity of females in the prime reproductive age classes (5–12), but males were better represented in all age classes. Males may have been overrepresented in the sample because their larger home ranges made them more likely to encounter a trap. However, males' tendencies to roam may have given them more varied foraging experiences and enhanced their chances of survival.

Mean litter sizes were smaller (1.86, 1975–80) than those observed during the dump years (2.18, before 1970). Smaller litter sizes may have resulted from the overall younger ages of females, a less nutritious diet, or a combination of both. Bears habitually foraging at garbage dumps weigh more and possess higher reproductive rates than their counterparts relying on native foods (Rogers 1976, 1978, Russell et al. 1979, Blanchard 1987, Stringham 1990). Nutritional condition of females has also been positively linked to reproductive success (Jonkel and Cowan 1971, Rogers 1987, Bunnell and Tait 1981, Stringham 1990). The small sample ($n = 61$) collected by 1980 may not have been representative of the entire population, but it seems likely that the effect of removing a major food source, along with the large number of management removals, severely reduced the core park population.

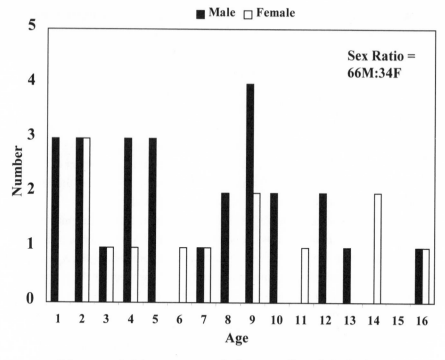

3.1. Yellowstone grizzly bear age structure for forty-one radio-marked animals, 1980.

By 1980 limited research results indicated that the reproductive rate was low (Blanchard and Knight 1980). By late 1982, with more data available, this finding was confirmed. Adult female mortality was excessively high, and the population was declining about 2 percent per year (Knight and Eberhardt 1985).

Consequences of Early Research on Federal Management Policy

Consultation by the senior author of this essay with sympathetic officials in Washington, D.C., resulted in the formation of the Interagency Grizzly Bear Committee (IGBC) in 1983. The IGBC comprises personnel from the wildlife agencies of Idaho, Montana, Washington, and Wyoming, Canadian provinces, the National Park Service, U.S. Forest Service, U.S. Fish and Wildlife Service, and Bureau of Land Management. This committee was charged with dictating management strategies designed to reverse the declining population trend. Adult female mortality was targeted as the most critical population parameter. This resulted in changes of policy regarding removal vs. relocation of problem bears. Major changes in food storage and garbage disposal were instituted on all federal lands within the recovery zone. Sheep allotments were eliminated or transferred to other areas. Law enforcement activities were increased. The Audubon Society offered a $10,000 reward for information resulting in the conviction of anyone illegally killing a grizzly bear. At this stage, it might be said that grizzly bear management became habitat management.

The beginning of habitat management for grizzly bears was a slowing of habitat degradation as federal agencies were required to comply with the Endangered Species Act of 1973. Under the act, agencies were directed to avoid destruction or adverse alteration of habitats of threatened and endangered species. Although the grizzly was declared threatened in 1975, little was accomplished toward recovery before formation of the IGBC.

By 1986 these policy changes resulted in progress toward slowing the population decline (Eberhardt and Knight 1987). An increasing population was recorded by 1993 (Eberhardt et al. 1994, Eberhardt 1995, Boyce 1995), and the population was estimated to be 344 in 1994, with a 90 percent confidence interval of 280–610 (Eberhardt 1995).

Estimating the size of the population has consistently proven a challenge because of the bears' solitary and secretive behavior and large home ranges covering mountainous and largely timber-covered terrain. Much of the study area is not easily accessible, resulting in difficulties obtaining adequate capture-recapture data. Difficulties also arise in keeping radio collars on individual females long enough to obtain reproductive data.

Given these complications, expectations of exact and frequent population size estimates are unrealistic and generally result in controversy and conflict among involved agencies, groups, and individuals. Estimates of population size must be analyzed in conjunction with population trend and habitat availability-suitability data to obtain an accurate description of the present and future situations. For these estimates and databases to remain viable, radio tracking of individuals must continue to assess reproductive and survival rates and monitor habitat use and food habits. To date, no other technology can supply this information.

How the Population Has Changed

POPULATION PARAMETERS

Adult females (< 4 yrs) weighed more during the dump years ($\bar{X}$ = 152 kg, before 1970) compared with this study ($\bar{X}$ = 106 kg, 1975–95). Adult females weighed more and were in better condition (C, calculated by dividing weight by front paw width^3) in the early part of this study (1975–82) compared with the more recent years (1983–94; table 3.1). Before 1983 only six of forty-one cap-

Table 3.1

Comparison of adult female grizzly bear weights and conditions during three periods

	Weight (kg)			Condition		
	mean	**SE**	*n*	**mean**	**SE**	*n*
1959–70	152		72			
1975–82[a]	121	30	41	.58	.14	33
1983–94	98	23	82	.51	.15	68

Notes: Adult is defined as > 4 years. Condition = (weight/front pad width)^3.
[a]Weight: $P = 0.000, t = 4.801$, df = 121; condition: $P = 0.018, t = 2.119$, df = 99.

tures were management actions, compared with thirty of eighty-two after 1982. One might speculate that research bears would weigh more than those caught in management actions, but this was not true. After 1982 management bears actually weighed more than research bears and were in better condition (table 3.2). Management bears were usually those that had been feeding on improperly stored human garbage and food with high nutritive values.

During 1975–82, the average age at which females had their first cubs was 6.0 years, but after 1983 it was 5.7 years. A higher proportion of five- and six-year-old bears were having first cubs after 1982 (ten of eighteen, 56 percent) compared with the earlier years (four of ten, 40 percent). Reasons for this include: (1) bears were exposed to lower nutritional conditions following closures of dumps until they adjusted to foraging only on native foods, and (2) more bears were captured inside the park before 1983 than later, when efforts were increased to capture bears outside the park, where the population was at lower densities. Lower densities would result in less competition for preferred food sources and increased individual nutritional condition.

The frequency of three-cub litters is higher now than during 1975–82, and

Table 3.2

Comparison of weights and conditions of radio-marked research vs. management grizzly bears, 1975–94

	Weight (kg)			Condition		
	mean	**SE**	*n*	**mean**	**SE**	*n*
Research[a]	104	30	87	.53	.15	76
Management[b]	110	22	36	.56	.14	25

Note: Condition = (weight/front pad width)^3. Weight: $P = 0.003, t = 2.870$, df = 80; condition: $P = 0.039, t = 1.796$, df = 66.
[a]Research bears were animals captured for research purposes.
[b]Management bears were animals captured in a conflict situation with humans.

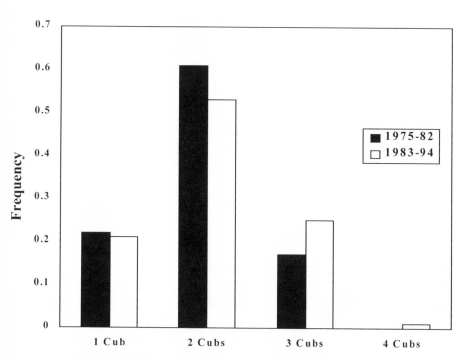

3.2. Yellowstone grizzly bear litter size frequency, 1975–82 and 1983–94. For 1983–94, $P = 0.069$, $X^2 = 3.310$, df = 1.

two four-cub litters were seen 1983–94 (figure 3.2). Larger litters are produced by females in better nutritional condition and by "prime" age females. Females produced most (70 percent) three-cub litters between ages eight and twelve (figure 3.3). Mean litter size of radio-marked females increased from 1.9 in 1975–82 to 2.3 in 1983–94. During 1959–67, mean litter size was 2.2. Litters contained 32 percent female cubs in 1975–82, compared with 61 percent female cubs in 1983–94.

Female reproductive rates—the number of cubs produced per year for completed cyles, or the span of time between production of two litters—were 0.648 during 1959–67, 0.777 during 1975–82, and 0.745 during 1983–94. Mean cycle lengths also changed for those periods: 3.4 years, 2.5, and 3.1, respectively. More marked females weaned cubs as yearlings before 1970 (thirteen of twenty-nine) compared with only two of ten after 1975.

GRIZZLY BEAR FORAGING STRATEGIES AND HABITAT USE

Most, if not all, animal species have a large repertoire of behaviors to deal with various environmental changes. A very abrupt environmental change

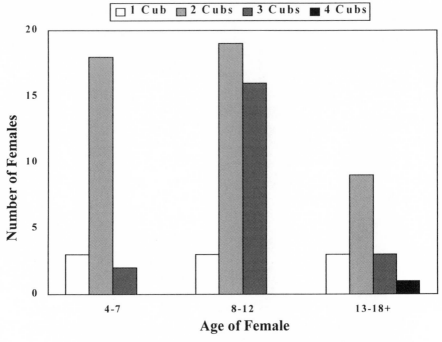

3.3.   Correlation between litter size and age of female Yellowstone grizzly bears.

may dictate a dramatic behavioral change for an entire population. A long-lived species such as the grizzly bear has a "population intelligence" regarding its use of available habitat. This intelligence consists of many learned behaviors accumulated over time and passed from one generation to another. Several decades of dump feeding by a significant portion of the Yellowstone grizzly bear population probably resulted in some loss of this population intelligence, especially as it related to natural summer foods. Thus, with the abrupt closing of the dumps, the bears were left with little knowledge of what to eat instead of garbage, so the remaining bears were compelled to begin accumulating entirely new or lost foraging strategies.

Lack of experience can be fatal to cubs that are inadequately prepared for weaning. For example, in 1978 there was an exceptionally abundant seed crop of whitebark pine *(Pinus albicaulis)*. Much of the crop remained in 1979, when pine seeds accounted for more than 40 percent of the annual diet. This would at first seem to be an advantage for new cubs, and in fact cubs born in 1979 gained more weight than average for cubs and were large and independent enough to be weaned as yearlings in 1980. (Grizzly bears typically wean young in the spring of their two-year-old year.) However, they had fed so exclusively on pine seeds that they lacked the necessary knowledge to find an adequate variety of foods, and they died of apparent malnutrition later in the year they were weaned.

Grizzly bears are omnivorous, and their role in affecting prey population numbers as predators is not as clear and definable as for true carnivores (Noss et al. 1996). Vegetation makes up approximately 80 percent of their annual diet (Mattson et al. 1991). The most stable of these foods are graminoids and forbs. Roots and whitebark pine seeds are consumed in proportion to their availability. Pine seeds are eaten nearly exclusively when available in sufficient quantities. Yellowstone grizzlies exhibit large-scale annual and seasonal diet switching, depending on availability of preferred food items.

As opportunistic carnivores, grizzly bears consume mammals, cutthroat trout, and insects during certain seasons. They feed extensively on winter-killed and weakened ungulates, mostly elk and bison, during March through May, and are efficient predators of newborn elk calves during May and June. Individual bears are known to be effective predators on healthy adult elk during summer, although it is not a widespread behavior in the population. Bull elk become susceptible to predation during the fall rut. The northern Yellowstone elk herd was reduced to approximately three thousand animals by 1968. Their numbers subsequently increased to more than 19,000 by 1995. Predation on elk by bears could be expected to increase with this increase in the prey population. Before 1968 documentation was lacking of substantial predation on elk, but recent observations showed that this behavior has become more common and widespread within the grizzly population.

Grizzly and black bears are known to prey on domestic sheep in the Yellowstone area. Sheep were commonly killed by grizzlies before 1980, when summer grazing allotments on federal land existed within the recovery zone. Black bears caught killing sheep are most frequently killed by federal Animal Damage Control agents. Predation on cattle has been less frequent, and observations suggest that it is done by grizzlies more than by black bears. Cattle are most frequently killed by adult male bears, and most predations involve calves and yearlings on summer pasture on federal allotments. Predation by a limited number of talented individual bears can reach economically significant levels at specific locations (Anderson et al. 1997).

A significant number of grizzly bears catch cutthroat trout during their spawning runs in Yellowstone Lake tributaries from May through July. The number of trout taken depends largely on stream conditions and the density of spawners. Fishing was not documented during the Craighead study, most likely because of the availability of garbage and subsequent changes in fisheries management that resulted in more large fish in the population. Larger fish swim farther up spawning streams, resulting in a wider distribution of trout of larger average size available to bears. Spawning cutthroat trout provide a substantial food source to bears that fish. One radio-marked female grizzly caught and consumed more than one hundred fish in one day (S. P. French, personal communication).

Yellowstone bears typically eat ants annually, but they increase consumption substantially during years when rainfall is low and succulent vegetation is limited. Both grizzlies and black bears raid unsecured bee hives on private land, most typically during late summer and fall of years when other native foods are scarce. In recent years substantial numbers of grizzlies have searched for army cutworm moths *(Eoxoa auxiliaris)* on alpine talus slopes when moths aggregate during August through October (French et al. 1994). Smaller numbers of black bears have also been observed feeding on the moths. Moths are high in fat (72 percent) and may be a preferred food item but are not available every year. Moths depend on snow-free talus slopes with sufficient quantities of nearby flowering alpine flowers to provide nectar, the moths' food supply.

Grizzly bear–habitat relations were gradually revealed by radio-telemetry. Meadows were documented as sources of succulent vegetation, pocket gophers *(Thomomys talpoides)* and their food caches in spring, and root crops during summer and fall. Telemetry revealed the importance of various timber cover types by successional stages, including old growth, not only for cover but also for certain types of foraging (Knight et al. 1984). Preference for dense, high, open-timber edges was also documented (Blanchard 1983).

The importance of maintaining adequate habitat and a level of freedom from human disturbance to preserve the viability of the population became clear to managers. As a result, grizzly bear management guidelines were implemented in 1979, establishing standardized protocols for the national forests, national parks, and Bureau of Land Management lands occupied by Yellowstone grizzlies (Mealey 1986). Grizzly bear management areas were created in Yellowstone Park in 1983. These were areas that could be closed temporarily to humans. Responsible habitat management has become increasingly sophisticated, with the development of cumulative effects analysis (CEA) using GIS and computer modeling. CEA was first applied for grizzly bear management in 1986 in the Cabinet-Yaak area of Montana, and it has subsequently been applied to site-specific cases within Yellowstone. The analyses assist in evaluating existing habitat and associated human activities in an area, and they predict the subsequent effects on grizzly bears inhabiting the area.

During 1988 wildfires burned approximately 12 percent of the Yellowstone ecosystem. Dramatically and widely publicized, these fires were perceived by the public to have devastated the ecosystem and its wildlife, including the threatened grizzly bear. There was no basis for the common perception that many grizzlies within the fire perimeter vacated their traditional home ranges for areas outside the burn perimeter. No radio-tagged bears changed home range location because of the fires. Some grizzlies avoided burned habitats during the year following the fires, but not during any subsequent years (Blanchard and Knight 1996). In fact, postfire effects (1990–95) have been beneficial to grizzly bears through increased production of such foods as forbs and tuberous root crops, largely because of increased soil nutrients (released by the ash) and reduced forest canopy. Clover and fireweed both proliferated following the fires, and grizzly bears dramatically increased consumption of these high protein (> 20 percent) forbs. Annual range sizes of males and females remained similar to those recorded before the fires (cf. Blanchard and Knight 1991, Blanchard and Knight 1996).

Approximately 30 percent of the most productive whitebark pine habitat was burned within Yellowstone Park in 1988. Natural regeneration of whitebark pine following fire is approximately fifty years, and significant cone crops are not produced until trees are about one hundred years old. The long-term effects of reduced cone-producing whitebark pine trees may prove to be deleterious to grizzly bears, especially during years when alternate native foods are also in short supply.

Evidence of Population Saturation

The nutritional condition of female bears has been positively linked to reproductive success, expressed as age of first reproduction, litter size, and length of breeding interval (Jonkel and Cowan 1971, Rogers 1987, Bunnell and Tait 1981, Stringham 1990). Body mass has most frequently been used to express nutritional condition in grizzly bears (Blanchard 1987, Stringham 1990). Female grizzly and black bears that achieved heavier weights at an earlier age have produced first litters at an earlier age (Bunnell and Tait 1981, Blanchard 1987, Rogers 1987, Schwartz and Franzmann 1991). By inference, lighter body weights and later age of first litter production would indicate poorer nutrition, which in turn would reflect quality and/or quantity of food. Food quantity might be affected by individual and cohort competition where food supplies are limited. In other words, competition varies with density of bears. Given this scenario, subordinate bears would do poorly at ecological carrying capacity (KCC) (McNab 1985) and, conversely, would achieve higher levels of nutrition at densities below KCC.

Table 3.3

Comparison of weights and conditions of three- and four-year-old female grizzly bears living inside and outside Yellowstone Park, 1975–94

Location	Weight (kg)			Condition		
	mean	SE	n	mean	SE	n
Inside YNP	66	10	4	.45	.06	3
Outside YNP[a]	100	31	9	.61	.16	8

Note: Condition = (weight/front pad width)^3.

[a]Weight: $P = 0.030$, $t = 2.100$, df = 11; condition: $P = 0.062$, $t = 1.700$, df = 9.

We observed five females with cubs-of-the-year (COY) at four years of age after 1975. All lived primarily outside Yellowstone Park—that is, more than 90 percent of all radio locations were beyond park boundaries. No four-year-olds were reported with COY during 1959–70 (Craighead et al. 1995). We also observed two females with four-cub litters living primarily outside the park. There is slight anecdotal evidence of four-cub litters in the early 1900s (Schullery 1991), and three litters of four cubs were observed inside the park between 1959–70, when garbage dumps were available. Garbage is generally considered to be a high-quality bear food (Herrero 1983), and those bears habitually foraging at dumps weighed more and possessed higher reproductive rates than their counterparts relying on less nutritious native foods (Blanchard 1987, Stringham 1990).

Three- and four-year-old females living outside Yellowstone Park weighed more and were in better condition than those inside the park (table 3.3), although sample sizes were small.

Within the past decade in the park, grizzlies have been more frequently observed feeding on native foods along roadsides and in developed areas in close proximity to people. Most are subadults and females with COY. It is well documented that grizzlies tend to avoid people when possible. This tolerance of human presence suggests greater competition from more dominant bears for similar foods in backcountry areas. This behavior has been noted for only one individual outside the park, an adult female raising litters from 1986 to 1995 near the east entrance.

Population Modeling of Yellowstone Grizzly Bears

In 1995 Robert Barbee, outgoing superintendent of Yellowstone National Park, expressed a view of population models that surprised some biologists but should have warned them of the current state of public and professional attitudes toward modeling as a management tool:

There's this whole notion of cumulative effects modeling in grizzly bear management, for example. Scientists tell us about the quantification of all the complex sets of variables out there, that leads to overlays in a formula that ultimately could be factored down to a certain kind of actionable result in the loss of one tenth of a bear or half a bear in the system in ten years. I am highly skeptical of that, and I think most people of the management mentality are also. . . . There's great skepticism in offices like mine, and a great infatuation with the science of modelling among the scientific people. . . .

It's a perpetual motion machine. It's a big money sump. I think it's gimmicky and it may deserve more respect, but I'm telling you that it doesn't have that respect in the management community. A lot of managers might keep their mouths shut because they don't want to admit how absolutely ignorant they are about all of this new technology. (Barbee 1995)

These views came from a manager who by most standards must be regarded not only as enlightened but as a world leader in the practical application of ecosystem management, and they provide some insight into the views of many managers about population modeling.

Population modeling of Yellowstone grizzly bears began with Craighead et al. (1974), who presented a data set that was repeatedly worked over and remodeled in subsequent years (Gross 1973, Peterson 1973, NAS 1974, McCullough 1981, 1983), in an exercise known cynically as "the numbers game." The approach used by the Craigheads and followed by several subsequent modeling efforts represented an attempt to combine age-specific survival estimates with age-structured reproductive information to estimate population growth.

If this initial round of modeling based on the Craighead data set had a long-term effect among the public and the management community, it may have been to strengthen the viewpoint described by Barbee (1995). By revealing a wide variety of modeling interpretations of the same data set, these modelers may inadvertently have helped create this management skepticism toward models.

Despite the complications of stochastic modeling (Boyce 1995) and bootstrapped growth-rate estimates (Eberhardt et al. 1994), all that the structured population models have revealed is that populations were relatively constant 1959–70, declined 1980–83, and increased 1983–96. Given the uncertainties associated with estimating bear populations in the field, these efforts have been important because they reinforce the same patterns observed in the trend counts based on observations of unduplicated females with cubs-of-the-year.

Another body of bear modeling literature has focused on estimating the probability of extinction for bears. The first such attempt used a demographic model with random births and deaths to anticipate the likelihood of future extinction (Shaffer 1983, Shaffer and Samson 1985). Generally, large populations are less likely to go extinct, so the focus has been on trying to estimate how many bears would be necessary to ensure persistence of a bear population for some arbitrarily chosen length of time into the future—that is, a minimum viable population (MVP). Several such efforts followed Shaffer's pioneering work (Suchy et al. 1985, Dennis et al. 1991, Foley 1994), but all are plagued by limitations in our ability to foretell future demographic environments (Boyce 1992, 1993).

Population modelers often give insufficient attention to connections with habitats. Ultimately the quality and extent of habitats will be fundamental to any long-term population viability for grizzly bears. Thus any useful model for grizzly bears must attempt a link between demography and habitats. There are signs that eventually models will be able to address a larger part of the breathtakingly complex suite of factors that make up a grizzly bear's environment.

But in the political present, the majority view of grizzly bear managers in Greater Yellowstone today is that models, though interesting exercises, still lack the sophistication and precision to be directly applied in management decisions. One reason the cumulative-effects model for the Yellowstone grizzly bear has so little credibility and has advanced so slowly is the lukewarm regard in which it is held by managers. A comment we heard following the sequential presentation of competing grizzly bear population models at the third biennial scientific conference on the Greater Yellowstone Ecosystem, in September 1995, was that though one model predicted population decline and the other increase, they were separated by a statistically trivial spread in data points. Given that models are to date inherently simplistic and unable to reflect even a modest percentage of the environmental variables in the grizzly bear's world, such narrow differences between models seem hardly worth debating and may serve to worsen the public image not only of the models but also of the scientists and managers who pay so much attention to them.

It might even be feared that modeling exercises tend to obscure the fundamental message of grizzly bear conservation behind an opaque screen of computerese. Many managers and political decision makers may presume (or wishfully hope) that somehow models can allow them to reduce the broad tolerances we now recognize as essential to preserving grizzly bears. The models may thus be used to suggest that we can determine exactly the smallest number of grizzly bears a recalcitrant manager must put up with in order to ensure their survival.

In its early days, the cumulative-effects model was presented to the public as a kind of Holy Grail of grizzly bear conservation, a device that would answer all our questions and settle many difficult management issues. In fact, given the realities of grizzly bear population ecology and the changing conditions across the millions of acres of Greater Yellowstone grizzly bear habitat, such a powerful and reliable tool may exist in some remote future, but its appearance is hardly imminent.

The Forgotten Yellowstone Black Bears

Black bears have shared Yellowstone with grizzly bears since before the establishment of the park, and presumably for much longer. An analysis of 168 accounts of Yellowstone wildlife between 1806 and 1882 yielded sixteen mentions of black bears, thirty-three mentions of grizzly bears, and 112 mentions of bears of unidentified species (Schullery and Whittlesey in press). It is difficult to judge from these accounts whether one species was more common. There was probably also a bias among unskilled observers toward assuming that any bears seen were grizzlies. Most modern visitors hopefully assume that any bear they see is a grizzly because of the greater prestige associated with sightings of grizzly bears. Many Yellowstone black bears have brown, cinnamon, and gold coat colors, a fact that causes most novice observers to conclude that they are not "black" bears.

Black bears were the first readily seen bears within the park. By 1889 they were commonly seen at garbage containers near hotels, and after about 1900 a few roadside panhandlers appeared. Roadside begging by black bears (mostly females with young and subadults) was common during the 1920s and through the 1960s. Adult male black bears typically frequented the open pit dumps within the park during hours that grizzly bears were not using them. By 1975 panhandling black bears were virtually eliminated with changes in park sanitation policies, including bear-proofing of all garbage cans, strict enforcement of the policy against feeding the bears, and removal of habituated roadside bears. Because they were no longer visible, people often concluded that the black bear population had been decimated.

Undoubtedly black bears were more numerous when their food supply was augmented by garbage and "hand-outs." Black bears typically prefer the spruce-fir habitats and interspersed meadows as foraging sites (Jonkel and Cowan

1971, Barnes and Bray 1967), but during the 1950s and 1960s, they were commonly observed in lodgepole pine forests along roadsides because of supplemental feeding (Barnes and Bray 1967).

A population estimate of one hundred Yellowstone black bears was first reported in 1920 in the superintendent's annual report. By the end of the decade, the estimate was about five hundred, and it ranged between 360 and 650 through the late 1970s. We cannot determine how these estimates were obtained, but they were very likely extrapolations from numbers seen at dumps and along roadsides. The artificially high densities of roadside-fed black bears extrapolated to the park's extensive backcountry appear to have contributed to highly inflated population estimates.

The only black bear population study in the park calculated a minimum density of one bear per 13.5 square kilometers in part of the Gallatin Mountain Range (Barnes and Bray 1967). Cole (1976) extrapolated this density to obtain a population estimate of 650 black bears within Yellowstone Park. This number appears inflated, but no further research has been conducted on black bears. The disappearance of black bears from the roadsides has led to widespread public belief that black bears are rare or nonexistent. However, sightings of black bears are numerous today, and they are routinely observed on IGBST grizzly bear radio-tracking flights over the Yellowstone ecosystem. Black bears are more commonly observed in certain areas, especially northern Yellowstone Park and northward, and along the Snake River drainage in the south and southwest portions of the park and southward. Gunther et al. (1995) reported that in 1994 visitors and park staff reported 703 observations of grizzly bears, 791 of black bears, and 135 of unknown species. There is no effort to screen or quantify these observations, but they do suggest the possibility of a roughly equal ratio of grizzly to black bears. Further studies are needed to better quantify the population status of black bears in the Greater Yellowstone Ecosystem and how they interact with grizzlies.

Conclusions

The Yellowstone grizzly bear population currently appears to be increasing as both reproductive and survival rates indicate a positive trend. More bears are being observed farther to the south and east of the park, outside designated wilderness and the recovery zone. Habitat degradation appears to be most severe in Montana and Idaho, with resulting low numbers and higher mortality rates in those areas. The future of the Yellowstone population appears to lie in the park and Wyoming. The extent of occupancy in Wyoming outside of wilderness will depend on the level of tolerance by managing agencies and local inhabitants.

Chances of grizzlies coming into conflict with people also increase as they move closer to larger concentrations of private land where garbage and other human foods are unsecured. Depending on circumstances, conflicts on private land often result in either the death of a bear or transport away from the situation (and, in many cases, the subsequent death of the bear). Mortalities will certainly escalate as the population continues to increase. To date, this problem has not been adequately addressed by managers, and if not reconciled may well result in a backlash of anti–grizzly bear sentiment among private landholders.

It is inevitable that in some years native food production is low as a result of annual weather variation. As bears enter hyperphagia and attempt to accumulate sufficient fat to prepare for denning, they seek food wherever they can find it. As population densities increase, subordinate animals are forced into nonpreferred situations, including human developments. In recent years this problem has become apparent as bears seek food during fall, and conflicts increase with humans on nearby national forests and private lands. These individuals are most frequently captured and transported to what we *hope* is unfamiliar territory where we *hope* they will find sufficient native foods to prepare for hibernation. Unfortunately, most adult grizzlies return to their original home range no matter where they are moved in the Yellowstone ecosystem (Blanchard and Knight 1995). Generally, most conflicts have not been resolved by the time the bear returns, primarily because the problems occur predominantly on private lands. Traditionally, the park has nearly always accepted these individuals from adjacent state wildlife agencies, except in instances of demonstrated aggression toward people or incorrigible food-rewarded behavior. If the park is at or near carrying capacity for grizzly bears, managers must ask what effects these transported individuals have on the resident bears and their social structure.

Other threats to the population loom on the horizon. The greatest single threat is the ever increasing development of private lands in grizzly bear range. Development not only decreases available habitat but also increases the potential for human-bear conflicts, especially during years of limited native food availability.

Whitebark pine in Yellowstone is threatened by white pine blister rust (*Cronartium rubicola*), which has killed nearly all whitebark pine in northwestern Montana and has been moving south. Infections in seedlings and the cone-bearing branches of mature trees were observed in 1995 in Yellowstone and Grand Teton parks (K. C. Kendall, personal communication). Grizzly bears eat whitebark pine seeds to the near exclusion of any other foods when they are available. However, cone production is unpredictable, and bears in Yellowstone have experienced many years when they had to search for other foods during fall to accumulate fat for hibernation. Unfortunately, some bears look to people and their developments for these foods. A late frost in the spring of 1993 killed the terminal reproductive organs on whitebark pine trees in most of the Yellowstone ecosystem except south of the park. This eliminated any cone production for three years—the time it takes whitebark cones to mature. In response, bears resorted to succulent vegetation in seep areas, root crops (such as yampa [*Perideridia gairdneri*]), ants, mushrooms, elk and moose *(Alces alces)* susceptible to predation, and food associated with people. Grizzly bears may not need whitebark pine seeds to survive. In 1995, after three years during which there were no whitebark pine seeds in most of the Yellowstone Ecosystem, bears captured for research purposes during late summer and fall were in excellent condition. The largest number of females with cubs-of-the-year ever recorded was in 1996 (thirty-three females with seventy cubs).

The discovery of lake trout *(Salvelinus namaycush)* in Yellowstone Lake in 1994 potentially threatens the cutthroat trout population and may compromise or destroy an important food source for bears that feed heavily on spawners. Lake trout have the potential to reduce the cutthroat trout population drastically through competition and predation, and they have the ability to become a keystone predator, an organism that greatly influences its ecosystem.

Global warming may affect many important bear foods, especially whitebark pine (> 2500 m elevation) and cutworm moths (> 3100 m). In worst-case scenarios, these two food sources would disappear. However, any warming will be gradual and give this extremely versatile forager time to adapt. The worst-case scenario for global warming does not project Yellowstone to become as warm as northern Mexico, where grizzly bears flourished into this century.

The status and trend of habitat available to grizzly bears appears to be the most critical determinant of viability of the population in the long term. A high degree of uncertainty exists concerning the quantification and monitoring of adequate habitat and associated human activities. Managers are relying on the Cumulative Effects Analysis (CEA) to produce the habitat and activities thresholds necessary to reduce that uncertainty. A flaw with the CEA is that the habitat coefficients (relative value of habitat types to grizzly bears) have not yet been validated as to their importance to grizzlies. We recommend the telemetry data gathered in this study be used to derive coefficients and determine probability of use by season. These coefficients should be incorporated in the CEA model before any CEA output values are used.

Currently the Yellowstone grizzly bear population is protected under the Endangered Species Act. Is this status good or bad for the long-term future of the population? Under this mandate, the Forest Service is directed not to adversely affect grizzly bear habitat. At this time there are no assurances that this direction would continue if the population were delisted and no longer protected under the act. Agencies are at this time operating under a "no net loss" premise. On the other hand, under the act, state agencies are severely restricted in their abilities to deal with an expanding population and more numerous human-bear conflicts outside the recovery zone. Inability to deal with problem situations has the very real potential of creating a backlash of reduced tolerance of grizzlies by local residents.

The Yellowstone grizzly bear population is currently isolated from other populations. Can it remain viable without injections of genetic diversity through transplants of genetically different individuals from other populations or establishment of safe travel corridors to other ecosystems (see Chapter 11)? DNA research has shown that Yellowstone grizzly bears have twice experienced a population bottleneck, a type of genetic drift resulting from a severe reduction in population size such that genetic material of the surviving population is a subset of that of the original population (Waits and Ward in press). The first occurred in 1850–1920, when grizzlies were exterminated to the south, and the second came in the early 1970s, when the garbage dumps were closed. It is unclear when Yellowstone grizzlies became isolated from northern and western populations; however, anecdotal information on sightings indicates this occurred approximately fifty to sixty years ago. More research is needed before infusion of possibly alien genetic strains by indiscriminately transporting individuals from other ecosystems. The possibility of establishing corridors among populations is currently being investigated through satellite imagery and GIS technology. Corridor possibilities are severely hampered by the predominantly private ownership of land in the most desirable locations. Bears living on Kodiak Island have been isolated from other populations for approximately ten thousand years and show the least genetic variability of any brown or grizzly bear populations (Waits and Ward in press). There is no indication that that population is suffering from loss of variability, as its reproductive rate is one of the highest of all populations (IGBC 1987). The Yellowstone population shows microsatellite DNA genetic diversity within the range of other Rocky Mountain grizzly bear populations (Waits and Ward in press). We see no reason to be concerned about lack of genetic diversity of the Yellowstone population until further research demonstrates a need for apprehension.

Expectations that DNA research will produce an accurate population estimate appear to be premature at best. Costs of sampling the entire ecosystem in a statistically rigorous procedure would be excessive even if logistically possible. Any population estimate gives a number of animals for that point in time and gives no information about the trend. Marking individuals with transmitters is currently the only technology that can provide information from which to calculate survival and reproductive rates and therefore population trends. We recommend that capture and marking continue throughout the ecosystem in an organized monitoring effort.

Acknowledgments: Rod Drewien and Chris Servheen contributed to this chapter through their constructive reviews, as did Mark Boyce through his insights on modeling.

Literature Cited: Albright, H. 1920. Annual report for Yellowstone National Park, 1920. National Park Service files, Yellowstone National Park Research Library, Yellowstone National Park.

Anderson, C. R., M. A. Ternent, D. F. Moody, M. T. Bruscino, and D. F. Miller. 1997. Grizzly bear–cattle interactions on two cattle ranges in northwestern Wyoming. Wyoming Game and Fish Department, Cheyenne.

Barbee, R. 1995. Yellowstone Science interview: A bee in every bouquet. Yellowstone Science 3(1):8–14.

Barnes, V. G., and O. E. Bray. 1967. Population characteristics and activities of black bears in Yellowstone National Park. Colorado Cooperative Wildlife Research Unit, Colorado State University, Fort Collins.

Blair, N. 1987. The history of wildlife management in Wyoming. Wyoming Game and Fish Department, Cheyenne.

Blanchard, B. M. 1983. Grizzly bear–habitat relationships in the Yellowstone area. International Conference on Bear Research and Management 5:118–23.

———. 1987. Size and growth patterns of the Yellowstone grizzly bear. International Conference on Bear Research and Management 7:99–107.

Blanchard, B. M., and R. R. Knight. 1980. Status of grizzly bears in the Yellowstone System. Proceedings of the 45th North American Wildlife Conference 45:263–67.

———. 1991. Movements of Yellowstone grizzly bears. Biological Conservation 58:41–67.

———. 1995. Biological consequences of relocating grizzly bears in the Yellowstone ecosystem. Journal of Wildlife Management 59(3):560–65.

———. 1996. Effects of wildfire on grizzly bear movements and food habits. Pp. 117–22 *in* J. Greenlee, ed., Ecological implications of fire in Greater Yellowstone. International Association of Wildland Fire, Fairfield, Washington.

Boyce, M. S. 1992. Population viability analysis. Annual Review of Ecology and Systematics 23:481–506.

———. 1993. Population viability analysis: Adaptive management for threatened and endangered species. Transactions of the North American Wildlife and Natural Resources Conference 58:520–27.

———. 1995. Population viability for grizzly bears *(Ursus arctos horribilis):* A critical review. College of Natural Resources, University of Wisconsin, Stevens Point.

Bray, O. E., and V. G. Barnes. 1967. A literature review on black bear populations and activities. National Park Service and Colorado Cooperative Wildlife Research Unit, Fort Collins.

Bunnell, F., and D. Tait. 1981. Population dynamics of bears: Implications. Pp. 75–98 *in* C. W. Fowler and T. D. Smith, eds., Dynamics of large mammal populations. John Wiley and Sons, New York.

Cannon, K. P. 1992. A review of archaeological and paleontological evidence for the prehistoric presence of wolf and related prey species in the northern and central Rockies physiographic province. Pp. 1-175 to 1-265 *in* J. Varley and W. Brewster, eds., Wolves for Yellowstone? A report to the United States Congress, vol. 4, research and analysis. National Park Service, Yellowstone National Park.

Cole, G. F. 1976. Management involving grizzly and black bears in Yellowstone National Park, 1970–75. Natural Resources Report no. 9, National Park Service, Yellowstone National Park.

Craighead, J., and F. Craighead. 1967. Management of bears in Yellowstone National Park. Unpublished report, National Park Service files, Yellowstone National Park.

Craighead, J., K. R. Greer, R. R. Knight, and H. I. Pac. 1988. Grizzly bear mortalities in the Yellowstone ecosystem, 1959–87. Report of Montana Department of Fish, Wildlife and Parks, Craighead Wildlife-Wildlands Institute, Interagency Grizzly Bear Study Team, and National Fish and Wildlife Foundation, Helena.

Craighead, J., J. S. Sumner, and J. A. Mitchell. 1995. The grizzly bears of Yellowstone. Island Press, Washington, D.C.

Craighead, J., J. Varney, and F. Craighead. 1974. A population analysis of the Yellowstone grizzly bears. Bulletin 40, Montana Forest and Conservation Experiment Station, University of Montana, Missoula.

Dennis, B., P. L. Munholland, and J. M Scott. 1991. Estimation of growth and extinction parameters for endangered species. Ecological Monographs 61:115–43.

Eberhardt, L. L. 1995. Population trend estimates from reproductive and survival data. Pp. 13–19 *in* Yellowstone grizzly bear investigations 1994, annual report of the Interagency Study Team. National Biological Service, Bozeman.

Eberhardt, L. L., B. M. Blanchard, and R. R. Knight. 1994. Population trend of the Yellowstone grizzly bear as estimated from reproductive and survival rates. Canadian Journal of Zoology 72:360–63.

Eberhardt, L. L., and R. R. Knight. 1987. Prospects for Yellowstone grizzly bears. International Conference on Bear Research and Management 7:45–50.

Foley, P. 1994. Predicting extinction times from environmental stochasticity and carrying capacity. Conservation Biology 8:124–37.

Franklin, W. S. 1913. The Yellowstone Park. Science 38:127–29.

French, S. P., M. G. French, and R. R. Knight. 1994. Grizzly bear use of army cutworm moths in the Yellowstone ecosystem. International Conference on Bear Research and Management 9:389–99.

Gross, J. 1973. A computer re-analysis of a computer analysis of the Yellowstone grizzly bear population by Craighead, Varney, and Craighead. Unpublished report, National Park Service files, Yellowstone National Park.

Gunther, K. A., M. J. Biel, S. D. Rice, and H. H. Hoekstra. 1995. Yellowstone National Park bear management summary, 1994. Report of the National Park Service, Yellowstone Center for Resources Bear Management Office, Yellowstone National Park.

Hadly, E. A. 1995. Evolution, ecology, and taphonomy of Late-Holocene mammals from Lamar Cave, Yellowstone National Park, Wyoming, USA. Ph.D. diss., University of California, Berkeley.

Haines, A. 1977. The Yellowstone story. 2 vols. Yellowstone Library and Museum Association and Colorado Associated University Press, Boulder.

Herrerro, S. 1983. Bear attacks, their causes and avoidance. Nick Lyons Books, New York.

Interagency Grizzly Bear Committee. 1987. Grizzly bear compendium. U.S. Fish and Wildlife Service, Missoula.

Jonkel, C. J., and I. McT. Cowan. 1971. The black bear in the spruce fir forest. Wildlife Monographs no. 27.

Knight, R. R., and L. L. Eberhardt. 1985. Population dynamics of Yellowstone grizzly bears. Ecology 66:323–34.

Knight, R. R., D. J. Mattson, and B. M. Blanchard. 1984. Movements and habitat use of the Yellowstone grizzly bear. Interagency Grizzly Bear Study Team Report, National Park Service, Bozeman.

Mattson, D. J., R. R. Knight, and B. M. Blanchard. 1991. Food habits of Yellowstone grizzly bears, 1977–87. Canadian Journal of Zoology 69:1619–29.

McCullough, D. R. 1981. Population dynamics of the Yellowstone grizzly bear. Pp. 173–96 *in* C. W. Fowler and T. D. Smith, eds., Dynamics of large mammal populations. J. Wiley and Sons, New York.

——. 1983. The Craigheads' data on Yellowstone grizzly bear populations: Relevance to current research and management. International Conference on Bear Research and Management 5:21–32.

McNab, J. 1985. Carrying capacity and slippery shibboleths. Wildlife Society Bulletin 13:403–10.

Mealey, S. P. 1986. Interagency grizzly bear guidelines. USDA Forest Service, USDI National Park Service, Bureau of Land Management, Idaho Fish and Game Department, Montana Department of Fish, Wildlife and Parks, Washington Game Department, Wyoming Game and Fish Department.

National Academy of Sciences. 1974. Report of committee on the Yellowstone grizzlies. National Academy of Sciences, Washington.

Noss, R. F., H. B. Quigley, M. G. Hornocker, T. Merrill, and P. C. Paquet. 1996. Conservation biology and carnivore conservation in the Rocky Mountains. Conservation Biology 10:949–63.

Peterson, S. 1973. Comments on the Craighead computer evaluation. Unpublished report, National Park Service files, Yellowstone National Park.

Rogers, L. L. 1976. Effects of mast and berry crop failures on survival, growth, and reproductive success of black bears. Transactions of the North American Wildlife and Natural Resources Conference 41:431–38.

——. 1978. Effects of food supply, predation, cannibalism, parasites, and other health problems on black bear populations. Pp. 194–211 *in* F. Bunnell, D. Eastman, and J. Peek, eds., Symposium on natural regulation of wildlife populations, March 10, 1978, Vancouver.

——. 1987. Effects of food supply and kinship on social behavior, movements and population growth of black bears in northern Minnesota. Wildlife Monographs no. 97.

Russell, R., J. Nolan, N. Moody, and G. Anderson. 1979. A study of the grizzly bear in Jasper National Park, 1975–1978. Canadian Wildlife Service, Edmonton.

Schullery, P. 1991. Yellowstone bear tales. Roberts Rinehart, Niwot, Colorado.

——. 1992. The bears of Yellowstone. 3d ed. High Plains Publishing Company, Worland, Wyoming.

——. 1995. Yellowstone's ski pioneers: Peril and heroism on the winter trail. High Plains Publishing Company, Worland, Wyoming.

Schullery, P., and L. Whittlesey. 1992. The documentary record of wolves and related wildlife species in the Yellowstone National Park area prior to 1882. Pp. 1-3 to 1-173 *in* J. D. Varley and W. G. Brewster, Wolves for Yellowstone? A report to the United States Congress, vol. 4, research and analysis. National Park Service, Yellowstone National Park.

——. In press. Summary of the documentary record of wolves and other wildlife species in the Yellowstone National Park area prior to 1882. *In* L. Carbyn, ed., Ecology and conservation of wolves in a changing landscape. Canadian Circumpolar Institute, University of Alberta, Edmonton.

Schwartz, C., and A. Franzmann. 1991. Interrelationships of black bears to moose and forest succession in the northern coniferous forest. Wildlife Monographs no. 113.

Shaffer, M. L. 1983. Determining minimum viable population sizes for the grizzly bear. International Conference on Bear Research and Management 5:133–39.

Shaffer, M. L., and F. B. Samson. 1985. Population size and extinction: A note on determining critical population sizes. American Naturalist 125:144–52.

Skinner, M. P. 1925. Bears in the Yellowstone. A. C. McClurg and Company, Chicago.

Smith, J. 1913. The Yellowstone Park. Science 37:941.

Stringham, S. 1990. Grizzly bear reproductive rate relative to body size. International Conference on Bear Research and Management 8:433–43.

Suchy, W., L. L. McDonald, M. D. Strickland, and S. H. Anderson. 1985. New estimates of minimum viable population size for grizzly bears of the Yellowstone ecosystem. Wildlife Society Bulletin 13:223–28.

Waits, L., and R. H. Ward. In press. A comparison of genetic variability in brown bear populations from Alaska, Canada, and the lower 48 states. Tenth International Conference on Bear Research and Management, July 16–20, 1995, Fairbanks, Alaska.

Wright, W. 1908. Letter to Acting Superintendent S. B. Young. Yellowstone National Park Archives, Army Records, Letters Received September–October 1980, vol. 7, pp. 341–49.

Cougar investigating a bighorn sheep carcass (P. Ian Ross)

Chapter Four

The Ecology of Anthropogenic Influences on Cougars

Kerry M. Murphy, P. Ian Ross, and Maurice G. Hornocker

Cougars *(Felis concolor)* play an important role in ecosystems, perhaps more than any other large carnivore in the Western Hemisphere. This influence stems from the cougar's extraordinary capability as a predator, its wide geographic distribution in both South and North America, and the broad diversity of habitats that it uses. Cougar predation helps structure ecological communities with direct effects on the relative abundance of different prey species (Terborgh 1988, 1990, Berger and Wehausen 1991). This influence may extend to vegetation if predation by cougars alters patterns of herbivory in its prey (Terborgh 1988). Cougar predation may also dampen prey oscillations or alter characteristics of individual prey populations by selectively removing individuals in vulnerable sex or age classes (Hornocker 1970, Ross et al. 1997, Chapters 10 and 12 this volume).

Cougars are highly specialized predators that hunt alone and often kill prey as large or larger than themselves. Because several days are required to consume large items, cougar food caches are often sites of interactions between cougars and scavengers. In the northern Greater Yellowstone Ecosystem (GYE), carrion provided by cougars may be a significant food source for bears (*Ursus americanus, U. arctos;* Murphy et al. 1998), coyotes *(Canis latrans),* or birds (for example, magpies [*Pica pica*]).

Although historically cougars were widely distributed across North America (Young 1946), habitat destruction, indiscriminate killing of prey, and direct persecution eventually reduced their known range to portions of southern Florida and mountainous regions of western Canada and the United States. Agricultural, industrial, and residential settlement continues to influence the range and abundance of cougars in the New World. Currently, the degradation of cougar habitat that results from human development and disturbance is the most serious long-term threat to cougar conservation. Human-caused mortality associated with hunting, control actions, or accidents may further threaten the persistence of small cougar populations in habitats fragmented by human development (e.g., Beier 1993).

Our goal is to review and synthesize knowledge of human effects on cougars over the entire range of scales at which the two species interact—from individual animals to populations whose persistence may be tied to processes that transcend ecosystem boundaries. This chapter is intended to be used by laymen and resource managers as a practical source of information on current issues in cougar conservation.

Cougar Life History and Habitat

LIFE HISTORY

Cougars regularly interact with members of their own species and are more aptly described as noncooperative than solitary (see Leyhausen 1965, Sandell 1989). Their social interactions communicate the status of important resources that pertain to conspecifics, such as the presence of potential mates or competitors for food. Most communication among cougars occurs passively, in absentia, through visual and olfactory markers. Common markers, particularly among males, are combinations of feces or urine deposited on mounds of soil and debris scraped from forest litter. Scrapes (or scratches) are left under large prominent trees or under rock overhangs at the mouths of canyons, on ridges, or along other natural travel routes of cougars. Vocalizations, facial and body posturing, and fighting may be used to communicate directly and to settle disputes over access to resources.

Adult male cougars (typically more than twenty-four months old) actively defend territories through both direct and indirect communication. Females occupy home ranges that are not defended. Territories and home ranges provide resources to the individuals that use them, at least partially to the exclusion of other cougars of the same sex. Cougars occupy territories and home ranges based primarily on land tenure—that is, their prior use of the space (Seidensticker et al. 1973). Direct interactions or indirect communications mediate the initial acquisition of a range or territory and may lead subsequently to modification of spatial and dominance relations among resident adults. Communication mechanisms used by cougars serve in part to facilitate the spacing of individuals at levels consistent with the prevailing density of critical resources. Although land tenure provides for consistency in spatial relations, social ties remain dynamic because of deaths and changes in the presence and dominance status among adults and because the spatial distribution of critical resources may change as well.

The dispersion of prey and the cover needed to stalk them may determine the spacing patterns of female carnivores (Seidensticker et al. 1973, Sandell 1989). For female cougars, these resources largely determine the sizes of home ranges (about 100 square kilometers in northern Yellowstone) and the extent that ranges overlap (Seidensticker et al. 1973). Home range sizes and spatial relations among resident females may change with increasing food requirements and the mobility of growing kittens. Territories of males in the northern GYE (about 425 square kilometers) are largely exclusive of other males and typically overlap those of up to three adult females. For males, adult female cougars, as well as food and cover, are important resources (Sandell 1989), and they may be aggressively defended. Adult male cougars are intensely competitive, and resident males may kill subadult males if they linger in their territories. Males provide no nutritional support to their mates and offspring, but through their larger territories may provide a spatial umbrella of protection for maternal females (that is, females supporting kittens) against subadult and intruding adult males that could be infanticidal (Seidensticker et al. 1973).

In southwestern Alberta and the northern GYE, cougars give birth primarily during late spring, summer, or early fall (figure 4.1). Litter sizes at birth are as high as four or possibly five kittens, but they frequently decline through natural mortality. Even during periods of population growth, fewer than one-half of the kittens born in the northern GYE may survive to independence (eleven to twenty-two months old). About two-thirds of the kittens that survive to reproductive age (nineteen to thirty-six months) in the northern GYE disperse to other cougar habitats in the ecosystem. These other areas also support local cougar populations that provide reciprocal contributions of subadult cougars, particularly males, to northern Yellowstone. A subadult male cougar marked in southern Idaho dispersed to northern Yellowstone, a distance of at least 380 kilometers (J. Laundré, personal communication). Similarly, a subadult male cougar in the northern GYE dispersed 174 kilometers to southern Idaho. Two female cougars from southwestern Alberta each moved more than two hundred kilometers from their natal areas. At least one of these was a "successful" disperser (Weaver et al. 1996), in that she survived long enough to reproduce. Immigration of cougars substantially increases recruitment of subadult and adult cougars to the northern GYE. During some years, immigration may add

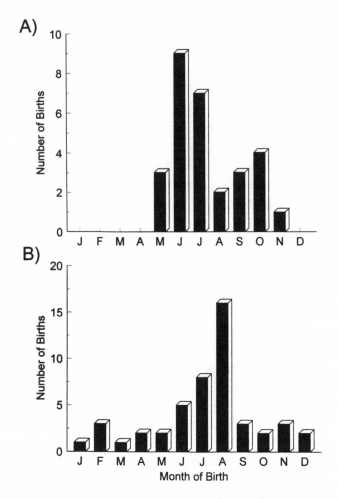

4.1.   Birth months of cougar litters in (A) the northern Yellowstone Ecosystem, 1987–95, and (B) southwestern Alberta, 1981–92.

more subadult cougars than the local production of kittens. The importance of immigration in enhancing population stability is documented in other ecosystems as well (Seidensticker et al. 1973, Logan et al. 1986, Beier 1993, Laing and Lindzey 1993, Logan et al. 1996).

THE HABITAT CONCEPT APPLIED TO COUGARS

The reproductive fitness of an individual cougar depends on its acquisition of food and mates, maintenance of home ranges or territories, and avoidance of enemies. Biologists are interested in the amounts, sizes, and arrangements of the components of habitat that support these activities. Although habitat components for cougars are often thought of as vegetation or physiographic types, they may also be such mobile aspects of habitat as prey, other cougars that are potential mates or aggressors, or large carnivores of other species that compete for food or that may pose a mortal threat. Because habitat components affect such vital processes as births, deaths, immigration, and emigration, these features affect the persistence of cougar populations (Ruggiero et al. 1988).

The cougars' relation to habitat is dynamic. The forces that affect cougar behavior change daily, seasonally, and annually, as influenced by the individuals' basic needs, physical maturation, and experience, as well as environmental variation. For example, such natural forces as fire or weather influence prey and cover. For cougars to survive and reproduce, their changing needs must be met by environments that also vary in their output of resources (Ruggiero et al. 1988).

Steep mountain slopes or abundant horizontal and vertical cover provided by vegetation and topography are important habitat components for cougars (Seidensticker et al. 1973, Logan and Irwin 1985, Van Dyke et al. 1986b, Belden et al. 1988, Koehler and Hornocker 1991, Laing and Lindzey 1991, Maehr et al. 1990, 1991, Williams et al. 1995). Cover facilitates their ambush-style hunting techniques and provides security from enemies. Such habitats as grassland steppe or agricultural fields provide abundant forage for ungulates in the GYE, and ungulates are common in these habitats, but cougars infrequently kill prey in these areas, presumably because stalking cover is inadequate. Cougars apparently also use cover for security from predators, including other cougars, when resting, traveling, denning, or feeding (Logan and Irwin 1985, Laing and Lindzey 1991). Maternal females may select dens and nursery sites that allow kittens to escape predators. Cougar dens and nurseries in the northern GYE and southwestern Alberta are invariably located in rock outcrops, in dense shrub fields, or under downed conifers. These sites contain small passages used by kittens that are difficult for larger predators to negotiate successfully. Laing and Lindzey (1991) documented that habitat selections of maternal females were significantly different from those of nonmaternal females. Maternal females preferred habitats with an understory that contained large boulders. Maehr et al. (1989, 1990) and Beier et al. (1995) also found that maternal females used natal dens in dense vegetation with abundant vertical and horizontal cover.

Such characteristics of prey as density, distribution, and behavior influence predator-prey relations (Sunquist and Sunquist 1989). These variables interact with physiographic features of habitat to influence cougar predation (Hornocker 1970, Seidensticker et al. 1973). Because prey are mobile, their distribution can change in response to their own density (Fretwell and Lucas 1970), seasonally variable climatic conditions, natural disturbance such as fire, and temporally fluctuating distributions of predators (Rogers et al. 1980).

Human Effects on Cougars

Human cultural, economic, and social systems affect cougars both directly and indirectly. Direct effects influence the lives of individuals as well as population survival rates. Cougars are affected indirectly by losses in habitat quality and by human-caused losses of habitat effectiveness. Because both direct and indirect effects influence vital population processes, they may strongly influence cougar population size and persistence.

INDIRECT EFFECTS

Humans alter the physical structure, biotic composition, and arrangement of cougar habitat components. The principal effects of human activity are (1) habitat destruction that reduces and fragments cougar populations, (2) modifications to the physical structure of cougar habitat when land-use practices reduce cover for hunting or security, (3) reductions in habitat effectiveness resulting from human disturbance, (4) habitat modifications that alter predator-prey relations, and (5) the modification of competitive relations between cougars and other predators and scavengers.

Habitat destruction. Development for industry, roads, or intensive agriculture can modify habitat so that it no longer provides the cover or security needed by cougars. Human population growth and resulting subdivision of cougar habitat for housing also contribute to habitat reduction and fragmentation.

Human populations and the number of households in counties making up the GYE increased 7.4 percent and 8.5 percent, respectively, from 1980 to 1990 (Feigley 1993). Riley and Malecki (1996) note that, with respect to human populations, seven of the ten fastest-growing states are located in the West, which also supports the majority of occupied cougar habitat. Maehr (1990b) found that "natural" cougar habitats in part of south Florida were reduced by more than 50 percent from 1900 to 1973 by urban and agricultural development of private land. This private land may also provide the most productive habitat for prey and cougars (Roelke 1987, Maehr et al. 1989), so development of these areas may have a disproportionately negative influence on cougar populations.

Environmental variation, demographic variation, and inbreeding effects may interactively cause extirpations of small wildlife populations. Vacant habitats may be naturally recolonized only if they remain geographically linked to viable populations that provide a source of immigrants. Cougars use travel corridors, habitats that facilitate movement of wildlife between habitat patches and conspecific populations (Beier 1995, Maehr 1990b). Cougar populations may become isolated because travel corridors are occluded by human development or disturbance. Destruction of travel corridors and core habitat severely endangers the Florida panther (Maehr 1990b) and currently threatens at least one cougar population in southern California (Beier 1993).

Although cougar habitat in the GYE is still largely contiguous, the open shrub and grassland steppe that occurs in its large valleys were probably historically avoided by cougars, except to the extent that riverine and riparian zones provided food and cover. Cougars still cross valley bottoms in the northern GYE, particularly where valleys are naturally constricted by mountainous terrain. But housing and agricultural developments exacerbate cougars' avoidance of open habitats in lowlands, thus helping to define travel corridors for cougars and to increase the ecological value of corridors that remain passable.

Travel corridors identified by Beier (1995) in southern California were typically drainage washes or ridges with abundant native woody vegetation that provided security to traveling cougars. Cougars avoided corridors characterized by noise, night lighting, domestic dogs, or other human disturbances (Beier 1995). Roadways with high-speed automobile traffic reduced cougar safety. Cougars may use highway underpasses or overpasses if they occur in travel corridors (Beier 1995).

Land-use practices in wildlands. In national forests timber harvest and road building alter wildlife habitat more than any other activity (Thomas 1979). Logging, burning, grazing, or other physical modifications of landscapes may reduce the cover needed by cougars (Logan and Irwin 1985, Van Dyke et al. 1986b, Laing and Lindzey 1991). Smallwood (1994) reported a 61 percent decrease in cougar track density in timber harvest areas of California from 1986 to 1992. Van Dyke et al. (1986b) concluded that in general logging would have negligible effects on resident cougars if logged areas were small relative to area requirements of cougars and if unlogged areas remained on the periphery of their home ranges. However, vegetative modifications are likely to have strong spatial and temporal aspects. Important variables are vegetative type, the extent to which vegetation structure is changed, and the short-term and long-term vegetative responses. An adverse temporary effect on cougar hunting cover might be compensated by positive long-term effects on prey abundance. For example, small-scale prescribed burns may locally reduce cover in fire-adapted vegetative types yet benefit prey by increasing habitat diversity and the nutritive quality of forage for prey (e.g., Schortemeyer et al. 1991). Beier and Barrett (1993) observed that three cougars avoided for only two to six weeks recently burned habitats characterized by chaparral, oak *(Quercus agrifolia* and *Q.*

engelmannii) woodlands, coastal scrub, and conifer forests. Williams et al. (1995) noted that two cougars used burned conifer forests that provided cover from rock outcrops or fireweed *(Epilobium angustifolium)*. In contrast, catastrophic fires that eliminate canopy and understory cover in conifer forests over areas exceeding three square kilometers may reduce cougar habitat quality until vegetative cover improves and prey populations again use these sites.

The effect of habitat alterations may also vary with climate. For instance, logging that opens forest canopies may lead to snow accumulations that, at least over the short-term, make the use of these sites energetically cost-ineffective for prey and cougars.

Reductions in habitat effectiveness from human disturbance. Human disturbance, particularly when associated with roads, can cause habitat to be less effective in supporting cougars, even when the habitat may be of high quality in other respects. Cougars may avoid such disturbances as mining, logging, or recreational machinery because they associate them with human presence, especially if these are novel stimuli and are interpreted as threats. They may continue to alter their behavior in anticipation of human presence, even after these stimuli are removed. Regardless, disturbed habitat is often less effective for cougars because its resources are used less optimally than undisturbed habitat.

Van Dyke et al. (1986a) found that cougars in Utah shifted toward nocturnal activity patterns in the presence of human disturbance and that cougars in Arizona crossed hard-surfaced and improved dirt roads less frequently than less-traveled, unimproved dirt roads. In Arizona resident cougars selected home ranges that were largely free from human disturbance and avoided timber sale sites (Van Dyke et al. 1986b). Maehr (1990a) documented that Florida panthers used a portion of the Big Cypress National Preserve less than expected during a hunting season for white-tailed deer *(Odocoileus virginianus)*, presumably because of disturbance by hunters. Belden et al. (1991) reported that disturbance associated with a hunting season displaced translocated cougars from newly occupied habitat.

Human activity, however, does not always lead to strong patterns of continued avoidance. If repeated exposure to humans has no strong negative consequences, cougars may continue to use habitat where human activity is a regular feature of their environment. For example, cougars in Alberta did not abandon core home ranges that included human presence, as long as human activity was intermittent (Jalkotzy and Ross 1995). Similarly, concentrated recreational activity at campground and picnic sites did not always deter cougars from using habitats within 250 meters. Beier (1995) found that dispersing cougars used areas and bedded near trails that were heavily used by hikers, bicyclists, and equestrians. Dispersers that traveled at night showed little aversion to parked vehicles with quiet motors and isolated residences without lighting. Halfpenny et al. (1991) and Ruth (1991) documented habituated cougars where prey concentrated near humans in Boulder County, Colorado, and Big Bend National Park, Texas, respectively. In these cases, the distribution of prey was strongly related to seasonal climatic conditions. Acquiring prey near human activity probably encourages cougars to habituate to human presence, because nutritional rewards may override the animal's opposing concern for its security (Ruth 1991).

Human effects on cougar prey. Humans also alter cougar habitat quality by influencing the availability of prey. For instance, the cost-effectiveness with which cougars hunt in different habitats may be affected if the distribution of their prey is altered by human disturbance or road building. Ungulates that are unhabituated to humans adjust their activity patterns in time and space to avoid human activity (Lyon 1979, Rost and Bailey 1979, Edge et al. 1985, Kufeld et al. 1988, Cassirer et al. 1992). If prey animals alter their habitat selection to favor environments that provide cover from human hunters (e.g., Kufeld et al. 1988), then cougars may also benefit if that cover improves their concealment and hunting success.

Alternatively, reductions in numbers of prey animals caused by recreational or subsistence hunting may also reduce cougar numbers (Jorgenson and Redford 1993). Competition between humans and cougars occurs when: (1) humans and cougars hunt the same prey, or (2) humans and cougars exhibit spatial overlap or are locally allopatric but hunt the same migratory prey, and (3) prey mortality from humans is at least partially additive to deaths caused by sources other than cougars. For example, the northern Yellowstone elk *(Cervus elaphus)* herd is reduced 1–14 percent by hunters outside Yellowstone National Park during the fall and early winter seasons (our calculations from Mack and Singer 1992). This fraction of the total herd is subsequently lost to cougars during the late winter and early spring. Human hunting pressure also modifies the sex and age structures of prey populations, influencing the availability of specific classes of prey that cougars favor or avoid.

Land-use practices also affect cougar prey at a variety of spatial and temporal scales. For example, settlement in the 1800s and 1900s dramatically improved habitat for the white-tailed deer in central and eastern Montana (Dusek et al. 1989). Agricultural and irrigation systems were introduced, frequent wildfire was eliminated, and white-tailed deer were released from competition after elk and bison *(Bison bison)* were reduced. These factors contributed to the development of diverse and extensive riverine habitats that promoted range and population expansions of white-tailed deer (Dusek et al. 1989). Similarly, fire suppression and the introduction of livestock and wild horses during the late 1800s and early 1900s led to habitat changes that favored mule deer *(O. hemionus)*. Gruell (1986) and Berger (1986) concluded that successional advance of shrubs and trees prompted large increases in the abundance and distribution of Great Basin mule deer from the early 1930s to the mid-1960s. Cougar distribution and abundance also apparently improved dramatically in response to increased numbers of mule deer (Berger and Weyhausen 1991).

Such landscape alterations in the GYE as logging or livestock grazing may radically alter the dynamics of cover, an important habitat component for cougar prey. The quality and spatial-temporal distribution of cover mediate forage accessibility, security, and thermoregulatory opportunities for ungulates (Peek et al. 1982). Logging reduces habitat quality for elk and deer if posttreatment moisture, soil, or temperature conditions preclude rapid vegetative responses where security cover is limiting. However, where vegetation regrowth is rapid, as in aspen *(Populus tremuloides)* forests, elk and deer may quickly reoccupy disturbed sites. In general, landscape alterations that enhance the vital rates and selection for a habitat by prey will also likely enhance prey acquisition by cougars, given that human disturbance is minimal and that cover is adequate for cougars.

Relations between cougars and their carnivore competitors. The reduction, augmentation, or reintroduction of such large carnivores as wolves *(Canis lupus)*, coyotes, or bears may influence cougars if there is significant spatial and dietary overlap. Koehler and Hornocker (1991) documented dietary overlap among cougars, coyotes, and bobcats *(Felis rufus)* during winter in central Idaho. Coyotes or bears are known to displace cougars from cougar food caches in the northern GYE and elsewhere (Spreadbury 1989, Harrison 1990, Green 1994, Murphy et al. 1998).

Wolf restoration in the GYE could affect cougars in three ways. First, wolf predation may alter population characteristics (for example, age ratios), behaviors (for example, vigilance or gregariousness), and the distribution of prey. Elk calves are the primary source of food for cougars in Yellowstone National Park (Murphy 1998). Reductions in their abundance could adversely influence prey acquisition by cougars. Second, the two predators may compete for carcasses of ungulates killed by cougars. Wolves in Alberta and northwestern Montana occasionally usurp cougar kills (Ruth and Hornocker 1996, I. Ross, unpublished data), but the converse is apparently not likely. Cougars in the northern GYE and in southwestern Alberta seldom scavenge ungulates killed by other sources and are not known to contest wolf packs for wolf kills. Third, cougars and wolves may kill each other (Schmidt and Gunson 1985, White and

Boyd 1989, Boyd and Neale 1992, U.S. Fish and Wildlife Service 1995, I. Ross, unpublished data). We presume that direct cougar-wolf interactions will typically favor wolves because wolf group sizes are usually larger than those of cougars and because cougars appear to be timid in defending their kills.

DIRECT EFFECTS

Road access and survival of hunted cougars. Besides providing a source of disturbance to cougars, road access increases the vulnerability of cougars to hunters and consequently may strongly affect cougar survival. Murphy (1983) and Barnhurst (1986) found that automobiles were the primary source of transportation for cougar hunters using dogs in Montana and Utah. Road type, density, and distribution affect the probability that houndsmen will detect a cougar track and reach a treed cougar. Cougar home range size and location, the extent of daily movement, and the cougars' use of topography (for example, travel in drainage bottoms with roads vs. without roads) interact with weather and the availability of escape terrain to influence cougar vulnerability (Murphy 1983, Barnhurst 1986).

Sources of human-caused cougar mortality. Most cougar populations are strongly influenced by human-caused mortalities. These losses primarily result from regulated hunting, removals by personnel of state or provincial agencies to control negative cougar-human interactions and predation on livestock, and accidental human-caused mortality. Legal hunting is usually the single greatest source of mortality among adult and subadult cougars in hunted populations (e.g., Murphy 1983, Logan et al. 1986, Anderson et al. 1992, Ross and Jalkotzy 1992), including southwestern Alberta and the northern GYE outside Yellowstone National Park (table 4.1). Illegal shooting occurs but apparently accounts for a low proportion of mortality (e.g., Maehr et al. 1991, Anderson et al. 1992, Ross and Jalkotzy 1992, Beier and Barrett 1993).

Deaths of maternal female cougars can result in the deaths of their orphaned kittens. Hunters cannot always distinguish lactating from nonlactating females, and tracks of kittens may only occur with their mother 25 percent of the time (Barnhurst and Lindzey 1989). Maternal females may be killed during periods when hunters are allowed to take cougars coincident to autumn hunting seasons for ungulates. Because hounds are often used to hunt cougars, kittens may also be mauled incidental to pursuit of their mothers (Barnhurst and Lindzey 1989).

Accidental cougar deaths result from collisions with vehicles (Maehr et al. 1991, Beier 1995) or incidental kills by fur trappers. In Alberta one radio-collared cougar died in a leg-hold trap set for lynx *(Lynx canadensis)*, and cougars

Table 4.1

Sources of death among adult and subadult radio-collared cougars in south-western Alberta and the northern Yellowstone ecosystem

		Number of cougar mortalities (%)			
Study area	N[a]	Sport hunting	Miscellaneous human[b]	Natural causes	Unknown causes
Sheep River, Alta	37	17 (46)	6 (16)	11 (30)	3 (8)
Northern Yellowstone ecosystem	25	12 (48)	1 (4)	12 (48)	0 (0)

Notes: Subadult is defined as 11–24 months old. Southwestern Alberta study was 1981–94, northern Yellowstone ecosystem study 1988–94.
[a]Does not include fourteen collared cougars at Sheep River and nine in the northern Yellowstone ecosystem that either died, moved out of the monitoring range, or had their transmitters fail.
[b]Miscellaneous deaths resulted from accidental trapping, collisions with vehicles, illegal kills, or control actions by management agencies.

are commonly killed in wolf or coyote snares (I. Ross, unpublished data). In northern Yellowstone, no human-related accidental deaths were recorded for eighty-five radio-marked cougars. Anderson (1983) summarized the sources of mortality to 115 cougars monitored for research studies in three western states and identified only one death from an automobile collision. On the other hand, accidental losses can be important to small insular populations (less than fifteen to twenty adult cougars) if habitats are fragmented by roads and human development (Beier 1993). Beier and Barrett (1993) found that vehicles accounted for ten of thirty-one (32 percent) deaths to radio-tagged cougars or their offspring in southern California where urban and highway development was extensive. Maehr et al. (1991) attributed fifteen fatalities of collared and uncollared Florida panthers to automobiles in southern Florida from 1979 to 1990. However, natural mortality accounted for a much greater percentage of deaths than automobiles when only radio-collared individuals were considered (Maehr et al. 1991).

Cougar deaths are also brought about by people protecting pets against cougar attacks, agency-control actions prompted by livestock predation by cougars, and other human-related causes. In Montana combined cougar deaths from these sources averaged 9 percent of mortality from 1988 to 1994 (our calculations from Montana Fish, Wildlife and Parks 1996), with the remainder caused by sport hunting. Numbers of cougar deaths unrelated to legal hunting increased markedly from 1989 to 1994 in Montana. In unhunted cougar populations or where cougar depredation on livestock is substantial, control actions may be the largest source of human-caused mortality (e.g., Cunningham et al. 1995).

The Effects of Human-Caused Mortality on Cougar Populations

ASSESSMENT OF EFFECTS AND SIGNIFICANCE

An important concern of wildlife biologists is that mortality caused by humans does not increase the probability that "managed" cougar populations will become extinct. The size of a cougar population affects its chances of persistence, particularly if it contains few breeding females (e.g., fewer than eleven, Beier 1993). In general, large populations that are well connected by cougar dispersal are expected to persist longer than small, insulated ones. New breeding individuals added via immigration numerically augment populations, increasing their resilience to decimating factors such as catastrophes. This positive effect of immigration on population persistence is the "rescue effect" (Brown and Kodric-Brown 1977).

Adding human-caused mortality to a cougar population can limit or depress its size if net losses from other sources of mortality remain unchanged. This additive, human-caused mortality is documented in cougar populations (Lindzey et al. 1992, Ross and Jalkotzy 1992). Because persistence and population size are linked (Beier 1993), additive mortality may reduce the chances that small cougar populations will persist. Where cougar populations exceed two hundred adults and are well linked to other local populations, light additive losses caused by sport hunting probably have little influence on extinction probabilities. In addition, immigration may help ameliorate the effects of hunting on cougars in the GYE (and other ecosystems), where cougar habitat remains largely contiguous and exceeds 2,200 square kilometers (Beier 1993) and where travel corridors allow the free exchange of dispersers among its subpopulations. Relations among dispersal, immigration, and human-caused mortality appear similar in other ecosystems that are characterized by large, relatively intact landscapes with high-quality cougar habitat where human development does not preclude successful dispersal (e.g., Lindzey et al. 1992, Logan et al. 1996). The strong dispersal capability of cougars, however, is not a panacea for conservation ills. A multitude of threats may imperil potential dispersers, particularly if the habitats between the source and destination populations are fragmented (Noss et al. 1996).

Because data are inadequate concerning the extent to which cougar mortality sources compensate each other, few generalizations concerning the effects of hunting on large cougar populations are possible. The processes that cause changes in cougar populations are complex and vary over time and space. For example, vital rates of birth, death, immigration, and emigration can be expected to vary relative to cougar population size (that is, exhibit density dependence) and to such environmental factors as prey density. To achieve management goals and protect cougars from overexploitation, management prescriptions must be based on knowledge of the vital factors that characterize individual populations.

In evaluating the effects of hunting on cougars, managers usually focus on such numerical aspects of cougar populations as sex, age, and harvest-effort ratios. Human-caused mortality, however, may influence social relations among the adult and subadult cougars that remain after hunting seasons end, although little information is available concerning this influence and its long-term consequences to population welfare. The behavioral strategies and social organization of cougars evolved over thousands of years, and disruptions to populations by hunting may have long-term consequences (Sweanor 1990). Cougar deaths associated with hunting may not be similar to natural mortality patterns because turnover of experienced breeding individuals is often accelerated by hunter preferences for larger cougars, particularly males. Experienced adults, males or females, that are removed by hunters may be individuals that contribute disproportionately to the production of offspring that survive to breed. Logan et al. (1996) found that 26 percent of the reproducing female cougars they studied produced 50 percent of the kittens, and only sixteen of thirty-four (47 percent) of the adult males sired litters. In the northern GYE, only nine adult males, 37 percent of those present, sired all of twenty-three litters. Two males alone sired fourteen litters (Murphy 1998).

Cougar deaths may destabilize social relations among adults that remain after hunting seasons close, at least until new breeding and dominance relations are established. For a resident male, familiarity with the females within his own territory and knowledge of their current reproductive status may enhance his own reproductive success (Seidensticker et al. 1973). Presumably, estrous females can be located more efficiently by long-standing resident males.

Artificially high turnover rates for resident adult males could also decrease kitten survival (Ross and Jalkotzy 1992). Removal of territorial males creates vacancies for immigrants, who may be more infanticidal than resident males. This phenomenon has been demonstrated in African lions (*Panthera leo;* Packer and Pusey 1984) and grizzly bears (Wielgus and Bunnell 1994) but remains undocumented for cougars.

EXAMINING BIOLOGICAL JUSTIFICATIONS FOR COUGAR HUNTING

Because most cougar populations in western North America are hunted, we examine some of the common biological arguments used to justify this activity: (1) cougar predation reduces prey available to hunters, (2) cougar population reduction achieved through hunting or trapping alleviates livestock depredation, and (3) hunting decreases negative cougar-human interactions by aversively conditioning cougars, or by creating mortality sinks for dispersing cougars that would otherwise be a source of problems for humans.

As previously discussed, cougars and humans that hunt the same prey in a common geographic area compete if either humans or cougars reduce the abundance of prey available to the other. The effects of cougars on prey available to humans, and vice versa, would be largely determined by the total number of prey killed annually by each predator. Cougar or human predation may reduce numbers of prey if predation is additive to other sources of prey mortality. However, cougars or humans may not affect the abundance of prey available to each other if reproduction offsets these mortalities, or if other mortality sources are entirely compensatory to cougar and human predation.

Currently, the lack of detailed, long-term studies of cougar-prey relations preclude conclusions about the extent to which cougars reduce the number of prey available to hunters. However, some data are available on the fraction of prey removed by cougars from prey populations and trends in ungulate populations that are subject to cougar predation. Murphy (1998) found that cougars remove only about 3 percent of the elk and 4 percent of the mule deer in the northern GYE. The total kill by cougars included about 12 percent of the buck mule deer, 1 percent of the bull elk, 9 percent of the calf elk, and less than 5 percent of the remaining sex-age classes of elk and deer. By comparison, human hunters annually remove about 72 percent of the buck mule deer, 22 percent of the bull elk, 3 percent of the calf elk, and less than 8 percent of all other sex-age classes of elk and deer (Murphy 1998). Cougars had little direct effect on the size of elk or deer populations (Murphy 1998). Interactions between weather and intraspecific competition for forage largely determined the size and regulation of elk populations in the northern GYE (Houston 1982, Merrill and Boyce 1991, Coughenour and Singer 1996).

In southwestern Alberta cougars annually killed 0–13 percent of a wintering bighorn sheep herd (Ross et al. 1997) and about 16–30 percent of wintering moose calves *(Alces alces)* (Ross and Jalkotzy 1996). Shaw (1980) estimated that cougars removed less than 20 percent and less than 10 percent of the North Kaibab mule deer herd during two different years of study. Anderson et al. (1992) estimated that resident cougars killed about 8–9 percent of a mule deer population in Colorado annually. In California cougar predation caused precipitous declines in small bighorn sheep populations where few alternate prey were available (Wehausen 1996). However, Hornocker (1970), Shaw (1977, 1980), and Lindzey et al. (1994) found that cougar populations in Idaho, Arizona (two study sites), and Utah did not prevent their prey (elk or mule deer) from increasing. Logan et al. (1996) concluded from their ten-year study that mule deer increased despite cougar predation when deer were below carrying capacity, but that cougar predation depressed deer numbers when drought conditions reduced the carrying capacity for deer. Habitat quality and quantity were the ultimate limiting factors for mule deer (Logan et al. 1996).

The potential to reduce livestock depredation is also used to justify hunting that might reduce cougar populations. In California cougar depredation on domestic sheep is hypothesized to reflect the distribution and abundance of cougars or differences in native prey (Torres et al. 1996). Although cougar hunting seasons are typically not structured to target offending individuals, cougar hunting could potentially reduce livestock depredation if hunting losses led to a sustained decrease in the numbers of adult or subadult cougars. Sustained reductions in numbers of cougars occur when adult mortality is not rapidly compensated by (1) simultaneous reductions in other sources of cougar mortality (for example, fighting), and (2) rapid recruitment (within twelve months) of subadults from local females or from peripheral subpopulations. Per capita rates of livestock predation by established resident cougars vs. those of new recruits (typically subadults) should also be considered. For such a program to work, livestock losses to newly recruited cougars should not exceed those caused by former resident cougars that were removed by hunters. Lindzey et al. (1992) found that a single simulated cougar harvest (42 percent of adults) in southern Utah reduced adult cougar numbers for up to nine months, but that sustained reductions of adult cougars for longer periods would likely reduce the adult population. Logan et al. (1996) found that removal of 53 percent of adult cougars in south-central New Mexico reduced the adult population for a period of thirty-one months. The effect of sustained hunting on cougar population sizes is probably enhanced where hunting mortality is at least partially additive to natural adult deaths (Lindzey et al. 1992, Ross and Jalkotzy 1992).

In the northwestern United States and western Canada, including the GYE, rates of cougar depredation are generally low (e.g., Alberta Fish and Wildlife 1992). In Montana from 1984 to 1993, only 8.2 predation incidents involving cougars occurred annually, and monetary losses averaged only $3,825 (Montana Fish, Wildlife and Parks 1996). However, depredation rates may surge locally if individual cougars develop a habit of preying on livestock. In this circumstance, selective removal of offending individuals by management action is more certain to reduce livestock losses and carries less risk of extirpation for small cougar populations than a nonselective removal strategy associated with hunting seasons.

For hunting seasons or control actions to reduce cougar attacks, they should result in fewer absolute numbers of cougars in a population that would potentially prey on humans. This is particularly desirable in areas where humans and cougars interact most frequently. With few exceptions, hunting seasons in western North America are not specifically structured to focus cougar hunting at the human-wildlands interface, where cougar-human or cougar-pet interactions are more likely to occur (see Torres et al. 1996 for relevant data). Hunters may find access to cougar habitats difficult here because (1) these lands are typically fragmented by multiple private ownerships, and (2) landowners with property along the human-wildlands interface are less likely to accommodate hunters than those in more remote habitats. Hunting programs also do not focus on individuals that are one to three years of age, the age class that is involved more frequently in incidents with humans than older cougars (Beier 1991, Aune 1991, but see Torres et al. 1996). These factors are unlikely to change, and in many areas they preclude the effective use of cougar hunting as a tool to reduce cougar attacks.

The common belief that cougar hunting can reduce negative human-cougar encounters is often based on the presumption that aggressive cougar behavior is a consequence of habituation. Although cougars habituate to human activity, there is no evidence that the increased number of attacks by cougars in recent decades is related to habituated cougars (Beier 1992). However, encounters between cougars conditioned by human food and humans or pets are noted in the literature (Aune 1991). Young cougars that are conditioned by anthropogenic foods may be more likely to experiment with humans as prey than unconditioned individuals (Aune 1991).

Coping with the Uncertainty

SOURCES OF UNCERTAINTY

Cougar conservation is characterized by uncertainty that arises from (1) inadequate knowledge of cougars and their ecological relations with humans, (2) the inherent variability of cougar populations and their habitats, and (3) the volatility of human tolerance for undesired cougar behavior. Relations between cougars and their environment and anthropogenic influences are complex, and collecting the necessary information is and will remain difficult and costly. Cougar populations typically occur at low densities and in habitats where field studies are logistically difficult. Cougars are also normally cryptic and secretive. These factors render many study techniques (for example, aerial surveys) largely ineffectual and detract from our ability to solve problems by gaining reliable knowledge.

Wildlife populations live in environments that are inherently variable. Landscape-scale environmental variation reduces the certainty with which spatially and temporally specific data can be reliably extrapolated. The production of kittens, their survival to breeding age, and other characteristics of cougar populations routinely fluctuate about mean values because of chance. Consequently, population parameters may exhibit short-term (for example, under five years) trends that are not characteristic of means calculated from longer time periods. Finally, values (for example, means) calculated from samples are usually only estimates of population parameters with some imprecision and some bias. In a practical sense, imprecision and bias in estimates of parameters reduce the certainty that "management" will achieve its objectives.

Public tolerance of cougars is also volatile and related to trends in cougar behavior toward people or related to current perceptions of the extent cougars compete for prey with human hunters. Public perceptions can also be modified by recreational use (e.g., cougar hunting), property relations, and management status (Kellert et al. 1996). Cougars attack humans and pets more frequently than do wolves, and their distribution in settled portions of North America is far greater than that of grizzly bears. Therefore, more than for these other large carnivores, attitudes toward the cougar may be affected by real rather than perceived relations with people.

Public opinion strongly influences the leniency with which managers resolve conflicts between cougars and human interests. Volatility in public opinion compounds the difficulties of cougar management, particularly because research and management information and the process of regulatory change are unlikely to keep pace with real or perceived changes in cougar behavior and populations. A "frontier mentality" still persists in many rural communities, wherein the majority vision of nature is one where humans have dominion over all wildlife. Although this view has softened, an unsympathetic perspective can be restored quickly with a few negative interactions between cougars and people. Because the principal "management" activity by many agencies is the regulation of mortality, the behavior of cougars toward people can profoundly influence levels of human-caused mortality resulting from the feedback between public opinion and the management policy of regulatory agencies.

MANAGEMENT CONSIDERATIONS

Hunters and wildlife managers should recognize that research and management data sets for evaluating the effects of hunting on cougars are typically inadequate. Biologists should manage hunter-caused mortality conservatively, particularly in small and fragmented cougar populations, because additive hunting mortality may significantly increase the probability of extinction. Managers should use nonlethal approaches to dealing with cougars involved in livestock depredation or cougar-human conflicts wherever possible (Ross and Jalkotzy 1995), while recognizing that some individuals may not be suitable candidates for translocation (Ruth et al. 1998).

Stochastic influences may result in trends in cougar populations that temporarily mislead managers to believe that their corrective management efforts have achieved results (or have failed) well before any such conclusion is warranted. Our limited knowledge of the species and uncertainty resulting from stochastic influences are usually linked and further warrant a conservative approach to cougar population and habitat management.

Although regulatory agencies should use regionally specific approaches to achieving management goals (Torres et al. 1996), the importance of such landscape-scale processes as dispersal in cougar population dynamics should be recognized in population and habitat management. Cougar dispersal frequently operates at larger spatial scales than are typically covered by the administrative units that states and provinces use to regulate cougar hunting. Efforts to increase cougar populations may be unsuccessful if adjacent units are managed as mortality sinks that recruit dispersers but provide few in return. Conversely, attempts to decrease populations may be unsuccessful because adjacent units contribute dispersers that are recruited into the focal unit. The dynamics of source and sink populations are particularly applicable to the GYE, where three states regulate hunting. Cougars that were radio-marked in the northern GYE dispersed across contiguous wildlands to distant parts of the ecosystem in Idaho, Wyoming, and Montana (Murphy 1998). The GYE actually supports a single large cougar population over seven to eight million acres, rather than a complex of interacting local populations.

Although cougar habitats in the GYE are largely contiguous, habitat managers should proactively identify and protect the dispersal and travel corridors that may be critical to populations that could become occluded by human development of wildlands. The management of landscapes as cougar habitat should be based on integrated, cooperative, long-range efforts by planners employed at the county, state, and federal levels. Managers and planners should remember that ecological principles ultimately determine the persistence of large carnivore populations. Administrative boundaries are irrelevant if the ecological needs of carnivores are neglected.

RESEARCH AND INFORMATION NEEDS

Successful management of any species requires basic information on its needs and role in ecological systems. Research on large carnivores should now focus particularly on how humans influence the viability of carnivores on scales of one hundred to one thousand years. For cougars, we need reliable knowledge of the effects of human activity on their habitat. The destruction and degradation of cougar habitat from unchecked human demand for space and natural resources is the most significant impediment to cougar conservation, and it is the most difficult influence to reverse. Specifically, habitat managers need to know more about interactions between cougar habitat components such as cover, prey, other carnivores, and human effects. Habitat quality is important to cougar viability because it determines breeding cougar densities and such vital characteristics of populations as birth and survival rates. Habitat needs of cougars may differ at life-history stages and for different processes (Ruggiero et al. 1988). For example, habitat needs for subadults and family groups may differ from those for solitary adults.

Managers need better information on which habitats constitute travel or dispersal corridors and the extent to which humans influence their use by cougars. Although most ecosystems that support cougars in western North America are currently connected by cougar dispersal (Beier 1993), their degree of connectivity continues to decline, in large part because of human expansion into wildlands. The variables that drive cougar selection of corridors should be identified across the full range of habitats in which cougars live.

Knowledge of cougar-prey, cougar-livestock, and cougar-human relations is crucial to cougar conservation. These aspects of the species' ecology influence public opinion and ultimately determine what management alternatives should be used to address perceived conflicts between cougars and people. Because management policy strongly influences cougar mortality, reliable knowledge in these areas could lead managers to develop ecologically based solutions that prevent conflicts and unnecessary cougar deaths.

To date, research and management programs have not been well matched to the temporal and spatial scales at which population dynamics function. For example, the selection of boundaries for cougar population studies is typically influenced heavily by financial constraints. Ideally, boundaries for population studies should be delimited by topographic features (for example, the interface between a montane forest and a desert) that are associated with natural declines in rates of successful dispersal of cougars. We are also unaware of a single study of cougar population dynamics that has spanned even one full cycle of major fluctuation in its primary prey.

Conclusion

Among North America's large carnivores, cougars demonstrate comparatively high resilience because of their broad plasticity in prey selection, relatively high female survivorship and reproductive capacity, and strong dispersal ability (Weaver et al. 1996). The intrinsic rate of increase in cougar populations can exceed that of mule deer (Tanner 1975), explaining why cougars recover quickly from suppression in some ecosystems (Ross and Jalkotzy 1992). Cougar populations appear to have stabilized or even increased in many areas of the northern Rocky Mountains under management as a game animal. Thus human-caused mortality associated with hunting is not the primary long-term impediment to cougars where it is conservatively regulated and applied to large populations that are well connected through exchange of dispersers. The indirect effects of humans on habitat components—cover, prey, security from human disturbance, and the connectivity of populations—will remain the most significant long-term threat to cougars.

Ensuring the persistence of large carnivores requires a commitment that extends far beyond regulatory agencies to encompass societal and cultural values at large. All wild populations need room. We must allocate the space required to sustain large carnivores and remain tolerant when they pursue their needs in conflict with our own. This will be the litmus test of our resolve to conserve not only large carnivores, but much of the diversity inherent in ecological systems.

Literature Cited: Alberta Fish and Wildlife. 1992. Management plan for cougar in Alberta. Wildlife Management Planning Series, no. 5. Edmonton.

Anderson, A. E. 1983. A critical review of literature on puma *(Felis concolor)*. Special Report no. 54. Colorado Division of Wildlife, Denver.

Anderson, A. E., D. C. Bowden, and D. M. Kattner. 1992. The puma on Umcompahgre Plateau, Colorado. Technical Publication no. 40. Colorado Division of Wildlife, Denver.

Aune, K. 1991. Increasing mountain lion populations and human–mountain lion interactions in Montana. Pp. 86–94 *in* C. L. Braun, ed., Mountain Lion–Human Interaction Symposium. Colorado Division of Wildlife, Denver.

Barnhurst, D. 1986. Vulnerability of cougars to hunting. M.S. thesis, Utah State University, Logan.

Barnhurst, D., and F. G. Lindzey. 1989. Detecting female mountain lions with kittens. Northwest Science 63:35–37.

Beier, P. 1991. Cougar attacks on humans in the United States and Canada. Wildlife Society Bulletin 19:403–12.

———. 1992. Cougar attacks on humans: An update and some further reflections. Proceedings of the Vertebrate Pest Conference 15:365–67.

———. 1993. Determining minimum habitat areas and habitat corridors for cougars. Conservation Biology 7:94–108.

———. 1995. Dispersal of juvenile cougars in fragmented habitat. Journal of Wildlife Management 59:228–37.

Beier, P., and R. H. Barrett. 1993. The cougar in the Santa Ana Mountain Range, California. Final Report, Orange County Cooperative Mountain Lion Study, University of California, Berkeley.

Beier, P., D. Choate, and R. H. Barrett. 1995. Movement patterns of mountain lions during different behaviors. Journal of Mammalogy 76:1056–70.

Belden, R. C., W. B. Frankenberger, R. T. McBride, and S. T. Schwikert. 1988. Panther habitat use in southern Florida. Journal of Wildlife Management 52:660–63.

Belden, R. C., B. W. Hagedorn, and W. B. Frankenberger. 1991. Responses of translocated mountain lions to human disturbance (abstract). P. 26 *in* C. L. Braun, ed., Mountain Lion–Human Interaction Symposium. Colorado Division of Wildlife, Denver.

Berger, J. 1986. Wild horses of the Great Basin: Social competition and population size. University of Chicago Press, Chicago.

Berger, J., and J. D. Wehausen. 1991. Consequences of a mammalian predator-prey disequilibrium in the Great Basin Desert. Conservation Biology 5:244–48.

Boyd, D. K., and G. K. Neale. 1992. An adult cougar, *Felis concolor,* killed by gray wolves, *Canis lupus,* in Glacier National Park, Montana. Canadian Field-Naturalist 106:524–25.

Brown, J. H., and A. Kodric-Brown. 1977. Turnover rates in insular biogeography: Effect of immigration on extinction. Ecology 58:445–49.

Cassirer, E. F., D. J. Freddy, and E. D. Ables. 1992. Elk response to disturbance by cross-country skiers in Yellowstone National Park. Wildlife Society Bulletin 20:375–81.

Coughenour, M. B., and F. J. Singer. 1996. Elk population processes in Yellowstone National Park under the policy of natural regulation. Ecology Applications 6:573–93.

Cunningham, S. C., L. A. Haynes, C. Gustavson, and D. D. Haywood. 1995. Evaluation of the interaction between mountain lions and cattle in the Aravaipa-Klondyke area of southeast Arizona. Arizona Game and Fish Department Technical Report no. 17, Phoenix.

Dusek, G. L., R. J. Mackie, J. D. Herriges Jr., and B. B. Compton. 1989. Population ecology of white-tailed deer along the lower Yellowstone River. Wildlife Monographs no. 104.

Edge, W. D., C. L. Marcum, and S. L. Olson. 1985. Effects of logging activities on home-range fidelity of elk. Journal of Wildlife Management 49:741–44.

Feigley, H. P. 1993. Comments on the large carnivore conservation problem. Pp. 43–46 *in* T. W. Clark, A. P. Curlee, and R. P. Reading, Conserving threatened carnivores: Developing interdisciplinary problem-oriented strategies. Northern Rockies Conservation Cooperative, Jackson, Wyoming.

Fretwell, S. D., and H. L. Lucas. 1970. On territorial behavior and other factors influencing distribution in birds. Acta Biotheoretica 19:16–36.

Green, G. I. 1994. Use of spring carrion by bears in Yellowstone National Park. M.S. thesis, University of Idaho, Moscow.

Gruell, G. E. 1986. Post-1900 mule deer irruptions in the Intermountain West. U.S. General Technical Report INT-206:1–37.

Halfpenny, J. C., M. R. Sanders, and K. A. McGrath. 1991. Human-lion interactions in Boulder County, Colorado: Past, present, and future. Pp. 10–16 *in* C. L. Braun, ed., Mountain Lion–Human Interaction Symposium. Colorado Division of Wildlife, Denver.

Harrison, S. 1990. Cougar predation on bighorn sheep in the Junction Wildlife Management Area, British Columbia. M.S. thesis, University of British Columbia, Vancouver.

Hornocker, M. G. 1970. An analysis of mountain lion predation upon mule deer and elk in the Idaho primitive area. Wildlife Monographs no. 21.

Houston, D. B. 1982. The northern Yellowstone elk. Macmillan, New York.

Jalkotzy, M., and I. Ross. 1995. Cougar responses to human activity at Sheep River, Alberta. Arc Wildlife Services Ltd., Calgary.

Jorgenson, J. P., and K. H. Redford. 1993. Humans and big cats as predators in the Neotropics. Symposia of the Zoological Society of London 65:367–90.

Kellert, S. R., M. Black, C. Reid Rush, and A. J. Bath. 1996. Human culture and large carnivore conservation in North America. Conservation Biology 10:977–90.

Koehler, G. M., and M. G. Hornocker. 1991. Seasonal resource use among mountain lions, bobcats, and coyotes. Journal of Mammalogy 72:391–96.

Kufeld, R. C., D. C. Bowden, and D. L. Schrupp. 1988. Influence of hunting on movements of female mule deer. Journal of Range Management 41:70–72.

Laing, S. P., and F. G. Lindzey. 1991. Cougar habitat selection in southern Utah. Pp. 27–37 *in* C. L. Braun, ed., Mountain Lion–Human Interaction Symposium. Colorado Division of Wildlife, Denver.

———. 1993. Patterns of replacement of resident cougars in southern Utah. Journal of Mammalogy 74:1056–58.

Leyhausen, P. 1965. The communal organization of solitary mammals. Symposia of the Zoological Society of London 14:249–63.

Lindzey, F. G., W. D. Van Sickle, B. B. Ackerman, D. Barnhurst, T. P. Hemker, and S. P. Laing. 1994. Cougar population dynamics in southern Utah. Journal of Wildlife Management 58:619–24.

Lindzey, F. G., W. D. Van Sickle, S. P. Laing, and C. S. Mecham. 1992. Cougar population response to manipulation in southern Utah. Wildlife Society Bulletin 20:224–27.

Logan, K. A., and L. L. Irwin. 1985. Mountain lion habitats in the Big Horn Mountains, Wyoming. Wildlife Society Bulletin 13:257–62.

Logan, K. A., L. L. Irwin, and R. Skinner. 1986. Characteristics of a hunted mountain lion population in Wyoming. Journal of Wildlife Management 50:648–54.

Logan, K. A., L. L. Sweanor, T. K. Ruth, and M. G. Hornocker. 1996. Cougars of the San Andres Mountains, New Mexico. Final Report, Project W-128-R. New Mexico Department of Game and Fish, Sante Fe.

Lyon, L. J. 1979. Habitat effectiveness for elk as influenced by roads and cover. Journal of Forestry 77:658–60.

Mack, J. A., and F. J. Singer. 1992. Population models for elk, mule deer, and moose on Yellowstone's winter range. Pp. 4-5 to 4-42 *in* J. D. Varley and W. G. Brewster, eds., Wolves for Yellowstone? A report to the United States Congress, vol. 4, research and analysis. National Park Service, Yellowstone National Park.

Maehr, D. S. 1990a. Florida panther movements, social organization, and habitat utilization. Final Report, Florida Game and Fresh Water Fish Commission, Tallahassee.

——. 1990b. The Florida panther and private lands. Conservation Biology 4:167–70.

Maehr, D. S., E. D. Land, and M. E. Roelke. 1991. Mortality patterns of panthers in Southwest Florida. Proceedings of the Annual Conference of Southeast Association of Fish and Wildlife Agencies 45:201–7.

Maehr, D. S., E. D. Land, and M. E. Roelke, and J. W. McCown. 1989. Early maternal behavior in the Florida panther *(Felis concolor coryi)*. American Midland Naturalist 122:34–43.

——. 1990. Day beds, natal dens, and activity of Florida panthers. Proceedings of the Annual Conference of Southeast Association of Fish and Wildlife Agencies 44:310–18.

Merrill, E., and M. S. Boyce. 1991. Summer range and elk population dynamics in Yellowstone National Park. Pp. 263–74 *in* R. B. Keiter and M. S. Boyce, eds., The Greater Yellowstone Ecosystem: Redefining America's wilderness heritage. Yale University Press, New Haven.

Montana Fish, Wildlife, and Parks. 1996. Final environmental impact statement: Management of mountain lions in Montana. Montana Fish, Wildlife and Parks, Helena.

Murphy, K. M. 1983. Relationships between a mountain lion population and hunting pressure in western Montana. M.S. thesis, University of Montana, Missoula.

——. 1998. The ecology of the cougar *(Puma concolor)* in the northern Yellowstone ecosystem. Ph.D. diss., University of Idaho, Moscow.

Murphy, K. M., G. S. Felzien, Hornocker M. G., and T. K. Ruth. 1998. Encounter competition between bears and cougars: Some ecological implications. Ursus 10:55–60.

Noss, R. F., H. B. Quigley, M. G. Hornocker, T. Merrill, and P. C. Paquet. 1996. Conservation biology and carnivore conservation in the Rocky Mountains. Conservation Biology 10:949–63.

Packer, C., and A. E. Pusey. 1984. Infanticide in carnivores. Pp. 31–42 *in* G. Hausfater and S. B. Hardy, eds., Infanticide: Comparative and evolutionary perspectives. Aldine Publishing, New York.

Peek, J. M., M. D. Scott, L. J. Nelson, D. J. Pierce, and L. L. Irwin. 1982. Role of cover in habitat management for big game in northwestern United States. Transactions of the North American Wildlife Conference 47:363–73.

Riley, S. J., and R. A. Malecki. 1996. Developing an adaptive management program for mountain lions *(Felis concolor)* in Montana (abstract). Fifth Mountain Lion Workshop, California Department of Fish and Game, Sacramento.

Roelke, M. E. 1987. Panther health and reproduction. Annual performance report, E-11-11. Florida Game and Fresh Water Fish Commission, Tallahassee.

Rogers, L. L., L. D. Mech, D. K. Dawson, J. M. Peek, and M. Korb. 1980. Deer distribution in relation to wolf pack territory edges. Journal of Wildlife Management 44:253–58.

Ross, P. I., and M. G. Jalkotzy. 1992. Characteristics of a hunted population of cougars in southwestern Alberta. Journal of Wildlife Management 56:417–26.

——. 1995. Fates of translocated cougars, *Felis concolor*, in Alberta. Canadian Field-Naturalist 109:475–76.

——. 1996. Cougar predation on moose in southwestern Alberta. Alces 32:1–8.

Ross, P. I., M. G. Jalkotzy, and M. Festa-Bianchet. 1997. Cougar predation on bighorn sheep in southwestern Alberta during winter. Canadian Journal of Zoology 74:771–75.

Rost, G. R., and J. A. Bailey. 1979. Distribution of mule deer and elk in relation to roads. Journal of Wildlife Management 43:634–41.

Ruggiero, L. F., R. S. Holthausen, B. G. Marcot, K. B. Aubry, J. W. Thomas, and E. C. Meslow. 1988. Ecological dependency: The concept and its implications for research and management. Transactions of the North American Wildlife and Natural Resources Conference 53:115–26.

Ruth, T. K. 1991. Mountain lion use of an area of high recreational development in Big Bend National Park, Texas. M.S. thesis, Texas A&M University, College Station.

Ruth, T. K., and M. G. Hornocker. 1996. Interactions between cougars and wolves (and a bear or two) in the North Fork of the Flathead River, Montana (abstract). Fifth Mountain Lion Workshop, California Department of Fish and Game, Sacramento.

Ruth, T. K., K. A. Logan, L. L. Sweanor, M. G. Hornocker, and L. J. Temple. 1998. Evaluating cougar translocation in New Mexico. Journal of Wildlife Management 62:1264–75.

Sandell, M. 1989. The mating tactics and spacing patterns of solitary carnivores. Pp. 164–82 *in* J. L. Gittleman, ed., Carnivore behavior, ecology, and evolution. Cornell University Press, Ithaca.

Schmidt, K. P., and J. R. Gunson. 1985. Evaluation of wolf-ungulate predation near Nordegg, Alberta: Second year progress report, 1984–85. Alberta Energy and Natural Resources, Fish and Wildlife Division, Edmonton.

Schortemeyer, J. L., D. S. Maehr, J. W. McCown, E. D. Land, and P. D. Manor. 1991. Prey management for the Florida panther: A unique role for wildlife managers. Transactions of the North American Wildlife and Natural Resources Conference 56:512–26.

Seidensticker, J. C., IV, M. G. Hornocker, W. V. Wiles, and J. P. Messick. 1973. Mountain lion social organization in the Idaho Primitive Area. Wildlife Monographs no. 35.

Shaw, H. G. 1977. Impact of mountain lion on mule deer and cattle in northwestern Arizona. Pp. 17–32 *in* R. L. Phillips and C. Jonkel, eds., Proceedings of the 1975 Predator Symposium. Montana Forest and Conservation Experiment Station, University of Montana, Missoula.

——. 1980. Ecology of the mountain lion in Arizona. Final Report, P-R Proj. W-78-R, Work Plan 2, Job 13. Arizona Game and Fish Department, Phoenix.

Smallwood, K. S. 1994. Trends in California mountain lion populations. Southwestern Naturalist 39:67–72.

Spreadbury, B. 1989. Cougar ecology and related management implications and strategies in southeastern British Columbia. M.E.D. thesis, University of Calgary, Calgary.

Sunquist, M. E., and F. C. Sunquist. 1989. Ecological constraints on predation by large felids. Pp. 283–301 *in* J. L. Gittleman, ed., Carnivore behavior, ecology, and evolution. Cornell University Press, Ithaca.

Sweanor, L. L. 1990. Mountain lion social organization in a desert environment. M.S. thesis, University of Idaho, Moscow.

Tanner, J. T. 1975. The stability and intrinsic growth rates of prey and predator populations. Ecology 56:855–67.

Terborgh, J. 1988. The big things that run the world: A sequel to E. O. Wilson. Conservation Biology 2:402–3.

——. 1990. The role of felid predators in neotropical forests. Vida Silvestre Neotropical 2:3–5.

Thomas, J. W. 1979. Introduction. Pp. 10–21 *in* J. W. Thomas, ed., Wildlife habitats in managed forests: The Blue Mountains of Oregon and Washington. USDA Agricultural Handbook no. 533.

Torres, S. G., T. M. Mansfield, J. E. Foley, T. Lupo, and A. Brinkhaus. 1996. Mountain lion and human activity in California: Testing speculations. Wildlife Society Bulletin 24:451–60.

U.S. Fish and Wildlife Service. 1995. Annual report of the Rocky Mountain Interagency Wolf Recovery Program. U.S. Fish and Wildlife Service, Helena.

Van Dyke, F. G., R. H. Brocke, and H. G. Shaw. 1986a. Use of road track counts as indices of mountain lion presence. Journal of Wildlife Management 50:102–9.

Van Dyke, F. G., R. H. Brocke, H. G. Shaw, B. B. Ackerman, T. P. Hemker, and F. G. Lindzey. 1986b. Reactions of mountain lions to logging and human activity. Journal of Wildlife Management 50:95–102.

Weaver, J. L., P. C. Paquet, and L. F. Ruggiero. 1996. Resilience and conservation of large carnivores in the Rocky Mountains. Conservation Biology 10:964–76.

Wehausen, J. D. 1996. Effects of mountain lion predation on bighorn sheep in the Sierra Nevada and Granite Mountains of California. Wildlife Society Bulletin 24:471–79.

White, P. A., and D. K. Boyd. 1989. A cougar, *Felis concolor,* kitten killed and eaten by gray wolves. Canadian Field-Naturalist 103:408–9.

Wielgus, R. B., and F. L. Bunnell. 1994. Dynamics of a small, hunted brown bear *Ursus arctos* population in southwestern Alberta, Canada. Biological Conservation 67:161–66.

Williams, J. S., J. J. McCarthy, and H. D. Picton. 1995. Cougar habitat use and food habits on the Montana Rocky Mountain Front. Intermountain Journal of Sciences 1:16–28.

Young, S. P. 1946. History, life habits, economic status, and control, Part 1. Pp. 1–173 *in* S. P. Young and E. A. Goldman, eds., The puma, Mysterious American cat. American Wildlife Institute, Washington.

Male gray wolf (Franz J. Camenzind)

Chapter Five

Wolves in the Greater Yellowstone Ecosystem: Restoration of a Top Carnivore in a Complex Management Environment

Douglas W. Smith, Wayne G. Brewster, and Edward E. Bangs

Restoration of the Rocky Mountain wolf *(Canis lupus)* in the Greater Yellowstone Ecosystem (GYE) has been a controversial challenge, as are most large carnivore reintroductions (Reading and Clark 1996). Wolves are well studied and we know a lot about managing them (Fuller 1995, Mech 1995), but successfully negotiating this wolf restoration policy process has been arduous and has embodied many nonbiological dimensions, including strong emotions (Gasaway et al. 1983, Mech 1995, Clark et al. 1996). Practically, this has made for a complex, dynamic policy arena. The decision to reintroduce wolves was bogged down by twenty years of acrimonious debate (Varley and Brewster 1992, Fritts et al. 1997), but despite numerous obstacles and close calls, wolves were reintroduced to Yellowstone National Park (YNP)—arguably some of the best wolf habitat in the world (Varley and Brewster 1992)—and central Idaho in 1995 and 1996. The questions remaining now are, How will wolves be managed? What will be their ecosystem effects? Will there be wolf control? How will the various interest groups position themselves regarding future wolf management?

Our goal in this chapter is to address the critical management questions concerning wolves in the northern Rocky Mountains to the extent possible now and to provide the reader with the necessary backdrop. We will describe the animal and some aspects of its biology and role in ecosystems. Then we will summarize aspects of the GYE wolf restoration and the goals of the program. Finally, we will conclude with some recommendations for wolf management in

the Rockies and our prediction that it will continue to be a conflictual sociopolitical process more than a biological problem.

Wolf Biology

Wolves are large (eighteen to sixty kilograms) social members of the dog family (*Canidae*, Mech 1970). They live in family groups called packs and defend an
area called a territory from other packs of wolves. Territory sizes vary with food
availability (ungulate prey, Fuller 1989) but can be from one hundred (in Wisconsin) to ten thousand square kilometers (in Alaska). Territories in the northern Rockies range from 350 to six thousand square kilometers (Ream et al. 1991).
The social structure of the pack is elaborate, with a dominant male and female
called the alpha wolves; usually only the alphas breed, although polygamous
matings have been reported (Mech and Nelson 1989). Mating depends on latitude
and can occur from February through April (Mech 1970), but at midlatitudes
such as the Rockies it occurs in February (Ream et al. 1991). Pups are born about
sixty-three days after mating, and litter size averages five pups (Mech 1970,
Pletscher et al. 1997). Wolves do not become sexually mature until twenty-two
months old (Rausch 1967, Rabb 1967) but may disperse as early as ten months old
or delay dispersal for years (Gese and Mech 1991, Boyd et al. 1995). Age of dispersal probably depends on the availability of unoccupied habitat (habitat saturation). Dispersal distance varies but can be extremely far, up to eight hundred kilometers (Fritts 1983, Boyd et al. 1995), although it is usually much less.

Wolves are a primary carnivore on ungulates in North America (Mech 1970),
and hence their ecosystem effects are nothing less than spectacular. They share
the top predator position with other carnivores like cougars *(Puma concolor)* and
grizzly bears *(Ursus horribilis)*, but it is arguable that other carnivores have less
influence on ecosystems than wolves. In any case, wolves affect prey density,
distribution, and population structure, which can impact vegetation (Gasaway
et al. 1983, McLaren and Peterson 1994). Wolves also affect smaller carnivores
like coyotes *(Canis latrans)*, typically keeping coyote density low. This has ripple
effects on the prey of coyotes (small mammals) and a competitor of coyotes, the
red fox *(Vulpes vulpes)*, a smaller canid kept at low density by coyotes but not by
wolves (Peterson 1977, Allen 1979).

There are few places where wolves, cougars, grizzly bears, and coyotes coexist (for example, northwest Montana, Kunkel 1997) with so abundant and diverse a prey base. Studying this complex of predators and prey is a great challenge for research in the Yellowstone ecosystem; any studies completed will no
doubt add to a literature dominated by northern studies with fewer predators
and prey (Gasaway et al. 1992, Carbyn et al. 1993). Indeed, the effect of wolves

and other predators on prey has been one of the most hotly debated ecological issues in wildlife management. Data from Yellowstone will not only aid wildlife management in the West but will shed light on enigmatic debates about interactions between wolves and prey (Bergerud and Ballard 1988, Thompson and Peterson 1988, Van Ballenberghe and Ballard 1994).

The Role of Wolves in Ecosystems

Most studies of wolves have focused on predator-prey interactions, ecology, behavior, genetics, or physiology (Harrington and Paquet 1982, Carbyn et al. 1993). This is interesting because wolf predation on the major herbivores in the ecosystems where they co-occur (historically most of the Northern Hemisphere) has the potential to affect vegetation and primary productivity, yet this potential impact has generally not been the focus of research. More studies have focused on the predator-prey aspect of wolf ecology (see Gasaway et al. 1983, 1992, Van Ballenberghe and Ballard 1994, and Carbyn et al. 1993 for reviews), usually without assessing the effects of this relation on broader ecosystem events, including other animals and plants (see also Chapter 12). When moose (*Alces alces*) are preyed on by both wolves and grizzly bears, they occur at lower densities than when they are preyed on by neither predator or only one predator (Peterson 1977, Gasaway et al. 1992). The ecosystem-wide impacts of these lower moose densities has received far less research attention (McLaren and Peterson 1994).

One problem of studying wolves experimentally in a very large ecosystem context is that there is no practical way to set up the controls and replications necessary for experimental science—or if there were, it would be too expensive to conduct (Boutin 1992). This is currently a problem in the GYE studies of the impacts of wolf restoration on the Northern and Madison-Firehole elk herds. No comparable situation exists outside the YNP to serve as a control for elk migration behavior, habitat, predators, human hunting, and weather. The complexity of such studies is another problem; the ecological complexity of wolf studies in the field is much greater than lab studies, prohibiting any kind of unifying laws from emerging (see Chapter 12).

An ecosystem study involving wolves, their prey, and vegetation has been conducted on Isle Royale (McLaren and Peterson 1994). It is also the longest running predator-prey study in the world (Peterson 1995). Isle Royale is a 544–square kilometer island in northwestern Lake Superior. Moose reached the island early in the twentieth century, erupted to high population levels by the 1930s in the absence of predation, and then suffered catastrophic starvation (Mech 1966, Peterson 1977, Allen 1979). Wolves did not make it to the island until the late 1940s, when they eliminated the coyotes and began preying on the moose and beavers (*Castor canadensis*, Allen 1979).

The value of the Isle Royale research, from an ecosystem perspective, is that the island confines the boundaries of the ecosystem and reduces the number of interacting species. Isle Royale has only sixteen mammal species, but all the major boreal forest players are there—wolves, moose, beavers, and snowshoe hares (*Lepus americanus*, Peterson 1977). This has simplified the research, making a one-predator, two-prey system rewarding for study (Mech 1966, Peterson 1977, Allen 1979, Peterson and Page 1988).

A pressing question in ecosystem studies has been top-down vs. bottom-up influences (Hairston et al. 1960, McLaren and Peterson 1994, Chapters 8 and 10 this volume). Does primary productivity through plant growth determine herbivore biomass, which in turn determines predator numbers? Or do predators control herbivores, which in turn affect consumption of plants? On Isle Royale, for example, do wolves control moose and thereby affect browsing on woody plants, or does plant availability determine moose population size, which ultimately determines the wolf numbers? Peterson et al. (1984) determined on Isle Royale that wolves and moose cycle with a periodicity of twenty-five to thirty years and that the wolf population lags behind moose by seven to ten years, suggesting bottom-up regulation.

With continued study, McLaren and Peterson (1994) found that low wolf numbers were followed by high moose numbers, which were followed by heavy plant browsing and vice versa, which is consistent with top-down regulation. High moose numbers also had an effect on balsam fir, a key winter moose browsing plant, eliminating it where it occurred at low density (Brandner et al. 1990). The interesting aspect to Isle Royale, made clear only through long-term research, is that top-down regulation is not always the key influence. A crash in the moose population in 1996 was precipitated by severe winter weather and heavy tick infestations. Because tick outbreaks are weather dependent, the moose population crash was caused not by wolves but by a combination of density-dependent and density-independent factors (DelGiudice et al. 1997). This also illustrates the dynamic nature of ecosystem processes and the challenge for managers to allow for such natural fluctuations (see Chapters 2 and 12).

Another ecosystem study in Algonguin Park, Ontario, of wolves, their prey, and plants, also emphasizes the paramount importance of long-term ecosystem studies. Douglas Pimlott began wolf studies in Algonquin in 1959 (Rutter and Pimlott 1968) and passed along supervision of the project to John and Mary Theberge, who are continuing the research. It is thus the second-longest predator-prey study in the world, although there was a brief hiatus from intensive fieldwork in the late 1980s. Unlike Isle Royale, Algonquin is large, about 7,700 square kilometers, and not an island. The forest type there, however—northern hardwood and southern boreal—is similar to Isle Royale's (Theberge 1989). Since the study's inception, wolf numbers have remained remarkably stable, at around 150 animals in thirty to thirty-five packs (possibly the result of immigration because mortality at times has been high), while prey populations have fluctuated. Moose have increased, but white-tailed deer *(Odocoileus virginianus)* have declined, although wolf predation is not implicated in the deer decline (Forbes and Theberge 1996). Beavers have become a more common prey of wolves in recent years, but quantitative data are lacking.

An unusual feature of Algonquin is that logging has been permitted since the park was established. Where logging has occurred in stands dominated by pine *(Pinus* spp.), moose and deer browsing has suppressed regeneration of deciduous vegetation. In mixed stands of pine and northern hardwoods, deciduous vegetation is too lush for moose and deer to suppress hardwood regeneration, and young pines are not becoming established because they are being outcompeted by deciduous trees. Wolves do not seem to play a role in controlling herbivore numbers and hence are not playing a role in forest regeneration. Bottom-up influences then would seem to be more important in Algonquin (Forbes and Theberge 1994).

The important ecosystem aspect of both these studies is that three trophic levels have been the focus of examination—primary producers (plants), primary consumers (deer and moose), and secondary consumers (wolves). Wolves were not implicated in controlling the prey populations as strongly as they have been in other predator-prey studies, which have suggested that predation does control prey populations and that top-down influences are more important (Gasaway et al. 1992, Van Ballenberghe and Ballard 1994). These studies are from areas where more than one predator utilizes the prey population—in the north, for example, where sympatric wolves and grizzly bears have been found to control prey populations (Gasaway et al. 1992). This difference in results—and explanations for it—should aid in generating hypotheses for study in the GYE.

With wolf restoration, the GYE offers an unparalleled opportunity to study the impacts of a major predator on an ecosystem. Despite the persistent problems of scale and indefinite ecosystem boundaries (which are often arbitrarily defined by researchers), we have valuable pre- and postwolf reintroduction data on a number of species, including the vegetation. Population sizes of most of the ungulates in the ecosystem are known, and the wolf's major prey, elk *(Cervus elaphus,* Phillips and Smith 1996), is well studied both in and outside the park (Houston 1982, Singer 1991). Major long-term studies are under way on the other carnivores: grizzly bears (Servheen and Knight 1990), cougars (Murphy et al. 1998), and coyotes (Chapter 6). Although not as complete, data also exist on small carnivores (Chapter 7). Finally, studies of Yellowstone's grasslands are complete (Tracy 1997), and research on woody vegetation is under way (J. Varley, personal communication). Indeed, the array of studies of both people and animals in the GYE is unique. It is likely our best shot at unraveling many interesting ecosystem-level processes.

Another heartening factor for research (but not for conservation) is that, like Isle Royale, the GYE is a gigantic "island" and hence definable. But this island is surrounded not by water but by unsuitable habitat in the form of ranch land. So far wolves have not fared well at this interface, and the national park–national forest boundary has been a filter that wolves have not succeeded in penetrating.

Wolf Eradication and Restoration in the Greater Yellowstone Ecosystem

WOLF ERADICATION

The history of wolves in North America and the West is straightforward: we killed them (Young and Goldman 1944, McIntyre 1995). Wolves were considered competition for wild prey and destroyers of domestic animals and therefore obstacles to the purposes of civilization, which was the main objective of settlers of the New World (Lopez 1978, McIntyre 1994). European folklore and mythology had seen the wolf as evil, and settlers, familiar with wolves from their homeland, wasted no time in transplanting this view to an American wilderness that was full of wolves (Lopez 1978, Casey and Clark 1996). This view of wolves ran counter to that of the Native Americans already inhabiting this wilderness, to whom wolves were often considered sacred. Unfortunately, settlers all too often lumped the two together, wolves and Native Americans, with the goal of annihilating them both.

The result was the elimination of wolves from the continental United States except for the northern portion of Minnesota, and the reduction of their numbers over much of Canada and Alaska (Young and Goldman 1944). In the West, many wolves made their last stand in YNP, but they were shown no mercy and persecuted within park boundaries. Weaver (1978) concluded from the historical record that no wolf packs persisted after the 1930s and that, other than sporadic singles and pairs, no wolves have existed in the GYE since that time (Meagher 1986).

THE ENDANGERED SPECIES ACT AND THE ADMINISTRATION
OF RESTORATION

In 1974 wolves in the northern Rocky Mountains were listed as endangered under the Endangered Species Act (ESA). Returning the wolf to at least portions of its former range was therefore federally mandated. For many (Pimlott 1967, Cole 1969, Mech 1970, 1991, Weaver 1978, Peek et al. 1991), beginning with Aldo Leopold (1944), reintroduction was the best strategy for restoring wolves. Natural recolonization was unlikely because of the insularity of the GYE; unsuitable and inhospitable habitat was an effective barrier for dispersing wolves to settle the ecosystem. Economics also played a part: reintroduction was cheaper. Recolonization of the GYE island through natural dispersal would have involved so few animals that population growth to levels adequate for delisting would almost surely have taken decades. Through this period, the costs of monitoring and nurturing this nascent population would have far exceeded the costs of reintroducing wolves. Finally, under provisions stated in Section 10(j) of the ESA, reintroduction allows much greater management flexibility than natural recolonization. This is important because wolves are controversial (Mech 1995, Bangs and Fritts 1996), and many local people were not in favor of the return of wolves. For example, under current provisions private landowners are allowed to haze wolves from their land as long as nonlethal means are employed. Further, they can kill a wolf if it is caught in the act of killing livestock and the kill is reported within twenty-four hours (U.S. Fish and Wildlife Service 1994a).

Regardless of management flexibility and law requiring restoration, wolf reintroduction or restoration has been controversial. It is a highly emotional issue tied to control of the West and to people's attitudes toward the federal government. Westerners would like the federal government out of their lives. Wolves occasionally kill livestock (Gunson 1983, Paul and Gipson 1994), and it is within the memory of many westerners that their parents and grandparents worked hard to eradicate wolves. To them, few efforts could be more ludicrous, more immoral, and more antithetical to their heritage than using taxpayer money to bring the wolf back.

As a result, an extensive administrative process had to be followed. Federal legislation directing that wolves be reintroduced to Yellowstone was introduced into Congress in 1987, 1989, and 1990, but none of these bills was passed into law (Schullery 1996). In 1990 Congress tried another approach to break the stalemate by directing the secretary of the interior to create a Wolf Management Committee to prepare a report on wolf restoration for both the GYE and the central Idaho wilderness. After months of deliberation, the Wolf Management Committee submitted its report and recommendations to the Congress, which chose not to act on the report. In 1991 Congress directed the U.S. Fish and Wildlife Service, in cooperation with the National Park Service and Forest Service, to prepare an environmental impact statement on the reintroduction of gray wolves to Yellowstone National Park and central Idaho.

On a parallel but contrary course, Congress had prohibited reintroduction through language in appropriations bills for the Department of the Interior. However, in 1988 through 1990 Congress appropriated funds to the National Park Service and U.S. Fish and Wildlife Service to investigate the social, ecological, legal, and economic effects of reintroducing wolves to YNP. Through these investigations, more than 1,300 pages were written on the potential effects of wolf reintroduction and presented to Congress (Yellowstone National Park et al. 1990, Varley and Brewster 1992). These investigations and other literature formed the scientific basis for the environmental impact statement (EIS) directed by Congress. The results of these investigations were invaluable in presenting accurate information to the public, evaluating the potential effects of various alternatives, and defending against legal challenges to the reintroduction.

There was never a law passed that directed the eradication of wolves. This policy was established through the Department of the Interior's appropriations in 1915 to the Bureau of Biological Survey, the precursor of the U.S. Fish and Wildlife Service. It is somewhat ironic that restoration was accomplished in the same manner.

Because the reintroduction of wolves to YNP had such high public and political interest, it was determined that extensive public involvement in the planning process was required. Similarly, because of uncertainty about the techniques that would be used in the project, extensive involvement was required by the scientific community and those experienced in wolf behavior, ecology, management, capture and handling, and confinement.

Preparation of the EIS was done by an interagency team led by the U.S. Fish and Wildlife Service. The process involved the public in identifying issues, developing potential alternatives, and reviewing draft documents. Between April 1992 and April 1994, the EIS team hosted more than 130 public meetings, distributed about 750,000 documents, and received comments from more than 170,000 people from all fifty states and forty countries (Bangs and Fritts 1996, Fritts et al. 1997). The final EIS was approved April 14, 1994, and a record of decision was signed by the secretary of the interior in June 1994 (U.S. Fish and Wildlife Service 1994a). Special regulations for reintroducing the wolves as nonessential, experimental populations under provisions of Section 10(j) of the Endangered Species Act were completed in November 1994 (Federal Register, November 22, 1994). Throughout the administrative process, extensive efforts were made to keep the public informed of the process and to provide factual information on a consistent and regular basis. Additionally, agency decision makers, congressional delegations, state elected officials, and state agency personnel were regularly informed of the progress. Finally in January

1995 and 1996, wolves were captured in Alberta and British Columbia, respectively, and transported to YNP.

CAPTURE, ACCLIMATION, AND RELEASE

Packs of wolves were helicopter darted and translocated to YNP. Wolf packs were chosen because it was believed that they would remain together after release and breed sooner than individual wolves. The packs were acclimated in one-acre (0.4-hectare) pens for ten weeks before release to break the homing desire (Fritts et al. 1984). Acclimation was the chosen strategy for Yellowstone because of the tremendous logistical support of the National Park Service. Yellowstone had year-round access to pens either by plowed road or groomed snowmobile trails, the ranger and maintenance staff assisted in myriad ways, and housing facilities allowed a labor force to be located on site. Unlike YNP, the central Idaho translocation effort captured not family groups but individuals, and reintroduced wolves were released immediately on site instead of being acclimated in pens. The central Idaho wilderness was big enough (and without roads) to allow wolves to wander, and there was no logistical support similar to that offered by the National Park Service. YNP wolves were "soft" released—that is, they were gradually introduced to the new environment via temporary enclosures and supplemental feeding, and finally, the gates or other sections of the pens were simply left open for the wolves to leave. Idaho wolves were "hard" released from their transport containers (Bangs and Fritts 1996, Phillips and Smith 1996).

YNP received fourteen wolves in three packs in 1995 and seventeen wolves in four packs in 1996 (Phillips and Smith 1996). In September 1996 the park also received ten wolf pups (five months old) from a depredating pack in northern Montana. Implementation of the reintroduction project involved innovative work by many biologists. Last-minute decisions were made and "adaptive management" was employed freely. Selecting specific individuals and sifting through packs of wolves while flying in a helicopter at minus 45°C was difficult, to say the least. Field personnel in Canada often had to make decisions about what wolves to ship where, after the animals were already at the compound. The goal of moving intact packs to YNP was achieved in only three of seven cases; for the others, packs of wolves were created (Phillips and Smith 1996).

Upon release, and contrary to initial expectations, the wolves did not bolt from the acclimation pens as soon as the gates were opened. After periods ranging up to twelve days, the wolves exited the pens. We considered this a fortunate circumstance because they could leave confinement in the absence of humans in an exploratory mode rather than an escape mode (Phillips and Smith 1996). Postrelease movements of four of five released groups consisted of three distinct periods of behavior: (1) restricted movements around the pen, (2) exploration, and (3) return and settlement near the acclimation pen. Two groups are not considered because mortality shortly after release precluded any kind of postrelease behavioral analysis. The fifth group fragmented at release, ranged widely, and did not reunite.

THE NEW GREATER YELLOWSTONE ECOSYSTEM WOLF ERA

As of July 1997 there were nine wolf packs and forty-four adult (yearling or older) free-ranging wolves and an unknown number of pups in thirteen litters (figure 5.1). Three packs have had more than one litter of pups, and one litter of pups perished because of the death of the loner mother. At the end of 1996 there were forty free-ranging wolves in nine packs, and four of those packs gave birth to a total of fourteen pups. At the end of 1995 there were thirteen adults in three packs, and two of those packs gave birth to a total of nine pups.

Twenty-six wolves have died since release. Nine (35 percent) of these are from the original thirty-one wolves translocated from Canada, and the other seventeen mortalities are from animals either introduced from Montana or born in YNP. Human-caused death was the leading cause of mortality for GYE wolves: four were illegally shot, four were legally shot, four were killed by vehicles, three were killed by other wolves, one was killed by a M-44 "coyote getter" device, and the other ten died from either natural or unknown causes, including four pups that died from exposure when their mother was killed by other wolves.

Although wolves have killed every ungulate species in the GYE except bighorn sheep (Phillips and Smith 1996), elk constituted 85 percent and moose made up 11 percent of the wolf kill during late winter 1997. Two bison were killed during the exceptionally severe winter of 1996–97. During winter, the rate of wolf kills of elk were higher than that reported for rates of killing elk elsewhere, and utilization and handling time (time spent feeding on a kill) has been less than that reported in other studies (Carbyn 1983, Kunkel 1997). Consequently, wolf territory size is at the low end of sizes reported for North America (Fuller 1989).

Besides elk, wolves have been killing coyotes at a high rate, not to acquire food but to eliminate competitors. Researchers know of fifteen coyote kills by wolves (National Park Service and R. Crabtree, unpublished data). In all but one case, coyotes were killed while scavenging on wolf kills before the wolves had left the carcass. Coyote pack size, litter size, territorial boundaries, and behavior have all changed in response to wolves (R. Crabtree, personal communication, Chapter 6 this volume).

Grizzly bears have not been killed by wolves. In most cases grizzly bears have

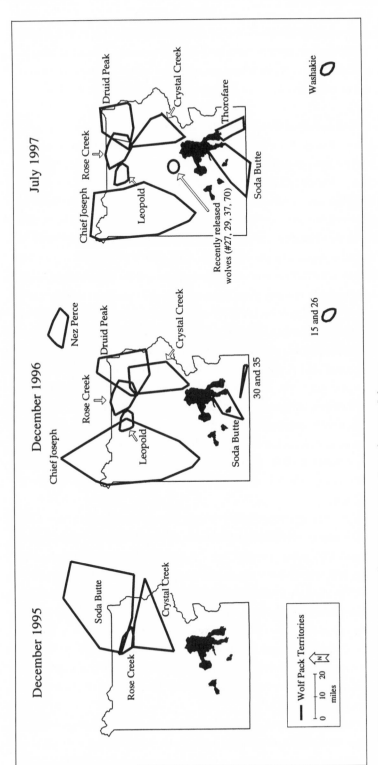

5.1. Wolf pack territories in the Greater Yellowstone Ecosystem, December 1995–July 1997.

displaced wolves from carcasses that wolves have killed. No case of a grizzly killing a wolf has been documented, despite several close approaches by grizzlies to wolf den sites. All black bear encounters so far have involved wolves chasing the bears away. We do not have data on wolf-cougar interactions.

Predictions and Appraisals About Wolf Restoration

Extensive research on wolves and on the processes of reintroduction and recovery was carried out before the YNP reintroduction (Yellowstone National Park et al. 1990, Varley and Brewster 1992, U.S. Fish and Wildlife Service 1994b). During this planning phase, numerous objectives were identified for GYE wolf recovery. The overall goal of the program—to remove wolves from the endangered species list—can be accomplished by having ten pairs of wolves breed in three successive years with the stipulation that two pups from each pack for that year's reproduction survive until December 31. When this objective of ten pairs in each area is reached simultaneously in the GYE, central Idaho, and northwestern Montana, then the gray wolf in the northern Rockies will be taken off the endangered species list (U.S. Fish and Wildlife Service 1994b).

As part of this process, predictions, or projections, were made about the effects of a restored wolf population on wildlife management, big game populations, livestock, land use, local economies, and visitor use (U.S. Fish and Wildlife Service 1994b). Assessment of the accuracy of these predictions will reveal how much we know about reintroducing wolves, about the GYE, and about what worked and what did not. This knowledge may be of value to other reintroduction programs. After two and a half years of free-ranging wolves (through September 1997), we can begin to assess the accuracy of these predictions (table 5.1).

The first prediction states that there were few or no wolves in the GYE. Contrary to this, some people asserted that naturally occurring wolf populations remained undetected in the ecosystem. The best way to find any animal anywhere, especially wolves, is to release conspecifics (Gese and Mech 1991). New packs are formed by lone wolves locating each other and breeding (Mech 1970). All the reintroduced wolves were radio-collared so that a collared wolf could "lead" us to any naturally occurring wolves. To date, no wolf has been located in the GYE with a non-reintroduced wolf, nor has any individual been found in a pack who was not the offspring of reintroduced wolves. Four packs have formed naturally since reintroduction, and they have all been collared wolves with other collared wolves.

It was predicted that some wolves would travel outside the experimental area. To date, no wolves have traveled outside the experimental area. Of nine packs currently in the ecosystem, eight are located within YNP (figure 5.1). Six

Table 5.1

Predictions and appraisals of Yellowstone National Park Wolf restoration

Predictions[a]	Appraisals[b]
No wolves existed in the Yellowstone ecosystem	None were found after reintroduction, except for reintroduced wolves
Wolves would travel outside the experimental area	None traveled outside the experimental area
Wolf packs would kill 120 ungulates per year	Wolf packs have killed about 130 ungulates per year
Wolves would kill 19 cattle and 58 sheep each year	No cattle killed, approximately 10 sheep killed per year
Land-use restrictions would occur around dens only	Land-use restrictions have occurred around dens only
State management would occur before delisting	States will not manage wolves before delisting
Few visitors to Yellowstone National Park would see wolves	About 20,000 people had seen wolves as of July 1997

[a]Made before wolf reintroduction.
[b]Made two and a half years after wolves were reintroduced to YNP.

instances of wolves using areas on private land, in livestock areas, or close to concentrated human populations have been recorded. In each case the offending wolves were captured and relocated or legally killed (U.S. Fish and Wildlife Service 1994a).

Based on an estimated ten wolves per pack, wolf packs were predicted to kill 120 ungulates per year (U.S. Fish and Wildlife Service 1994b). Based on an actual average of five wolves per pack, GYE's wolf packs have been killing at a slightly higher rate of about 130 ungulates per year, most of which have been elk. This may be explained by the fact that the EIS calculated a typical ungulate as slightly larger than an elk (because bison and moose are larger than an elk). Moreover, wolves kill many calves, which are considerably smaller than the typical ungulate, thus increasing the actual number of ungulates killed.

About nineteen cattle and sixty-eight sheep were predicted to be killed annually by a recovered wolf population of ten breeding pairs (U.S. Fish and Wildlife Service 1994b). No cattle have been killed and fifty-five sheep have been killed in two and a half years. One lion-hunting dog was killed on private land after the dog followed a wolf trail into a pack of wolves. A private group has compensated owners for all verified wolf kills. In all cases the offending wolf was removed or killed legally.

Land-use restrictions were predicted to occur around den sites from April 1 through June 30, curtailing some Department of Agriculture Wildlife Services predator control activities (primarily use of M-44 "coyote getter" devices) on private lands because of the presence of wolves. No restrictions have occurred on sites other than den sites or acclimation pens. Two areas around dens were closed later than June 30 in 1997 because of proximity to a road. In one case the rationale was to preserve viewing opportunities for the public, and in the other case it was to protect a litter of pups that was not being fed regularly because their mother had been struck by a vehicle. Two groups of wolves moved off public land into areas where Wildlife Services was using M-44 devices and neck snares for preventive control of coyotes. These devices were removed from several ranches while wolves were in the vicinity.

During the EIS process Idaho, Montana, and Wyoming were asked for the flexibility to manage wolves in their states. At the start of the reintroduction program in 1995, all three states, while not supporting reintroduction, did consider themselves partners in the effort. It was hoped that the states would be the primary managers of wolves by autumn 1995. None of the states has chosen to manage wolves before they are delisted. Wolf management has been carried out by the U.S. Fish and Wildlife Service, the National Park Service, Wildlife Services, and the Nez Perce tribe in Idaho.

The EIS predicted that few people would see wolves in the wild and that local economies would benefit by up to $23 million annually. From April 1995 through September 1997 about twenty thousand people have observed free-ranging wolves in YNP. No quantitative information is available on the economic impacts, but interviews in the local communities indicate that business has increased because of the presence of wolves (R. McIntyre, personal communication).

In general the predictions made about wolf reintroduction to the GYE were accurate. Surveys before reintroduction did not find evidence of any existing wolves, nor did our reintroduced wolves. The distances traveled by the reintroduced wolves were similar to those traveled by non-reintroduced wolves and less than those traveled by relocated wolves in other areas (Fritts 1982, Fritts et al. 1984). Livestock depredations were rare, as predicted. Predictions about wolf kill rates were based on wolf packs of ten individuals; we observed that packs of five wolves were able to kill at approximately the rate predicted for packs of ten. This prediction may have been the most difficult because of the large number of factors that determine wolf kill rates. Tremendous variability has been found seasonally and annually within the same wolf population, so this is not a surprising result (Peterson 1977). Land-use restrictions have been limited as predicted, and in the two cases where restrictions were extended, human activity and benefit were the reasons. State management of wolves did not proceed as predicted, reflecting the immense complexity of the policy process (predictions about wolf biology being considerably more certain than predictions about policy processes). Tourism did increase and associated economic benefits to communities occurred because of the high, and unpredicted, visibility of the wolves.

Rocky Mountain Wolf Management and the Future of Wolves in the West

The future of wolves in the Rocky Mountain West looks promising relative to the record of the past six decades. However, given the dynamic nature of the wolf policy and management process, including a U.S. District Court ruling in late 1997 that the reintroduction violated the intent of the Endangered Species Act, details of the wolf's future are not clear at present.

WOLF MANAGEMENT IN THE WEST

The goal of Rocky Mountain wolf restoration and management is to remove the wolf from the endangered species list and turn management over to the states (Idaho, Montana, and Wyoming). Wolves would eventually be common enough to be managed as a game animal, and a wolf harvest would almost surely take place (U.S. Fish and Wildlife Service 1994b). YNP will always be managed as a wolf sanctuary. There will be no harvest or wolf control inside the park. (This may seem to be a ridiculous prospect, but it has been proposed at some public meetings.) When ten breeding pairs have two pups that survive until December 31 for three successive years in each recovery area simultaneously (U.S. Fish and Wildlife Service 1987), then the recovery program goals will have been met. Wolves can then be delisted, with a five-year oversight period during which the states and the U.S. Fish and Wildlife Service will monitor wolves. If the wolf population in any of the three recovery areas falls below the minimum of ten breeding pairs, then wolves would be relisted as an endangered species (U.S. Fish and Wildlife Service 1994b).

Until wolves are delisted there will be no sport harvest (U.S. Fish and Wildlife Service 1994a). The special regulations under the experimental, nonessential status do allow for wolves to be killed when they are in the act of killing livestock on private land (U.S. Fish and Wildlife Service 1994a). It is also legal under the special regulations to haze a wolf nonlethally off private land (U.S. Fish and Wildlife Service 1994a). Other management actions for problem wolves are permitted on public land but will involve federal personnel. Citizens can take action after federal efforts to remove a problem wolf have proven unsuccessful (U.S. Fish and Wildlife Service 1994b). These and other provisions are possible because of the experimental, nonessential status of the reintroduced population—provisions that would not have been available if wolves had naturally recolonized and were classified as an endangered species.

When a wolf kills livestock in the experimental area, the Department of

Agriculture Wildlife Services unit will remove the offending wolf or wolves. These animals will be relocated and rereleased. If any of these animals kills livestock again, it will be removed from the wild (killed or placed into captivity). So far this management scheme has been highly effective and gained the restoration effort tremendous credibility with the ranching community.

Recently, Idaho, Montana, and Wyoming have all decided not to manage wolves before they are delisted. Each state has prepared a draft wolf management plan for the post-delisting era, but coordination of wolf management among the states seems likely, and one management plan may be developed. All state plans must be approved by the U.S. Fish and Wildlife Service.

In short, wolves will exist where humans will let them (Kellert et al. 1996). Human values will regulate wolf density and distribution. Wolves will be absent from areas where livestock is raised, they will exist at low densities in roaded areas, and they will be naturally regulated in parks and wildlife refuges. This management scenario is similar to zoned wolf management used in other areas and proposed for widespread recovery of wolf populations (Mech et al. 1988, Mech 1995, Clarkson 1995). Some of these predictions depend on how much money is available to pay for wolf control and management. No matter what happens, it is safe to say that wolves will remain the center of a very lively public policy debate about the role of carnivores in ecosystems and their relation to humans.

LESSONS FROM MINNESOTA AND ALASKA

This process of taking the wolf from endangered to recovered and delisted will not be easy. Delisting is likely to be more controversial than reintroduction and restoration (Clark and Brunner 1996). We know from experiences in Minnesota, and to some extent in Alaska (where wolves are not endangered or rare), that wolf management is fraught with enormous difficulties and national debates (Anonymous 1993, Mech 1995). Wolf issues in those areas and in Canada have progressed through a series of lawsuits, and this will probably be unavoidable in the West as well (Keiter and Locke 1996). Wolves will continue to be used to advance points of view on other environmental or western issues— trust or mistrust of the government is one such issue.

People's core beliefs about the relation of humans to nature and about environmental issues in general influence their views of the wolf debate— whether there should be control or no control, harvest or total protection, and so on. Because these core beliefs vary widely among Americans, the only way to deal with the nearly guaranteed volatility surrounding wolf issues is to continue to involve as many people as possible (Clark 1992, 1993). There was unprecedented public outreach and communication with all concerned parties before wolf reintroduction actually occurred (Fritts et al. 1997), and this approach is the only hope in a democracy for reducing inaction and litigation during delisting conflicts. If not, the courts will influence wolf delisting as they did the reintroduction.

Ways to avoid this controversy can be learned by studying the Alaskan and Canadian situations. Wolves have never been endangered in these two places, and both have always had legal harvests (Hayes and Gunson 1995). Wolf control, aimed at eliminating most of the wolves in a particular area to enhance ungulate populations, has been employed at times (Gasaway et al. 1983, Bergerud and Elliot 1986, Cluff and Murray 1995). This practice has been extremely controversial and has attracted national attention. As a result, wolf management in Alaska has become an issue for the courts and political arenas. Killing wolves as a management practice in the western United States, especially for GYE wolves that leave the park, will likely be even more distasteful to the American public.

Alaska ended the widespread persecution that dominated wolf management in the 1950s and 1960s, and many wolf populations throughout the state grew. However, after ending wolf persecution, the state chose not to protect any wolf population aggressively. Denali National Park is the only area in Alaska where wolves are protected. A wolf pack that roams in and out of the park can be killed when it leaves the boundaries of the park. If the state had identified the Denali ecosystem as a wolf sanctuary, state managers would have gained more credibility for wolf management elsewhere in the state (W. Gasaway, personal communication). A more balanced approach, aimed at protecting and managing wolves, would have gained greater acceptance and might not have attracted national controversy.

Alaska also did not pay enough attention to nonhunters (W. Gasaway, personal communication). Alaska estranged nonhunters by not protecting wolves in some areas, and nonhunters far outnumber hunters, especially in the lower forty-eight states. Wildlife is a resource for all the state's people, not just the hunters. Management of wolves should consider people who want to view wolves and those who value the existence of free-ranging, unharassed wolves. Social questions for the Yellowstone restoration project then include: Should wolf packs that range inside and outside of Yellowstone be protected from harvest? Should the GYE, and not just YNP, be our wolf sanctuary? Certainly, with the Ministry of Natural Resources in Ontario leading the way in protecting wolves that migrate outside Algonquin Provincial Park (Forbes and Theberge 1996), the management precedent has been set. A large volume of literature advocates an ecosystem approach, irrespective of political boundaries, in the GYE (Keiter and Boyce 1991, Clark and Minta 1994). Wolves are recently new to the GYE; the time is right to try new approaches.

Finally, if we learn little else from Alaska and Canada, we should recognize that wolf control by aerial gunning will not be accepted. If this is attempted, court battles will surely ensue. Wolf management will be much more successful if it is done through "normal" harvest strategies similar to the practices employed for other wildlife species (Clarkson 1995, Mech 1995).

THE FUTURE OF WOLVES IN THE YELLOWSTONE ECOSYSTEM

The reintroduction and subsequent restoration of wolves to the GYE can be likened to recovery after severe predator control (W. Gasaway, personal communication). As the years pass, the full ecological significance of the wolf will become clearer. When wolves are delisted, there will be approximately three hundred wolves over a 155,000-square kilometer area, or an extremely low density of one wolf per 517 square kilometers (Fuller 1989). Their distribution will perhaps be more important than their density, for there will be pockets of high wolf density. But this underscores the point that wolves in the West will not be like wolves in the North: wolves will never be common in the West because there are too many people and too much development. Rather than a widespread distribution at low to moderate density, wolves will be unevenly distributed with high densities in some places. Understanding this fact will affect the design of both research and management of wolves in the West.

Understanding, however, has not come easily. It is important to recognize that it was the federal government that brought the wolves back, and it will be the state governments that finish the job. Wolf restoration will take time, and all levels of government, the public, and special interest groups need to acknowledge that much patience will be necessary. In addition, because wolf recovery and restoration are complex enough, the simplest and least expensive solutions should always be sought. Our overriding point in this chapter is that it will be the human-biological milieu that will determine the future of wolves. The whole issue revolves around politics, public relations, education, and constant dialogue between conflicting groups. If we do not develop an effective, cooperative management process, then the legal system will significantly determine wolf management (see Clark 1997). The courts are ill-suited to guide wolf management because of the adversarial proceedings, limited issues under adjudication, and constraints on how the proceedings are actually conducted. As a result, many interested parties are likely to find that their concerns are inadequately addressed. Despite the magnitude of the wolf recovery project and the widespread support for GYE wolf restoration, it almost failed. It was one legal decision away from success or failure.

Besides the political aspects, the science of GYE wolf restoration is an invaluable ecological lesson. Most predator-prey studies have looked at two or sometimes three species in a management unit (not an ecosystem) for a few years. Five large predators exist in the Yellowstone ecosystem—wolves, grizzly bears, black bears *(Ursus americanus)*, cougars, and coyotes—and data are available on all of them. One other situation like this exists in northwestern Montana, but studies are not ongoing, and far fewer researchers are involved. Long-term studies are now possible—and, in fact, imperative—in the GYE. It would be a crime against history and science if we did not take advantage of this opportunity.

Wolf conservation biology in the GYE is important not only to wildlife management in the West, combining the world's first and most famous national park with North America's most controversial predator will clearly garner worldwide visibility and attention. Conservation biology and maintenance of biodiversity in ecosystems are pressing issues worldwide. Yellowstone's wolf recovery program has proven that species restoration is politically difficult, biologically uncertain, costly, and protracted—even in the world's richest nation with broad pubic support for conservation and in one of the world's most visible ecosystems (Weber and Rabinowitz 1996). And yet, GYE wolf restoration took more than twenty years and was (and still is) one court case away from failing. What fate will befall the many imperiled species and ecosystems in tropical and other fragile ecosystems managed by cash-strapped, politically unstable, developing countries where conservation may be seen as a low priority or a costly extravagance?

A key question currently in the GYE is who will have the most to say about wolf management. Based on the record to date, it looks as if biologically based management of wolves will be difficult and that actual management will mirror the past policy dynamic. For example, the decision to reintroduce wolves was delayed by twenty years of harsh debate, with the final decision left up to a judge (Varley and Brewster 1992, Fritts et al. 1997). In other words, wildlife and ecosystem management was not in the hands of wildlife professionals (see Kellert and Clark 1991, Primm and Clark 1996). This has made current and future implementation of GYE wolf restoration difficult. But as professionals we must follow the best principles and practices of modern wildlife management as we manage and study wolves—keeping in mind the wolf's powerful public images and the diverse human perspectives at play, and remaining open to change and improvements.

Literature Cited: Allen, D. L. 1979. The wolves of Minong: Their vital role in a wild community. Houghton Mifflin Company, Boston.

Anonymous. 1993. Alaska wolf plan defeated by communications failures. International Wolf 3:10–15.

Bangs, E. E., and S. H. Fritts. 1996. Reintroducing the gray wolf to central Idaho and Yellowstone National Park. Wildlife Society Bulletin 24:402–13.

Bergerud, A. T., and W. B. Ballard. 1988. Wolf predation on caribou: The Nelchina herd case history, a different interpretation. Journal of Wildlife Management 52:251–59.

Bergerud, A. T., and J. P. Elliot. 1986. Dynamics of caribou and wolves in northern British Columbia. Canadian Journal of Zoology 64:1515–29.

Boutin, S. 1992. Predation and moose population dynamics: A critique. Journal of Wildlife Management 56:116–27.

Boyd, D. K., P. C. Paquet, S. Donelon, R. R. Ream, D. H. Pletscher, and C. C. White. 1995. Transboundary movements of a recolonizing wolf population. Pp. 135–40 *in* L. N. Carbyn, S. N. Fritts, and D. R. Seip, eds., Ecology and conservation of wolves in a changing world. University of Alberta Press, Edmonton.

Brandner, T. A., R. O. Peterson, and K. L. Risenhoover. 1990. Balsam fir in Isle Royale National Park, Michigan: Effects of moose herbivory as influenced by density. Ecology 71:155–64.

Carbyn, L. N. 1983. Wolf predation on elk in Riding Mountain National Park, Manitoba. Journal of Wildlife Management 47:963–76.

Carbyn, L. N., S. M. Oosenbrug, and D. W. Anions. 1993. Wolves, bison . . . and the dynamics related to the Peace-Athabasca delta in Canada's Wood Buffalo National Park. University of Alberta Press, Edmonton.

Casey, D., and T. W. Clark. 1996. Tales of the wolf. Homestead Publishing, Moose, Wyoming.

Clark, T. W. 1992. Practicing natural resource management with a policy orientation. Environmental management 16:423–33.

———. 1993. Creating and using knowledge for species and ecosystem conservation: Science, organizations, and policy. Perspectives in Biology and Medicine 36:497–525.

———. 1997. Averting extinction: Reconstructing endangered species recovery. Yale University Press, New Haven, Connecticut.

Clark, T. W., and R. D. Brunner. 1996. Making partnerships work in endangered species conservation: An introduction to the decision process. Endangered Species Update 13:1–5.

Clark, T. W., A. P. Curlee, and R. P. Reading. 1996. Crafting effective solutions for large carnivore conservation. Conservation Biology 10:940–48.

Clark, T. W., and S. C. Minta. 1994. Greater Yellowstone's future: Prospects for ecosystem science, management, and policy. Homestead Press, Moose, Wyoming.

Clarkson, P. L. 1995. Recommendations for more effective wolf management. Pp. 527–34 *in* L. N. Carbyn, S. H. Fritts, and D. R. Seip, eds., Ecology and conservation of wolves in a changing world. University of Alberta Press, Edmonton.

Cluff, H. D., and D. L. Murray. 1995. Review of wolf control methods in North America. Pp. 491–504 *in* L. N. Carbyn, S. N. Fritts, and D. R. Seip, eds., Ecology and conservation of wolves in a changing world. University of Alberta Press, Edmonton.

Cole, G. F. 1969. The elk of Grand Teton and southern Yellowstone national parks. Yellowstone Library and Museum Association, Yellowstone National Park.

DelGiudice, G. D., R. O. Peterson, and W. Samuel. 1997. Trends of winter nutritional restriction, ticks, and numbers of moose on Isle Royale. Journal of Wildlife Management 61:895–903.

Forbes, G. J., and J. B. Theberge. 1994. Multiple landscape scales and winter distribution of moose *(Alces alces)* in a forest ecotone. Canadian Field Naturalist 107:201–7.

———. 1996. Cross-boundary management of Algonquin Park wolves. Conservation Biology 10:1091–97.

Fritts, S. H. 1982. Wolf depredation on livestock in Minnesota. U.S. Fish and Wildlife Service, Resource Publication no. 145.

———. 1983. Record dispersal by a wolf from Minnesota. Journal of Mammalogy 64:166–67.

Fritts, S. H., E. E. Bangs, J. A. Fontaine, M. R. Johnson, M. K. Phillips, E. D. Koch, and J. R. Gunson. 1997. Planning and implementing a reintroduction of wolves to Yellowstone National Park and central Idaho. Restoration Ecology 5:7–27.

Fritts, S. H., W. J. Paul, and L. D. Mech. 1984. Movements of translocated wolves in Minnesota. Journal of Wildlife Management 48:709–21.

Fuller, T. K. 1989. Population dynamics of wolves in north-central Minnesota. Wildlife Monographs no. 105.

———. 1995. Guidelines for gray wolf management in the northern Great Lakes region. International Wolf Center Technical Publication 271.

Gasaway, W. C., R. D. Boertje, D. V. Grangaard, D. G. Kelleyhouse, R. O. Stephenson, and D. G. Larsen. 1992. The role of predation in limiting moose at low densities in Alaska and Yukon and implications for conservation. Wildlife Monographs no. 120.

Gasaway, W. C., R. O. Stephenson, J. L. Davis, P. E. K. Shepherd, and O. E. Burris. 1983. Interrelationships of wolves, prey, and man in interior Alaska. Wildlife Monographs no. 84.

Gese, E. M., and L. D. Mech. 1991. Dispersal of wolves *(Canis lupus)* in northeastern Minnesota, 1969–1989. Canadian Journal of Zoology 69:2946–55.

Gunson, J. 1983. Wolf predation of livestock in western Canada. Pp. 102–5 *in* L. N. Carbyn, ed., Wolves in Canada and Alaska: Their status, biology, and management. Canadian Wildlife Service Report Series no. 45.

Hairston, N. G., F. E. Smith, and L. B. Slobodkin. 1960. Regulation in terrestrial ecosystems and the implied balance of nature. American Naturalist 94:421–27.

Harrington, F. H., and P. C. Paquet. 1982. Wolves of the world: Perspectives of behavior, ecology, and conservation. Noyes Publications, Park Ridge, New Jersey.

Hayes, R. D., and J. R. Gunson. 1995. Status and management of wolves in Canada. Pp. 21–33 *in* L. N. Carbyn, S. N. Fritts, and D. R. Seip, eds., Ecology and conservation of wolves in a changing world. University of Alberta Press, Edmonton.

Houston, D. B. 1982. The northern Yellowstone elk. Macmillan Publishing, New York.

Keiter, R. B., and M. S. Boyce. 1991. The Greater Yellowstone Ecosystem: Redefining America's wilderness heritage. Yale University Press, New Haven.

Keiter, R. B., and H. Locke. 1996. Law and large carnivore conservation in the Rocky Mountains of the U.S. and Canada. Conservation Biology 10:1003–12.

Kellert, S. R., and T. W. Clark. 1991. The theory and application of a wildlife policy framework. Pp. 17–36 *in* W. R. Mangun, ed., Public policy issues in wildlife management. Greenwood Press, New York.

Kunkel, K. E. 1997. Predation by wolves and other large carnivores in northwestern Montana and southeastern British Columbia. Ph.D. diss., University of Montana, Missoula.

Leopold, A. 1944. Review of *The wolves of North America* by S. P. Young and E. H. Goldman. Journal of Forestry 42:928–29.

Lopez, B. H. 1978. Of wolves and men. Charles Scribner's Sons, New York.

McIntyre, R. 1995. War against the wolf. Voyageur Press, Stillwater, Minnesota.

McLaren, B. E., and R. O. Peterson. 1994. Wolves, moose, and tree rings on Isle Royale. Science 266:1555–58.

Meagher, M. 1986. Summary of possible wolf observations, 1977–1986. Mimeo. Yellowstone National Park, Wyoming.

Mech, L. D. 1966. The wolves of Isle Royale. U.S. National Park Service Fauna Series 7.

———. 1970. The wolf: The ecology and behavior of an endangered species. Natural History Press, Doubleday, New York.

———. 1991. Returning the wolf to Yellowstone. Pp. 309–22 *in* R. B. Keiter and M. S. Boyce, eds., The Greater Yellowstone Ecosystem: Redefining America's wilderness heritage. Yale University Press, New Haven.

———. 1995. The challenge and opportunity of recovering wolf populations. Conservation Biology 9:270–78.

Mech, L. D., S. H. Fritts, G. L. Radde, and W. J. Paul. 1988. Wolf distribution and road density in Minnesota. Wildlife Society Bulletin 16:85–87.

Mech, L. D., and M. E. Nelson. 1989. Polygyny in a wild wolf pack. Journal of Mammalogy 70:675–76.

Murphy, K. M., G. S. Felzien, M. G. Hornocker, and T. K. Ruth. 1998. Encounter competition between bears and cougars: Some ecological implications. Ursus 10:55–60.

Paul, W. J., and P. S. Gipson. 1994. Wolves. Pp. 123–29 *in* Prevention and control of wildlife damage. University of Nebraska, Lincoln.

Peek, J. M., L. D. Mech, D. E. Brown, J. H. Shaw, S. R. Kellert, and V. Van Ballenberghe. 1991. Restoration of wolves in North America. Wildlife Society Technical Review 91-1.

Peterson, R. O. 1977. Wolf ecology and prey relationships on Isle Royale. U.S. National Park Service Scientific Monograph Series no. 11.

———. 1995. The wolves of Isle Royale: A broken balance. Willow Creek Press, Minocqua, Wisconsin.

Peterson, R. O., and R. E. Page. 1988. The rise and fall of Isle Royale wolves, 1975–1986. Journal of Mammalogy 69:89–99.

Peterson, R. O., R. E. Page, and K. M. Dodge. 1984. Wolves, moose, and the allometry of population cycles. Science 224:1350–52.

Phillips, M. K., and D. W. Smith. 1996. The wolves of Yellowstone. Voyageur Press, Stillwater, Minnesota.

Pimlott, D. H. 1967. Wolf predation and ungulate populations. American Zoologist 7:267–78.

Pletscher, D. H., R. R. Ream, D. K. Boyd, M. W. Fairchild, and K. E. Kunkel. 1997. Population dynamics of a recolonizing wolf population. Journal of Wildlife Management 61:459–65.

Primm, S. A., and T. W. Clark. 1996. Making sense of the policy process for carnivore conservation. Conservation Biology 10:1036–45.

Rabb, G. B. 1967. Social relationships in a group of captive wolves. American Zoologist 7:305–11.

Rausch, R. A. 1967. Some aspects of the population ecology of wolves, Alaska. American Zoologist 7:253–65.

Reading, R. P., and T. W. Clark. 1996. Carnivore reintroductions: An interdisciplinary examination. Pp. 296–336 *in* J. L. Gittleman, ed., Carnivore behavior, ecology and evolution. Cornell University Press, Ithaca.

Ream, R. R., M. W. Fairchild, D. K. Boyd, and D. H. Pletscher. 1991. Population dynamics and home range changes in a colonizing wolf population. Pp. 349–66 *in* R. B. Keiter and M. S. Boyce, eds., The Greater Yellowstone Ecosystem: Redefining America's wilderness heritage. Yale University Press, New Haven.

Rutter, R. J., and D. H. Pimlott. 1968. The world of the wolf. J. B. Lippincott, Philadelphia.

Schullery, P. 1996. The Yellowstone wolf: A guide and sourcebook. High Plains Publishing Co., Worland, Wyoming.

Servheen, C. W., and R. R. Knight. 1990. Possible effects of a restored wolf population on grizzly bears in the Yellowstone area. Pp. 4-35 to 4-49 *in* National Park Service, Wolves for Yellowstone? A report to the United States Congress, vol. 2, research and analysis. National Park Service, Yellowstone National Park.

Singer, F. J. 1991. The ungulate prey base for wolves in Yellowstone National Park. Pp. 323–48 *in* R. B. Keiter and M. S. Boyce, eds., The Greater Yellowstone Ecosystem: Redefining America's wilderness heritage. Yale University Press, New Haven.

Theberge, J. B. 1989. Legacy: The natural history of Ontario. McClelland and Stewart and Co., Toronto.

Thompson, I. D., and R. O. Peterson. 1988. Does wolf predation alone limit the moose population in Pulaskwa Park? A comment. Journal of Wildlife Management 52:556–59.

Tracy, B. 1997. Fire effects in Yellowstone's grasslands. Yellowstone Science 5:2–5.

U.S. Fish and Wildlife Service. 1987. Northern Rocky Mountain wolf recovery plan. U.S. Fish and Wildlife Service, Denver.

———. 1994a. Establishment of a nonessential experimental population of gray wolves in Yellowstone National Park in Wyoming, Idaho, and Montana and central Idaho and southwestern Montana. Final Rule, November 22, Federal Register, 59:60252–81.

———. 1994b. The reintroduction of gray wolves to Yellowstone National Park and central Idaho. Final Environmental Impact Statement. U.S. Fish and Wildlife Service, Helena.

Van Ballenberghe, V., and W. B. Ballard. 1994. Limitation and regulation of moose populations: The role of predation. Canadian Journal of Zoology 72:2071–77.

Varley, J. D., and W. G. Brewster, eds. 1992. Wolves for Yellowstone? A report to the United States Congress, vol. 4, research and analysis. National Park Service, Yellowstone National Park.

Weaver, J. L. 1978. The wolves of Yellowstone. U.S. National Park Service Natural Resources Report no. 14.

Weber, W., and A. Rabinowtiz. 1996. A global perspective on large carnivore conservation. Conservation Biology 10:1046–54.

Yellowstone National Park, U.S. Fish and Wildlife Service, University of Wyoming, University of Idaho, Interagency Grizzly Bear Study Team, and University of Minnesota Cooperative Parks Studies Unit, eds. 1990. Wolves for Yellowstone? A report to the United States Congress, vol. 1, research and analysis. National Park Service, Yellowstone National Park.

Young, S. P., and E. A. Goldman. 1944. The wolves of North America. Vols. 1 and 2. Dover Publications, New York.

Coyote visiting a winter-killed elk carcass on the National Elk Refuge (Franz J. Camenzind)

Chapter Six

Coyotes and Canid Coexistence in Yellowstone

Robert L. Crabtree and Jennifer W. Sheldon

Adolph Murie's (1940) pioneering work on the ecology of the coyote *(Canis latrans)* in Yellowstone National Park was a landmark of predator research in North America. The Greater Yellowstone Ecosystem (GYE) was to become a major center of carnivore research, and many classic studies followed his lead (e.g., Craighead 1979, Clark 1994), including additional research on coyotes (Robinson and Cummings 1951, Camenzind 1978, Bekoff and Wells 1986). By the late 1980s Yellowstone National Park had undertaken long-term studies of ungulate-killing carnivores, such as the grizzly bear *(Ursus horribilis,* Chapter 3) and mountain lion *(Felis concolor,* Chapter 4), but not the coyote. Because gray wolves *(Canis lupus)* were about to be restored to the park and because the coyote is the most abundant ungulate predator and a major competitor with the wolf, an intensive long-term study of coyotes began in 1989 on the Northern Range of Yellowstone National Park.

Coyotes are ideal carnivores to study because of their ability to adapt and thrive in diverse environments and because of their variable social behavior (Bekoff and Wells 1986). From loose pairs (Berg and Chesness 1978) to packs of ten or more (Crabtree and Varley in press), the midsized coyote displays many of the behavioral characteristics seen among the thirty-five species within the family Canidae (Sheldon 1992). Canids themselves are instructive groups through which to examine the community structure of carnivores because of their wide distribution and variable behavioral and ecological adaptations (Johnson et al. 1996).

This chapter is divided into two sections. The first is a review of major ecological studies of the coyote. Based on these studies, we develop a synthetic view of the coyote from an ecosystem perspective focusing on controversial themes, recent findings, and sociodemographic population limitation issues. The second section examines canid coexistence and competition because it is the least understood, least studied, and possibly most important aspect of coyote ecology. After a brief historical review of Yellowstone's three canids, we discuss the coyote's ecological role in Yellowstone. In striving to understand how canids, and ultimately carnivores, coexist, we focus on the coyote because of its pivotal role, competitively positioned in size between the red fox *(Vulpes vulpes)* and gray wolf. In order to develop a general theory of canid coexistence, we then review sympatric studies of two or more canid species. Our findings are placed in a theoretical framework and applied to a general scenario of three different-sized coexisting canid species typical on other continents. We end with conservation, management, and research recommendations.

The Ecology of Coyotes from an Ecosystem Perspective

In review of past and current field studies of the coyote, we chose eleven study sites based on both study duration and their inclusion of both social-behavioral aspects and population dynamics (table 6.1). Although autecological in nature, most of these studies include the work of two or more projects over two or more time periods, sometimes in two or more areas within the study sites. This allows us to take more of an ecosystem perspective by treating these studies as a continental metapopulation through space and time. For example, the pioneering work of F. Camenzind, M. Bekoff, J. Weaver, and W. Tzilkowski all occurred in close proximity to one another and is lumped together under the heading Jackson Hole, Wyoming.

Carnivores are notoriously difficult to study, so the emphasis of a particular project is often technique dependent. Studies focused on behavioral ecology (for example, Jackson Hole, Wyoming, and in Yellowstone) utilize direct observation, while studies of spatial organization and estimation of demographic parameters (for example, south Texas brushland and northern Utah) usually require intensive capture and radio-tagging efforts. A thorough understanding of long-lived carnivore species and, more important, the coexistence of carnivore communities requires a long-term approach, including direct observations of these behaviorally complex species (Frame 1986). Without such observations, inferences regarding social organization, social structure, and social interactions are suspect (Bekoff and Wells 1986) and can produce erroneous conclusions (Waser 1974). Extensive and systematic behavioral observations were conducted only in the Jackson Hole, Wyoming, and Yellowstone field studies.

Another hurdle in synthesizing information across North American coyote studies is the variable and sometimes large effect of human exploitation on the results of field studies. Human exploitation is here defined as human-caused mortality. Although several studies appear to have light exploitation rates, only two (southeast Washington and Yellowstone) examine the unexploited condition. Because of the substantial impact of exploitation on coyote behavior and demographics, we provide subsections to help interpret how coyote populations actually operate under unexploited and undisturbed conditions, thereby gaining valuable insight into the conditions under which they evolved.

FOOD-PREY RELATIONS

Predator-prey relations are best understood from the simultaneous study of both predator and prey (Errington 1935). Fortunately, the first monographical studies of coyotes did focus on prey populations, including a twenty-four-year study in northern Utah and a twelve-year study in central Alberta (table 6.1). Both sets of studies demonstrated the significant influence of coyotes on cyclic populations of snowshoe hares *(Lepus americanus)* and black-tailed jackrabbits *(Lepus californicus)*. Even more profound were the functional and the numerical effects of varying prey abundance on coyotes. Strong functional responses are expected when a classic habitat generalist–feeding generalist like the coyote preys on a relatively large, easy-to-handle prey source (see Chapter 9). Strong numerical responses should also occur, given the behavioral plasticity and high reproductive capability of coyotes. When examined, reproductive parameters do respond to major increases or decreases in food supply (Clark 1972, Knudsen 1976, Todd et al. 1981). Few vertebrate mammals the size of coyotes and wolves have the reproductive potential to produce five to eight young per year.

Recent studies tend to focus more on population and behavioral ecology and less on food-prey relations. However, current work in Yellowstone has examined the ecology of coyote predation on small mammal populations (Gese et al. 1996a,b), as did studies conducted in Jackson Hole, Wyoming, during the 1970s (table 6.1). Studies of coyote food habits are numerous but provide unreliable information regarding prey impacts because of sampling biases related to differential digestibility and inconsistent analysis methods (Kelly 1991). Nevertheless, food habit studies indicate that coyotes rely primarily on small mammal prey. Major exceptions appear to be seasonally abundant foods such as fruits in some southern regions, and carrion in northern regions.

Table 6.1

Comparison of coyote studies

Study site	Period	Food-prey relations
Northern Utah	1963-86	Functional and numerical response to cyclic prey abundance (jackrabbits). Coyote abundance correlated with jackrabbit abundance.
Central Alberta	1964-75	Functional and numerical response to cyclic prey abundance (snowshoe hare). Overall positive correlation between coyote and snowshoe hare abundance.
Jackson Hole, Wyoming	1970-82	Spring abundance correlated with carrion availability. Various aspects of behavior and population demography related to prey resources (carrion and rodent availability).
Southwest Alberta	1974-77	Pack size related to percent of mule deer (not elk) in the winter diet. Functional feeding response with mule deer density and winter killed elk.
South Texas Brushland	1974-82	Not addressed in this study
Southeast Idaho	1975-86	Territory size not related to food abundance but higher proportion of transients during prey scarcity.
South Texas Plains	1978-82	Found no relations between group size and prey size.
Maine	1979-84	Low prey densities may preclude delayed dispersal, resulting in lower pack size.
Southeast Washington	1974-88	Heavy predation on mule deer fawns, while rodents accounted for a large portion of seasonal diets.
Southeast Colorado	1983-86	Prey density and habitat features affect territory size.
Yellowstone	1937-39 1946-49 1989-95	Functional response to seasonally available prey. Coyotes not limiting ungulate populations but are major elk predator due to high density. Coyotes take large portion of rodent prey compared with other predators.

Behavioral ecology	Demographic limitation
Territory size and percent of transients are related to prey density and exploitation. More juvenile females dispersed, and dispersed farther, than did juvenile males.	Litter size and pregnancy rates correlated with jackrabbit abundance. Winter food limits coyote density during the low period of the jackrabbit abundance cycle.
Not addressed in this study.	Snowshoe hare abundance correlated with litter size and reproduction. Carrion abundance related to ingress and egress rates.
Wolflike social, spatial, and breeding system, including a pack structure. Group size benefits scent-marking and active defense of territories.	Lightly exploited, with territoriality providing spatial limitation of coyotes. Demographic parameters related to food availability, disease, and human-caused mortality.
Indicates pack formation is an adaptation for efficient capture and defense of ungulate prey. Found a division of labor and evidence of territorial avoidance.	Not addressed.
Nonoverlapping territorial core areas. Transients also avoided core areas. Many females ovulate and become pregnant, but only territorial females produce pups.	Exclusive and successful breeding by territorial females. Immigration of territorial and breeding replacements due to population reduction. Juvenile females dispersed more than males.
Spatial use of home range area related to behavior mode, season, temperature, and prey. Avoidance of novel items inside territory but not outside and peripheral.	Lightly exploited population had low recruitment but high dispersal rates. Spring density not affected by mortality in previous fall–winter period. More females dispersed.
Exclusive breeding by a territorial female. Alloparental care observed. Movements related to breeding season and not group size or foraging behavior.	Stable, habitat-saturated population. High pup mortality observed that may be related to disease. A reservoir of transients to replace breeding goups.
Survival rate lower for dispersers than for residents belonging to social groups.	No difference between male and female dispersal characteristics.
Exclusive breeding by alpha female. High spatially structured social classes. Transient avoided packs. Survival was a function of social class, not age class.	Unexploited, habitat-saturated population. High pup mortality and dispersal related to social and nutritional stress. Adult survival is 9% for residents and higher for dispersers.
Coyote group size increased with amount of large prey (deer) in the diet. Increased cohesiveness of social groups during breeding season.	Implied relation between coyote density and prey abundance. Population is habitat-saturated, with large territory size in low prey abundance areas. High juvenile dispersal.
Division of labor between alpha male and female. Helping behavior results in more food and defense of pups and higher litter size and survival in good food years. Dispersal by subordinate pack members.	Unexploited, stable, dense, habitat-saturated population. Pack size regulated by neonatal pup mortality and dispersal of subordinates, factors that are related to prey abundance and availability. High adult survival.

continued

Table 6.1

Continued

Study site	Period	Intraspecific competition
Northern Utah	1963–86	Not addressed.
Central Alberta	1964–75	Not addressed.
Jackson Hole, Wyoming	1970–82	Habitat saturation with direct evidence of conspecific killing without consumption. Scent-marking, vocalization, and active defense involved in territorial maintenance.
Southwest Alberta	1974–77	Intrapack dominance hierarchy and interpack spatial avoidance.
South Texas Brushland	1974–82	Implied intraspecific competition based on intrapack spatial avoidance.
Southeast Idaho	1975–86	Implied intraspecific competition based on nonoverlapping home ranges of adults.
South Texas Plains	1978–82	Implied intraspecific competition based on nonoverlapping territorial packs.
Maine	1979–84	Implied intraspecific competition based on nonoverlapping home ranges of adults.
Southeast Washington	1974–88	Intraspecific strife inferred from non-overlapping territories, active defense, and vocalization playbacks. Intruders often identified from olfaction and vision.
Southeast Colorado	1983–86	Implied intraspecific competition based on contiguous, nonoverlapping territories.
Yellowstone	1937–39 1946–49 1989–95	Intraspecific strife inferred by defended, nonoverlapping territories. In poor food years, larger packs produce fewer pups. Territories maintained by physical presence, scent-marking, evictions, and vocalizations.

Interspecific competition	Sources
Not found or addressed.	Clark 1972, Knudsen 1976, Hoffman 1977, Hibler 1977, Davison 1980, Harris 1983, Mills and Knowlton 1991.
Exploitative competition implied between coyotes, lynx, and raptors. A community approach reaching across three trophic levels.	Nellis and Keith 1976, Todd and Keith 1976, Keith et al. 1977, Todd et al. 1981, Todd and Keith 1976.
Not found or addressed.	Weaver 1977, Camenzind 1978, Tzilkowski 1980, Bekoff and Wells 1981, Wells and Bekoff 1982, Bekoff and Wells 1986.
Some evidence of interspecific competition.	Bowen and Cowan 1980, Bowen 1981, Bowen 1982.
Not found or addressed.	Windberg et al. 1985, Knowlton et al. 1986, Windberg and Knowlton 1988.
High diet overlap with red fox but spatial segregation by habitat type. Possible exploitative competition with bobcats.	Davison 1980, Laundré and Keller 1984, Laundré 1981, Green and Flinders 1981, Harris 1983, Mills and Knowlton 1991.
Not addressed but potential exploitative competition with two scavenging vulture species.	Andelt 1985.
Interference competition with red fox. Spatial segregation occurred as coyotes relegated fox to inferior habitats. High diet overlap and exploitative competition with bobcats.	Harrison and Gilbert 1985, Major and Sherburne 1987, Harrison et al. 1989, Litvaitis and Harrison 1989, Harrison 1992.
Not found or addressed.	Stoel 1976, Steigers and Flinders 1980, Springer 1982, Crabtree 1989, Fulmer 1990, Blatt 1994.
Not found or addressed.	Gese 1988, 1989, Gese et al. 1988.
Interference competition with wolves, grizzlies, and mountain lions at carcasses. Mountain lions also act as predators. Spatial and temporal segregation between coyotes and red fox. Coyote is the major scavenger.	Murie 1940, Robinson and Cummings 1951, Crabtree and Sheldon 1995, Gese 1995, Hatier 1995, Crabtree and Varley in press.

The role of ungulate neonates in the early summer diet of the coyote has received surprisingly little attention, given its apparent impacts on the demography of coyotes and its potential to affect ungulate populations. Ungulate neonates are available during early summer, when pups are at maximum growth rates. The major period of pup mortality from disease and starvation occurs immediately after this period, in July and August (Crabtree and Varley in press). Central-place foragers (for example, den-attending adults) are constrained by time—in addition to the energetic demands of provisioning the young with food—and this should result in prey specialization. Thus adult coyotes would be predicted to specialize on large food items (energy maximizers) according to optimal foraging theory (Pyke et al. 1977). Althoff and Gipson (1981) and Till and Knowlton (1983) indicate that provisioning of pups stimulated adults to prey on domestic ungulates. Behavioral observations in Yellowstone and the preponderance of elk calf remains at den sites corroborate the occurrence of prey specialization in June during pup provisioning. Because differential reproductive success and pup survival are apparently directly linked to the timing and availability of ungulate neonates (or other food sources), evolutionary consideration should be given to the related topics of coyote group formation, the timing of reproduction, and antipredator behavior of both wild and domestic ungulates.

Bekoff and Wells (1986) estimated that about 90 percent of the coyote's diet is mammalian flesh. Because ungulate flesh in the diet is usually carrion (Bekoff 1977, Weaver 1977, Houston 1978) and seasonal in occurrence, few studies have investigated coyote impacts on ungulate populations (but see Messier et al. 1986). Coyotes usually kill ungulates that are weak, impaired, domesticated, or starving, but they are certainly capable of killing healthy adults, even elk in Yellowstone (Gese and Grothe 1995, Crabtree unpublished data). Impacts of coyotes on ungulate populations appear to be mainly via predation on ungulate neonates during pup rearing. We know of no study that indicates significant impacts on the adult segment of an ungulate population. However, predation rates on young ungulates can be high (Hamlin et al. 1984). In Yellowstone, coyotes kill more elk calves (neonates and older calves in winter) than do grizzly bears and mountain lions combined (table 6.2) and inflict heavy predation (greater than 80 percent) on radio-tagged antelope fawns (D. Scott 1994, personal communication). Till and Knowlton (1983) experimentally demonstrated that coyotes kill domestic sheep to provide food for young pups.

BEHAVIORAL AND SOCIAL ECOLOGY

The belief that coyotes are more solitary than other similar-sized canids seems to be the result of cultural folklore and biases in field sampling. In con-

Table 6.2

Comparison of predation rates by large predators in Yellowstone's Northern Range

| Species | n | Elk predation | | | | | Elk biomass | Per capita kill rate |
		Neonate calves	Short yearling	Adult winter	Adult non-winter	Total		
Mountain lion[a]	17	35	313	70	193	611	76,150	36
Grizzly bear[b]	60	750[c]	0	0	few	750+	13,500	13
Coyote[d]	400	750	360–626	20–35	0	1130–1411	66,760	3

[a]Data from Kerry Murphy, Hornocker Wildlife Research Institute.
[b]Bonnie Blanchard, personal communication, estimated ~60 grizzlies using the Northern Range.
[c]Francis Singer, National Biological Service.
[d]This study, projected estimates.

trast to gray wolves, coyote pack members rarely travel all together, and field counts of pack size are usually derived from visual counts of traveling individuals in winter. In addition, human exploitation lowers group size and may cause coyotes to become more secretive and nocturnal. The coyote actually fits nicely into the ecological/body weight relations described by Moehlman (1986) for many canid species. Larger canids, like coyotes, are more social and tend to form packs.

Social organization. Coyote populations are explicit in their spatial arrangement and have well-defined social classes. In synthesizing the results of the studies in table 6.1, which include two recent studies (conducted by the authors) of unexploited populations, we modify the classification of adult coyotes originally proposed by Bowen (1978). His classification was adopted by Bekoff and Wells (1986) to describe two distinct adult social class categories— territorial *pack members* (members of a social group) and nonterritorial *loners.* Adults in territorial packs are further divided into the dominant *alphas* or *breeding pair* and their subordinate *betas.* Betas are pups born in previous years that stay in their natal territory (Crabtree 1989). Beta pack members can either be *helpers,* which help with pup-rearing, or *slouches,* which occupy the territorial area and interact with the breeding pair but seldom contribute toward pup feeding, pup rearing, and den guarding (Hatier 1995).

Loners are subdivided into *solitary residents* and *nomads.* The term *transient* has been used in previous studies to describe all nonterritorial coyotes. However, this is inappropriate because a significant portion of coyotes express site fidelity but do not defend the area they occupy. Solitary residents have levels of site fidelity and home range size similar to that of pack members. In contrast, nomads have low site fidelity and range over large areas (fifty to three hundred square miles), presumably in search of a mate and a territorial vacancy (Crabtree 1989).

Solitary residents generally make up less than 15 percent of the population and are the most heterogeneous social class. They are divided into two subclasses: *floaters* and *former alphas.* Floaters tend to be younger (one to three years old), show weak fidelity to an area, and range over a larger area than most older solitary residents. They spend substantial amounts of time on the periphery of several territories and are suspected to be outcasts from one of the adjacent territories (Crabtree 1989). The characteristics of this subclass match those of the roamers described by Bekoff and Wells (1986) and of individuals described by Messier and Barrette (1982).

The second subclass of solitary residents, *former alphas,* consists of older adults (age three and a half to thirteen and a half), with a degree of site fidelity close to a territorial pack member. Evidence presented by Crabtree (1989) indicates that many of these individuals are former territorial alphas. They have noticeable head and facial scars, which are common on breeding males, and evidence of former reproduction is seen on females. A vocalization study by Fulmer (1990) indicated that older solitary residents occasionally respond to territorial group yip-howls.

Only Bekoff and Wells (1986) described a social category called *resident mated pairs* that do not defend a territory. We believe that this situation arises infrequently, though we have observed it several times in Yellowstone as a direct result of wolf disturbance (Crabtree and Sheldon unpublished data) or as loose, short-term social bonds between male and female solitary residents (observed twice in southeast Washington, table 6.1).

Territories. From 65 to 90 percent of individuals in a coyote population belong to territorial social groups or packs. Territories are defended and are stable in unexploited and lightly exploited areas. They are typically around ten square kilometers and range from about two square kilometers in southern regions (south Texas, table 6.1) to around twenty square kilometers in northern regions (Bowen 1978). Home-range analysis indicates some overlap between territories, but we believe that this could simply be an artifact of the statistical method employed. Observation of scent-marking and territorial defense in Yellowstone indicates relatively little, if any, overlap between groups (Crabtree unpublished data). Statistically defined territorial core areas do not show any overlap in the studies reviewed.

Territory size can vary inversely with prey abundance in other species (Hixon et al. 1983 for birds, Mares et al. 1982 for small mammals), but this relation is not consistent in coyotes (Mills and Knowlton 1991). It is implied for coyotes in southeast Colorado (Gese et al. 1988), but habitat saturation of contiguous, nonoverlapping territories (for example, Yellowstone, south Texas, southeast Washington, table 6.1) may not allow territories to expand and contract with changing prey densities. In addition, the Yellowstone and southeast Washington studies showed a six-year average period for alpha pairs residing in territories.

Sociality, cooperative foraging, and delayed dispersal. The formation of packs, or sociality, in coyotes has been attributed to increased foraging efficiency (see Bekoff and Wells 1986), but this relation, though it has received significant attention, remains unclear. Messier and Barrette (1982) provide an excellent discussion of the subject. Bowen (1981) concludes that group formation in coyotes is an adaptation for the efficient capture of ungulates or economic defense and consumption of carcasses. This hypothesis appears to be supported by other fieldwork (Camenzind 1978, Bekoff and Wells 1986, Bowyer 1987, and for gray wolves, Packard and Mech 1980). Sheldon (1992) questions whether successful hunting is a secondary effect, as implied by Gese et al. (1988), while Messier and Barrette (1982) provide criticism and compelling alternative hypotheses.

We agree with Messier and Barrette (1982) and find no empirical evidence to demonstrate that larger social groups lead to increased per capita food intake in coyotes. In fact, single individuals and groups of two commonly take down and kill both deer (*Odocoileus* spp.) and elk *(Cervus elaphus)*, but larger groups can also be involved (Gese and Grothe 1995, Crabtree unpublished data). This illustrates one of the major problems in evolutionary ecology: are the observations an effect or an evolutionary cause?

Messier and Barrette (1982) propose that group formation in coyotes is the result of delayed dispersal and that exploitation of ungulate prey is a secondary effect. Juveniles that forgo dispersal accrue a variety of benefits, such as secure foraging, increased survival, continued learning, and the attainment of alpha status within or adjacent to their own territory. In addition, delayed dispersal may also be the result of habitat saturation—no vacancies for dispersing juveniles (Davison 1980). Movements of radio-tagged juveniles in Yellowstone and southeast Washington indicate that some juveniles disperse in fall or early winter, but return to their natal territories later in the winter or in the spring before whelping (Crabtree unpublished data). The additional contention by some authors that delayed dispersal may be additionally related to delayed sexual maturity appears weak (Bekoff 1977, Messier and Barrette 1982). Evidence suggests that all females are capable of breeding at ten months of age and that variation in reproductive statistics (for example, age at breeding, percentage of yearling females reproducing) is caused by socially mediated breeding suppression and exploitation effects.

Could delayed dispersal, along with cooperative foraging, be an effect rather than the cause of sociality? We believe that the existence of coyote packs may have evolved along two lines. First, the delayed dispersal hypothesis described above does not necessarily convey a benefit to other related pack members and also delays the time of first reproduction. In fact, delayed dispersers or betas come with an inherent cost to the reproducing pair and their pups—they deplete food resources within the territorial foraging area. Data from southeast Washington and Yellowstone clearly indicate the existence of slouches, beta individuals that do not appear to help the alpha pair with pup rearing. Betas (especially juveniles) are often denied access to winter food and subsequently disperse (Gese et al. 1996b). However, if staying in the natal territory increases a beta's chance of survival and later reproductive success, then from an inclusive fitness standpoint, the dominant alpha parents should tolerate it. Older juveniles and young yearlings (ages nine to fourteen months) have been observed successfully eliciting regurgitations from the alphas with pups at the den. Thus delayed dispersal may be a form of extended neotony in coyotes when certain ecological conditions dictate a "staying" strategy. Furthermore, a dominance hierarchy provides an efficient social mechanism to regulate a subordinate beta's dispersal. Dominant individuals, i.e., alphas, may assess ecological information, such as low prey abundance or large pack size, and force dispersal of subordinates.

We term the second catalyst for pack formation the pup protection hypothesis, which is similar to that described for lions in Africa (Packer et al. 1990). Under this hypothesis genetic fitness (and a beta's inclusive fitness) is increased if helping actually increases pup survival. Selection pressure comes in the form of pup predation by conspecifics (neighboring packs), predators (golden eagles and bears), and competitors (wolves and mountain lions). In Yellowstone we have observed golden eagles capturing coyote pups at the den, wolves killing coyote pups, and nearly forty instances of den-guarding adult coyotes chasing off both conspecifics and other large carnivores that approach dens. We have also observed adult coyotes chase and attack bears and mountain lions. Camenzind (1978) twice saw adult coyotes from an adjacent territory killing pups at the den site while attending adults were gone. Bekoff and Wells (1986) report that larger groups of coyotes (additional betas) were more effective at chasing off intruders. Hatier (1995) found that larger pack sizes (those with beta helpers) resulted in more den guarding and more food provisionings for pups.

The delayed dispersal and pup protection hypotheses are not mutually exclusive. Because some betas help and others are slouches, it could be that helpers tend to be dominant and add to pack size via pup protection and delayed dispersal, whereas slouches tend to be subordinate and add to sociality via delayed dispersal only. Under this unified hypothesis, a core of one or two helpers is essential. Additional pack members tend to be slouches and are added if beneficial ecological conditions prevail (probably prey abundance). Additional field data are needed to test these hypotheses.

DEMOGRAPHIC LIMITATION

Among the numerous studies of coyote population demographics, only Knowlton (1972) and Knowlton and Stoddart (1983) have attempted a synthetic review regarding the regulation and limitation of coyote populations. In addition to the population mechanics they describe, it is clear that incorporation of coyote social class dynamics, behavior, human exploitation, and competition is essential to any synthesis.

Reproduction and neonatal survival. Female coyotes are monestrous, and the alpha pair mates once a year. Like wolves, coyote packs occasionally produce a double litter (two breeding females in one pack). We have observed this several times in Yellowstone and estimate that double litters occur 4 percent of the time. In one case, an eleven-year-old alpha female had seven pups, while her daughter, a two-year-old beta female, had a litter of five pups. All pups were communally nursed and reared. The beta female had been a den helper the previous year and appeared closely associated with her alpha female mother.

Litter size at birth appears relatively invariant with respect to changes in prey abundance. Litter size averages about six pups per year and an even sex ratio is common (Bekoff 1977). Numerous studies, spanning a variety of habitats, prey abundances, and exploitation rates, report litter sizes, taken from den counts, between five and seven (Sheldon 1992, Crabtree unpublished data, and see various studies, table 6.1). Even with drastic, ten- to fortyfold changes in jackrabbit and hare abundance, litter size varied only from 6.6 to 7.6 (unborn fetus counts, Clark 1972) and 4.3 to 6.0 (placental scar counts, Todd et al. 1981) in northern Utah and central Alberta, respectively.

Contrary to Knowlton (1972), litter size at birth also appears largely unaffected by levels of human exploitation. He reported an inverse relation between an abundance index (number of coyotes caught per standard trap line in the fall) and litter size varying from 4.3 to 6.9. However, the litter sizes reported from den counts varied only from 5.0 to 5.7. Based on examination of female reproductive organs, he then inferred litter sizes in seven south Texas counties to be 2.8 and 4.2 in a lightly exploited area, 3.7 and 5.3 in a moderately exploited area, and 6.2, 6.3, and 8.9 in an intensively controlled area. It appears that these data have led to the commonly held notion that litter size at birth increases when populations are exploited.

We disagree with the contention of density-dependent adjustments in litter size for several reasons. First, the average litter sizes (den counts) for the two unexploited studies (southeast Washington and Yellowstone, respectively) were 5.6 and 5.4 for successful females. However, 27 percent and 14 percent of alpha females (mostly old-aged) failed to produce pups successfully, and thus the corrected values are 4.1 and 4.5 pups per alpha female in the southeast Washington and Yellowstone studies, respectively. Interestingly, both studies found that radio-tagged alpha females without pups at den emergence time (den counts) all showed localized movements near traditional denning areas at the time of birth (early April). Furthermore, intensive visual observation and capture of two of the reproductively failed alpha females in May revealed evidence of lactation. Thus unsuccessful alpha females appear to have become pregnant and probably lost entire litters shortly after birth (see Sayles 1984). This was first suspected by Knudsen (1976) in northern Utah.

Second, reproductive data gathered from the examination of female carcasses of unknown social status appear to be misleading. In a particularly enlightening study, Knowlton et al. (1986) radio-tagged sixty-five females to determine age and to classify them as territorial or transient based on intensive monitoring. Females were then collected to examine their reproductive organs. Yearlings, both territorial and nonterritorial, did not ovulate or implant. Fifty percent of the nonterritorial females age two and older ovulated, and 25 percent implanted. None of these ovulating and implanted females successfully whelped. In addition, beta females may have also ovulated and implanted. Thus a much higher percentage of females initiate reproduction than are ever successful. Obviously, this casts serious doubt on the reliability of unborn fetus and placental scar counts being used to infer litter size, let alone successful reproduction. It is also interesting to note that two of four solitary resident females recovered by the authors during spring periods had four and eight embryos in the process of resorption. We have also repeatedly seen various loner females and beta females in copulatory ties during February.

Third, it appears that litter survival (mortality from birth to early winter), not litter size at birth, is the major reproductive parameter that responds in a density-dependent manner to human exploitation. Canids place relatively little energetic investment in gestation, compared with lactation and provisioning of pups during the pups' maximum growth period. Field data collected by the senior author suggest that intraspecific strife leads to alpha females that are in poor condition. Consequently, pups that are born underweight may have inadequate food-provisioning rates, and are predisposed to disease, the proximate cause of mortality (Crabtree 1989 and unpublished data). In Yellowstone mortality appears to affect pups according to a dominance hierarchy already formed by ten weeks of age (Knight 1978). This results in a skewed sex ratio favoring males because males tend to be dominant over females and may gain more access to nursing and regurgitated food.

In studies of unexploited and lightly exploited populations, females first attain alpha status and initiate reproduction when two to five years of age. The probability of successful litters decreases around age seven (Crabtree 1989, Crabtree and Varley in press). Socially sterile beta females, combined with a reservoir of loner females, provide a high potential for replacement of breeding females in exploited populations. In Yellowstone only 35 percent of the females in the population are alphas, of which 86 percent successfully have pups. As a population becomes exploited, it has been reported (Connolly and Longhurst 1975) that as many as 90 percent of females breed. Unless territory size significantly shrinks, these levels are inconsistent with the coyote's classic land tenure system of nonoverlapping territories with one female breeding per territory. Unless exploitation levels are so high as to break down this system, such as that described by Berg and Chesness (1978), it is difficult to imagine levels above 66 percent. All studies examining female reproduction have reported territorial and nonterritorial females that do not breed.

Dispersal. We agree with Knowlton and Stoddart (1983) that dispersal is the primary mechanism for maintaining population densities near saturation levels. Davison (1980) and the two studies of unexploited populations clearly indicated emigration from habitat-saturated areas. This serves to reduce population density or to lessen intraspecific competition in the case of a food shortage (Harrison 1992). These studies also indicated lower survival rates for dispersers than for residents. Low survival of dispersers, locally abundant prey, and lack of territorial vacancies in local and nearby areas appear to be selective conditions favoring delayed dispersal (natal philopatry).

In the case of a density-reducing event, a reservoir of loner replacements colonizes vacancies. There is intense competition for vacant territorial areas because newly mated pairs are able exclusively to reproduce successfully, thereby directly increasing fitness. Although loner and beta females occasionally become pregnant, it appears that at least 95 percent are unsuccessful in producing pups. This, combined with the dispersal potential of coyotes (see Robinson and Cummings 1951), suggests that immediate colonization occurs whereby loners and betas intensively compete for territorial residency and the chance to successfully breed. Observations in southeast Washington and Yellowstone indicate that vacated territories act as immigration sinks. Immediate colonization occurred by loners and betas from adjacent packs. In four of ten replacement events, it took one and a half to two years for a new breeding pair to emerge and reproduce. This delay appeared to be the result of intense competition between adults and adult pairs trying to establish pair bonds. In contrast, some territorial vacancies are immediately replaced by a new breeding pair without skipping a successful breeding season.

Evidence gathered in southeastern Washington and Yellowstone (Crabtree 1989, Crabtree and Sheldon unpublished data) indicates that some loners are semidispersed pack members that occasionally visit their natal pack. These individuals are one to four years old and float throughout a large area (for example, 50 km^2) surrounding their natal pack. We suspect that intraspecific competition for food within the natal territory results in the semi- or distant dispersal of nonalpha pack members. This allows for efficient reduction of pack size, and more food is available for the remaining dominant adults. Reduction in pack size can result in increased survival for pups as demonstrated by Crabtree and Varley (in press). This loose affiliation may also explain the high proportion of loners in some populations. Mills and Knowlton (1991) proposed that food shortages cause an increase in the proportion of nonterritorial individuals (loners). A mechanistic explanation of how these loners might be derived is provided by Gese et al. (1996b), who describe the dispersal of subordinate pack members that had decreased access to food sources.

Adult survival. Data from the two unexploited coyote populations reported a 9 percent and 10 percent annual adult mortality rate, with occasional vehicle collisions accounting for half of these losses. This is far less than the 40 percent mortality rate assumed by Knowlton (1972) in the absence of exploitation. With Knowlton and Stoddart (1983) calculating that 89 percent of mortalities are attributable to humans, they must have been assuming that human exploitation largely substituted for natural mortality. Even some exploited populations have annual adult mortality rates less than 40 percent (e.g., Windberg et al. 1985). In the absence of exploitation either from humans, predators, or lethal competitors, the major mortality period for coyotes is from birth to six months old. Once a coyote reaches its first winter, the probability is high that it will reach old age.

ASSESSING HUMAN EXPLOITATION

Frank (1979) called for studies of unexploited coyote populations in order to understand the evolutionarily significant situation. Unfortunately, only a few studies, like Camenzind (1978), have been conducted on lightly exploited populations, and only two studies have documented unexploited populations (Crabtree 1989, Crabtree and Varley in press). Nearly all field studies of coyotes have been conducted on populations subjected to substantial levels of exploitation. The results of these studies are thus biased because of the effects of exploitation. For example, coyotes are reported to show flexibility in their social system (Lott 1984). Could such social flexibility be the result of human exploitation, and to a lesser extent, unnatural and human-disturbance conditions? Various studies of coyote populations, all subjected to various levels of exploitation, report significant variation in both social and spatial organization (Berg and Chesness 1978, Camenzind 1978, Danner and Smith 1980, Bowen 1981, Messier and Barrette 1982, Andelt 1985, Bekoff and Wells 1986).

We conducted a survey of coyote radio-telemetry studies in order to assess the effects of human exploitation on coyote populations and the extent to which exploitation clouds our understanding of how coyote populations operate under natural conditions. From this review (R. Crabtree and M. Matteson unpublished data), we constructed sociodemographic scenarios according to three levels of adult coyote exploitation. This subject is obviously complex, and we present these levels so that scientists, managers, and conservationists can understand and interpret these effects better.

Level 1. Unexploited to lightly exploited (0–24 percent annual human-related mortality). These populations are characterized by stable to fairly stable, habitat-saturated, nonoverlapping territories with boundaries that are stable across time. Pack formation and delayed dispersal is apparent with one to eight subordinate betas depending on prey abundance and proximity to exploited areas that function as dispersal sinks. Pup survival is low (20–60 percent), and few, if any, yearlings reproduce. Average age of adults is three to four years, and the territorial residency time for alpha pairs is three to six years.

Level 2. Moderately exploited populations (25–49 percent annual human-related mortality). These populations are characterized by high turnover of alpha pairs (one to three years). There is a 10–20 percent reduction in population density resulting from decreases in pack size, rather than decreases in number of territories. The land tenure system is intact but has an unstable system of nonoverlapping territories with shifting boundary areas—the population is in a state of semicolonization from immigrants. Some yearling females successfully breed because of a shortage of older, mature females. Pup survival varies from 50–90 percent according to prey resources. Average age of adults is about two years old, and the territorial residency time for alphas is one to three years.

Level 3. Highly exploited populations (50 percent or greater annual human-related mortality). These populations are characterized by an unstable social and spatial system. Individuals have a low probability or surviving until age two. Many yearling females breed because of low competition. Litter size is slightly elevated and pup survival is high, averaging 70–100 percent annually. Packs of three adults are uncommon, with most breeding groups made up of the single breeding pairs. Evidence of loose pair bonds and occasional polygynous breeding events may occur. The age structure includes more than 50 percent yearlings, and the population is in a constant state of colonization with high immigration rates.

Sociodemographic population regulation. Various sociodemographic factors that can regulate coyote populations have been identified—territoriality, dominance hierarchies, breeding longevity, subordinate dispersal, reproductive failure, double litters, and pup mortality. Most studies reviewed indicate direct or indirect evidence of intraspecific competition. Unexploited and habitat-saturated populations indicate intraspecific strife (see Packard and Mech 1980) and a higher level of intraspecific competition. Manifestations include low pup weights, scarring, reproductive failure, frequent territorial disputes, and high pup mortality, including the probable loss of entire litters shortly after birth.

The abundance and availability of prey is certainly a major limiting factor, but the extent to which it is involved in population regulation remains uncertain. Access to prey has been linked to every major social and demographic event. Crabtree (1989) reported that the population unit on which natality and mortality primarily act is not age but social class. We agree with Knowlton and Stoddart (1983) that social intolerance, mediated by the abundance and availability of food, is the primary determinant of coyote density. However, recent research indicates that the role played by social behaviorally mediated access to prey has been previously underestimated and that human exploitation common in most field studies has severely confounded our understanding of coyote populations. Long-term studies of coyotes in unexploited and lightly exploited areas, especially when combined with systematic visual observations, have proven irreplaceable in their contribution to a general understanding of coyote ecology.

The one factor that has largely been ignored in coyote research agendas is the role of interspecific competition (but see Keith et al. 1977, Paquet 1989, and Maine, table 6.1). Coyotes coexisted with the competitive pressures of gray wolves over much of their distribution before European settlement. Hence inference regarding the evolutionary mechanisms of coyotes must be taken with caution because most studies were conducted in areas without wolves and other coevolved competitors. An example of the importance of interspecific competition in limiting coyote populations comes from recent data in Yellowstone (Crabtree and Sheldon 1996, Crabtree and Sheldon unpublished data). Wolf killing of coyotes in Yellowstone's Lamar Valley from 1996 through 1998 has resulted in a 50 percent sustained reduction in coyote density. Numerically, this reduction is accounted for by a decrease in pack size as well as a reduction in the number of territorial coyote packs. Either wolves kill alpha coyotes, causing pack disintegration and dispersal, or packs are relegated to an adjacent area. In either case, coyote packs fail to recolonize vacated territories in the high-use areas of a wolf territory. For the above reasons, interspecific competition, specifically canid coexistence, is thoroughly examined in the second half of this chapter.

Coyotes and Canid Competition

In contrast to intraspecific competition, we know very little about the effects of interspecific competition on coyote populations. In order to understand the role of interspecific competition in coyote and canid communities, we provide a brief review of the history of canids in the GYE and the general ecological role of the coyote before wolf restoration. We then examine sympatric canid studies and place them into a theoretical framework applicable to the three canid species present in Yellowstone.

Interference competition—including fighting, killing, direct displacement, and relegation to inferior habitats—has been clearly demonstrated in previous studies, yet wolves, coyotes, and red fox persist in sympatry. Wolves are able to exclude coyotes (Peterson 1996), and coyotes are able to exclude red foxes (Harrison et al. 1989, Sargeant et al. 1987) at various scales, from individual encounters and territories, to entire regions, yet they coexist in many regions of North America. We are therefore particularly interested in the mechanisms of coexistence.

BACKGROUND AND HISTORY OF CANIDS IN THE YELLOWSTONE ECOSYSTEM

Based on the pre-European distribution of the red fox (see Aubry 1983), coyote (Bekoff 1977), and gray wolf (Mech 1974), the red fox and coyote have greatly increased their range across North America while wolves declined substantially during the century-long predator eradication era beginning in the 1860s. All three species occurred naturally in the GYE and the Northern Range of Yellowstone (see also Chapter 2) and coexist once again with the reintroduction of gray wolves in 1995 (see Chapter 5).

The increase in red fox distribution into the lower forty-eight states has been attributed to intentional introduction from Europe and widespread habitat changes accompanying agricultural development (Sheldon 1992). The European red fox inhabits agricultural and human-disturbed habitat at lower elevations, while the red fox endemic to North America resides in either the high-elevation montane-alpine zones of the Rockies, Sierra, and Cascade mountain ranges or the boreal forests of Canada and the Great Lakes states (Crabtree 1993).

The majority of the coyote's range expansion occurred during the predator eradication era (1860s to 1960s), which resulted in drastic reductions in the range distributions of many carnivores. Besides its famed resiliency to predator control techniques, the coyote's range expansion has been attributed to widespread reduction in the distribution of the gray wolf and the clearing of forests. The coyote now occupies most habitat types in North America, although it is best adapted to the arid and open shrub–grassland areas of the West (Bekoff 1977).

The red fox, adapted to alpine tundra and boreal forest, was certainly present in the Yellowstone region during, before, and after the Wisconsin glaciation. The coyote, however, was probably present in the Yellowstone region only during interglacial periods and possibly during glacial periods in the lower elevation areas surrounding the ice-capped mountains. The gray wolf probably inhabited only areas of boreal forest to the south of ice sheets, where ungulate populations supported breeding populations, and moved into the higher elevations once the continental ice sheet began to retreat.

Schullery and Whittlesey (1992) reviewed historical sightings of canids prior to 1890 and found that sightings of wolves and foxes were common, while coyote sightings were rather infrequent. Although this could partly be the result of misclassifying coyotes as wolves, it is clear that several park officials were very adept at distinguishing species, even color morphs of red fox (see Norris 1881). The surprising lack of coyote sightings stands in sharp contrast to the trapping records of Skinner (1927) from 1906 to 1927 when the last wolves were extirpated from the Northern Range of Yellowstone. While 127 wolves and 134 mountain lions were killed, 4,356 coyote mortalities were recorded. Even if 80 percent were pups killed at dens (which is doubtful), coyotes appear much more abundant that wolves. This leaves us with a peculiar paradox—where did all the coyotes come from? Were they already there? Could coyote numbers, once released from wolf pressure, have quickly rebounded?

Distribution and abundance of coyotes in the Greater Yellowstone Ecosystem. Currently, the coyote inhabits all vegetation communities below 8,000 feet in the GYE, except for areas of contiguous deep snow and steep rocky areas. Based on extensive winter surveys conducted from 1992 through 1995 (Gehman et al. 1997), the coyote uses all elevations above 8,000 feet, but only on a transitory basis. The typical coyote behavior of territorial establishment, courting, pair-bonding, and breeding that takes place from December through February is absent above 8,000 feet. However, coyotes are commonly observed from 8,000 to 11,000 feet in meadow and mixed forest-meadow habitats from May to July. These coyotes are seen in male-female pairs engaged in courtship and pair-bonding. Nonterritorial coyotes that reside below 8,000 feet during winter apparently travel to these areas in late April and May as snow melts and prey becomes available. These pairs vocalize, scent-mark, and defend territories just like lower-elevation coyotes in winter, but no successful reproduction has been detected.

The estimated density of adult coyotes on the Northern Range averages 0.45 per square kilometer (Crabtree and Varley in press) based on total counts and capture-recapture estimates (Crabtree et al. 1989). In the open, shrub-steppe and mesic grasslands of the GYE, coyotes can reach local densities exceeding one per square kilometer. However, across much of the mixed meadow-forest habitat types of the GYE, coyote densities range from 0.1 to 0.4 coyotes per square kilometer.

Prewolf ecological role of the coyote. The high density of coyotes on the Northern Range adds a strong numerical component to an already broad functional role as a generalist consumer. In this section we compare historical data to current information, estimate biomass consumption of different prey species, and calculate the percentage of prey species removed by coyotes. These data further underline the coyote's stable and broad ecological role on Yellowstone's Northern Range.

Table 6.3

Coyote food habits as indicated by analysis of scats (% of biomass intake)

Prey species	Murie 1940 (5,086 scats)	This study (1995; 500 scats)
Microtus spp.	42.4	41.3
Pocket gopher	27.0	24.5
Ground squirrel	0.6	3.0
Snowshoe hare	4.3	4.4
Elk	20.3	21.2

The ecological role of the coyote was defined by Murie (1940) from analysis of 5,086 scats. We summarized these data and compared them to a subsample of five hundred scats collected in 1991 and 1992 (table 6.3). The data sets were remarkably similar, indicating stable resource use. This, combined with the apparent stability in the location of traditional den sites, suggests that prey availability and preferences have also been relatively stable. Based on our results, the observations of Murie (1940), and the similarity of den site locations during 1946–49 (Robinson and Cummings 1952) and 1990–94, we believe that the number and location of coyote packs on the Northern Range became stable shortly after the extirpation of the gray wolf in 1927.

The coyote is the major elk predator on the Northern Range, killing an estimated 1,276 elk annually, the majority of which are neonates (table 6.2). The coyote population accomplishes this not by specialization but by sheer numbers (450 coyotes), along with a propensity for killing mostly young neonates (an estimated 750 annually) in June and weak or starving adults during winter. Although coyotes are capable of killing healthy adult elk during winter, they seldom do so (Crabtree unpublished data, Gese and Grothe 1995). In comparison, mountain lions kill around six hundred elk (only thirty-five neonates) and grizzlies kill an estimated 750 neonates and a few adults (B. Blanchard, personal communication). The estimated per capita annual kill rates for an estimated 450 coyotes, sixty grizzly bears, and seventeen adult mountain lions on the Northern Range are three, thirteen, and thirty-six elk, respectively.

Estimates of the biomass of various prey species consumed by coyotes were based on: (1) independent estimates of predation rates (Gese et al. 1996a), (2) carcass consumption rates (S. Grothe unpublished data), (3) observed predation rates on ungulates (Gese and Grothe 1995, R. Crabtree unpublished data), and (4) seasonal estimates of the fresh weight of prey consumed, from scat analysis incorporating differential digestibility corrections (Kelly 1991). Nearly 50 percent of the annual biomass intake came from small mammals (taken mostly in summer and fall), and nearly 45 percent from ungulates taken mostly in winter and spring (figure 6.1). In the seven nonwinter months, microtines made

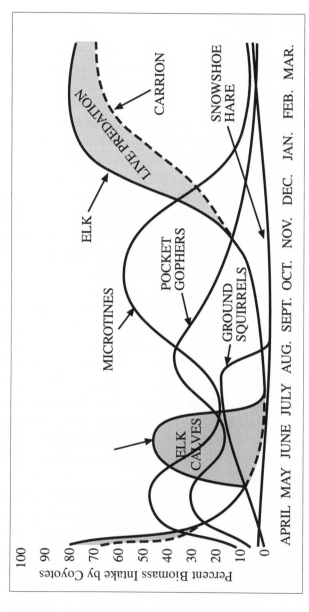

6.1. Seasonal biomass intake by coyotes in Yellowstone's Northern Range.

up 41 percent of prey biomass consumed, pocket gophers 25 percent, ground squirrels 3 percent, snowshoe hares 4 percent, and elk (calves and carrion) 21 percent. An estimated 74 percent of the biomass intake during the five winter months came from elk (primarily carrion), and 26 percent from small-mammal prey.

The coyote's estimated take of small-mammal prey on the Northern Range is indicative of the high density of the prey and of the coyote's broad functional role. Among fifteen species of carnivores that rely on small-mammal prey as their major food source, the coyote accounts for an estimated removal of 76 percent of the estimated population of microtines, 24 percent of pocket gophers, 35 percent of ground squirrels, and 10 percent of neonate elk calves (Chapter 3, Singer unpublished data). Nearly three-fourths of the elk biomass consumed during the five winter months is carrion, and at least two-thirds of the live predation on elk involves old or starving individuals. As Murie (1940) suspected, harsh winters result in large numbers of carcasses for scavenging coyotes. In mild winters with little carrion, coyotes can lose up to 30 percent of their body weight. The availability of carrion directly affects litter size (Crabtree and Varley in press).

CANID SYMPATRY AND COEXISTENCE

Because they are remarkably diverse, canids are an instructive group through which to examine carnivore competition and the resultant mechanisms of coexistence. Members of this family range in weight from one and a half to eighty kilograms, with variable diets and feeding strategies ranging from insectivory to omnivory to almost complete carnivory. For example, the largest canid, the gray wolf, is an obligate carnivore that specializes in killing ungulates but can seasonally utilize a substantial portion of nonungulate, small-mammal prey in its diet (Crabtree 1992). Life-history traits and social organization are highly variable among the Canidae, as are their adaptability and behavioral plasticity (Sheldon 1992).

In a comprehensive review of canid sympatry, Johnson et al. (1996) examined how resources are partitioned among potential competitors. These authors assessed sympatric canid pairs throughout the world in terms of diet breadth, as well as temporal, spatial, and habitat overlap. A summary follows of canid ecological relations described in Johnson et al. (1996) and a multitude of other studies of sympatric canids (e.g., Chambers 1987, Chypher and Scrivner 1992, Dekker 1983, Dibello et al. 1990, Engelhardt 1986, Gittleman 1986, Green and Flinders 1981, Jimenez 1993, Johnson 1992, Klett 1987, Moehlman 1986, Sargeant and Allen 1989, Sargeant et al. 1987, Theberge and Wedeles 1989, Voigt and Earle 1983, White et al. 1994, Wooding 1984).

Review of the above papers gives rise to generalizations grouped into the topics presented in the following sections.

Size dominance. Species dominance among sympatric canids is universally a function of size: in competitive interactions, larger canid species are dominant over smaller ones. However, numerical advantage in smaller species can sometimes temporarily reverse the direct dominance of the larger species (Litvaitis 1992, Davis 1980, Schamel and Tracy 1986, Dekker 1983, 1989, 1990, Johnson 1992, Jimenez 1993, Paquet 1991, Crabtree and Sheldon personal observations).

Three-species canid pattern. According to Johnson et al. (1996), a pattern of three sympatric canid species occurs in North America, Eurasia, and Africa. This consistent pattern of partitioning based on ecological roles seems to permit canid coexistence (Rosenzweig 1995), but these three species are not the only canids or competing carnivores in many of these regions. In each region, even though the species are different, the three types of canids occur in congruent patterns of functional roles, consisting of: (1) a large (more than twenty kilograms), obligate ungulate killer—for example, African wild dog *(Lycaon pictus)*, dhole *(Cuon alpinus)*, or gray wolf; (2) a medium-sized omnivore of ten to twenty kilograms—for example, golden jackal *(Canis aureus)* or coyote; and (3) a small, highly omnivorous species—for example, corsac fox *(Vulpes corsac)*, fennec fox *(Fennecus zerda)*, or red fox.

Disruptions to predator assemblages brought about by the human removal of the largest canid and obligate carnivore are widespread. In Africa the African wild dog, in Asia the dhole, and in North America the gray wolf have been eradicated from regions where human populations conflict with these species. These large, highly social, low density, obligate ungulate predators have proven relatively easy to eradicate. Smaller, omnivorous, generalist canid species are generally more resilient and tolerant of human presence. Their populations often increase following removal of the largest canid, suggesting that competition between the canid species had depressed populations of the smaller canid. Rapid range expansion of generalist canids such as coyotes, red foxes, and jackals often follows these eradications (Sheldon 1992). The ecosystem-wide effects of this rapid range expansion of generalist canids need to be assessed. In some ways these ultrageneralists can be seen as weedy species, the gulls of the dog world.

The predominant diet pattern among sympatric canids is the three-species canid pattern described above; alternatively, in a two-canid system the larger canid is more strictly carnivorous while the smaller is an opportunistic omnivore with significantly greater niche breadth. Diet studies are numerous (e.g., Lamprecht 1978, Dibello et al. 1990, Paquet 1992, Fuller et al. 1989, Johnson and Franklin 1994, Nel 1984, Jimenez 1993, Jaksic et al. 1983, Andriashek et al. 1985,

Brown 1990, Bothma 1971, Bothma et al. 1984, Smits et al. 1989, Hockman and Chapman 1983, Small 1971, Green and Flinders 1981). Niche breadth, prey biomass, prey diversity, and habitat heterogeneity all seem to be important in determining the economies of resource partitioning. In general, little information is available on numerical aspects of canid sympatry; for instance, density or relative abundance ratios. Most of the information on population numbers is derived indirectly from trapping indices.

Spatial and temporal partitioning. There is clear evidence of temporal and spatial partitioning between sympatric canids. In general, the smaller canids avoid the larger ones (Theberge and Wedeles 1989, Paquet 1992, Ingle 1990, Follman 1973, Bailey 1992, Davis 1980, Hersteinsson and Macdonald 1982, Dekker 1983, 1989, 1990, Paquet 1992, Moehlman 1983). Differences in activity patterns do not usually result in decreased diet overlap, but rather may represent behavioral avoidance of potential agonistic interactions with competitors. The same is true for spatial avoidance patterns. The rule of thumb on the part of the smaller canid seems to be avoidance of potentially lethal encounters rather than maximization of energetic intake.

Interference competition. Interference competition between sympatric canid species is common (O'Farrell 1984, Paquet 1992, Berg and Chesness 1978, Carbyn 1982, Egoscue 1956), often culminating in the killing of one canid species by another (Murie 1944, Berg and Chesness 1978, Carbyn 1982, O'Farrell 1984; see Peterson 1996 for a recent review). In fact, the first kill that reintroduced wolves made in Yellowstone was a red fox that climbed into a wolf enclosure. In general, one canid species does not completely exclude another regionally (Johnson et al. 1996). Exceptions to this general rule occur on Isle Royale, Michigan, where the last coyote from an originally robust population was seen seven years after the first gray wolf crossed over to the island on an ice bridge (Krefting 1969), and in California, where coyotes effectively seem to exclude kit foxes across large areas (O'Farrell 1984).

The extent to which high interspecific kill rates among canids are an artifact of colonizing or perturbed populations is not clear. Examination of wolf-caused coyote mortality in Yellowstone National Park should provide insight into accommodations by the two species as wolves approach habitat saturation. As Frame (1986:xxi) points out: "Only the direct observation of interference competition in long-term field studies identified where competitive interactions are occurring with sufficient intensity to provide a numerical response in a population." Observations of direct competitive interactions provide valuable insight into the behavioral and demographic components of coexistence.

Competition imposes energetic costs, affects survival and fecundity, and may force relegation to less optimal habitats (with lower resource availability or higher risks). The manifestations of interspecies competition can be quite subtle. Follman's (1973) study of sympatric red and gray foxes found little overt evidence of competition, and the two species were segregated by habitat. It can also be less subtle: in South America where the chilla (*Dusicyon griseus*, five kilograms) and culpeo (*Dusicyon culpaeus*, five to thirteen kilograms) are sympatric, the larger canid excluded the smaller from optimal habitat via interference competition, and the chilla was relegated to suboptimal habitats (Johnson 1992). This pattern of spatial relegation of the smaller canid is commonly found in other sympatric canid pairs. Temporal segregation reduces competition for food only if different activity patterns result in access to different prey populations (Pianka 1974) or if food resources are renewed within the period of temporal segregation (Litvaitis 1992).

Character displacement. The carnivore complex in East Africa contains five sympatric canids, the highest density of canid sympatry anywhere, a phenomenon that may in part result from the enormous diversity and biomass of available prey. Van Valkenburgh and Wayne's (1994) study of three sympatric jackals in East Africa brings to light some difficulties with the traditional view of species overlap and its relation to morphological character displacement. The three sympatric jackal species in East Africa *(Canis mesomelas, C. adustus,* and *C. aureus)* appear to have converged rather than diverged morphologically and seem to segregate ecologically (Fuller et al. 1989). Similarly, sympatric coyotes and wolves in Ontario appear to be converging in body weight and length (Schmitz and Lavigne 1987).

Home-range interspersion. The geometry of canid home ranges—that is, how they are interspersed and spaced—is important. The few concurrent telemetry studies of canids have shown how the home ranges of smaller canids and their prey fit in between the home ranges of the larger canids. Deer densities are highest in the gaps between the home ranges of different wolf packs (Mech 1977). Similarly, the red fox home ranges in Yellowstone are located between and on the periphery of coyote territories (Fuhrmann 1998). Essentially all of the sympatric studies reviewed that had concurrently tracked two species of radio-tagged canids indicated some degree of home range interspersion or fine-scale allopatry.

Scavenging. In ecosystems where large predators kill more prey than they can consume in one feeding bout, smaller coexisting canids may benefit (see Paquet 1989). This opportunistic scavenging behavior is a function of group size, with unpredictable outcomes. Large social groups of smaller-sized scavengers can usurp kills, and large social groups of top carnivores can keep scavengers away more effectively. Both satiation level and stage of carcass consumption appear to influence the outcomes of interspecific interactions at kills (S. Grother personal communication). Among canids it is clear that sociality functions to protect against scavengers stealing carcasses. Preliminary data from observed wolf-killed ungulate carcasses on the Northern Range suggest that bears, eagles, ravens, and coyotes all benefit from the addition of carrion biomass via wolf reintroduction.

MECHANISMS OF COEXISTENCE IN THE GREATER
YELLOWSTONE ECOSYSTEM

Based on the review of sympatric canid studies, we hypothesize that canid coexistence is primarily a function of avoiding fatal encounters while attempting to secure or defend the prey necessary to survive and successfully reproduce. Canids therefore employ a variety of behaviors that result in spatial and temporal resource partitioning. Given the historic and prehistoric sympatric distributions, the abundance and consistency of observations reporting aggressive and fatal interactions between coyotes and gray wolves and between coyotes and red foxes imply that they are not rare events and that strong selection pressures occur.

In the Yellowstone ecosystem, as well as other areas of North America, we propose four mechanisms that mediate canid coexistence: (1) simple geometry, (2) individual behavioral avoidance as mediated by sensory perception, (3) Paquet's (1992) scavenger potential hypothesis, and (4) effective social group size.

Simple geometry contends that the disproportionate relation between body size (four, twelve, and thirty-six kilograms, respectively) and territory size (two, twelve, and two hundred square kilometers, respectively) in the red fox, coyote, and gray wolf further facilitates the efficient interspersion of the home ranges of two or more canid species. The smaller canid has relatively more space in between and on the periphery of the larger canid's territory. In addition, the smaller canid likely has a lower probability of encounter with the larger canid. This may be especially true for the red fox, which is relegated to suboptimal habitats by the dominant coyote (Harrison et al. 1989). Red foxes can avoid the core of coyote territories and still survive on relatively dense rodent and insect prey populations.

Behavioral avoidance is based on the highly evolved senses of members of the family Canidae. Avoiding fatal encounters requires detection of the larger canid through visual, auditory, or olfactory cues. In addition, the red fox is highly nocturnal in Yellowstone compared with the more diurnal and crepuscular coyote (Fuhrmann 1998). Both red foxes and coyotes have been observed to wait at distances of two hundred meters to over one kilometer until wolves have left their kills, thereby temporally partitioning a high-energy food resource during winter.

Scavenger potential was first described by Paquet (1989). This hypothesis is based on the observation of wolves consuming large ungulates over multiple feeding bouts. The smaller canid scavenges on the ungulate carcass between feeding bouts when wolves (or coyotes) are resting for several hours (Paquet 1992). For example, in areas where white-tailed deer *(Odocoileus virginianus)* are the primary prey, packs of wolves can consume a whole deer in one feeding bout, leaving little for coyotes.

Effective group size appears to be an important means of avoiding fatal encounters. Of more than two hundred wolf-coyote encounters observed in Yellowstone from 1996 through 1998 (Crabtree and Sheldon 1996, unpublished data), fifteen resulted in coyote deaths. All deaths involved three or more wolves and a single coyote. Groups of coyotes were also observed attacking single wolves and usurping their kills. Genetically related individuals engage in group defense of a carcass, den site, or open-field encounter with a larger canid. In addition, vigilance (and detection) is often more efficient in a social group. Preliminary analysis of data collected in Yellowstone (A. Gladwin unpublished data) indicate a per capita decrease in vigilance behaviors with increasing group size.

Conservation, Management, and Research Recommendations

COYOTE POPULATIONS

The coyote's distribution across North America has tripled during the past century. Clearly, the coyote does not represent a small-population conservation problem. Nevertheless, restoration of coyotes to some areas may serve a functional role in the conservation of other species. For example, it may be useful to reassess the functional role of coyotes where deer are overabundant, or where red fox populations inflict heavy predation on ground-nesting birds, livestock, or poultry. We believe that conservation science can learn important lessons from understanding a successful and ubiquitous species like the coyote, in addition to the important lessons learned from declining species.

Because of the numerous sociodemographic and density-dependent processes that take place in response to widespread population reductions and indiscriminate killing, federal programs to limit coyote numbers have proven ineffective and costly. Field research is needed to experimentally examine the effectiveness of control and the responses of coyote populations to exploitation. We believe that the coyote's famed behavioral plasticity and demographic resiliency to exploitation is an evolutionary product of coexisting with competing species, mainly the gray wolf. The proposed fifteen-year study in Yellowstone (pre- and postwolf) should provide this needed information. So far, gray wolves have inflicted heavy mortality on coyotes during the winter in wolf core-use areas (see Crabtree and Sheldon 1996), and the sex and age structure of twenty-four wolf-killed coyotes appears to be relatively indiscriminate, with a possible bias toward younger individuals (much like human trapping efforts).

CANID COEXISTENCE

It is clear that interspecific relations between canid species competing for limited resources can have significant impacts. This has major implications for conservation and management. Medium- and smaller-sized canids have shown a release (Soulé et al. 1988) following a reduction or extirpation of the larger canid (Hersteinsson et al. 1989, Lewis et al. 1993, Peterson 1996). Thus future reintroduction efforts should consider these and other community effects before implementation. Canid hybridization has occurred repeatedly (Lehman et al. 1991, Wayne and Jenks 1991, Boitani 1982, Gottelli and Sillero-Zubiri 1992) and can hamper or complicate expensive reintroduction efforts (Gittleman and Pimm 1991).

We agree with Johnson et al. (1996) that monitoring the effects of canid reintroduction programs provides a splendid opportunity to examine canid community structure and competition experimentally. Furthermore, Yellowstone's wolf reintroduction will provide a unique opportunity to understand such community-wide effects as functional and numerical responses by other small-mammal predators (for example, red foxes, weasels, raptors) responding to significant reductions in coyote populations. Initial estimates of wolf numbers in northern Yellowstone made by Garton et al. (1990) suggest there will be a tripling of the amount of potential carrion available to scavengers, which could have significant impacts on, for example, threatened Yellowstone grizzly bear populations. Changes in elk behavior and distribution followed by subsequent changes in vegetation are yet other possible responses (Chapter 8).

Canids, especially coyotes, display wide-ranging variability in their behavioral ecology and population demography, and thus serve as excellent candidates for understanding carnivore communities. Long-term study, combined with systematic behavioral observation and experimental (and natural) manipulation is required in order to craft successful conservation strategies for communities in dire need of restoration. Such field studies are difficult and pose challenges to agencies, universities, and private organizations that must cooperate in order to plan, fund, and execute these invaluable studies.

Acknowledgments: We thank the staff of Yellowstone National Park, especially John Varley, chief scientist, for support, guidance, and timely advice. A special thank-you to the numerous field technicians, interns, and volunteers who worked on various aspects of the field research, especially graduate students R. Fuhrmann, E. Gese, S. Grothe, K. Hatier, W. Monroe, P. Moorcroft, Deborah Smith, and C. Wilmers. M. Phillips, Douglas Smith, K. Murphy, and W. Brewster of Yellowstone National Park's Wolf Ecology Project provided field and logistical support. We also thank Battelle Pacific Northwest Laboratories, Hanford, Washington, and E. Ables, University of Idaho, for support of the research conducted in southeast Washington under U.S. Department of Energy contract DE-AC06-76RLO-1830. The research in Yellowstone was funded primarily by the National Park Service (1989–94) under cooperative agreement 1268-1-9001 to R. L. Crabtree, and by Yellowstone Ecosystem Studies (1993–99). Additional funding and support was provided by Bob and Dee Leggett, Defenders of Wildlife, Bob Landis of Landis-Trailwood Films, National Geographic Society, World Wildlife Fund, Marian Probst, Earthwatch, Hornocker Wildlife Research Institute, Max McGraw Wildlife Foundation, Greenville Foundation, Turner Foundation, U.S. Fish and Wildlife Service, U.S. Department of Agriculture, U.S. National Biological Service, M. Johnson of Wildlife Veterinarian Services, Biology Department at Montana State University, Department of Wildlife Ecology at the University of Wisconsin–Madison, and J. Williams at University of North Carolina–Wilmington. P. Curlee and S. Minta provided valuable review and comments on earlier drafts.

Literature Cited: Althoff, D. P., and P. S. Gipson. 1981. Coyote family spatial relationships with reference to poultry losses. Journal of Wildlife Management 45:641–49.

Andelt, W. F. 1985. Behavioral ecology of coyotes in South Texas. Wildlife Monographs no. 94:1–45.

Andriashek, D., H. P. L. Kiliaan, and M. K. Taylor. 1985. Observations on foxes, *Alopex lagopus* and *Vulpes vulpes,* and wolves, *Canis lupus,* on the off-shore sea ice of northern Labrador. Canadian Field-Naturalist 99:86–89.

Aubry, K. B. 1983. The Cascade red fox: Distribution, morphology, zoogeography, and ecology. Ph.D. diss., University of Washington, Seattle.

Bailey, E. P. 1992. Red foxes, *Vulpes vulpes,* as biological control agents for introduced arctic foxes, *Alopex lagopus.* Canadian Field-Naturalist 106:200–205.

Bekoff, M. 1977. *Canis latrans.* Mammalian Species no. 79:1–9.

Bekoff, M., and M. C. Wells. 1981. Behavioral budgeting by wild coyotes: The influence of food resources and social organization. Animal Behaviour 29:794–801.

———. 1986. Social ecology and behavior of coyotes. Pp. 251–338 *in* J. S. Rosenblatt, C. Beer, M.-C. Busnel, and P. J. B. Slater, eds., Advances in the study of behavior. Academic Press, New York.

Berg, W. E., and R. A. Chesness. 1978. Ecology of coyotes in northern Minnesota. Pp. 229–48 *in* M. Bekoff, ed., Coyotes: Biology, behavior, and management. Academic Press, New York.

Blatt, J. W. 1994. Demographics and dispersal of coyote *(Canis latrans)* pups. M.S. thesis, Central Washington University, Ellensburg.

Boitani, L. 1982. Wolf management in intensively used areas of Italy. Pp. 158–72 *in* F. H. Harrington and P. C. Pacquet, eds., Wolves of the world: Perspectives of behavior, ecology, and conservation. Noyes Publications, Park Ridge, New Jersey.

Bothma, J. D. P. 1971. Food habits of some Carnivora (Mammalia) from Southern Africa. Annals of the Transvaal Museum 27(2):15–26.

Bothma, J. D. P., J. A. J. Nel, and A. Macdonald. 1984. Food niche separation between four sympatric Namib Desert carnivores. Journal of Zoology (London) 202:327–40.

Bowen, W. D. 1978. Social organization of the coyote in relation to prey size. Ph.D. diss., University of British Columbia, Vancouver.

———. 1981. Variation in coyote social organization: The influence of prey size. Canadian Journal of Zoology 59:639–52.

Bowen, W. D., and McT. I. Cowan. 1980. Scent marking in coyotes. Canadian Journal of Zoology 58:473–80.

Bowyer, R. T. 1987. Coyote group size relative to predation on mule deer. Mammalia 51:515–26.

Brown, G. W. 1990. Diets of wild canids and foxes in East Gippsland, 1983–1987, using predator scat analysis. Australian Mammalogy 13:209–13.

Camenzind, F. J. 1978. Behavioral ecology of coyotes on the National Elk Refuge, Jackson, Wyoming. Pp. 267–94 in M. Bekoff, ed., Coyotes: Biology, behavior, and management. Academic Press, New York.

Carbyn, L. N. 1982. Coyote population fluctuations and spatial distribution in relation to wolf territories in Riding Mountain National Park, Manitoba. Canadian Field-Naturalist 96:176–83.

Chambers, R. E. 1987. Diets of Adirondack coyotes and red foxes. Transactions of the Northeastern Section of the Wildlife Society 44:90.

Chypher, B. L., and J. H. Scrivner. 1992. Coyote control to protect endangered San Joaquin kit foxes at the Naval Petroleum Reserves, California. Pp. 42–46 in J. E. Borrecco and R. E. Marsh, eds., Proceedings of the 15th Vertebrate Pest Conference, Davis, California. University of California, Davis.

Clark, F. W. 1972. Influence of jackrabbit density on coyote population change. Journal of Wildlife Management 36:343–56.

Clark, T. W. 1994. Restoration of the endangered black-footed ferret: A twenty-year overview. Pp. 272–97 in M. L. Bowles and C. J. Whelan, eds., Restoration and recovery of endangered species. Cambridge University Press, Cambridge.

Connolly, G. E., and W. M. Longhurst. 1975. The effects of control on coyote populations. University of California, Davis, Division of Agricultural Science, Bulletin no. 1872.

Crabtree, R. L. 1989. Social, spatial, and demographic characteristics of an unexploited coyote population. Ph.D. diss., University of Idaho, Moscow.

———. 1992. A preliminary assessment of the non-ungulate mammal prey base for wolves in Yellowstone National Park. Pp. 5-122 to 5-138 in J. D. Varley and W. G Brewster, eds., Wolves for Yellowstone? A report to the United States Congress, vol. 4, research and analysis. National Park Service, Yellowstone National Park.

———. 1993. Gray ghost of the Beartooth. Yellowstone Science 1(3):13–16.

Crabtree, R. L., F. G. Burton, T. R. Garland, D. A. Cataldo, and W. H. Rickard. 1989. Slow-release radioisotope implants as individual markers for carnivores. Journal of Wildlife Management 53:949–54.

Crabtree, R. L., and J. W. Sheldon. 1996. Summary of the interactions of gray wolves and coyotes on Yellowstone's Northern Range. Special report to Yellowstone National Park, Mammoth.

Crabtree, R. L., and J. D. Varley. In press. Ecological and sociodemographic role of the coyote on Yellowstone's Northern Range. Proceedings of the 3d biennial science conference on the Greater Yellowstone Ecosystem, Greater Yellowstone's predators, September 24–27, 1995. Yellowstone National Park and Northern Rockies Conservation Cooperative, Mammoth.

Craighead, F. C., Jr. 1979. Track of the grizzly. Sierra Club Books, San Franscisco.

Danner, D. A., and N. S. Smith. 1980. Coyote home range, movement, and relative abundance near a cattle feedyard. Journal of Wildlife Management 44:484–87.

Davis, G. K. 1980. Interaction between bat-eared fox and silver-backed jackal. East Africa Natural History Society Bulletin, September–October, 79.

Davison, R. P. 1980. The effect of exploitation on some parameters of coyote populations. Ph.D. diss., Utah State University, Logan.

Dekker, D. 1983. Denning and foraging habits of red foxes, *Vulpes vulpes,* and their interactions with coyotes, *Canis latrans,* in central Alberta, 1972–1981. Canadian Field-Naturalist 97:303–6.

———. 1989. Population fluctuations and spatial relationships among wolves, *Canis lupus,* coyotes, *Canis latrans,* and red foxes, *Vulpes vulpes,* in Jasper National Park, Alberta. Canadian Field-Naturalist 103:261–64.

——. 1990. Population fluctuations and spatial relationships among wolves, coyotes, and red foxes in Jasper National Park. Alberta Naturalist 20:15–20.

Dibello, F. J., S. M. Arthur, and W. B. Krohn. 1990. Food habits of sympatric coyotes, *Canis latrans*, red foxes, *Vulpes vulpes*, and bobcats, *Lynx rufus*, in Maine. Canadian Field-Naturalist 104:403–8.

Egoscue, H. J. 1956. Preliminary studies of the kit fox in Utah. Journal of Mammalogy 37:351–57.

Engelhardt, D. B. 1986. Analysis of red fox and coyote home range use in relation to artificial scent marks. M.S. thesis, University of Maine, Orono.

Errington, P. L. 1935. Over-populations and predation: A research field of singular promise. Condor 37:230–32.

Follman, E. H. 1973. Comparative ecology and behavior of red and gray foxes. Ph.D. diss., Southern Illinois University, Carbondale.

Frame, G. W. 1986. Carnivore competition and resource use in the Serengeti ecosystem of Tanzania. Ph.D. diss., Utah State University, Logan.

Frank, L. 1979. Review of *Coyotes: Biology, behavior, and management*, by Marc Bekoff. Journal of Mammalogy 60:658–59.

Fuhrmann, R. T. 1998. Distribution, morphology, and habitat use of the red fox in the northern Yellowstone ecosystem. M.S. thesis, Montana State University, Bozeman.

Fuller, T. K., A. R. Biknevicius, P. W. Kat, B. Van Valkenburgh, and R. K. Wayne. 1989. The ecology of three sympatric jackal species in the Rift Valley of Kenya. African Journal of Ecology 27:313–23.

Fulmer, K. F. 1990. Characterizing the functions of coyote vocalizations through the use of playback. M.S. thesis, University of Idaho, Moscow.

Garton, E. O., R. L. Crabtree, B. B. Ackerman, and G. L. Wright. 1990. The potential impact of a reintroduced wolf population on the northern Yellowstone elk herd. Pp. 3-59 to 3-91 *in* Wolves for Yellowstone? A report to the United States Congress, vol. 2, research and analysis. National Park Service, Yellowstone National Park.

Gehman, S., R. Crabtree, E. Robinson, M. Harter, and S. Consolo-Murphy. 1997. Comparison of three methods for detecting mammalian carnivores. Final Report to Yellowstone National Park, Mammoth.

Gese, E. M. 1988. Relationship between coyote group size and diet in southeastern Colorado. Journal of Wildlife Management 52:647–53.

——. 1989. Population dynamics of coyotes in southeastern Colorado. Journal of Wildlife Management 53:174–81.

——. 1995. Foraging ecology of coyotes in Yellowstone National Park. Ph.D. diss., University of Wisconsin, Madison.

Gese, E. M., and S. Grothe. 1995. Analysis of coyote predation on deer and elk during winter in Yellowstone National Park, Wyoming. American Midland Naturalist 133:36–43.

Gese, E. M., O. J. Rongstad, and W. R. Mytton. 1988. Home range and habitat use of coyotes in southeastern Colorado. Journal of Wildlife Management 52:640–46.

Gese, E. M., R. L. Ruff, and R. L. Crabtree. 1996a. Intrinsic and extrinsic factors influencing coyote predation of small mammals in Yellowstone National Park. Canadian Journal of Zoology 74:784–97.

——. 1996b. Social and nutritional factors influencing the dispersal of resident coyotes. Animal Behaviour 52:1025–43.

Gittleman, J. L. 1986. Carnivore life history patterns: Allometric, phylogenetic, and ecological associations. American Naturalist 127:744–71.

Gittleman, J. L., and S. L. Pimm. 1991. Crying wolf in North America. Nature 351:524–25.

Gottelli, D., and C. Sillero-Zubiri. 1992. The Ethiopian wolf: An endangered endemic canid. Oryx 26:205–14.

Green, J. S., and J. T. Flinders. 1981. Diets of sympatric red foxes and coyotes in southeastern Idaho. Great Basin Naturalist 41:251–54.

Hamlin, K. L., S. J. Riley, D. Pyrah, and R. J. Mackie. 1984. Relationships among mule deer fawn mortality, coyotes, and alternate prey species during summer. Journal of Wildlife Management 48:489–99.

Harris, C. E. 1983. Differential behavior of coyotes with regard to home range limits. Ph.D. diss., Utah State University, Logan.

Harrison, D. J. 1992. Dispersal characteristics of juvenile coyotes in Maine. Journal of Wildlife Management 56:128–38.

Harrison, D. J., J. A. Bissonette, and J. A. Sherburne. 1989. Spatial relationships between coyotes and red foxes in eastern Maine. Journal of Wildlife Management 53:181–85.

Harrison, D. J., and J. R. Gilbert. 1985. Denning ecology and movements of coyotes in Maine during pup rearing. Journal of Mammalogy 66:712–19.

Hatier, K. G. 1995. Effects of helping behaviors on coyote packs in Yellowstone National Park, Wyoming. M.S. thesis, Montana State University, Bozeman.

Hersteinsson, P., A. Angerbjorn, K. Frafjord, and A. Kaikusalo. 1989. The arctic fox in Fennoscandia and Iceland: Management problems. Biological Conservation 49:67–81.

Hersteinsson, P., and D. W. Macdonald. 1982. Some comparisons between red and arctic foxes, *Vulpes vulpes* and *Alopex lagopus*, as revealed by radio tracking. Symposia Zoological Society of London 49:259–89.

Hibler, S. J. 1977. Coyote movement patterns with emphasis on home range characteristics. M.S. thesis, Utah State University, Logan.

Hixon, M. A., F. L. Carpenter, and D. C. Paton. 1983. Territory area, flower density, and time budgeting in hummingbirds: An experimental and theoretical analysis. American Naturalist 122:366–91.

Hockman, J. G., and J. A. Chapman. 1983. Comparative feeding habits of red foxes *(Vulpes vulpes)* and gray foxes *(Urocyon cinereoargenteus)* in Maryland. American Midland Naturalist 110:276–85.

Houston, D. B. 1978. Elk as winter-spring food for carnivores in northern Yellowstone National Park. Journal of Applied Ecology 15:653–61.

Ingle, M. A. 1990. Ecology of red foxes and gray foxes and spatial relationships with coyotes in an agricultural region of Vermont. M.S. thesis, University of Vermont, Burlington.

Jaksic, F. M., J. L. Yanez, and J. R. Rau. 1983. Trophic relations of the southernmost populations of *Dusicyon* in Chile. Journal of Mammalogy 64:697–700.

Jimenez, J. E. 1993. Comparative ecology of Dusicyon foxes at the Chinchilla National Reserve in northeastern Chile. M.S. thesis, University of Florida, Gainesville.

Johnson, W. E. 1992. Comparative ecology of two sympatric South American foxes, *Dusicyon griseus* and *Dusicyon culpaeus*. Ph.D. diss., Iowa State University, Ames.

Johnson, W. E., and W. L. Franklin. 1994. The role of body size in the diets of sympatric grey and culpeo foxes. Journal of Mammalogy 75:163–74.

Johnson, W. E., T. K. Fuller, and W. L. Franklin. 1996. Sympatry in canids: A review and assessment. Pp. 189–218 *in* J. L. Gittleman, ed., Carnivore behavior, ecology, and evolution, vol. 2. Cornell University Press, Ithaca.

Keith, L. B., A. W. Todd, C. J. Brand, and R. S. Adamcik. 1977. An analysis of predation during a cyclic fluctuation of snowshoe hares. Proceedings of the International Conference on Game Biology 13:151–75.

Kelly, B. 1991. Analysis of coyote scat and estimation of diet. M.S. thesis, University of Idaho, Moscow.

Klett, S. 1987. Home ranges, movement patterns, habitat use, and interspecific interaction of red foxes and coyotes in northwest Louisiana. M.S. thesis, Southeast Louisiana University, Hammond.

Knight, S. W. 1978. Dominance hierarchies of captive coyote litters. M.S. thesis, Utah State University, Logan.

Knowlton, F. F. 1972. Preliminary interpretations of coyote population mechanics with some management implications. Journal of Wildlife Management 36:369–82.

Knowlton, F. F., and L. C. Stoddart. 1983. Coyote population mechanics: Another look. Pp. 93–111 *in* F. L. Bunnell, D. S. Eastman, and J. M. Peek, eds., Symposium on natural regulation of wildlife populations. Forest, Wildlife, and Range Experiment Station, University of Idaho, Moscow.

Knowlton, F. F., L. A. Windberg, and C. E. Wahlgren. 1986. Coyote vulnerability to several management techniques. Pp. 165–76 *in* D. B. Fagre, ed., Proceedings of the seventh Great Plains wildlife damage control workshop, San Antonio. USDA Forest Service, Rocky Mountain Forest and Range Experiment Station, Fort Collins, Colorado.

Knudsen, J. J. 1976. Demographic analysis of a Utah-Idaho coyote population. M.S. thesis, Utah State University, Logan.

Krefting, L. W. 1969. The rise and fall of the coyote on Isle Royale. Naturalist 20(4):24–31.

Lamprecht, J. 1978. On diet, foraging behaviour and interspecific competition of jackals in the Serengeti National Park, East Africa. Zeitschrift für Säugetierkunde 43:210–33.

Laundré, J. W. 1981. Home range use by coyotes in Idaho. Animal Behaviour 29:449–61.

Laundré, J. W., and B. L. Keller. 1984. Home-range size of coyotes: A critical review. Journal of Wildlife Management 48:127–39.

Lehman, N., A. Eisenhawer, K. Hansen, L. D. Mech, R. O. Peterson, P. J. P. Gogan, and R. K. Wayne. 1991. Introgression of coyote mitochondrial DNA into sympatric North American gray wolf populations. Evolution 45:104–19.

Lewis, J. C., K. L. Sallee, and R. T. Goligtly. 1993. Introduced red fox in California. California Resources Agency, Department of Fish and Game, Nongame Bird and Mammal Section Report 93-10, Sacramento.

Litvaitis, J. A. 1992. Niche relations between coyotes and sympatric Carnivora. Pp. 73–85 *in* A. H. Boer, ed., Ecology and management of the eastern coyote. Wildlife Research Unit, University of New Brunswick, New Brunswick.

Lott, D. F. 1984. Intraspecific variation in the social systems of wild vertebrates. Behaviour 88:266–325.

Major, J. T., and J. A. Sherburne. 1987. Interspecific relationships of coyotes, bobcats, and red foxes in western Maine. Journal of Wildlife Management 51:606–16.

Mares, M. A., T. E. Lacher, M. R. Willig, N. A. Bitar, R. Adams, A. Klinger, and D. Tazik. 1982. An experimental analysis of social spacing in *Tamias striatus*. Ecology 63:267–73.

Mech, L. D. 1974. *Canis lupus*. Mammalian Species no. 37.

———. 1977. Wolf pack buffer zones as prey reservoirs. Science 198:320–21.

Messier, F., and C. Barrette. 1982. The social system of the coyote *(Canis latrans)* in a forested habitat. Canadian Journal of Zoology 60:1743–53.

Messier, F., C. Barrette, and J. Huot. 1986. Coyote predation on a white-tailed deer population in southern Quebec. Canadian Journal of Zoology 64:1134–36.

Mills, L. S., and F. F. Knowlton. 1991. Coyote space use in relation to prey abundance. Canadian Journal of Zoology 69:1516–21.

Moehlman, P. 1983. Socioecology of silverbacked and golden jackals *(Canis mesomelas* and *Canis aureus)*. Pp. 423–53 *in* J. F. Eisenberg and D. G. Kleiman, eds., Recent advances in the study of mammalian behavior. Special Publication no. 7, American Society of Mammalogists, Pittsburgh.

———. 1986. Ecology of cooperation in canids. Pp. 64–86 *in* D. I. Rubenstein and R. W. Wrangham, eds., Ecological aspects of social evolution: Birds and mammals. Princeton University Press, Princeton.

Murie, A. 1940. Ecology of the coyote in the Yellowstone. National Park Service Fauna Series no. 4. U.S. Government Printing Office, Washington.

———. 1944. The wolves of Mount McKinley. National Park Service Fauna Series no. 5. U.S. Government Printing Office, Washington.

Nel, J. A. J. 1984. Behavioural ecology of canids in the south-western Kalahari. Koedoe 1984:229–35.

Nellis, C. H., and L. B. Keith. 1976. Population dynamics of coyotes in central Alberta, 1964–68. Journal of Wildlife Management 40:389–99.

Norris, P. W. 1881. Annual report of the superintendent of the Yellowstone National Park to the secretary of the interior for the year 1880. U.S. Government Printing Office, Washington.

O'Farrell, T. P. 1984. Conservation of the endangered San Joaquin kit fox *Vulpes macrotis mutica* on the Naval Petroleum Reserves, California. Acta Zoologica Fennica 172:207–8.

Packard, J. M., and L. D. Mech. 1980. Population regulation in wolves. Pp. 135–50 *in* M. N.

Cohen, R. S. Malpass, and H. G. Klein, eds., Biosocial mechanisms of population regulation. Yale University Press, New Haven.

Packer, C., D. Scheel, and A. E. Pusey. 1990. Why lions form groups: Food is not enough. American Naturalist 136:1–19.

Paquet, P. 1989. Behavioral ecology of wolves *(Canis lupus)* and coyotes *(C. latrans)* in Riding Mountain National Park, Manitoba. Ph.D. diss., University of Alberta, Edmonton.

——. 1991. Winter spatial relationships of wolves and coyotes in Riding Mountain National Park. Journal of Mammalogy 72:397–401.

——. 1992. Prey use strategies of sympatric wolves and coyotes in Riding Mountain National Park. Journal of Mammalogy 73:337–43.

Peterson, R. O. 1996. Wolves as interspecific competitors in canid ecology. Pp. 315–23 *in* L. N. Carbyn, S. H. Fritts, and D. Seip, eds., Wolves in a changing world. Canadian Circumpolar Institute, University of Alberta, Edmonton.

Pianka, E. R. 1974. Niche overlap and diffuse competition. Proceedings of the National Academy of Sciences 71:2141–45.

Pyke, G. H., H. R. Pulliam, and E. L. Charnov. 1977. Optimal foraging: A selective review of theory and tests. Quarterly Review of Biology 52:137–54.

Robinson, W. B., and M. W. Cummings. 1951. Movements of coyotes from and to Yellowstone National Park. U.S. Fish and Wildlife Service Special Scientific Report on Wildlife no. 11. USDI Fish and Wildlife Service, Washington.

Sargeant, A. B., and S. H. Allen. 1989. Observed interactions between coyotes and red foxes. Journal of Mammalogy 70:631–33.

Sargeant, A. B., S. H. Allen, and J. O. Hastings. 1987. Spatial relations between sympatric coyotes and red foxes in North Dakota. Journal of Wildlife Management 51:285–93.

Sayles, N. D. 1984. The effect of nutrition on maternal behavior among captive coyotes. M.S. thesis, Utah State University, Logan.

Schamel, D., and D. M. Tracy. 1986. Encounters between arctic foxes, *Alopex lagopus,* and red foxes, *Vulpes vulpes.* Canadian Field-Naturalist 100:562–63.

Schmitz, O. J., and D. M. Lavigne. 1987. Factors affecting body size in sympatric Ontario canids. Journal of Mammalogy 68:92–99.

Schullery, P., and L. Whittlesey. 1992. The documentary record of wolves and related wildlife species in the Yellowstone National Park area prior to 1882. Pp. 1-3 to 1-174 *in* J. D. Varley and W. G. Brewster, eds., Wolves for Yellowstone? A report to the United States Congress, vol. 4, research and analysis. National Park Service, Yellowstone National Park.

Sheldon, J. W. 1992. Wild dogs: The natural history of the nondomestic Canidae. Academic Press, New York.

Skinner, M. P. 1927. The predatory and fur bearing animals of the Yellowstone National Park. Roosevelt Wildlife Bulletin 4:163–281.

Small, R. L. 1971. Interspecific competition among three species of Carnivora on the Spider Ranch, Yavapai Co., Arizona. M.S. thesis, University of Arizona, Tucson.

Smits, C. M. M., B. G. Slough, and C. A. Yasui. 1989. Summer food habits of sympatric arctic foxes, *Alopex lagopus,* and red foxes, *Vulpes vulpes,* in the northern Yukon Territory. Canadian Field-Naturalist 103:363–67.

Soulé, M. E., D. T. Bolger, A. C. Alberts, M. Wright, M. Sorice, and S. Hill. 1988. Reconstructed dynamics of rapid extinctions of chaparral-requiring birds in urban habitat islands. Conservation Biology 2:75–92.

Springer, J. T. 1982. Movement patterns of coyotes in south-central Washington. Journal of Wildlife Management 46:191–200.

Steigers, W. D., and J. T. Flinders. 1980. Mortality and movements of mule deer in Washington. Journal of Wildlife Management 44:381–88.

Stoel, P. F. 1976. Some coyote food habit patterns in the shrub-steppe of south-central Washington. M.S. thesis, Portland State University, Portland, Oregon.

Theberge, J. B., and C. H. R. Wedeles. 1989. Prey selection and habitat partitioning in sympatric coyote and red fox populations, southwest Yukon. Canadian Journal of Zoology 67:1285–90.

Till, J. A., and F. F. Knowlton. 1983. Efficacy of denning in alleviating coyote depredations upon domestic sheep. Journal of Wildlife Management 47:1018–25.

Todd, A. W., and L. B. Keith. 1976. Responses of coyotes to winter reductions in agricultural carrion. Wildlife Technical Bulletin no. 5, Alberta Department of Recreation, Parks, and Wildlife, Edmonton.

Todd, A. W., L. B. Keith, and C. A. Fischer. 1981. Population ecology of coyotes during a fluctuation of snowshoe hares. Journal of Wildlife Management 45:629–40.

Tzilkowski, W. M. 1980. Mortality patterns of radio-marked coyotes in Jackson Hole, Wyoming. Ph.D. diss., University of Massachusetts, Amherst.

Van Valkenburgh, B., and R. K. Wayne. 1994. Shape divergence associated with size convergence in sympatric East African jackals. Ecology 75:1567–81.

Voigt, D. R., and B. D. Earle. 1983. Avoidance of coyotes by red fox families. Journal of Wildlife Management 47:852–57.

Waser, P. M. 1974. Spatial associations and social interactions in a "solitary" ungulate: The bushbuck *Tragelaphus scriptus* (Pallas). Zeitschrift für Tierpsychologie 37:24–37.

Wayne, R. K., and S. M. Jenks. 1991. Mitochondrial DNA analysis implying extensive hybridization of the endangered red wolf *Canis rufus*. Nature 351:565–68.

Weaver, J. L. 1977. Coyote-food base relationships in Jackson Hole, Wyoming. M.S. thesis, Utah State University, Logan.

Wells, M. C., and M. Bekoff. 1982. Predation by wild coyotes: Behavioral and ecological analyses. Journal of Mammalogy 63:118–27.

White, P. J., K. Ralls, and R. A. Garrott. 1994. Coyote-kit fox spatial interactions as revealed by telemetry. Canadian Journal of Zoology 72:1831–36.

Windberg, L. A., H. L. Anderson, and R. M. Engeman. 1985. Survival of coyotes in southern Texas. Journal of Wildlife Management 49:301–7.

Windberg, L. A., and F. F. Knowlton. 1988. Management implications of coyote spacing patterns in southern Texas. Journal of Wildlife Management 52:632–40.

Wooding, J. B. 1984. Coyote food habits and the spatial relationship of coyotes and foxes in Mississippi and Alabama. M.S. thesis, Mississippi State University, State College.

Adult female and two young badgers (Franz J. Camenzind)

Chapter Seven

Mesocarnivores of Yellowstone

Steven W. Buskirk

Most species of carnivores (Mammalia: Carnivora) of the Greater Yellowstone Ecosystem (GYE) are medium-sized (adult male body weights of one to fifteen kilograms). Here I call them mesocarnivores, adapting the terminology of Soulé et al. (1988), who showed the strong community-shaping relations involving carnivores and their prey (see also Terborgh and Winter 1980). The mesocarnivores of the GYE, following the taxonomy of Jones et al. (1986), comprise the red fox *(Vulpes vulpes)*, coyote *(Canis latrans)*, raccoon *(Procyon lotor)*, marten *(Martes americana)*, fisher *(Martes pennanti)*, mink *(Mustela vison)*, wolverine *(Gulo gulo)*, badger *(Taxidea taxus)*, striped skunk *(Mephitis mephitis)*, river otter *(Lutra canadensis)*, lynx *(Felis lynx)*, and bobcat *(Felis rufus)* (table 7.1). Although it falls within the size range of this group, the coyote is covered in detail in Chapter 6. Mink are at the lower limit of this size range; I have included them, though little GYE-specific information exists. Before being taken into captivity, the black-footed ferret *(Mustela nigripes)* was last known as a free-ranging population (Richardson et al. 1987) within the eastern boundary of the GYE (Harting and Glick 1994), but it is not treated here.

In this chapter I describe the composition and dynamics of the mesocarnivore community of the GYE. I address the functional roles of this group that have been recognized and propose others that have not. I explore the factors that structure communities of mesocarnivores, especially those in the GYE, and

Table 7.1

Articles, monographs, and theses dealing with mesocarnivores, except coyotes, specific to the Greater Yellowstone Ecosystem

Species	General	Habitat	Feeding Ecology	Populations
Red fox	Fichter and Williams 1967, Alfred 1979, Crabtree 1993			
Marten	Clark et al. 1989a, b; Negus and Findley 1959	Sherburne and Bissonette 1993, 1994, Coffin 1994	Murie 1961	Lacy and Clark 1993, Fager 1991, Coffin 1994, Clark and Campbell 1976
Mink	Mitchell 1961			Mitchell 1961
Wolverine	Newby and McDougal 1964, Hoak et al. 1982			
Badger	Minta 1990		Minta et al. 1992	Minta 1993
River otter	Rudd et al. 1986, Zackheim 1982			
Lynx	Halloran and Blanchard 1959, Reeve et al 1986			
Bobcat	Bailey 1979, Knick 1990		Knick 1990	

Note: Publications incidentally mentioning these species are not included. Skinner (1927), Bailey (1930), and Negus and Findley (1959) provide natural history accounts for most or all species. No publications deal primarily with the raccoon, striped skunk, or river otter in the GYE.

consider some aspects of the relation between phylogeny, morphology, and community structure. I discuss the role that protected areas of the GYE have played in the conservation of this group and propose measures that should enhance that role.

Significance of Mesocarnivores

Mesocarnivores are important ecologically because they affect the behaviors and demography of prey (Johnson and Sargeant 1977, Krebs et al. 1995). They also cycle nutrients by scavenging carrion (Hornocker and Hash 1981, Martin 1994), they affect plant fitness and likely landscape patterns through dispersal and predation of seeds (Willson 1992), and they influence distributions and abundances of nonprey vertebrates, especially each other. Mesocarnivores complete or interrupt the life cycles of pathogens or parasites of other animals, including humans (Thorne et al. 1982). Some mesocarnivores inflict harm on human economies, ranging in magnitude from the minor damage caused by martens in wilderness cabins to the depredations on livestock caused by coyotes and to locally severe losses to hatchery fish stocks caused by river otters (Toweill and Tabor 1982). In parts of the GYE not managed by the National Park Service, some mesocarnivores are trapped or shot for their furs (Crowe 1986). They contribute esthetically and emotionally to human experiences, particularly for recreationists in the GYE, to an unmeasured degree. In many parts of the world, carnivores of this size are the largest-bodied, most wilderness-dependent, and most ecologically important Carnivora persisting in landscapes fragmented and modified by humans (Buskirk 1994a, Newmark 1995).

Relative to size-delimited subsets of other mammalian orders, mesocarnivores are remarkable for their diversity of taxa, form, and function. They include generalist runners (coyote), swimmers (river otter), and climbers (marten). Some are habitat generalists (striped skunk) while others are specialists (marten and river otter). Some species coexist closely with humans (coyotes), whereas others are "wilderness species" (wolverine and lynx). Diets range from the highly omnivorous (striped skunk) to the narrowly specialized (lynx). Their unusual physiological adaptations include facultative heterothermy (badger), reflex ovulation (mink), and embryonic diapause (marten, fisher, wolverine). Resting metabolic rates vary widely among species, depending on diet (McNab 1989, Harlow 1994) and season. Their behaviors vary widely among species, remain plastic within species, and involve high levels of learning. They employ half-sibling care of young (red fox), behavioral thermoregulation (marten), and complex chemical signals (most species) to maintain spatial systems that vary within and among species.

In general, predators occur at lower densities than herbivores, a consequence of the loss of energy and biomass as food ascends trophic levels. Wolverines, for example, occur at minuscule densities, 0.006 per square kilometer to 0.02 per square kilometer (Banci 1994). The allometric equation of Peters (1983) predicts population density of a temperate carnivorous mammal to be about 6 percent that of an herbivore of the same size. These typically low densities of carnivores have important conservation implications. Low densities mean small populations, and small populations are predisposed to the stochastic and genetic changes that lead to extinction (Gilpin and Soulé 1986). It is not coincidental that the three species of mammals native to the GYE that are listed pursuant to the Endangered Species Act of 1973 (grizzly bear, wolf, black-footed ferret) are all predators. A correlate of their low densities is that mesocarnivores range very widely. Dispersing lynx have been reported to travel more than four hundred kilometers in each of five studies that used telemetry to follow them (Koehler and Aubry 1994).

As a policy concern in the western contiguous United States, mesocarnivores are important because of their general scarcity and the habitat specialization and dependency of some species on wilderness. These species are predisposed to impacts from humans (Finch 1992). Some species have undergone distributional losses (reviewed by Ruggiero et al. 1994) that have been attributed to humans. Examples of regional extinctions include the apparent disappearance of the wolverine from the Sierra Nevada Mountains since 1980 (Maj and Garton 1994), the disappearance of the fisher from the area between Mount Shasta and the Yosemite Valley since the 1930s (Zielinski et al. 1995), and the apparent extinction since 1950 of an entire subspecies, the Humboldt marten *(Martes americana humboldtensis)*, from northern California (Kucera et al. 1995). The lynx in the contiguous United States has been proposed for listing under the Endangered Species Act (Federal Register 63[130]:36993–37013); it is a protected species in Wyoming, although still trapped for its fur in Montana (Koehler and Aubry 1994). The river otter was eradicated from many temperate states (Toweill and Tabor 1982) and is protected in Wyoming and Idaho, though still trapped in Montana.

Zoogeography and Historical Occurrence

Our knowledge of the mesocarnivore community in the GYE has changed little since Ernest Thompson Seton (1913) provided species portraits and a species list based on his 1897 visit. Only a few species, the marten, for example, have received fairly detailed hypothesis-driven study, mostly in relation to habitat (table 7.1). By contrast, our knowledge of the raccoon, wolverine, river otter, mink, lynx, and bobcat is almost entirely from anecdote. The very existence of a population of fishers in the GYE is a matter of conjecture; a recent photograph taken in the northeastern GYE (Gehman 1995) was the first plausible physical evidence of the fisher in more than ninety years (Long 1965). Our ignorance of mesocarnivores of the GYE is profound and long-standing.

Mesocarnivores of the GYE can be logically divided into two groups based on the relation of the GYE to their continental distributions. The coyote, raccoon, mink, badger, striped skunk, and bobcat are widespread, and the GYE represents but a small part of their continental range. For the fisher, wolverine, and lynx, however, the GYE represents a southern peninsular extension of a vast geographic range centered in Canada and Alaska (Ruggiero et al. 1994). The marten is intermediate between these conditions. The river otter, though originally widely distributed at temperate latitudes, has been reduced to such a degree that the GYE now represents an important refugium. The red fox is a special case discussed below.

Some mesocarnivore distributions and abundances have changed perceptibly over the historical period. Most writers who have addressed the subject, lacking reliable population data, have subjectively estimated the abundance of these species at an ordinal scale (for example, "rare," "common," "numerous"). These characterizations, as well as those of population change, must therefore be interpreted with the greatest caution. The most likely factors to have contributed to long-term population fluctuations include the widespread use of predator poisons in the late 1870s (Murie 1940), which would have affected various scavenging species. The abundance of coyotes was mentioned as a problem in the writings of various superintendents around 1900 (Murie 1940). On the other hand, the eventual extermination of wolves in the region in the 1920s should have increased coyote populations and decreased red fox populations. Other potential influences, such as on densities of ungulates, have not been examined.

Red foxes were called common by Seton (1913) but rare by Skinner (1927) (see also Chapter 6). They were reported to have declined in numbers by Murie (1940) and to be uncommon in the Jackson Hole area. Fox numbers may have responded to those of coyotes and wolves. Raccoons generally occur at low densities in the GYE (Seton 1913, Bailey 1930). They are most likely to be found along the lower reaches of the major rivers, such as the Yellowstone and Snake, but have been reported from various other locations (Wyoming Game and Fish Department unpublished data). Martens are widespread in forested areas of the GYE and more (Skinner 1927) or less (Seton 1913) common. Murie (1940) considered martens "moderately common." In my experience, the east face of the Teton Mountains is one of the most likely places to casually observe them. Negus and Findley (1959) reported about four hundred taken annually from the Jackson Hole area. I can infer no long-term trends in distribution or abundance.

Fishers seldom have been reported in the GYE. Various authors (e.g., Skinner 1927) referred to a specimen taken from a poacher in Yellowstone National Park in the 1890s. Thomas (1954) described two specimens trapped on the Beartooth Plateau, and occasional unconfirmed reports have described fishers from there. The remote photography in 1995 of possible fishers, also in that area (Gehman 1995), has renewed belief in a fisher population persisting since the 1800s. This is the southernmost area of the Rocky Mountains with a pattern of repeated reports of fishers. Minks are found along the rivers and streams of the GYE but have been considered uncommon by most writers (Skinner 1927, Bailey 1930).

Wolverines were called "very numerous" during the period 1834–43 by Russell (1965), were mentioned "surprisingly often" in reports from 1872–81

(Schullery and Whittlesey 1992), and were "of general distribution, but not common" in the 1890s (Seton 1913:225). Skinner (1927) thought that they were especially common in the northern part of the GYE, but less so than twenty years before, and Bailey (1930) only called them "well-distributed." Thomas (1954) considered the wolverine a "vanishing species" in Wyoming outside Yellowstone National Park. Long (1965) reported only a few records from northwestern Wyoming for the period before 1961, but Hoak et al. (1982) reported fifty new records for the area south of Yellowstone National Park that suggested a population increase. Meagher (1986), however, reported very similar numbers of sightings of wolverines between the periods 1960–73 and 1974–85.

Badgers have been mentioned as common by various authors (Seton 1913, Skinner 1927, Bailey 1930), generally in sagebrush steppe habitats (Negus and Findley 1959). They are most commonly noted along the Yellowstone River from Gardiner to the Lamar Valley and on the valley floor of Jackson Hole. I can infer no patterns of long-term change in distribution or abundance. Striped skunks are common at lower elevations in the GYE, especially the Yellowstone and Snake River valleys (Seton 1913, Skinner 1927, Negus and Findley 1959), usually in riparian, meadow, or sagebrush habitats. River otters were common in the GYE, especially in Yellowstone Lake and the Snake River, before 1882 (Haines 1974, Schullery and Whittlesey 1992), in the 1890s (Seton 1913), and throughout the 1900s (Skinner 1927, Bailey 1930, Thomas 1954, Rudd et al. 1986).

Lynx were called common by Seton (1913), but less so by later writers (Skinner 1927, Bailey 1930, Halloran and Blanchard 1959, Negus and Findley 1959). Reeve et al. (1986) observed no striking changes in distributions in Wyoming between 1856–1973 and 1973–86. A similar absence of change through time (1961–82 vs. 1983–93) is seen in the distribution maps of Maj and Garton (1994). Lynx are generally reported from forested areas of the Yellowstone Plateau, the northern Wind River Mountains, and the Wyoming-Salt River Ranges. Bobcat records are, if anything, rarer than those of lynx (Seton 1913, Skinner 1927, Murie 1940, Negus and Findley 1959). Murie (1940:10) wrote that "bobcats, once common, now are apparently gone." Sightings tend to be in low-elevation, non-forested habitats, especially in the northern GYE. No long-term trends in distribution or abundance are apparent.

Factors That Structure Mesocarnivore Communities

The major factors that structure mesocarnivore communities are food abundance, habitat structure, interference competition, and humans, especially their trapping. Studies of numerical responses of mesocarnivore populations are few (marten, Thompson and Colgan 1987; lynx, reviewed by Koehler and Aubry 1994), and the results are intuitive: when food is scarce, survival, densities, and reproduction are low.

HABITAT STRUCTURE

Habitat structure shapes carnivore communities by meeting life needs and mediating trophic and competitive relations. At a guild level, Van Valkenburgh (1985) compared the composition and morphological adaptations of Carnivora more than seven kilograms in body weight in Yellowstone, three other modern sites (Serengeti, Malaysia, Chitawan), and Orellan (Oligocene) deposits of Colorado, Nebraska, Wyoming, and South Dakota. Of the modern ecosystems, Yellowstone has an intermediate number of carnivore species, but the lowest number of herbivore species more than five kilograms in weight. Compared with the Serengeti, Yellowstone carnivores comprise fewer species but manifest a wider range of locomotor adaptations, the latter reflecting the presence of ursids. This difference is consistent with habitat structure: Yellowstone predators, occupying a mosaic of forests and parklands, include a balance of ambushers, distance runners, and climbers, whereas the Serengeti guild, occupying savanna, is dominated by runners, and the guild of forested Malaysia is dominated by climbers.

At the level of species and individuals, habitat structure meets multiple needs that have been reviewed by Buskirk and Powell (1994) and Ruggiero et al. (1994). Structure provides access to prey and to thermal microenvironments for resting and reproduction, helps avoid predators, and meets a "psychological need" for overhead cover (Hawley and Newby 1957) that is not well understood. The importance of habitat structure seems to be body-size dependent. Large carnivores like brown bears *(Ursus arctos)*, cougars *(Felis concolor)*, and wolves *(Canis lupus)* have few or no competitors larger than themselves, have low mass-specific rates of heat loss, and can capture prey in a wide range of habitat types. Their need for habitat structure largely is limited to that required to protect neonates. This important hypothesized difference between large carnivores and mesocarnivores is perhaps the reason that the former have received relatively little study with regard to habitat, whereas martens, fishers, and raccoons, smaller by one to two orders of magnitude, have strong and well-studied ties to habitat structure. It is also the reason that large carnivores cannot serve as "umbrellas" for the conservation of all carnivores. Habitat generalists, by definition, saturate the landscape with their home ranges more thoroughly than do habitat specialists. So the minimum landscape sizes required to maintain populations of specialists are larger than for generalists, other factors being equal. Wolves pursue prey in taiga, subtropical forest, tundra, desert, and, as I once observed, in ocean surf. For wolves, habitat structure explains little of the variation in densities over a vast geographic range. The same cannot be said for mesocarnivores.

Habitat structure influences not only local distributions of mesocarnivores but also the evolution of their forms and behaviors. Species with highly derived forms or behaviors tend to occupy special habitats, whereas species with generalized forms or flexible behaviors (for example, red fox and striped skunk) tolerate many habitat types. The specialized structures and behaviors associated with narrow habitat niches include the streamlined body form of the river otter, which aids swimming but hinders running, and the "psychological avoidance" of areas without overhead cover by martens and fishers. Breadth of the feeding niche also is related to the structure of mesocarnivore communities. Morphological specializations for feeding include the loss of postcarnassial teeth in highly predaceous felids and the long front claws of fossorial badgers. Surprisingly, these two forms of specialization, habitat and feeding, each with its morphological and behavioral correlates, seem not to be correlated. Martens, for example, evolved narrow habitat niches, but wide feeding niches, whereas short-tailed weasels (Mustela erminea) became food specialists in a wide range of habitats. This would be difficult to test rigorously because feeding niche breadths of river otters and wolverines would be hard to compare. Its significance, however, is that understanding mesocarnivore distributions and abundances requires examination of both predation and habitat ecology. Among large carnivores, traditional emphasis has been on the former.

INTERSPECIES INTERACTIONS

Interspecific competition has a powerful structuring role in mesocarnivore communities that often has been recognized in individual interactions (table 7.2), but seldom as a broader pattern. And although we lack rigorous tests, accumulated anecdotes suggest that communities of mesocarnivores, more than those of herbivores, are structured by interference competition, as defined by Keddy (1989; see also Chapter 6). Body size is the primary determinant of success in these interactions; in case after case, a larger-bodied carnivoran species has been found to reduce or exclude populations of a smaller one (table 7.2), but never, to my knowledge, the reverse. This is not to say that carnivores are incapable of exploitation interactions in which the smaller-bodied participants better the larger (e.g., King and Moors 1979). The likelihood of an interference interaction has an interesting relation to the difference in body sizes between potential competitors (figure 7.1). Body size differences predisposing to interference are as high as a factor of seven, but more commonly one and a

Table 7.2

Observed or inferred interference competition involving mesocarnivores

Dominant species	Subordinate species	Type[a]	Reference
Coyote	Red fox	I	Voigt and Earle 1983
		I	Major and Sherburne 1987
		O	Gese et al. 1996
Coyote	Bobcat	I	Reviewed by Litvaitis 1992
Red fox	Arctic fox	O	Schamel and Tracy 1986
Red fox	Eurasian pine marten	I	Lindström et al. 1995
Foxes	Weasels	O,I	Latham 1952
Cougar	Bobcat	O	Koehler and Hornocker 1991
Cougar	Coyote	O	Boyd and O'Gara 1985
		O	Koehler and Hornocker 1991
Coyote	Lynx	O	O'Donoghue et al. 1995
Lynx	Red fox	O	Stephenson et al. 1991
Fisher	Marten	I	Krohn et al. 1995
		I	Thomasma 1996
		O	Hodgman et al. 1997

Notes: Mesocarnivores are defined as 1–15 kg in body weight. For all pairs, the larger-bodied species is dominant, although interactions of coyotes and bobcats, similar in size, vary among species.
[a]O = observed (aggressive behavior); I = inferred (local parapatry, habitat or temporal partitioning, inverse population fluctuations).
[b]However, see Sargeant et al. 1987 for local parapatry without predation.

half to four. This is the limiting similarity described by Fox (1982) for predatory dasyurids (Marsupialia) and clearly results from the positive prey size–predator size relation described by Rosenzweig (1966): excluding another predator of about one's own size and shape increases prey availability to oneself. The proximal mechanism of exclusion often is not known, but likely includes a wide range of chemical and behavioral signals and attack (Rosenzweig 1966), sometimes resulting in death. Ruggiero et al. (1994) reviewed these interactions for marten, fisher, and lynx.

Of course, interference with an allospecific has potentially high fitness costs as well. As a general model of how individuals "decide" whether to coexist with allospecifics, I propose the following: The probability of strife is positively related to rates of encounter and to the uncertainty of the outcome of potential agonism. Being of approximately equal size and shape (for example, coyote and bobcat) means that two species have similar prey. Hunting for common prey predisposes them to occupy similar habitats and increases encounter rates. Upon encounter, uncertainty about the outcome of strife (because of similar body size) would predispose animals to approach each other to perceive each other's size and thereby increase the likelihood of agonism. As animals age and learn, olfactory or other signaling may substitute for this process. Clearly, however, agonism risks one's own fitness, and more economical means

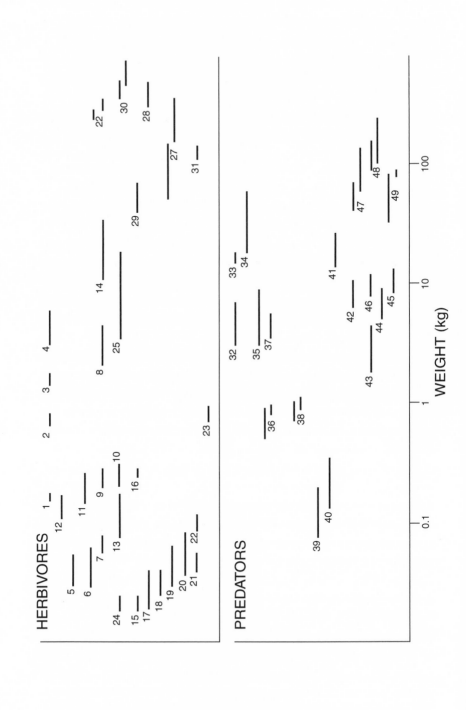

of resource partitioning (for example, chemical signaling) must be traded off against opportunities lost because of the more indirect information that they likely convey.

Because of the low densities and cryptic behaviors of the participants, human observations of aggressive interspecific interactions, especially killing, are extremely rare. So our understanding of this process in shaping carnivore community structure comes from accumulated anecdote, which diminishes, unfairly I believe, our sense of its importance.

Habitat structure and interspecific competition must not work in isolation; they interactively shape mesocarnivoran communities. For example, spatial analyses from Maine (Krohn et al. 1995) and California (Krohn et al. 1997) support the hypothesis that deep, soft snow limits the distribution of fishers, and fishers limit the distribution of martens. Martens, with their lower foot loadings, occur in areas where snow is too deep or soft for fishers. Snow depth, in turn, is affected by forest canopy characteristics, which we ordinarily suppose affects martens and fishers directly (Buskirk and Powell 1994).

Other interspecies interactions involving mesocarnivores have been shown to harm neither species and to benefit one or both (commensalism or mutualism). Minta et al. (1992) showed that coyotes and badgers preying on the same species *(Spermophilus armatus)* in Jackson Hole, Wyoming, interacted while foraging in ways that suggested benefits for both. The effect of large predators on the temporal-spatial distribution of carrion has not been systematically studied, but I hypothesize that large carnivores (wolves, bears, and cougars) make carrion available in patterns that benefit mesocarnivores, including red foxes,

---

7.1. (opposite) Body-mass distributions of herbivorous and predatory mammals of the Greater Yellowstone Ecosystem. Primarily invertibrativorous forms are excluded. Body masses are from Clark and Stromberg (1987) and Chapman and Feldhamer (1982). Herbivorous mammals are most species in the body-mass ranges of 0.02–0.3 kg and 40–600 kg. Predators are more evenly distributed across a narrower range (0.07–200 kg) of body masses. I attribute the more even body-mass distribution of predators than of herbivores to the size-structuring effects of interspecific interference competition observed in mammalian predators operating on communities over evolutionary time periods. 1 = pika, 2 = Nuttall's cottontail, 3 = snowshoe hare, 4 = white-tailed jackrabbit, 5 = least chipmunk, 6 = yellow pine-chipmunk, 7 = Uinta chipmunk, 8 = yellow-bellied marmot, 9 = Uinta ground squirrel, 10 = golden-mantled ground squirrel, 11 = red squirrel, 12 = northern flying squirrel, 13 = northern pocket gopher, 14 = beaver, 15 = deer mouse, 16 = bushy-tailed woodrat, 17 = southern red-backed vole, 18 = heather vole, 19 = meadow vole, 20 = montane vole, 21 = long-tailed vole, 22 = water vole, 23 = muskrat, 24 = western jumping mouse, 25 = porcupine, 26 = elk, 27 = mule deer, 28 = moose, 29 = pronghorn, 30 = bison, 31 = bighorn sheep, 32 = red fox, 33 = coyote, 34 = wolf, 35 = raccoon, 36 = marten, 37 = fisher, 38 = mink, 39 = short-tailed weasel, 40 = long-tailed weasel, 41 = wolverine, 42 = badger, 43 = striped skunk, 44 = river otter, 45 = lynx, 46 = bobcat, 47 = black bear, 48 = grizzly bear, 49 = cougar.

martens, and wolverines, and that would not occur otherwise. Specifically, I predict that predation-inflicted mortality on ungulates is more uniformly distributed in time and space than winter-kill mortality. This should result from the continuous food needs of large predators compared with the seasonal and highly episodic pattern of ungulate deaths resulting from starvation and hypothermia. Of course, winter-killed carcasses may be more intact and may present a richer food supply than predator-killed carcasses. For migratory ungulates, such as elk in the northern GYE (Chapter 8), strong temporal patterning of ungulate deaths should correspond with strong spatial patterning, although this hypothesis requires testing.

In at least a few cases, mesocarnivores also can benefit from nonpredatory interactions with vertebrates other than mesocarnivores. Martens associate closely with the middens of pine squirrels (*Tamiasciurus* spp.) to attain prey other than squirrels (Sherburne and Bissonette 1993) and to gain hypothesized thermal energetic benefits (Buskirk 1984). Throughout the range of the white-bark pine *(Pinus albicaulis),* martens may enjoy further benefits by eating squirrel-cached seeds of this tree. I have found the stomachs of martens killed in the Wind River Mountains in midwinter to be filled with these seeds. At that season, seeds would be impossible to gather from the forest floor, and martens are not equipped to remove them from cones. The relation of white-barked pine nuts to omnivorous carnivores like the marten are potentially important phenomena that are ideally studied in the GYE.

Radiation and Niche Breadth

Among carnivores, interspecific interactions and the radiation of lineages are mutually influential. Some lineages, such as the families Mustelidae and Viverridae, evolved diverse body sizes and shapes, whereas others, such as the Canidae, radiated in size but retained the ancestral shape. Ursids, one of the shortest branches on the carnivoran phylogenetic tree (Wozencraft 1989), have had insufficient time to radiate in either way: they are all large and similarly shaped. Carnivoran families with diverse sizes and shapes are species-rich both continentally and locally: as many as nine species of Mustelidae are locally sympatric in North America, and as many occur in the GYE (Skinner 1927, Bailey 1930). This is true in spite of the strong sexual dimorphism for body size shown by Mustelidae (Buskirk and Lindstedt 1989), which produces a greater variety of forms than species, perhaps making interspecific niche "fitting" more problematical.

By contrast, families with a common form are predisposed to interference competition with confamilials, as exemplified by the Canidae (Chapter 6). The

Table 7.3

Species richness, ranges of body weights, ratios of largest to smallest, and three measures of postcranial morphological diversity of some families of North American Carnivora

	Number of species	Range of body weight (kg)	Largest/ smallest	Morphological diversity		
				FMT[a]	MCP[b]	OLL[c]
Canidae	8	2–43	21	0.036	0.035	0.0004
Felidae	7	9–71	8	0.098	0.045	0.0022
Mustelidae	17	0.08–27	340	0.274	0.073	0.0025
Ursidae	3	78–360	5	0.003	0.018	0.0001

Notes: Species are from North America, north of Mexico. Numbers of species are from Jones et al. 1986. Body weights are from Chapman and Feldhamer 1982. Data on morphological diversity are from Van Valkenburg 1985 and from specimens in the Denver Museum of Natural History (DMNH) and the Wyoming State Archeological Laboratory (WSAL), Laramie. Measurements of specimens in DMNH and WSAL were restricted to adults as indicated by epiphyseal fusion and made with calipers to the nearest mm (available from the author). One member of each sex was measured where possible and the mean used; otherwise, one sex was sampled. Within-species replicates showed minimal variation among individuals. FMT and MCP are related to cursorial ability and predation type (e.g., pursuit, pounce, ambush). OLL is related to digging ability; fossorial species have high values. Because variability rather than location (i.e., mean) was of interest, variances were calculated for each ratio by carnivoran family. Nonzero correlations between the three variables reduced the inferential value of variable-by-variable tests of variance equality (i.e., F-tests). So a likelihood ratio test (Anderson 1984) was used to test the equality of the variance-covariance matrix between pairs of families. Such a test takes into account covariances between variables within families and allows for comparison with other families. Mustelids showed the greatest interspecies postcranial morphological diversity, followed by felids, canids, and ursids. Likelihood ratio tests showed that mustelids were more variable than both canids ($\chi^2 = 13.7$, 6 df, $P = 0.03$) and felids ($\chi^2 = 38$, 6 df, $P < 0.001$), whereas canids and felids do not differ ($\chi^2 = 8.4$, 6 df, $P = 0.21$). Although ursids consistently had the lowest variability, the low species richness of this family precluded hypothesis testing.
[a]Length of femur divided by length of third metatarsal.
[b]Length of third metacarpal divided by length of third proximal phalanx.
[c]Length of olecranon process of the ulna divided by the length of the shaft of the ulna.

likelihood of such coexistence is positively related to the difference in body sizes between any two species. Only three canid species occur in all of the GYE, and coyotes are predicted (Singer 1991) to be drastically reduced by the recent reintroduction of wolves. The interspecific range of body weights in mustelids, canids, and ursids is positively related to their range of morphologies (table 7.3). The effect of this radiation of body form and body size among Carnivora has been to separate niche spaces, enhancing coexistence, but its cause is much less clear. It should reflect the body size and degree of derivation of ancestors, their isolation, and time available for radiation.

Mesocarnivore communities can be thought of as occupying a series of niche spaces, the most important dimensions of which are body size (predator and prey) and body shape. Temporal partitioning of resources does not appear to be an important structuring mechanism among mesocarnivores. Body shape is a correlate of foraging style (for example, ambushing, digging) and of the use of habitat structure for other life needs, like predator avoidance. A niche space can be narrowed or eliminated under the influence of another carnivore species, especially one similar in shape and slightly larger in body size. I propose that this niche narrowing is accomplished in the following order: home range displacement, microhabitat avoidance, prey shifting.

The Role of Yellowstone in Conserving Mesocarnivores

The important role of the GYE in the conservation of mesocarnivores has been enhanced by several remarkable attributes:

Size. The nearly 59,000 square kilometers of wild lands (Glick et al. 1991)—more or less protected from conversion to development or agriculture, having a core on which extractive uses of terrestrial animals are prohibited, and occurring at a comparable geographic latitude—is remarkable or unique in the world. Although by no means a closed system, the GYE is an ecosystem, not a community or landscape.

Proscriptions on hunting and trapping. The relative absence of human-inflicted mortality from hunting, trapping, and poisoning, especially in Yellowstone National Park, has profoundly affected the behaviors and abundances of mesocarnivores.

Proscriptions on mining. Mining has the potential to alter mesocarnivore communities in ways that are powerful, yet poorly understood and difficult to test experimentally. Metal-bearing leachates from mines have contaminated some large river systems in the West. These metals contaminate benthic invertebrates as far as 380 kilometers downstream and reduce fish densities (Moore and Luoma 1990). Birds and mesocarnivores, such as river otters, mink, and raccoons, that feed on aquatic animals are susceptible to reduction or elimination by these toxic chemicals. With few exceptions (Meyer 1993), the GYE is relatively free of mines and mine-generated water contaminants.

Intact vertebrate fauna. The bird and mammal communities of the GYE are quite similar to those in presettlement times, so that most interactions among vertebrates, notwithstanding altered densities, reflect processes unaltered by man. Those mesocarnivores of GYE near the southern limits of their distributions (for example, wolverine, lynx, fisher) may possess unique adaptations to southern latitudes.

The GYE has served an important, even critical conservation function for mesocarnivores. Bailey (unpublished, cited in Skinner 1927:194–95) thought that trappers around Yellowstone National Park caught more wolverines than elsewhere and that "the park evidently serves as a breeding and recruiting ground which has kept this . . . animal from local extermination." Perhaps the river otter best illustrates how the GYE, specifically Yellowstone and Grand Teton national parks, has helped to conserve mesocarnivores more broadly. Otters have been locally common in what is now Yellowstone National Park throughout the historical period. Much of the plains and montane West outside the GYE was occupied by river otters in presettlement times. But during the early 1900s the species became extinct or nearly so over most of the West, including Colorado, Utah, North Dakota, the Southwest (Toweill and Tabor 1982), and Wyoming (Thomas 1954, Long 1965). By the 1970s the river otter was recovering in some of these areas (Toweill and Tabor 1982), primarily through reintroduction. In contrast, areas south of Yellowstone National Park appear to have been recolonized naturally from the Yellowstone source population (Rudd et al. 1986), so that otters now are common on the Snake, upper Green, and Wind Rivers. For the river otter, the GYE has represented a refugium from trapping and water pollution, from which outlying areas have been repopulated gradually, at zero management cost, and with no controversy.

The absence of trapping in Yellowstone National Park also may have helped conserve the mountain red fox and the fisher. It is generally believed that current red fox stocks in North America, especially those in the East and South, derived from transplants from Europe (Fichter and Williams 1967, Aubry 1983). In presettlement times, mountain red foxes in the western contiguous United States were restricted to montane habitats (Aubry 1983); lower elevations later were colonized by foxes of European descent. Remnants of the native stocks, however, can still be found at higher elevations in the Sierra Nevada and Cascade Mountains. These "mountain red foxes," thought to be phenotypically distinguishable by their grayish pelage, have been tentatively identified by Crabtree (1993) in the northeastern GYE. This hypothesized persistence likely has been facilitated by proscriptions on trapping.

The fisher, if it has persisted undetected in the GYE since the 1800s, would represent the longest reported persistence of a presumed small, insular population of that species. The lack of timber harvesting in the northern GYE seems likely to help account for it. Both of these examples, admittedly constructed from minimal data, illustrate the potentially important role of the GYE refugium to mesocarnivore conservation (Buskirk 1994b), and how sketchily we understand it. Still, this role is imperfect and does not assure long-term persistence of mesocarnivores, especially those with large spatial requirements.

Newmark (1995) showed that twenty-two of the twenty-nine extinctions of lago-morphs, carnivorans, and artiodactyls from national parks in western North America, after their establishment, involved Carnivora. Of the thirty-nine species considered, eight (red fox, raccoon, ringtail, mink, river otter, fisher, striped skunk, and lynx) accounted for seventeen of twenty-nine population extinctions. Clearly, mesocarnivores are predisposed to extinction in national parks, and the establishment of national parks of the size typical in western North America around mesocarnivore populations does not assure the persistence of these species. The GYE itself may not be large enough to assure persistence of some, especially larger-bodied species.

The Future

The GYE will continue to be important in the understanding, public enjoyment, and conservation of mesocarnivores, in part because of its large size and strong protection. Still, the GYE is undergoing changes with important implications:

The reintroduction of wolves. This action has been predicted to result in reduced coyote numbers, releasing red fox populations (Singer 1991). The literature endorses these predictions, as well as my prediction that the prey of coyotes, such as rabbits and squirrels, will increase when coyotes decline. Such a prey release effect was inferred by Palomares et al. (1995) for a lynx–mongoose *(Herpeoes ichneumon)*–rabbit *(Oryctolagus cuniculus)* system in Spain, though disputed by Litvaitis and Villafuerte (1996). The onset of "mesocarnivore suppression" in the GYE would be the opposite of the "mesopredator release" described by Terborgh and Winter (1980). Other mesocarnivores, such as martens and badgers, may be affected indirectly—martens because coyotes are their potential predators, and badgers because they prey on ground squirrels.

The introduction of lake trout (Salvelinus namaycush) *to Yellowstone Lake.* This presumably illegal introduction has been forecast to have profound effects on other fish species, especially Yellowstone cutthroat trout (*Oncorhynchus clarki bouvieri*, Kaeding et al. 1996). Changes in fish stocks can be predicted to affect densities and distributions of piscivorous mesocarnivores, especially river otters and mink, depending on the abundance and vulnerability of fish in future aquatic communities. Considering the apparently important past role of the GYE in conserving river otters, this potential impact should be studied carefully.

Increased development. Development, which I consider to include construction of roads, buildings, and other structures, could have important local effects on mesocarnivores, but I do not foresee systemwide changes that could be detected by the most sensitive existing methods. This is because so much of the GYE receives statutory protection that foreseeable developments do not seem capable of altering the land at a large scale.

Increased visitation. Visitation per se likely affects mesocarnivores primarily via vehicle traffic, which causes accidental deaths. In nature reserves, mesocarnivores, unlike bears and wolves, tend not to conflict with humans in ways that lead to the animals' removal. It is difficult to imagine visitation or traffic levels that could cause systemwide changes in mesocarnivore community composition, although extremely low-density species like wolverines could be vulnerable.

Timber cutting. Some parts of the GYE have undergone large-scale timber cutting, which removes large volumes of woody biomass, converts sites to early successional stages, and interrupts the recruitment of coarse woody debris to the forest floor. Studies in the GYE (Campbell 1979) and elsewhere (reviewed by Thompson and Harestad 1994) have been nearly unanimous in showing that martens avoid regenerating clearcuts for several decades. Fishers also have been shown to avoid regenerating clearcuts (Buck et al. 1994, Jones and Garton 1994). Other species, including lynx and river otters, have been little studied in relation to forest management in western North America, but have needs for structure that likely are not met by post-cutting seres in the Rocky Mountains. Timber cutting can alter mesocarnivore communities beyond the borders of the cut areas by affecting landscape attributes (Chapin et al. 1998). This is therefore a special concern for mesocarnivores of Yellowstone.

We can enhance the role of the GYE in the conservation of mesocarnivores if:

First, we know what mesocarnivore species are present and where they occur. No longer can we set aside areas the size of Yellowstone National Park and correctly assume that, without further action, all carnivore taxa will be protected in perpetuity. We require knowledge to plan for, actively conserve, and, if necessary, reintroduce mesocarnivores. For the GYE, adequate information is not available for most species, and for some there is almost no information. No published studies, theses, or comprehensive reports deal with fishers, river otters, wolverines, mountain red foxes, or lynx in the GYE. Major recent advances in detection methods (reviewed by Zielinski and Kucera 1995) make monitoring for some of these species feasible and cost-effective.

Second, water is protected from mine-generated pollution. The historical role of Yellowstone National Park in conserving river otters and other piscivorous vertebrates (for example, mink, bears, and white pelicans [*Pelecanus erythrorhynchus*]) could be drastically diminished if mine leachates were to contaminate important feeding or fish production areas.

Third, the herbivorous biota of the GYE remains more or less intact. Drastic alterations in the vertebrate fauna of the GYE have the potential to affect the predators that feed on them.

The community of mesocarnivores in the GYE is diverse, functionally important, relatively unaltered by humans, and relatively, but by no means completely, protected from human-caused change. This community has national significance, no less so than for large carnivores, but it lacks the public interest and support that have traditionally been directed toward bears and wolves. We have long presumed that if we succeeded at conserving large carnivores, smaller-bodied carnivores would be shielded under their "umbrella." This assumption is false because mesocarnivores in general are more food- and habitat-specialized as an indirect consequence of their smaller body sizes. Our steadfast acceptance of this assumption increases, perhaps to certainty, the likelihood that some of the products of evolution found in the GYE for millennia will be lost within decades.

Acknowledgments: J. Erb measured and analyzed morphological attributes. W. J. Zielinski, D. J. Harrison, and J. Halfpenny provided detailed and insightful comments on early drafts.

Literature Cited: Allred, E. M. 1979. Denning behavior of the red fox *(Vulpes fulva)* in eastern Idaho and western Wyoming. M.S. thesis, Idaho State University, Pocatello.

Anderson, T. W. 1984. An introduction to multivariate statistical analysis. John Wiley and Sons, New York.

Aubry, K. B. 1983. The Cascade red fox: Distribution, morphology, zoogeography and ecology. Ph.D. diss., University of Washington, Seattle.

Bailey, T. N. 1979. Den ecology, population parameters and diet of eastern Idaho bobcats. National Wildlife Federation Scientific Technical Series 6:62–69.

Bailey, V. 1930. Animal life of Yellowstone National Park. Charles C. Thomas, Baltimore.

Banci, V. 1994. Wolverine. Pp. 99–127 in L. F. Ruggiero, K. B. Aubry, S. W. Buskirk, L. J. Lyon, and W. J. Zielinski, eds., The scientific basis for conserving forest carnivores: American marten, fisher, lynx, and wolverine in the western United States. USDA Forest Service, General Technical Report RM-254.

Boyd, D., and B. O'Gara. 1985. Cougar predation on coyotes. Murrelet 66:17.

Buck, S. G., C. Mullis, A. S. Mossman, I. Show, and C. Coolahan. 1994. Habitat use by fishers in adjoining heavily and lightly harvested forest. Pp. 368–76 in S. W. Buskirk, A. S. Harestad, M. G. Raphael, and R. A. Powell, eds., Martens, sables, and fishers: Biology and conservation. Cornell University Press, Ithaca.

Buskirk, S. W. 1984. Seasonal use of resting sites by marten in southcentral Alaska. Journal of Wildlife Management 48:950–53.

——. 1994a. An introduction to the genus *Martes*. Pp. 1–10 in S. W. Buskirk, A. S. Harestad, M. G. Raphael, and R. A. Powell, eds., Martens, sables, and fishers: Biology and conservation. Cornell University Press, Ithaca.

——. 1994b. The refugium concept and the conservation of forest carnivores. Pp. 242–45 in Proceedings of the XXI International Congress of Game Biologists, Halifax, Nova Scotia.

Buskirk, S. W., and S. L. Lindstedt. 1989. Sex biases in trapped samples of Mustelidae. Journal of Mammalogy 70:88–97.

Buskirk, S. W., and R. A. Powell. 1994. Habitat ecology of fishers and American martens. Pp. 283–96 *in* S. W. Buskirk, A. S. Harestad, M. G. Raphael, and R. A. Powell, eds., Martens, sables, and fishers: Biology and conservation. Cornell University Press, Ithaca.

Campbell, T. M. 1979. Short-term effects of timber harvests on pine marten ecology. M.S. thesis, Colorado State University, Fort Collins.

Chapin, T. G., D. J. Harrison, and D. D. Katnik. 1998. Influence of landscape pattern on habitat use by American marten in an industrial forest. Conservation Biology 12:1327–37.

Chapin, T. G., D. J. Harrison, and D. M. Phillips. 1997. Seasonal habitat selection by marten in an untrapped forest preserve. Journal of Wildlife Management 61:707–17.

Chapman, J. A., and G. A. Feldhamer. 1982. Wild mammals of North America. Johns Hopkins University Press, Baltimore.

Clark, T. W., M. Bekoff, T. M. Campbell, T. Hauptman and B. D. Roberts. 1989a. American marten, *Martes americana*, home ranges in Grand Teton National Park, Wyoming. Canadian Field-Naturalist 103:423–25.

Clark, T. W., T. M. Campbell III, and T. N. Hauptman. 1989b. Demographic characteristics of American marten populations in Jackson Hole, Wyoming. Great Basin Naturalist 49:587–96.

Clark, T. W., and T. M. Campbell III. 1976. Population organization and regulatory mechanisms of pine martens in Grand Teton National Park, Wyoming. Pp. 293–95 *in* R. M. Linn, ed., Conference on scientific research in national parks, vol. 1. National Park Service Transactions and Proceedings Series no. 5, Washington.

Clark, T. W., and M. R. Stromberg. 1987. Mammals in Wyoming. University of Kansas Museum of Natural History, Lawrence.

Coffin, K. W. 1994. Population characteristics and winter habitat selection by pine marten in southwest Montana. M.S. thesis, Montana State University, Bozeman.

Crabtree, R. 1993. Gray ghost of the Beartooth. Yellowstone Science (spring): 13–16.

Crowe, D. M. 1986. Furbearers of Wyoming. Wyoming Game and Fish Department, Cheyenne.

Fager, C. W. 1991. Harvest dynamics and winter habitat use of the pine marten in southwest Montana. M.S. thesis, Montana State University, Bozeman.

Fichter, E., and R. Williams. 1967. Distribution and status of the red fox in Idaho. Journal of Mammalogy 48:219–30.

Finch, D. M. 1992. Threatened, endangered, and vulnerable species of terrestrial vertebrates in the Rocky Mountain region. USDA Forest Service General Technical Report RM-215.

Fox, B. J. 1982. A review of dasyurid ecology and speculation on the role of limiting similarity in community organization. Pp. 97–116 *in* M. Archer, ed., Carnivorous marsupials. Royal Zoological Society of New South Wales, Sydney.

Gehman, S. 1995. Stalking the elusive fisher. Yellowstone Science 3(4):2–3.

Gese, E. M., T. E. Stotts, and S. Grothe. 1996. Interactions between coyotes and red foxes in Yellowstone National Park, Wyoming. Journal of Mammalogy 77:377–82.

Gilpin, M. E., and M. E. Soulé. 1986. Minimum viable populations: Processes of species extinction. Pp. 19–34 *in* M. E. Soulé, ed., Conservation biology: The science of scarcity and diversity. Sinauer Associates, Sunderland, Massachusetts.

Glick, D., M. Carr, and B. Harting. 1991. An environmental profile of the Greater Yellowstone Ecosystem. The Greater Yellowstone Coalition, Bozeman, Montana.

Haines, A. L. 1974. Yellowstone National Park: Its exploration and establishment. National Park Service, Washington.

Halloran, A. F., and W. E. Blanchard. 1959. Lynx from western Wyoming. Journal of Mammalogy 40:450–51.

Harlow, H. J. 1994. Trade-offs associated with the size and shape of American martens. Pp. 391–403 *in* S. W. Buskirk, A. S. Harestad, M. G. Raphael, and R. A. Powell, eds., Martens, sables, and fishers: Biology and conservation. Cornell University Press, Ithaca.

Harting, A., and D. Glick. 1994. Sustaining Greater Yellowstone: A blueprint for the future. Greater Yellowstone Coalition, Bozeman, Montana.

Hawley, V. D., and F. E. Newby. 1957. Marten home ranges and population fluctuations in Montana. Journal of Mammalogy 38:174–84.

Hoak, J. H., J. L. Weaver and T. W. Clark. 1982. Wolverines in western Wyoming. Northwest Science 56:159–61.

Hodgman, T. P., D. J. Harrison, D. M. Phillips, and K. D. Elowe. 1997. Survival of American marten in an untrapped forest preserve in Maine. Pp. 86–99 in G. Proulx, H. N. Bryant, and P. M. Woodward, eds., *Martes:* Taxonomy, ecology, techniques, and management. Proceedings of the Second International Martes Symposium, Edmonton, Alberta. Provincial Museum of Alberta, Edmonton.

Hornocker, M. G., and H. S. Hash. 1981. Ecology of the wolverine in northwestern Montana. Canadian Journal of Zoology 59:1286–1301.

Johnson, D. H., and A. B. Sargeant. 1977. Impact of red fox predation on the sex ratio of prairie mallards. U.S. Fish and Wildlife Service, Wildlife Research Report 6.

Jones, J. K., D. C. Carter, H. H. Genoways, R. S. Hoffmann, D. W. Rice, and C. Jones. 1986. Revised checklist of North American mammals north of Mexico, 1986. The Museum, Texas Tech University, Occasional Papers no. 107.

Jones, J. L., and E. O. Garton. 1994. Selection of successional stages by fishers in north-central Idaho. Pp. 377–87 in S. W. Buskirk, A. S. Harestad, M. G. Raphael, and R. A. Powell, eds., Martens, sables, and fishers: Biology and conservation. Cornell University Press, Ithaca.

Kaeding, L. R., G. D. Boltz, and D. G. Carty. 1996. Lake trout discovered in Yellowstone Lake threaten native cutthroat trout. Fisheries 21(3):16–20.

Keddy, P. A. 1989. Competition. Chapman and Hall, New York.

King, C. M., and P. J. Moors. 1979. On co-existence, foraging strategy and the biogeography of weasels and stoats *(Mustela nivalis* and *M. erminea)* in Britain. Oecologia 39:129–50.

Knick, S. T. 1990. Ecology of bobcats relative to exploitation and a prey decline in southeastern Idaho. Wildlife Monographs 108.

Koehler, G. M., and K. B. Aubry. 1994. Lynx. Pp. 74–98 in L. F. Ruggiero, K. B. Aubry, S. W. Buskirk, L. J. Lyon, and W. J. Zielinski, eds., The scientific basis for conserving forest carnivores: American marten, fisher, lynx and wolverine in the western United States. USDA Forest Service, General Technical Report RM-254.

Koehler, G. M., and M. G. Hornocker. 1991. Seasonal resource use among mountain lions, bobcats, and coyotes. Journal of Mammalogy 72:391–96.

Krebs, C. J., S. Boutin, R. Boonstra, A. R. E. Sinclair, J. N. M. Smith, M. R. T. Dale, K. Martin, and R. Turkington. 1995. Impact of food predation on the snowshoe hare cycle. Science 269:1112–15.

Krohn, W. B., K. D. Elowe, and R. B. Boone. 1995. Relations among fishers, snow, and martens: Development and evaluation of two hypotheses. Forestry Chronicle 71:97–105.

Krohn, W. B., W. J. Zielinski, and R. B. Boone. 1997. Relations among fishers, snow, and martens in California: Results from small-scale spatial comparisons. Pp. 211–32 in G. Proulx, H. N. Bryant, and P. M. Woodard, eds., *Martes:* Taxonomy, ecology, techniques, and management. Proceedings of the Second International Martes Symposium, Provincial Museum of Alberta, Edmonton.

Kucera, T. E., W. J. Zielinski, and R. H. Barrett. 1995. The current distribution of American marten, *Martes americana,* in California. California Fish and Game 81:96–103.

Lacy, R. C., and T. W. Clark. 1993. Simulation modeling of American marten *(Martes americana)* populations: Vulnerability to extinction. Great Basin Naturalist 53:282–92.

Latham, R. M. 1952. The fox as a factor in the control of weasel populations. Journal of Wildlife Management 16:516–17.

Lindström, E. R., S. M. Brainerd, J. O. Helldin and K. O. Overskaug. 1995. Pine marten-red fox interactions: A case of intraguild predation? Annals Zoologica Fennici 32:123–30.

Litvaitis, J. A. 1992. Niche relationships between coyotes and sympatric Carnivora. Pp. 73–85 in A. H. Boer, ed., Ecology and management of the eastern coyote. Wildlife Research Unit, University of New Brunswick, Fredericton.

Litvaitis, J. A., and R. Villafuerte. 1996. Intraguild predation, mesopredator release, and prey stability. Conservation Biology 10:676–77.

Long, C. A. 1965. The mammals of Wyoming. University of Kansas Publications, Museum of Natural History 14:493–758.

Maj, M., and E. O. Garton. 1994. Fisher, lynx, wolverine summary of distribution information. Pp. 169–75 *in* L. F. Ruggiero, K. B. Aubry, S. W. Buskirk, L. J. Lyon, and W. J. Zielinski, eds., The scientific basis for conserving forest carnivores: American marten, fisher, lynx, and wolverine in the western United States. USDA Forest Service, General Technical Report RM-254.

Major, J. T., and J. A. Sherburne. 1987. Interspecific relationships of coyotes, bobcats, and red foxes in western Maine. Journal of Wildlife Management 51:606–16.

Martin, S. K. 1994. Feeding ecology of American martens and fishers. Pp. 297–315 *in* S. W. Buskirk, A. S. Harestad, M. G. Raphael, and R. A. Powell, eds., Martens, sables, and fishers: Biology and conservation. Cornell University Press, Ithaca.

McNab, B. K. 1989. Basal rate of metabolism, body size, and food habits in the Order Carnivora. Pp. 335–54 *in* J. L. Gittleman, ed., Carnivore behavior, ecology, and evolution. Cornell University Press, Ithaca.

Meagher, M. 1986. Cougar and wolverine in Yellowstone National Park. Unpublished report, Yellowstone National Park, Resource Management Office.

Meyer, G. A. 1993. A polluted flash flood and its consequences. Yellowstone Science 2(1):2–6.

Minta, S. C. 1990. The badger, *Taxidea taxus* (Carnivora: Mustelidae): Spatial-temporal analysis, dimorphic territorial polygyny, population characteristics, and human influences on ecology. Ph.D. diss., University of California, Davis.

———. 1993. Sexual differences in spatio-temporal interaction among badgers. Oecologia 96:402–9.

Minta, S. C., K. A. Minta, and D. F. Lott. 1992. Hunting associations between badgers *(Taxidea taxus)* and coyotes *(Canis latrans)*. Journal of Mammalogy 73:814–20.

Mitchell, J. L. 1961. Mink movements and populations on a Montana river. Journal of Wildlife Management 25:48–54.

Moore, J. N., and S. N. Luoma. 1990. Hazardous wastes from large-scale metal extraction. Environmental Science and Technology 24:1278–85.

Murie, A. 1940. Ecology of the coyote in Yellowstone. U.S. National Park Service Conservation Bulletin no. 4, Fauna Series no. 4. U.S. Government Printing Office, Washington.

———. 1961. Some food habits of the pine marten. Journal of Mammalogy 42:516–21.

Negus, N. C., and J. S. Findley. 1959. Mammals of Jackson Hole, Wyoming. Journal of Mammalogy 40:371–81.

Newby, F. E., and J. J. McDougal. 1964. Range extension of the wolverine in Montana. Journal of Mammalogy 45:485–87.

Newmark, W. D. 1995. Extinction of mammal populations in western North American national parks. Conservation Biology 9:512–26.

O'Donoghue, M., E. Hofer, and F. I. Doyle. 1995. Predator versus predator. Natural History 104:6–9.

Palomares, F., P. Gaona, P. Ferreras, and M. Delibes. 1995. Positive effects on game species of top predators by controlling smaller predator populations: An example with lynx, mongooses, and rabbits. Conservation Biology 9:295–305.

Peters, R. H. 1983. The ecological implications of body size. Cambridge University Press, Cambridge.

Reeve, A., F. G. Lindzey and S. W. Buskirk. 1986. Historic and recent distribution of the lynx in Wyoming. Unpublished report, Wyoming Cooperative Fish and Wildlife Research Unit, Laramie.

Richardson, L., T. W. Clark, S. C. Forrest, and T. M. Campbell III. 1987. Winter ecology of black-footed ferrets *(Mustela nigripes)* at Meeteetse, Wyoming. American Midland Naturalist 117:225–39.

Rosenzweig, M. L. 1966. Community structure in sympatric carnivora. Journal of Mammalogy 47:602–12.

Rudd, W., L. R. Forrest, F. G. Lindzey and S. W. Buskirk. 1986. River otters in Wyoming: Distribution, ecology and potential impacts from energy development. Unpublished report, Wyoming Cooperative Fish and Wildlife Research Unit, Laramie.

Ruggiero, L. F., K. B. Aubry, S. W. Buskirk, L. J. Lyon, and W. J. Zielinski, eds. 1994. The scientific basis for conserving forest carnivores: American marten, fisher, lynx, and wolverine in the western United States. USDA Forest Service, General Technical Report RM-254.

Russell, O. 1965. Journal of a trapper. University of Nebraska Press, Lincoln.

Sargeant, A. B., S. H. Allen, and J. O. Hastings. 1987. Spatial relations between sympatric coyotes and red foxes in North Dakota. Journal of Wildlife Management 51:285–93.

Schamel, D., and D. M. Tracy. 1986. Encounters between arctic foxes, *Alopex lagopus,* and red foxes, *Vulpes vulpes.* Canadian Field-Naturalist 100:562–63.

Schullery, P., and L. Whittlesey. 1992. The documentary record of wolves and related wildlife species in the Yellowstone National Park area prior to 1882. Pp. 1-3 to 1-175 *in* J. D. Varley and W. G. Brewster, eds., Wolves for Yellowstone? A report to the United States Congress, vol. 4, research and analysis. National Park Service, Yellowstone National Park.

Seton, E. T. 1913. Wild animals at home. Grosset and Dunlap, New York.

Sherburne, S. S., and J. A. Bissonette. 1993. Squirrel middens influence marten *(Martes americana)* use of subnivean access points. American Midland Naturalist 129:204–7.

———. 1994. Marten subnivean access point use: Response to subnivean prey levels. Journal of Wildlife Management 58:400–405.

Singer, F. J. 1991. Some predictions concerning a wolf recovery into Yellowstone National Park: How wolf recovery may affect park visitors, ungulates and other predators. Transactions of the North American Wildlife and Natural Resources Conference 56:567–83.

Skinner, M. P. 1927. The predatory and fur-bearing animals of the Yellowstone National Park. Roosevelt Wildlife Bulletin 4(2):163–281.

Soulé, M. E., D. T. Bolger, A. C. Alberts, R. Sauvajot, J. Wright, and S. Hill. 1988. Reconstructed dynamics of rapid extinctions of chaparral-requiring birds in urban habitat islands. Conservation Biology 2:75–92.

Stephenson, R. O., D. V. Grangaard, and J. Burch. 1991. Lynx, *Felix Lynx,* predation on red foxes, *Vulpes vulpes,* caribou, *Rangifer tarandus,* and Dall sheep, *Ovis dalli,* in Alaska. Canadian Field-Naturalist 105:255–62.

Terborgh, J., and B. Winter. 1980. Some causes of extinction. Pp. 119–33 *in* M. E. Soulé and B. A. Wilcox, eds., Conservation biology: An evolutionary-ecological perspective. Sinauer Associates, Sunderland, Massachusetts.

Thomas, E. M. 1954. Wyoming fur bearers. Wyoming Game and Fish Department, Bulletin no. 7, Cheyenne.

Thomasma, L. E. 1996. Winter habitat selection and interspecific interactions of American martens *(Martes americana)* and fishers *(Martes pennanti)* in the McCormick Wilderness and surrounding area. Ph.D. diss., Michigan State University, Houghton.

Thompson, I. D., and P. W. Colgan. 1987. Numerical responses of martens to a food shortage in northcentral Ontario. Journal of Wildlife Management 51:824–35.

Thompson, I. D., and A. S. Harestad. 1994. Effects of logging on American martens with models for habitat management. Pp. 355–67 *in* S. W. Buskirk, A. S. Harestad, M. G. Raphael, and R. A. Powell, eds., Martens, sables, and fishers: Biology and conservation. Cornell University Press, Ithaca.

Thorne, E. T., N. Kingston, W. R. Jolley, and R. C. Bergstrom, eds. 1982. Diseases of wildlife in Wyoming. Wyoming Game and Fish Department, Cheyenne.

Toweill, D. E., and J. E. Tabor. 1982. River otter. Pp. 688–703 *in* J. A. Chapman and G. A. Feldhamer, eds., Wild mammals of North America. Johns Hopkins University Press, Baltimore.

Van Valkenburgh, B. 1985. Locomotor diversity within past and present guilds of large predatory mammals. Paleobiology 11:406–28.

Voigt, D. R., and B. D. Earle. 1983. Avoidance of coyotes by red fox families. Journal of Wildlife Management 47:852–57.

Willson, M. F. 1992. Mammals as seed-dispersal mutualists in North America. Oikos 67:159–76.

Wozencraft, W. C. 1989. The phylogeny of the Recent Carnivora. Pp. 495–535 *in* J. L.

Gittleman, ed., Carnivore behavior, ecology, and evolution. Cornell University Press, Ithaca.

Zackheim, H. S. 1982. Ecology and population status of the river otter in southwestern Montana. M.S. thesis, University of Montana, Missoula.

Zielinski, W. J., and T. E. Kucera. 1995. American marten, fisher, lynx, and wolverine: Survey methods for their detection. USDA Forest Service, General Technical Report PSW-157.

Zielinski, W. J., T. E. Kucera, and R. H. Barrett. 1995. The current distribution of the fisher, *Martes pennanti*, in California. California Fish and Game 81:104–12.

Cow elk in Yellowstone National Park (Franz J. Camenzind)

Chapter Eight

Predicting the Effects of Wildfire and Carnivore Predation on Ungulates

Francis J. Singer and John A. Mack

Ungulates provide the single largest source of food for the large and mid-sized carnivores in the Greater Yellowstone Ecosystem (GYE). More than 80,900 individuals of eight species—elk *(Cervus elaphus)*, bison *(Bison bison)*, mule deer *(Odocoileus hemionus)*, moose *(Alces alces)*, bighorn sheep *(Ovis canadensis)*, pronghorn *(Antilocapra americana)*, white-tailed deer *(Odocoileus virginianus)*, and mountain goat *(Oreamnos americanus)*—are found within and immediately adjacent to Yellowstone National Park (YNP) (figure 8.1, table 8.1). This constitutes the largest and most spectacular concentration of hoofed animals in any area of comparable size in North America. The largest single concentration of ungulates in the GYE occurs on the northern Yellowstone winter range, where populations of six ungulates are found at densities of nineteen to twenty-one ungulates per square kilometer. By way of comparison on a biomass basis, ungulates (table 8.2) are twenty-two times more abundant (4,400 kilograms) than all seven species of nonungulate mammalian prey combined (200 kilograms; beaver [*Castor canadensis*], snowshoe hare [*Lepus americanus*], Uinta ground squirrel [*Spermophilus armatus*], northern pocket gopher [*Thomomys talpoides*], *Microtus* spp., red-backed vole [*Clethrionomys gapperi*], deer mouse [*Peromyscus*]). For every one hundred kilograms of biomass of ungulate prey, there are only four and a half kilograms of smaller mammalian prey on the Northern Range. Elk biomass alone exceeds the small mammal prey by a ratio of twenty to one.

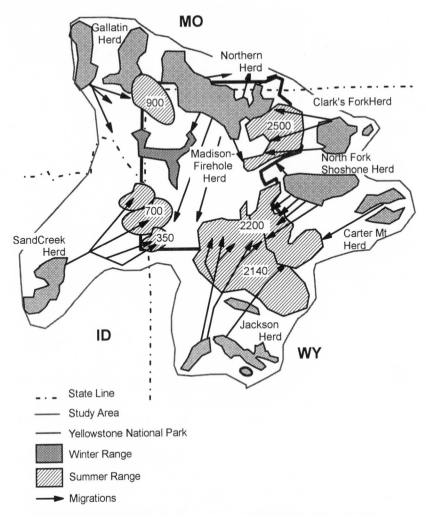

8.1.   Map of study area and elk summer and winter home ranges and migration routes.

In this chapter we first review the history of ungulate management in the GYE, as well as some general issues about population regulation that pertain in particular to Yellowstone-area ungulates. Second, we detail the ungulate numbers, distributions, biomass, interspecific interactions, and rates of predation on ungulates before the two major events of the past dozen years—the large fires of 1988 and the restoration of wolves. Third, we describe the short-term effects of the fires of 1988 through 1993 and project longer-term effects on the habitat and populations of ungulates. Fourth, we predict that the restoration of wolves into the northern portions of YNP during the winters of 1995 and 1996 is likely to have far-reaching effects on the ungulate-predator system. Although

Table 8.1

Numbers of ungulates available to carnivores

| | Winter ranges | | | | | Summer ranges | Entire |
Ungulate species	Northern Range	Other YNP winter ranges	Total	Visibility corrections[a]	Corrected total	Parkwide[b]	study area[c]
Elk	17,500	1,900	19,400	+0.23	23,900	31,000	46,000
Bison	533	~2,000	2,500	tr	2,600	2,600	2,900
Mule deer	1,814	<30	2,000	+0.40	2,800	>3,000	21,000
Moose	70	50–70	120	+1.00–1.50	240–300	>1,000	5,500
Bighorn	366	a few	350	+0.50	525	>300	4,000
Pronghorn	392	0	392	+0.02	400	500	600
White-tails/mountain goat	<100	0	<100	–	<100	<100	900
Total	20,875	3,980	24,900		30,600	>38,600	80,900

Note: The actual counts presented are the average of highest annual aerial counts (from Singer 1991a, Meagher 1989a, Keating 1985) of ungulates, visibility corrections, and estimated number of ungulates corrected for the animals missed during aerial counts in the Yellowstone and Grand Teton national parks and adjacent study area, 1980–88, before the fires of 1988.

[a]Visibility corrections are described in Singer et al. 1989, Singer and Garton 1994, Mack and Singer 1993b, and Scott unpublished data.

[b]Calculations for ungulate estimates in summer are provided in Singer and Mack 1993.

[c]Insufficient information for visibility corrections.

Table 8.2

Biomass of ungulates (kg $\times$ 10^3) available to predators on winter and summer ranges in Yellowstone National Park just before the fires of 1988

| | Winter ranges | | | |
Ungulate species	Northern Range	Other park winter ranges	Total	Summer ranges
Elk	3,930	430	4,360	15,500
Bison	240	1,000	1,240	3,000
Mule deer	100	2	102	350
Moose	55	30	85	185
Bighorn	17		17	42
Pronghorn	20		20	59
White-tailed deer/ mountain goat	5		5	<13
Total	4,400	1,500	5,900	19,200
Visibility-corrected estimate for ungulates[a]	5,400	1,500	7,000	23,100

Note: Biomass based on ungulate weights for the area provided in Houston 1982:71–72, 157.
[a]Based on counts and visibility corrections for elk, moose, bighorns, and deer provided in Singer 1991b, Singer et al. 1989, and Mack and Singer 1993a.

this restoration is too recent to result in any documented influences, we foresee cascading effects of the wolves on the numbers, distributions, and behavior of the ungulates, and also on other predators that share the ungulate prey base. Finally, we use a computer model to predict the effects of wolf predation on numbers, populations, and hunter harvests of ungulates, and to explore the role of compensation between mortality from winter starvation of elk and wolf predation on those same variables. We conclude with an ecosystem macrohypothesis that combines the predicted effects of climate change, the fires of 1988, wolf predation, and cascading effects of wolf restoration on the ungulate and predator guilds.

Ungulates as Prey

Populations of several carnivores are directly linked to ungulate abundance. Wolves and mountain lions are obligate predators on ungulates, although both occasionally consume smaller prey, such as rabbits, and wolves occasionally prey on beaver. Wolves and mountain lions rarely scavenge ungulate carrion. Grizzly bears, black bears, and coyotes also have a high proportion of ungulates in their diets, particularly as carrion, and neonatal ungulates are also captured and eaten (Murie 1940, French and French 1990, Mattson et al. 1991). Yellowstone grizzlies are unique in that they consume relatively more ungulates as prey or carrion than do most grizzly populations in North America (Mattson et al. 1991, Servheen et al. 1986). Mesopredators such as foxes, wolverines, fishers, and badgers also consume variable amounts of ungulate carrion, which is typically available in mid- to late winter and sometimes well into spring on the winter ranges of the GYE. In general, Yellowstone's predators fall into three groups relative to their consumption of ungulates. Mountain lions and wolves consume mostly ungulates, nearly all as live prey, at a very high rate (90–95 percent of their diet; Chapters 4 and 5). Both species of bears and coyotes consume a mix of ungulate food as live prey and carrion at a moderate rate (about 20–30 percent of their diets; Murie 1940, Mattson et al. 1991, Chapters 3 and 6 this volume). Mesopredators consume ungulate carrion locally and for brief periods (on a year-round basis ungulates constitute less than 10 percent of their diets), although the consumption of ungulates may be more important at certain times (Chapter 7).

Elk are the overwhelmingly dominant ungulate in the ecosystem. On a numerical basis within YNP, elk outnumber all other ungulates by a ratio of five to two. When considering the entire study area, including both Yellowstone and Grand Teton national parks (GTNP) and some adjacent national forest and other lands, elk outnumber all other ungulates by a ratio of four to three (table 8.1). On the basis of biomass, elk contribute one hundred kilograms to sixty-one kilograms for all other ungulates on YNP winter ranges, and one hundred kilograms to forty-nine kilograms for all other ungulates on all YNP summer ranges.

Because of their overwhelming abundance, elk are more likely than other ungulates to influence vegetation, regulate ecosystem processes, and provide food for large predators. Populations of the other large ungulates—moose and bison—influence vegetation in restricted locales (Chadde and Kay 1988, Frank and McNaughton 1992, 1993, Singer and Cates 1995), but these effects are not on the landscape scale of elk influences. Regardless of which species is driving the relation (and it may vary spatially and temporally), population dynamics of wolves, grizzlies, and coyotes are tied to elk (Singer and Mack 1993, Boyce 1993, Chapters 3, 6, and 10 this volume).

Elk are also predicted to be the dominant ungulate into the foreseeable future. They are likely to increase in the ecosystem in the next two decades as a result of the fires of 1988 (Boyce and Merrill 1991, Coughenour and Singer 1995). Wolves are not predicted to reduce elk numbers greatly or proportionately more than any other ungulate (Lime et al. 1993, Boyce 1993, Mack and Singer 1993a, Chapter 10 this volume), with the possible exception of mule deer, and wolves therefore are not predicted to alter ungulate abundance ratios substantially.

We restricted our observations to that major portion of the GYE that includes all the seasonal ranges and migration routes of the overwhelmingly dominant and abundant ungulate, elk. This study area includes all of YNP, GTNP, and parts of three adjacent national forests (figure 8.1). Dominant topographic features include a high, rolling central plateau and high mountain ranges in the central part of the study area, grading to sharply dissected or more gentle river valleys at lower elevations. Conifer forests dominate on the high elevation central plateau of YNP *(Pinus contorta, Pseudotsuga menziesii, Abies lasiocarpa)*, mixed with wet meadows *(Deschampsia cespitosa, Festuca idahoensis, Artemesia cana, Carex spp.)*. Drier forests (mostly Douglas fir, *Pseudotsuga menziesii)* predominate at lower elevations, as do dry grasslands *(Pseudoroegneria spicata, Festuca idahoensis, Poa sandberghii)* and dry shrublands *(Chyrosothamnus viscidiflorus, C. nauseousus, Artemisia tridentata;* Houston 1982, Steele et al. 1983, Despain 1991).

The northern winter range is emphasized in our review because it supports the single largest concentration and highest diversity of ungulates. The northern Yellowstone winter range consists of about twelve hundred square kilometers of rolling grasslands (55 percent) interspersed with conifer forests (41 percent). Average annual temperatures range from 1.8°C at high elevations to 6.6°C at lower elevations. Average annual precipitation is less than seventy-five centimeters, including about two hundred centimeters of annual snowfall. About 82 percent of the northern winter range is located within YNP, and the remaining 18 percent is on national forest and private lands outside the park. Elk, mule deer, pronghorns, and bison migrate outside the park each year, with larger migrations in years of heavy snowfall.

BRIEF HISTORY OF UNGULATE MANAGEMENT IN THE ECOSYSTEM

Historical management of ungulates can be divided into several fairly distinct periods following the reviews of Houston (1982) and Romme et al. (1995).

Before 1886: Early exploration. Early explorers reported all the ungulates present in the GYE today, with the exception of mountain goats (Haines 1965, Norris 1877, Houston 1982, Schullery and Whittlesey 1996). The accounts of the early explorers indicated that bighorn sheep, mule deer, pronghorn antelope, and bison were all common in the area (Houston 1982, Schullery and Whittlesey 1996). Moose were apparently rare, especially in the northern half of YNP, until the early 1900s (Walcheck 1976, Houston 1982). Considerably more controversy surrounds the historic abundance of elk in the GYE. Several authors have concluded that elk were abundant in the northern Yellowstone, Jackson Hole, Gros Ventre, and Gallatin valleys, based on accounts by early explorers

(Cole 1963, Lovaas 1970, Houston 1976, 1982, Boyce 1989, Schullery and Whittlesey 1996). Kay (1994), however, argued that elk were rare in the GYE because of overexploitation by Native Americans, citing as evidence the paucity of elk remains in archaeological sites in the area.

1886–94: Ungulate overexploitation and population reductions. A period of intense market hunting and overexploitation of elk populations followed the visits of early explorers. Elk and other ungulate populations were greatly reduced between 1886 and 1910. Poaching of elk within YNP continued until the 1890s, when the park staff finally brought it under control.

1895–1930: Protectionism. A period of more protective management followed, when elk were completely protected within YNP and GTNP. Large predators were controlled within YNP under the assumption that ungulates required protection from predators to recover their populations (Chapter 2). Ungulates, including elk, mule deer, and white-tailed deer, were artificially fed on the Northern Range near Mammoth Hot Springs, while bison were fed hay in the Lamar Valley. Elk were first fed hay during the winter in Jackson Hole in 1911 as growth in the town of Jackson encroached on their winter range.

1930–67: Modern game management with sport hunting of harvestable surplus outside parks and artificial reductions within the parks. The principles of traditional game management were also applied within the boundaries of YNP and GTNP from 1930 to 1967. Concerns about apparent overabundance of ungulates and severe hedging of willow, aspen, and big sagebrush were voiced as early as 1919. By 1930 staff of YNP embarked on a program of artificial control of elk, bison, and pronghorns. The goal of three thousand to six thousand elk, however, was set by using principles of economic carrying capacity, defined as the population size that results in highest possible yield and productivity of animals. Economic carrying capacity tends to be much less than the ecological carrying capacity (ECC) of the habitat. The intense trapping, live removal, and shooting program of the 1960s drew criticism from the public, and, following public hearings in 1967, the artificial reductions within YNP were terminated (USDI National Park Service 1968).

1967–present: Natural regulation of ungulates within YNP and the continuation of modern game management outside parks. Application of the concepts of economic carrying capacity to wildlife in national parks was questioned in 1963 in an advisory report to the National Park Service (Leopold et al. 1963) and by park managers and biologists (Cole 1971, Houston 1971).

GTNP never officially adopted a program of natural regulation of ungulates, and a winter hunt for elk has occurred along the park's edge (Cole 1969, Boyce 1989).

Management Philosophies for Migrating Ungulates

The fact that Yellowstone and Grand Teton national parks do not include all of the areas used by migrating ungulates creates challenges for agencies with different management philosophies. Natural regulation relies on density-dependent controls that operate as a population approaches the limits of the vegetative food base and snow cover.

Traditional game management strives to limit ungulate populations at about one-half ECC (usually), well below the point at which the vegetation is significantly influenced by ungulate herbivory (McCullough 1979, Wagner et al. 1995, Kay and Wagner 1994). Both natural regulation and economic carrying capacity philosophies have their supporters, but applying both management approaches to the same ungulate population is contradictory (Chase 1986, Boyce 1991, Wagner et al. 1995). The northern Yellowstone elk herd, for example, is subject to natural regulation within YNP, but outside the park other agencies attempt to limit elk numbers through hunting to minimize complaints from private landowners about damage caused by elk.

Elk are also protected within GTNP, except for the winter management hunt east of the Snake River, yet the herd exceeds the population goals set by the management agencies. The unharvested GTNP segments of the Jackson herd have grown considerably at the expense of other, more heavily hunted segments of the herd (Boyce 1991, Smith and Robbins 1994).

Regulation is the process by which a population returns to an equilibrium density point through density-dependent mortality and natality (Messier 1991). If natural regulation is sufficient, it obviates the need for culling or artificial regulation of populations within parks. A limiting factor, either density dependent or density independent (Ballard 1992), is any process that lowers an equilibrium point through a change in population production or loss (Sinclair 1989, Boutin 1992). Limiting factors are important to managers because they can suppress a population, limit its growth, and reduce the total number of ungulates available for harvest. Examples of potential regulatory factors include predation and sport hunting, but on closer scrutiny these factors are more often identified as limitations (Boutin 1992). Density-dependent factors may regulate a population with its habitat-limited resources or ECC. Climate factors such as snowfall and precipitation are considered part of the effective habitat limits.

Natural regulation of ungulates within YNP predicted that strong density-dependent mechanisms would regulate the ungulates around a single equilibrium point. The concepts of natural regulation were based on the best scientific information available at the time, but in the ensuing decades evidence has accumulated to indicate that expectation of a single equilibrium was naïve. Systems are more complex, stochastic events alter the equilibrium, and multi-equilibrium or nonequilibrium states are just as likely (May 1977, Wiens 1977, Connell and Sousa 1983). Evidence since 1970 has shown that unmanaged populations of wolves and bears can limit or regulate ungulate populations (Bergerud et al. 1983, Messier and Crete 1985, Gasaway et al. 1992). But Sinclair (1989) and Boutin (1992) concluded that regulation by predators is not well documented. The YNP natural regulation model predicted a similar new equilibrium between the plants and ungulates, as ungulate populations were released following cessation of controls, yet in this case no retrogression or major changes in soil and vegetation resources were predicted (Cole 1969, Houston 1976). Achieving both of these outcomes at the same time is unlikely.

Evaluation of the success of natural regulation management of ungulates in YNP is a contentious topic (Chase 1986, Despain 1991, Wagner et al. 1995, Singer et al. 1996). On the one hand, density-dependent regulation is well documented in both the northern Yellowstone and Jackson Hole elk herds (Sauer and Boyce 1983, Houston 1982, Boyce 1989, Dennis and Taper 1994, Coughenour and Singer 1996). Initial evaluations suggest that natural regulation has resulted in very little degradation of vegetation or soil within YNP (Cayot et al. 1979, Houston 1982, Coughenour 1991, Engstrom et al. 1991, Frank and Mc-Naughton 1992, 1993, Singer et al. 1996). On the other hand, the declines in willow, aspen, and cottonwood in the area suggest to some scientists that natural regulation of ungulates within YNP has been a failure (Kay 1984, 1990, Kay and Wagner 1994, Wagner et al. 1995).

There is a potential for compensation between mortality from carnivores and losses from hunting or winter starvation. Compensation is defined as the nonadditive effect of two or more forms of mortality when one form of mortality results in decreased mortality from another. Recreational hunting, for example, proposes to remove a surplus of animals that would otherwise die from other causes (Connolly 1981). Compensation between various sources of mortality is most likely to occur for populations closer to ECC, for populations where mortality rates are high, and for populations influenced by density-dependent regulation (Bartmann et al. 1992). For populations far below ECC, mortality is much more likely to be additive.

We hypothesize that compensatory mechanisms will occur, following wolf restoration to YNP, between winter starvation mortality of elk calves and wolf predation. Compensation has strong implications for management because the degree of compensation from wolf predation will greatly influence the allowable hunter harvest of the northern Yellowstone elk herd when animals migrate from the park. Our arguments that compensation will occur between wolf predation and overwinter juvenile starvation in elk is based upon the following evidence: (1) wolves kill mostly young of the year and adult ungulates that are older than prime age (Mech 1966, Pimlott et al. 1969, Peterson 1977, Carbyn 1983), and (2) winter malnutrition losses of juvenile elk and red deer is higher at higher cervid densities (Guiness et al. 1978, Houston 1982, Sauer and Boyce 1983, Clutton-Brock et al. 1982). Losses of elk-sized cervids from malnutrition in winter are essentially zero in heavily hunted populations and in populations with wolves present (Lonner and Schladweiler 1986, Adams et al. 1995). These statements are supported by a review of seven studies of first-year mortality of elk-sized cervids (table 8.3). Essentially no winter malnutrition losses of juveniles were observed in two populations with wolves and two other populations hunted at moderate rates. But in two other populations with no wolves and only light harvests by humans, loss of juveniles from winter malnutrition was extensive. These comparisons suggest that hunting and wolf predation are compensatory for winter malnutrition losses of juveniles.

Ungulates have traditionally been viewed as passive components of ecosystems, and managers believed effects of ungulates on the ecosystem needed to be held to a minimum (Dyskerius 1949, Wagner et al. 1995). The concept that ungulates should have no effect on the ecosystem persists in some circles even today (Wagner et al. 1995). Evidence has accumulated that ungulates may play a role in the regulation of ecosystem processes (McNaughton 1979, 1988, Pastor and Naiman 1992). This regulation has been referred to as control by components (McNaughton 1988). Optimal grazing intensities to stimulate production of grasses is apparently reached at 40–60 percent consumption of aboveground production. In Yellowstone, ungulate herbivory has been demonstrated to result in higher nutrient turnover, higher nutrient concentrations in steppe grasses and shrubs, and stimulation of net aboveground primary productivity (47 percent more production by grazed over ungrazed grasses; Frank and McNaughton 1992, 1993, Singer and Harter 1996). Aboveground consumption of herbaceous vegetation by ungulates averaged 45 percent in Yellowstone and 65 percent in African savannas, but native herbivory in other temperate grassland reserves averaged only 7.9 percent (Frank and McNaughton 1992).

The stimulation of aboveground production of grasses by ungulates in YNP may be related to the migratory behavior of ungulates and their ability to track young, high-quality forage as it shifts upward in elevation and across the Yellowstone ecosystem each growing season (Frank and McNaughton 1993). The removal of standing dead vegetation by ungulate herbivory can improve light and soil conditions that favor soil moisture and nutrient availability (McNaughton 1979, Knapp and Seastedt 1986). Also, urine and dung deposition

Table 8.3

Sources of elk and caribou calf mortality during the first year of life in seven case studies

| | Wolves absent | | | | | Wolves present | |
| | High elk densities | | Moderate elk densities | | Low elk densities | Caribou | Elk |
	Island Rhum	Yellowstone Park	Yellowstone Park	Gravelly Mountains	North-central Idaho[a]	Denali National Park	Riding Mountain National Park[b]
	Summer mortality (% of all mortalities)						
Bears	0	39		22	73	50	majority
Wolves	0	0		0	0	18[c]	5
Other predators	8	34		0	24	32	0
Accidents	11	5		0	0	0	0
Abandonment/ nursing deformities	36	0		78	0	0	0
Other/unknown	45	22		0	3	0	0
Summer mortality rate	18	22	49	16	59	43	60
	Winter mortality (% of all mortalities)						
Malnutrition	82	58		0		0	0
Hunting	0	15		0		0	8
Disease/deformities	9	4		0		0	0
Accidents	9	4		0		0	0
Predation	0	4		0		0	92
Unknown/other	0	15		0		0	0
Winter mortality rate	11	49	7	0	n.d.	0	26[d]
Years of data	6	4	4	3	4	4	–
Marked calves (n)	221	132	132	57	70	233	–

Sources: Schlegel 1976, Guiness et al. 1978, Singer 1987, Singer et al. 1997, Adams 1995.

Note: Caribou data for Denali National Park only.

[a]Winter data were not available (Schlegel 1976).

[b]Information based on studies of wolf and black bear predation rates (Paquet 1989).

[c]Wolf predation increased to 29 percent during latter part of study, when wolf densities increased (Adams et al. 1995).

[d]Estimated from population reconstruction (P. Paquet personal correspondence).

increases the fertility of grassland patches. Frank et al. (1994) concluded that Yellowstone's ungulates returned nitrogen to the soil at a rate four and a half times greater than did the decomposition of ungrazed litter.

Trends in Ungulate Prey and Interactions Within the Ungulate Guild

ELK

Eight elk herds winter and summer in or immediately adjacent to Yellowstone and Grand Teton national parks. We estimate that approximately 52,800 elk occur in these eight herds. The single largest concentration of summering elk is on the high central plateaus and mountain meadows of YNP, where about 38,000 elk gather (table 8.1). The remaining elk spend the summers within about forty kilometers of the park. About 22,500–25,000 elk—slightly fewer than one-half of the total—winter within YNP.

Elk in the YNP area are strongly migratory. Seven of the eight elk herds migrate an average of twenty kilometers (range: ten to fifty-five kilometers) between winter and summer ranges. Only the Madison-Firehole elk herd is completely nonmigratory (Craighead et al. 1972). Migrations of YNP elk coincide almost precisely with timing of peak net aboveground primary production and peak nitrogen concentrations in grasses (Frank and McNaughton 1992). Before winter, elk move off the high central plateau, where snow depths are prohibitive, to winter ranges at lower elevations outside or near the edges of both national parks. Fall migrations last only nineteen to twenty-seven days, while spring migrations are more protracted, lasting up to forty-six days. During the spring migration, elk are inhibited by melting snows, and they make extended use of transition ranges at intermediate elevations where newly emergent vegetation is at peak growth.

Six of the eight herds increased dramatically during recent decades. We estimated that total elk numbers in the study area increased about 75 percent during the two decades preceding the large fires of 1988 to more than 52,000 just before the fires (Singer 1991b, Singer and Mack 1993). We attributed these large-scale increases to the fruits of earlier elk restorations, to milder winters during the 1980s, and to conservative harvests of elk, combined with the inaccessibility to hunters of elk that migrate late from the large parks and remote wilderness areas. Only two elk populations did not increase during this period. First, the nonmigratory Madison-Firehole elk herd did not increase noticeably, although accurate counting of the herd is hampered by extensive conifer forest (Singer 1991b). This herd winters in a deep snow environment, where winter survival is closely tied to use of geothermal areas, and as a result it likely responds to different environmental and climatic factors than do the other seven herds. Second, the Gallatin elk herd increased very little during this period (λ = 1.07 during the 1980s) because of efforts by Montana Fish, Wildlife, and Parks to limit the herd through hunting, in order to protect vegetation and reduce complaints from private landowners.

Actual elk populations in the study area are likely 20–40 percent larger than the estimates based on counted elk, for many animals that are in conifer forest are missed during aerial surveys (Samuel et al. 1987, Vales and Peek 1993, Lockman et al. 1989, Singer and Garton 1994).

Natural regulation of elk within YNP did not, however, result in limitless growth of elk without any controls. Most of the increase in the northern Yellowstone elk occurred before 1981. Six elk populations, all more heavily hunted, increased more rapidly during the 1980s than did the only herd under natural regulation management, the northern Yellowstone herd. The northern Yellowstone herd was likely closer to ECC by 1980, and density-dependent mechanisms probably operated on this herd more than on the others. By 1988 four of the more extensively hunted herds were above population goals (Clarks Fork, North Fork Shoshone, Jackson, and Sand Creek). Late migrations from YNP are often cited as reasons for the less-than-desired harvest levels in these other herds (Boyce 1989, Rudd et al. 1983).

Total harvests of all sex and age classes averaged 6–36 percent of minimum population counts of the eight elk herds. Harvests of the northern Yellowstone elk herd averaged 6 percent of the herd per year up to 1988 (Singer and Mack 1993), but because of increased migrations after the 1988 fires, the annual elk harvest has nearly doubled since then. Because of the recent increased harvest in the northern herd, Boyce (1991) felt that the term *natural regulation* was no longer accurate, and more recently the terms *minimal management* of ungulates within the park (Yellowstone National Park 1996) and *natural process management* (Boyce 1991) have been used.

BISON

Three more or less distinct bison populations inhabit YNP—the Mary Mountain, Pelican, and Northern Range herds. A fourth bison herd occupies GTNP, the National Elk Refuge, and adjoining lands in Jackson Hole. YNP bison are progeny of native YNP and Montana populations that survived the large-scale slaughters in the western United States. A small group of native bison survived in Pelican Valley, and in 1902 this population was augmented with twenty-one animals from the National Bison Range, Moisee, Montana. Like several other ungulate species, bison increased during the mild winters of the 1980s and increased their migrations from the park (Meagher 1989a, b). The increased bison migrations have resulted in intense management outside of YNP, where bison are culled to prevent any transmission of brucellosis to domestic cattle.

The Jackson bison herd, originally from Theodore Roosevelt National Park and YNP herds, developed from animals that escaped from an enclosure in GTNP. Since their accidental release, they have established traditional use areas and migration routes. The Jackson herd grew slowly until 1975, when bison began to winter on the National Elk Refuge and use supplemental winter feed intended for elk. The Jackson bison population was estimated at 380 animals in the fall of 1994, and the intention of the various agencies is to limit the population at four hundred animals through sport hunting.

PRONGHORNS

Only the northern Yellowstone pronghorn herd inhabits the GYE on a year-round basis. Pronghorns spend the summer in four other locales in the study area—Jackson Hole; Cody, Wyoming; Henry's Lake area, Idaho; and the Duck-Cougar Creeks area on the western edge of YNP.

The northern Yellowstone pronghorn population ranged from five hundred to eight hundred animals between 1930 and 1947 (Houston 1982). With the aim of protecting big sagebrush on the boundary line area of the northern Yellowstone winter range, pronghorns were artificially reduced between 1947 and 1967. These efforts, combined with several severe winters, reduced pronghorn numbers to fewer than two hundred animals by 1968, and the herd remained under two hundred animals until 1980 (Houston 1982, Scott and Geisser 1996). Several hypotheses have been proposed for the lack of herd growth during this period, including severe competition with elk over forage resources (Chase 1986, Wagner et al. 1995), although elk and pronghorns have minor niche or diet overlaps (Singer and Norland 1994). Coyote predation on fawns also appears substantial on this small, isolated, mostly nonmigratory population (O'Gara 1968, Berger 1991, Scott and Geisser 1996). O'Gara reported a high rate of live fetuses in does ($\bar{X} = 1.96$), yet twelve weeks after fawning there were only thirty-three fawns per one hundred does. Authors attributed the losses to coyotes and, to a lesser extent, to golden eagles. D. Scott (personal correspondence) verified these speculations for 1988–93, when more than 90 percent of marked fawns succumbed to predation before their first fall, nearly all to coyotes. Yellowstone's pronghorns, one of the few western pronghorn populations to live year-round on a small range, might be vulnerable to coyote predation.

Apparently distinct nonmigratory and migratory segments of the northern Yellowstone pronghorn herd exist. About one hundred to two hundred animals migrate from the winter range on the boundary line area to higher-elevation summer range in Antelope Creek, Specimen Ridge, Lamar Valley, Gardners

Hole, and Junction Butte. The nonmigratory segment of the herd spends the summer on the winter range in the Mammoth and boundary line area of YNP from Reese Creek to Lava Creek.

The northern Yellowstone pronghorn antelope population increased dramatically during the 1980s to between six hundred and eight hundred animals, probably as a consequence of milder winters (Singer 1991a). Also, as the herd increased, relatively more animals spent the winter, the fawning season, and the August–September rut season on the irrigated alfalfa and crop fields of the Royal Teton Ranch. This increased use of a rich food source might also have contributed to the herd's growth (D. Scott personal correspondence). Complaints from the landowner, the Church Universal and Triumphant, resulted in a damage hunt on pronghorns from 1987, with a permit limit of twenty-five animals; these permits were reduced to five in 1996.

MULE DEER

An estimated 15,500 mule deer occur in eight discrete herds in the study area. Most of these deer migrate during the summer months into YNP (Hurley et al. 1989, Kuck et al. 1989, Lockman et al. 1989, Singer 1991b, P. Gogan personal correspondence), but few radiotelemetry studies have been conducted, and no accurate estimates can be made of the number of deer in YNP in summer. Mule deer tolerate less snow depth than elk, moose, and bison and are forced to migrate from YNP to lower elevations in winter.

Seven of the herds increased dramatically from 1970 to 1988. The average λ value for these herds was 1.16 (any λ > 1.0 indicates an increasing trend). Only the Gallatin herd did not increase during this period. Aerial helicopter counts for the northern Yellowstone herd more than doubled during this period, but this estimate of an increase is based on only one helicopter count in 1979 (Singer 1991a). Recent counts also suggest a generally increasing trend (Yellowstone National Park 1997). Most of the northern Yellowstone deer herd resides outside of YNP during the hunting and winter seasons, where it is available to hunters. Thus the northern Yellowstone deer herd is heavily harvested and should not be considered naturally regulated. Annual hunter harvests averaged 22 percent of the estimated population size (Singer and Mack 1993). We conclude that the mild winters of the 1980s probably explain the recent increases in the mule deer populations in the GYE.

MOOSE

Moose were apparently rare in the system when first visited by Euro-Americans, but moose had colonized the GYE by the 1870s (Houston 1968, Walcheck 1976). Whether this was a temporary absence is unknown, but moose were uncommonly slow in colonizing areas of western North America following glacial recession (Houston 1982). Moose have not turned up in archaeological or faunal sites from the GYE (Lahren 1976, Houston 1982).

Fourteen discrete herds of moose inhabit the study area. Only six of these winter totally or partially within the boundary of YNP (Ritchie 1978, Trent et al. 1984, Alt and Foss 1987, Hurley et al. 1989, Lockman et al. 1989, Tyers 1996). Moose prefer conifer forests in the study area (Tyers 1996), and, as a result, moose are among the most difficult ungulates to count. A visibility correction for aerial helicopter surveys has been worked out only for the Jackson moose herd (B. Smith personal correspondence). Population reconstruction for the northern Yellowstone moose herd suggested that fixed-wing (Super Cub) surveys underestimated moose populations by one-third to one-half (Singer 1991a, Mack and Singer 1993a). Conservative, largely uncorrected estimates suggest that at least 5,500 moose occur in the study area. Moose are able to cope with deeper snows than are the smaller ungulates, and they are relatively more abundant in the central plateau and other interior YNP and GTNP winter ranges. Despite their relative rarity in the ecosystem (table 8.2), moose are a significant source of potential prey to any predator inhabiting the interior of YNP during winter.

Most of the moose herds were either stable or increased between 1970 and 1988, but they did not show the large-scale increases that elk, mule deer, and bison did over the two decades up to 1988. Six moose herds for which population trend information was available increased an average of 24 percent through 1988. But moose seen on elk surveys declined 47 percent on Yellowstone's northern winter range. Moose seen on fixed-wing surveys for elk conducted over the same areas averaged 32 ± 16 ($\bar{X} \pm SD$) in the 1960s, but only 17 $\pm$ 9 moose in the 1980s (Barmore 1980, Singer 1991a). Moose are now found mostly at the upper fringes of the elk winter range (Singer and Norland 1994, Singer et al. 1994), including the Cooke City area, upper Slough Creek near Frenchy's Meadow, and upper Pebble and upper Hellroaring Creeks.

BIGHORN SHEEP

Approximately four thousand bighorn sheep occur in thirteen herds in the study area. Two herds increased from 1970 to 1988, while the remaining eleven were either stable or declining.

THE LESS COMMON UNGULATES: WHITE-TAILED DEER
AND MOUNTAIN GOATS

White-tailed deer are uncommon in the GYE. A few hundred animals occur in the lower elevations of the Yellowstone River and Gallatin River drainages.

White-tails prefer extensive riparian, woody-plant communities and shallow snow depths, both of which are not common in the study area. White-tails are quite common, however, immediately outside the study area, where snow depths are less and their preferred habitats are more prevalent, such as the lower Gallatin and lower Yellowstone River valleys.

Mountain goats were native in the local region but not in the GYE (Laundré 1990). Introductions have resulted in several populations that currently number about 800–830 animals in the GYE (Peck 1972, Hayden 1984, Laundré 1990, Singer and Mack 1993).

GENERAL TRENDS

Large increases occurred in nearly all the deer and elk herds in the GYE in recent decades. In particular, the release of the ungulate guild from artificial controls on the northern winter range provided a classic quasi-experiment (figure 8.2). Total ungulate numbers tripled from 1968 to 1988, and—fortuitously—the numbers, population growth rate, diets, habitat use, and niche and diet overlaps among five species were studied both in the 1960s (Barmore 1980) and again in the 1980s (Singer and Norland 1994). Unfortunately, moose and beaver were not studied in as much detail as the other large herbivores, although general information is available on their numbers and distributions (Warren 1926, Jonas 1955, Barmore 1980, Singer 1991a, Consolo and Hanson 1993, Tyers 1996, Consolo-Murphy and Tatum 1995). The overwhelmingly dominant ungulate, elk, increased 2.8 times during this period. Because elk are diet and habitat generalists, it was predicted that they would negatively influence the other ungulates (Pengelley 1963, Chase 1986, Kay 1990, Kay and Wagner 1994). From the comparisons before and after the increases, the following tentative conclusions can be made:

1. Moose declined on the main elk winter range during the period of elk increases (Singer 1991a, Tyers 1996). Currently, moose are abundant in winter only in areas with deep snows that inhibit the movements of elk (Tyers 1996). Tyers (1996) concluded that the current moose distribution suggests partial competitive exclusion by high densities of elk.

2. Beaver present a more complex situation. Beaver declined dramatically in the early decades of this century (Warren 1926, Jonas 1955), and beaver are mostly found in YNP in higher elevations, where elk densities are lower (Consolo and Hanson 1993). The current patchy beaver distribution prompted Kay (1990) and Kay and Wagner (1994) to conclude that elk have competitively excluded beaver. But Schullery and Whittlesey (1996) counter that beaver abundance in 1900–1920 might have been artificially high as a consequence of release from an intense period of trapping in the park during the late 1800s. The

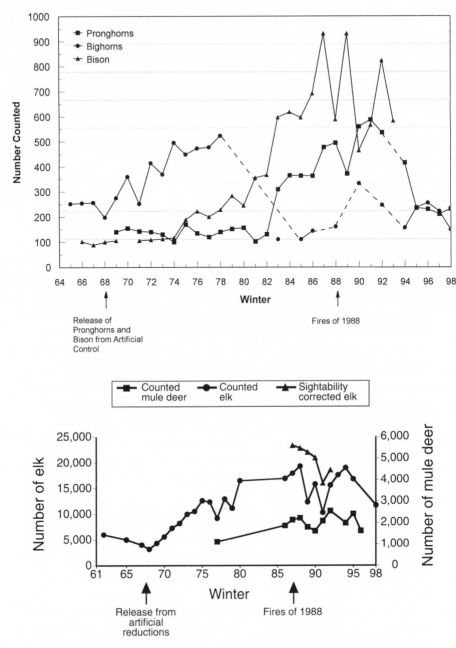

8.2. Increases in five species of ungulates in northern Yellowstone National Park following cessation of human controls in 1968.

1900–1920 period might not be a valid benchmark by which to compare later trends. Beaver on the Northern Range were more closely associated with aspen, not willow, for forage and dam construction, and the aspen decline has been attributed to multiple causes (Romme et al. 1995). The aspen decline might have caused the beaver decline, and that in turn might have caused the willow decline. Because factors other than elk abundance varied during the period that elk numbers increased, sorting out causes is difficult, if not impossible. For example, the climate of the area was more arid, and there were fewer fires concurrent with the elk increases. Local climates are cooler and wetter at the high elevations where beaver are currently found; higher precipitation supports more vigorous willow communities and more beaver and results in deep snows that inhibit elk. Our best conclusion at present is that the beaver declines did not coincide with but preceded the natural regulation management of ungulates in YNP, and thus events that preceded the new management model likely explain the beaver decline. Increasing elk densities in this century may be implicated in the beaver decline, but a multicausal explanation involving not only more elk but less woody riparian browse, fewer large fires, a more arid climate, and fewer wolves is more likely (Romme et al. 1995, Singer and Cates 1995).

3. All of the remaining five species of ungulates in the guild—elk, bison, pronghorn antelope, mule deer, and bighorn sheep—increased rapidly and simultaneously with the elk increase. The parallel increases resulted from cessation of artificial controls for elk, bison, and pronghorns and from milder winters (figure 8.2). We conclude that interspecific competition was not sufficient to stop the population growth of any species, but we cannot conclude unequivocally that the growth rates of the other ungulates were not slowed by the elk increase (Singer and Norland 1995). In particular, growth rates of bison, pronghorns, and bighorns were very rapid and suggest little slowing. The elk increase ($\lambda = 1.02$), however, was considerably less than the maximum observed ($\lambda = 1.20$) or maximum potential ($\lambda = 1.24$; Eberhardt et al. 1996), and intraspecific competition (for example, within the elk population) is suggested. As described earlier, the increase in mule deer is more tenuous, for early estimates are based on a single helicopter count. Changes in niche overlap between species pairs were unexpectedly minor during the population releases and the subsequent increases in the ungulate guild (figure 8.3). Only bison and bighorns increased their niche overlap with other species. The bison changes were attributed mostly to intraspecific competition and increased access to habitat and forage resources resulting from the sevenfold increase in numbers and the threefold increase in range occupied by bison. The changes in niche

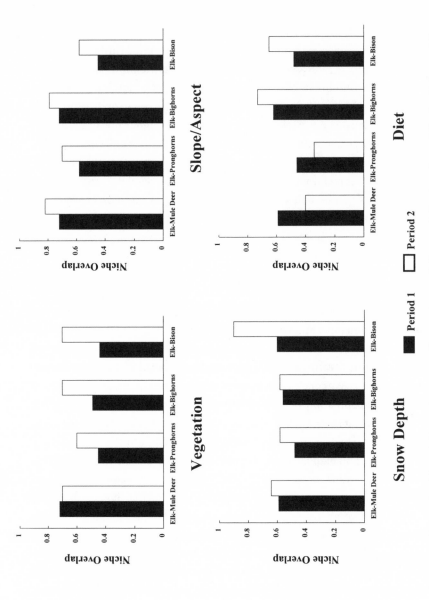

8.3. Changes in niche and diet overlaps among five ungulates on Yellowstone's Northern Range following their population release from artificial controls.

use by bighorn sheep might have been a consequence of increased competition with elk (Keating 1985, Keating et al. 1985, Houston 1982), but if so, the rate of population increase was still not significantly reduced following the elk increases (Singer and Norland 1994).

Effects of the 1988 Fires on Ungulates

COVER

YNP and adjacent areas were subjected to burning from several large fires in the late summer and fall of 1988, during unprecedented fire conditions. The fires followed on the heels of the most severe summer drought since the 1930s and a series of storm fronts that brought lightning and hot winds but no rain. The fires of 1988 were considered a several-century event for the area and constituted the largest single media event in the history of the National Park Service (Schullery 1989). Approximately one-half of YNP was included in fire perimeters, although only 20 percent of the surface area of the park actually burned. About 32 percent of the winter range within YNP and about 30 percent of the summer range for the northern Yellowstone elk herd burned. On the northern winter range, 13,700 hectares of grasslands and shrublands burned, along with 14,600 hectares of conifer forests. On the summer range for the northern Yellowstone elk herd, another 12,400 hectares of grasslands, meadows, and shrublands burned, as did 150,000 hectares of conifer forests. Cover for all ungulate species was dramatically affected. Vast areas of tall big sagebrush, aspen, and conifer forest burned, although much unburned conifer forest remained, and ungulates made some use of burned conifer stands.

SURVIVORSHIP

Direct mortality of ungulates during the fires was relatively rare. Our carcass surveys and monitoring of radio-collared elk during and after the fires suggest that only about 1 percent of the elk present in YNP at the time died as a result of the burning (Singer et al. 1989, 1997). Indirect mortality from the drought, fires, and the severe winter following the fires was much more substantial. The northern Yellowstone pronghorn and mule deer populations both declined about 21 percent, and the northern elk population declined about 24–37 percent by the end of that first winter after the fires (Singer et al. 1989). Hunter harvest increased the first winter as more elk migrated from the park; increased sport hunting accounted for losses totaling about 12 percent of the elk population. Winter malnutrition of elk accounted for about another 13–25 percent of the losses from the total elk population.

We observed migrating ungulates that moved long distances to unburned areas, and ungulates that occupied relatively unburned areas were more likely to survive the effects of the fires. Ungulates that were reluctant to leave their established home ranges, such as older bull elk and adult moose, and those whose ranges were extensively burned, died at high rates the first winter following the fires (Singer et al. 1989, Tyers 1996). For example, six of seven radio-collared bull elk that were seven years old or older did not leave their traditional home ranges and died that first winter (Dave Vales unpublished data). Adult bull elk and calf elk were particularly vulnerable to the immediate and secondary effects of the fires the first winter. Adult bull elk survived at only about one-half their normal rate (0.46 that winter vs. 0.82 normally), while calf survival dropped to less than one-fourth the normal rate (0.16 that winter vs. 0.65 normally). Three of fourteen radio-collared moose died the first winter following the fires, apparently of malnutrition (Tyers 1996). The moose that were reluctant to leave extensively burned home ranges were more likely to die. Radio-collared moose that survived the first postfire winter did so by making more concentrated use of unburned and lightly burned patches of mature conifer forest within their home ranges (Tyers 1996).

FORAGES

The ecological effects of the fires were greater where both fuel loads and relative fire intensities were greater. For example, effects of the fires were relatively minor in grasslands, where the fires burned rapidly, the fire heat intensity was relatively low, flame heights relatively long (about 0.8 m), and residual burning relatively low. Minor, but mostly positive and short-term (one to three years) responses to the fires were documented in grasslands. Aboveground herbaceous biomass in grasslands was only about 20 percent higher, protein was only 0–10 percent higher, and digestibility was not detectably influenced by the burning (Singer and Harter 1996). Fire influences in grasslands were relatively brief and occurred only the first and second summers following the fires. Thus, ungulates that feed mostly in grasslands—elk and bison—were generally positively influenced by the fires. But forage in burned conifer forests, where fires burned hotter and longer and where fuel loads were greater, was severely depleted for the first several years following the fires. Biomass of forage in forests declined to between one-fifth and one-fourth of preburn levels the first few growing seasons after the fires, although the quality of forage was much higher. But after five to seven years, burned forests are predicted to support severalfold more forage for ungulates than preburn levels (Norland et al. 1996). Also, protein concentrations increased

43 – 50 percent in forage in burned forests, compared with unburned forests, apparently because of the hotter fires and greater release of nutrients (Singer et al. 1996).

Our computer simulations predict that fewer elk will occur on Yellowstone's Northern Range for three years following the fires, but then elk numbers will increase approximately 20 percent over prefire conditions (figure 8.4, Coughenour and Singer 1996). The increases are predicted to persist for about fifteen to twenty-five years, but then about 2003 – 13 they will return to prefire levels (Coughenour and Singer 1996, Singer et al. 1997). Elk forage on summer ranges is predicted to take longer, eight to ten years, to recover to prefire conditions, but after that time, much larger increases in forage biomass and nutritional value are predicted to occur in burned forests. These effects on burned forests in the summer range are the result of the opening of conifer forest canopies, resultant warmer soil temperatures, and larger nutrient release from the burning of heavier fuel loads. As a result, summer range ECC for elk is predicted to increase 69 percent over prefire conditions for fifteen to twenty-five years after the fires (Coughenour and Singer 1996, Singer et al. 1997).

Researchers have concluded that moose were negatively influenced by the

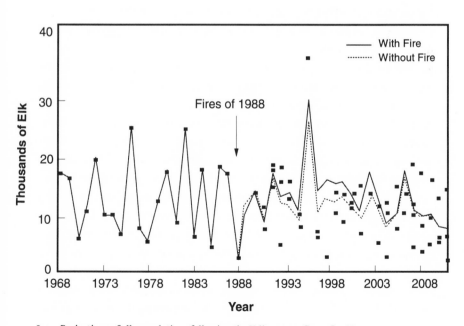

8.4.   Projections of elk populations following the Yellowstone fires of 1988.

fires of 1988, at least during the first three winters postfire (Tyers 1996). Moose counts from horseback on Yellowstone's Northern Range declined during this period (Tyers personal correspondence). Moose on the Northern Range prefer mature conifer stands with snow that is soft enough to permit moose movements but deep enough to discourage elk use of the same habitats (Tyers 1996). Some moose survived the secondary effects of fire by reducing their movements and home ranges and by concentrating feeding in the remaining unburned patches of mature forest; any subsequent increase in moose is likely to depend on succession and recovery of mature conifer forest canopies (Tyers 1996).

This prediction of a moose decline is counter to most observations of moose populations following fires. Typically, biomass production of woody browse is increased severalfold as a result of fires (Wolff 1978, MacCracken and Viereck 1990), and moose numbers increase dramatically (Franzmann and Schwartz 1986, LeResche et al. 1974). But moose on the Northern Range are found mostly in higher elevations, where the snows are deep. By mid- to late winter, moose are forced to leave open, tall-browse thickets because of deep snow and move to mature conifer forest canopies, where snows are shallow and soft enough to permit easy travel (Tyers 1996).

The postfire responses of the northern Yellowstone elk population to date have closely followed our model predictions (Coughenour and Singer 1996, Singer et al. 1996). Fewer elk were counted or estimated the first three years after the fires, but elk numbers apparently recovered to prefire levels by 1994 and 1995 and may be continuing to increase (Lemke et al. 1998).

Bison also increased dramatically after the fires. Aerial counts of bison in YNP averaged nearly one-third larger than prefire counts (Hodgson 1994, Yellowstone National Park 1996). There are two possible explanations for the increase. First, M. Meagher (cited in Hodgson 1994) reported that use of snow-mobile-packed interior park roads during winter allow bison greater access to feeding areas. She estimated that this road use by bison resulted in an increase of roughly one thousand bison in the population (Hodgson 1994). Second, the bison increase might have been a consequence of increased forage availability and quality as a result of the 1988 fires. Bison consume mostly grasses and sedges from grassland habitats, as do elk, and we predict that bison will experience the same increases predicted for elk unless significant hunting and removal occurs (Coughenour and Singer 1996, Singer et al. 1997).

Pronghorn antelope on the northern Yellowstone winter range declined dramatically after 1988, although the fires per se have not been implicated in their decline. This population declined from five hundred or six hundred to about four hundred animals the first winter after the fires, but unlike the elk and bison, pronghorns have continued to decline and in 1996 were reported to number fewer than 250 animals (Scott and Geisser 1996, D. Scott personal correspondence). All of the winter range for this herd remained unburned, and the increase in forbs on burned summer range (Singer and Harter 1996) should have benefited pronghorns. The pronghorn decline may be unrelated to the fires. First, winters have been more severe since 1988. Second, the harvest on Royal Teton Ranch has resulted in near abandonment of those lands by pronghorns in recent years (D. Scott personal correspondence). This has resulted in a loss of access to tall sagebrush winter range and irrigated alfalfa fields on the ranch. A third factor is the decline in big sagebrush, which has continued to occur on the pronghorn winter range within YNP (Singer and Renkin 1995). Fourth, coyote predation may also have contributed to the recent pronghorn decline.

During their population increase in the late 1980s, pronghorns were observed reexploring historic ranges. They were once again seen during summer in the meadows along Antelope Creek, in Hayden Valley, and along the Firehole River.

Live Prey and Carrion Biomass Available to Carnivores

Ungulates in the major portions of the GYE included in our study numbered at least 80,900 individuals of eight species before the fires of 1988 and before wolf restoration to the area. Nearly half of these ungulates spend the winters near large towns and concentrations of human settlement, where they will be relatively less available to several of the more secretive carnivores such as wolves. Yellowstone's northern winter range supported 24,900 counted ungulates, on the average, during winters in the 1980s and before the large fires. Thus approximately 4,400,000 kilograms of ungulates were available to carnivores (table 8.2). Based on visibility corrections and population modeling, we estimate that aerial counts of elk need to be corrected for animals not observed by 0.23, moose counts by 1.00, mule deer counts by 0.40, and bighorn sheep counts by 0.90 (Singer 1991a, Mack and Singer 1993b, Singer and Garton 1994, Bodie et al. 1995). Based on these corrections, we conclude that ungulate biomass available to carnivores in the summer exceeded twenty-three million kilograms within all of YNP and in the winter exceeded seven million kilograms on all YNP winter ranges. The largest concentration of potential prey during winter, about 5,500 kilograms of ungulates (corrected estimate), is available on the northern Yellowstone winter range.

Ungulate carrion can be very important as a localized and short-term food source for coyotes, bears, ravens, golden eagles, bald eagles, and black-billed magpies (Chapters 3 and 6). As many as seven species of passerine birds were observed feeding on insect larvae that use the carcasses (Houston 1978). Ungulate carrion is uniquely available on two winter ranges within YNP, the northern Yellowstone and the Madison-Firehole-Gibbon range. Mortality of elk and bison is attributed to density-dependent and weather-related losses. Most of these animals die of malnutrition (DelGuidice et al. 1991, 1994); rumens of dead elk contain food, but 88 percent of their marrows are gelatinous, indicating depletion of energy reserves (Houston 1978).

Available carrion in YNP in some years exceeds the consumption capability of the scavenger guild (Houston 1978, Blanchard and Knight 1990). But carrion availability varies greatly, and carrion is more abundant in severe winters and when ungulate densities are high. The largest documented source of carrion occurred following the dramatic events of 1988–89, including the drought, large fires, and severe winter. We estimated that 3,021 to 5,757 elk died on the northern Yellowstone winter range (Singer et al. 1989). Several large carcass piles were found during the fall of 1988 right after the fires (Singer et al. 1989). Spectacular concentrations of scavengers fed on the largest carcass piles, including more than six hundred ravens, thirty bald eagles, and six to twelve bears at a single carcass pile (Blanchard and Knight 1990, Singer et al. 1989). By December 1 more than 95 percent of the carcasses were consumed (Singer et al. 1989). Fall use of ungulates was 29 percent of the diet of grizzly bears in 1988, compared with 8 percent during a normal fall (Blanchard and Knight 1990).

During the typical April–May period, ungulates composed 40–60 percent of the diet of grizzlies in YNP (Knight et al. 1984). Spring availability of carrion can significantly influence production, condition, and survival of adult female grizzlies in the GYE (see Chapter 3).

Winterkill of ungulates is considerably less common in the more heavily hunted herds in the GYE, but less quantitative information has been gathered in those ranges. The Jackson elk herd, which is both heavily hunted and artificially fed, is subject to only about 1.4 percent winter mortality from malnutrition for those elk that use the National Elk Refuge (B. Smith unpublished data). Variance between winters in mortality from malnutrition is low. Mortality never exceeded 4.9 percent of the elk using the refuge even during the most severe winters (B. Smith unpublished data).

ELK

Predation is the second leading cause of mortality in the first year of life of an elk calf on Yellowstone's northern winter range (22.9 percent of marked calves), followed by winter malnutrition (12.6 percent of marked calves; Singer et al. 1997). Nearly all of the predation occurs during the first twenty-eight days of life. Bears, especially grizzlies, account for most of the predation (11.8 percent of marked calves), followed by coyotes (8.7 percent of marked calves). No calves are killed by bears after twenty-eight days of life. Coyotes occasionally kill calves or adult elk in poor condition during winter (Murie 1940, 1951, Geese and Grothe 1995), and mountain lions also kill elk during winter. Because mountain lions are so uncommon, population effects from mountain lion predation on an elk-population basis are minor (Chapter 4). We estimate that about sixty grizzlies present on the calving areas killed 1,688 elk calves, about four hundred coyotes killed 1,352 calves, 100–150 black bears killed 152 calves, and about seventeen mountain lions killed 120 calves out of a typical cohort of eight thousand calves born on the northern range.

Predators kill more late-born calves and calves with lower birth weights in YNP (Singer et al. 1997). This corroborates findings for red deer (the same species as elk, *Cervus elaphus*) on the Island of Rhum, where heavier calves are born to cows in better condition and these calves survive at higher rates (Clutton-Brock et al. 1982). A potential for compensation, or substitution, of mortality is suggested from this information. In other words, predators might kill lighter elk calves that otherwise would be more likely to die later from other sources of mortality, such as winter malnutrition.

Comparisons between YNP and Jackson Hole elk populations verified that, under natural regulation management in YNP, mortality of elk during the first year of life was a significant limiting, and potentially regulatory, factor on the population. The first-year mortality of elk calves in the Jackson elk herd (15.2 percent of marked calves) was only about a third of the total first-year mortality of calves in the northern Yellowstone herd (51.2 percent of marked calves; Singer et al. 1996, Smith and Anderson 1996). Black bear predation is the leading cause of elk calf mortality in the Jackson elk herd (Smith and Anderson 1996).

We classify grizzly bears and wolves (once they are fully recovered) as keystone predators on elk in the GYE—species whose impact on the community is far out of proportion to their abundance (Paine 1966, Power and Mills 1995). Dominant predators, such as coyotes, are those that are important but also abundant—that is, their large impact is in proportion to their high abundance.

Grizzlies are successful in pursuits of young elk calves up to twenty-eight days of age (Singer et al. 1996). Grizzlies are successful in 36 percent of pursuits, stalks, or searches for hidden elk calves (French and French 1990). They are often able to chase and kill more than one calf during a single episode. Multiple kills of elk calves by grizzlies are documented in 3 percent of predation episodes by French and French (1990) and in 7 percent of episodes by Singer et al. (1996); in one instance, a grizzly caught five elk calves in a fifteen-minute interval (French and French 1990). We estimate that all predators kill three thousand elk calves in a typical year on the Northern Range and that grizzlies kill 57 percent of that total.

Predators kill very few adult elk in the GYE study area. Predators killed only one of forty-seven radio-collared adult elk over a four-year period from the northern Yellowstone herd. The predator in this single case was a mountain lion. As a consequence, survival rates of adult elk in the northern herd were high, 0.96 for adult cows and 0.82 for adult bulls (Singer et al. 1996).

Similarly, predation on adult elk in Jackson Hole was relatively uncommon (Smith and Robbins 1994), which may change as grizzlies become more common and when wolves colonize the valley. Annual survival of adult elk in the Jackson Hole herd was 0.79 for adult cows and 0.63 for adult bulls (Smith and Robbins 1994). Eighty percent of all mortality of adult elk in the Jackson elk herd was from hunting.

PRONGHORNS

Predation, especially by coyotes, is apparently a significant limiting factor to increases in the northern Yellowstone pronghorn herd. In past decades, biologists speculated that coyote predation was very significant in the herd (O'Gara 1968, Barmore 1980, Houston 1982), but field evidence was limited. For example, physical condition of adult does is excellent and fetal fawn ratios are very high (two hundred fetuses per one hundred does; O'Gara 1968). In all probability, nearly two fawns are born each year to nearly every doe. But survival of fawns from birth to early winter was estimated to be only 17 percent, with most of the mortality probably from coyote predation (Houston 1982). A field study of fawn mortality from 1988 to 1991 verified that coyote predation on fawns was high. Fewer than 10 percent of fawns marked as neonates survived to their first October (D. Scott unpublished data).

No detailed study of survivorship or predation has yet been completed on other species, although an exhaustive study is in progress on mule deer of the Northern Range (P. Gogan personal correspondence).

Predicted Effects of Wolf Restoration on Ungulates

NUMBERS AND DISTRIBUTION

Wolves were restored to the northern Yellowstone range in February 1995, and their complete recovery into the area is expected within the next few years—barring unforeseen events, including actions by the court. Two widely disparate scenarios have been proposed for the number of wolves that will ultimately occupy the area:

Moderate wolf density. We predicted that seventy-five to one hundred wolves would ultimately occupy the northern winter range and 110–50 would occupy both the Northern Range and interior YNP winter ranges combined (Singer 1991b). Subsequent modelers either adopted these same estimates or reached similar conclusions (Garton et al. 1990, Boyce 1993, Mack and Singer 1993a). Biggs (1995) more rigorously analyzed snow depths and ungulate distributions and similarly concluded that seventy to ninety wolves would occupy the Northern Range. These wolf densities would result in the highest biomass of ungulates available per wolf ever reported and one of the highest densities of wolves ever reported (Singer 1991b). Our predictions are based on the assumption that wolf packs would adopt long, linear home ranges containing both summer and winter elk ranges and that upper limits of wolf densities would be set by the maximum territorial spacing observed in nonmigratory wolves elsewhere in North America (Singer 1991a). Conflicts with humans and the resultant high wolf mortality along the long northern boundary of the park (from road kills, poaching, and, ultimately, legal harvest) will likely limit the number of wolves that will occupy the long north, east, and west park boundary areas of Yellowstone National Park. Observations of wolf packs along the boundaries of Denali National Park, Alaska, and Riding Mountain National Park, Manitoba, confirm that wolves are reduced in boundary areas by human-caused mortality (Haber 1977, Carbyn 1983). YNP staff have verified this prediction to date. All seven dispersing adult wolves were either killed when outside of the park by humans or vehicles or were cap-tured and returned to the park. A pack that twice left the park for a few days but returned quickly to the park suffered no mortality.

High wolf density. Should the recovered wolves become migratory, this could result in a closer spacing of packs, and in that case we predict an eventual population of 150 to 200 wolves for the Northern Range, as Lime et al. (1993) and Messier et al. (1995) predicted. Migratory wolf packs in high arctic ecosystems move with caribou herds and establish temporary territories within howling distance of other packs. These packs may be found seasonally at very high densities of one pack every few square kilometers. The scenario of high wolf density is based on both the assumptions of migratory behavior and tighter spacing of wolf packs. As of early 1999, wolves had grown rapidly, as we predicted, and approached 75 animals on the Northern Range, but interpack strife and killings increased in 1998, and it is too early to determine whether the moderate or high density will prevail in the long term.

UNGULATES

Ungulate distribution and behavior. We predict that the distribution and behavior of adult female ungulates with young at heel will be most influenced by the introduction of wolves. Wolf avoidance behavior is documented in caribou and moose cows, including calving in such wolf-free areas as islands, mountaintops, or areas near human habitation (Bergerud 1980, Edwards 1983, Stephens and Peterson 1984).

Although adult female ungulates may further exhibit avoidance behaviors following wolf restoration, we do not predict large differences from the current selection of birthing sites and postnatal behavior. Yellowstone's elk, mule deer, moose, and pronghorns are already subjected to significant predation by coyotes and bears and probably already exhibit antipredatory behavior in their selection of birthing and neonatal hiding sites. For example, elk cows give birth near the administration buildings and government housing areas of Mammoth Hot Springs, presumably to avoid bears and coyotes. Similarly, a major fawning area for pronghorns is a field adjacent to the main street of Gardiner, Montana, presumably because fewer coyotes venture there. Bison have been little influenced by predators (Meagher 1973), and as a result their behavior may eventually change the most in response to the restoration of wolves.

Adult ungulates may alter their habitat use slightly in response to wolf restoration. Elk are predicted to increase use of forest cover after wolf restoration. Long-legged ungulates such as elk and moose seek such landscape features as heavy timber downfall, cliffs, and open water to escape pursuing wolves (Peterson and Allen 1974, Hatter 1982, Gunson 1986). During snow-free periods, bison also escape wolves by seeking tree cover (Carbyn and Trottier 1987, 1988). We predict that bighorn sheep will abandon some of the gentler slopes and stay closer to steep escape terrain following wolf recovery. J. Stelfox (personal correspondence) observed that bighorn sheep abandoned several gentle grass slopes near Jasper, Alberta, immediately following the return of wolves to the area after a control program.

Elk and bison that inhabit the thermal areas on the central plateau of YNP during winter might be particularly vulnerable to wolves. The thermal features are essential to ungulate occupation of the area in winter (Craighead et al. 1972, Meagher 1973, Singer 1991b). Wolves might chase ungulates into the deeper snows at the edges of the thermal areas and catch them there. We suspect that Yellowstone's ungulates might escape from wolves in the larger thermal areas but remain vulnerable in the smaller, peripheral thermal areas, where they can be chased into deep snow. Population consequences should not be large, however, because only 4 percent of the park's elk and 10 percent of its bison inhabit small, isolated thermal areas.

We developed a deterministic elk population model for the northern Yellowstone elk herd following Bartholow (1988). We incorporated the following data for 1975–92: (1) the number of age classes, (2) initial population proportions in various age classes (Houston 1982, Coughenour and Singer 1996, Singer et al. 1996), (3) beginning population size (Houston 1982), (4) pre- and postseason mortality rates gathered by Montana Fish, Wildlife, and Parks for 1976–92 (Alt and Foss 1987, Montana Fish, Wildlife, and Parks unpublished data), (5) winter severity indices (Farnes 1992), (6) age and sex of the harvest, (7) estimates of wounding losses (Swenson 1985), (8) reproductive rates by age class of elk, and (9) sex ratios at birth (Houston 1982).

Detailed information on pregnancy rates, age structure, and hunter harvest is obtained from the periods of artificial reductions of the northern Yellowstone elk herd, 1932–67 (Houston 1982). More recently, this information has came from a mandatory check station for harvested elk operated by the Montana Department of Fish, Wildlife, and Parks from 1976 to the present. Aerial counts of elk were obtained following procedures outlined in Houston (1982) and Singer (1991b) and Singer et al. (1989). Estimates of overwinter mortality of elk calves were obtained by herd classifications and population reconstruction described in Mack and Singer (1993b) and Coughenour and Singer (1996). We modeled an annual average harvest of elk by hunters of 404 bulls, 878 cows, and 226 calves, based on the average annual hunter harvest from 1984 through 1990. We projected calf ratios of thirty to thirty-three calves per one hundred cows in 1992–95 under the assumption that average calf ratios would increase following the density reductions and improved range conditions following the fires of 1988. We also reduced adult elk mortality 40 percent for the years 1992–95 for the same reasons.

We modeled a hypothetical initial transplant of six pairs of wolves that hypothetically occurred in 1983. This hypothetical population, increasing at a rate of $\lambda = 1.8$ in our model following the suggestion of Boyce (1993), grew to either seventy-eight wolves or one hundred wolves for just the northern winter range. We projected that wolves would most likely kill twelve ungulates per year based on an average kill rate in elk-dominated systems of fourteen ungulates per year (Carbyn 1983, Fuller 1989, J. Gunson unpublished data). We reduced the fourteen ungulates per year to twelve because in all those studies the kill rate was determined in midwinter, when kill rates are higher and wolves consume more meat. We recognize that kill rates can vary with prey availability and prey vulnerability, and we suggest the reader consult Boyce (1993, Chapter 10 this volume), Vales and Peek (1993), and Mack and Singer (1993a) for more information on the influence of variable kill rates for wolves on model predictions.

Table 8.4

Predicted number of ungulates preyed upon by wolves in Yellowstone's Northern Range following wolf recovery

	Relative abundance	Relative vulnerability	Ungulates killed	
Ungulate species			78 wolves	100 wolves
	Low kill rate (9 ungulates/wolf/year)			
Elk	100	1.0	534	685
Deer	21	1.3	128	164
Moose	3	0.7	11	14
Bighorn/pronghorn	2	0.3		
Bison	2.6	0.7		
	High kill rate (12 ungulates/wolf/year)			
Elk			712	913
Deer			171	219
Moose			15	19
Other ungulates			38	49

Note: Predicted number of ungulates killed from each species is corrected for relative ability times relative vulnerability as described in the text in the section on modeling.

We predicted the number of ungulates that wolves would kill based on relative vulnerability and relative availability of ungulates. Vulnerability was estimated from prey selection in multiungulate systems and from chest heights and weight loadings of ungulates (Telfer and Kelsall 1984). Elk vulnerability was set at 1.0. We estimated that mule deer would be more vulnerable than elk at 1.3 (Cowan 1947, Carbyn 1974), while other ungulates would be less vulnerable than elk. Moose vulnerability was set at 0.7 (Telfer and Kelsall 1984, Carbyn et al. 1987), bison at 0.7 (Telfer and Kelsall 1984, Carbyn and Trottier 1987, 1988), and bighorn sheep at 0.3 (Cowan 1947, Carbyn 1974). We estimated elk calves would be three times more vulnerable than cows, and bulls 1.3 times more vulnerable than cows, based on the work of Mech (1966), Pimlott et al. (1969), Carbyn (1983), Franzmann et al. (1980), and Ballard et al. (1987). Based on these assumptions, we estimated in our model that seventy-eight wolves would kill twelve ungulates per wolf per year, or 936 total ungulates per year, of which 712 ungulates will be elk (table 8.4). We predict a larger population of one hundred wolves using the same kill rate per wolf will kill twelve hundred ungulates per year.

We tested the accuracy of the elk model after it was first developed in 1990 by predicting the elk population size in 1991 and 1992 (see Mack and Singer 1993b). The model estimate approximated the elk population that was later observed those years (Singer and Mack 1993). Observed calf ratios were even higher postfire than we had first predicted in 1990 (forty-eight calves per one hundred cows in 1991, forty-nine calves per one hundred cows in 1992; Mack and Singer 1993b). For the current exercise, we modified the 1990 model with this new information. We also modeled wolf predation to be compensatory with juvenile winter starvation losses of elk calves at rates of 0 percent, 50 percent, and 80 percent compensation. For each model option, we simulated elk population recovery from the severe effects of the drought, fires, and severe winter of 1988–89, subsequent to the large-scale elk decline immediately following the fires (Singer et al. 1989). We assume managers will not permit wolves and hunters to greatly reduce elk populations without reducing the elk harvest. Therefore, we reduced the antlerless harvest of elk (mostly cow harvest) by 0 percent, 14 percent, and 27 percent in order to achieve recovering elk population trends following the fires.

Kill rates are influenced by a variety of factors, including the condition and vulnerability of ungulates. If the kill rate in the model is reduced to nine ungulates per wolf per year, a hypothetical population of seventy-eight wolves is predicted to reduce the northern Yellowstone elk herd by 5 percent compared with the model situation if no wolves were present. Modeled elk populations are predicted to decline 23 percent when a higher kill rate of twelve ungulates per wolf per year is used, and elk are predicted to decline 41 percent when the kill rate is fifteen ungulates per wolf per year.

Wolf densities are difficult to predict. If one hundred wolves eventually occupy the northern winter range, our model predicts that the elk population will decline and not recover under any of the kill rates modeled (nine, twelve, or fifteen ungulates per wolf per year), unless the antlerless harvest of elk is reduced by at least 27 percent. Similarly, moose populations decline under all model scenarios of wolf density and kill rates. We predict moose harvests will need to be reduced by half to avoid a decline in the northern Yellowstone moose herd. Mule deer model scenarios are even more pessimistic. Antlerless harvests of mule deer are completely eliminated in the model for hypothetical populations of seventy-eight and one hundred wolves at all the kill rates, with one exception: the scenario of seventy-eight wolves and the kill rate of nine ungulates per wolf per year is predicted to sustain an antlerless harvest of deer, but only if this antlerless harvest of deer is reduced by two-thirds under current levels.

Compensation between wolf predation and winter starvation of elk calves strongly influenced our model predictions. There is no way to predict the level of compensation that will occur, but three hypothetical scenarios serve to demonstrate the large effect that the level of compensation can have on the allowable harvest of elk. With no wolves present, the modeled hypothetical elk population recovers to prefire levels (23,000 estimated elk) by 1995 (figure 8.5).

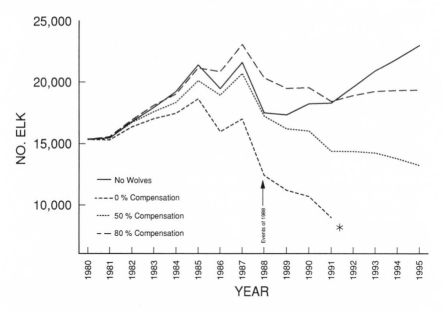

8.5. Compensation between winter malnutrition losses of elk calves and wolf predation in the presence of one hundred wolves.

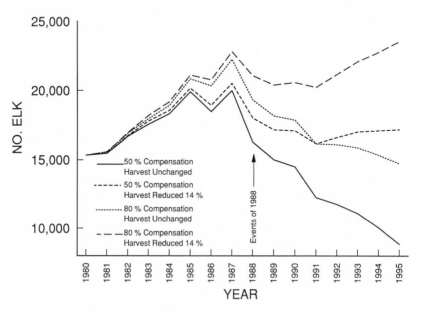

8.6. Scenarios of compensation among sources of mortality and reduced harvests of elk in the presence of one hundred wolves.

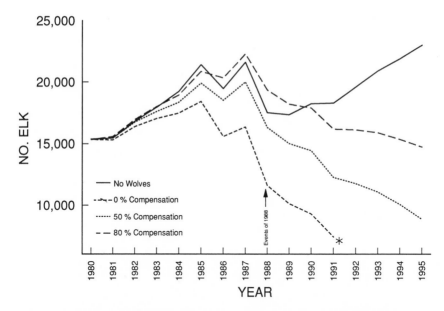

8.7. Compensation between winter malnutrition losses of elk calves and wolf predation in the presence of seventy-eight wolves.

With zero compensation, a population of one hundred wolves, and no change in the antlerless elk harvest, the elk population declines drastically by 1995 and will not recover from the fires of 1988. Even with scenarios of 50 and 80 percent compensation, the elk population still declines with wolf predation. In order for the hypothetical elk population to recover from the fires in the presence of 100 wolves on the Northern Range, the antlerless harvest of elk must be reduced 29 percent (from 878 to 629 elk) in the model (figure 8.6). The most recent counts, through 1998, suggest a declining elk population (figure 8.2), with increased hunter harvests (3,320 harvested in 1996–97) due to increased elk migrations out of the park and increased availability of elk to sport hunters outside the park (Lemke et al. 1998). Our models predict continued decreases in elk at current (moderate) wolf numbers and current high harvests of antlerless elk, and greater declines if high wolf densities occur.

For a hypothetical population of seventy-eight wolves, a kill rate of twelve ungulates per wolf per year, and zero change in antlerless harvest, the modeled elk herd still did not recover by 1995 (figure 8.7). But with compensation, we predicted recovery of the elk population following the fires. The model scenario of 80 percent compensation with seventy-eight wolves allowed for a con-

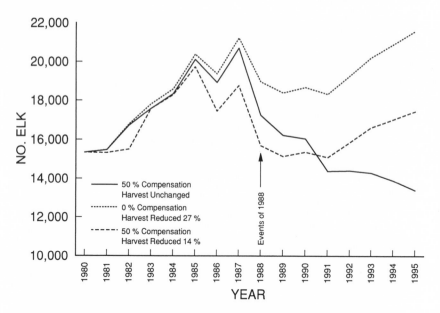

8.8. Scenarios of compensation among sources of mortality and reduced harvests of elk in the presence of seventy-eight wolves.

tinued high harvest of antlerless elk. In general, harvest reduction of antlerless elk of 14 percent provides more optimistic models for elk recovery (figure 8.8). But in the absence of compensation, a reduction of the antlerless elk harvest of 27 percent is required to recover the elk population.

CASCADING EFFECTS OF WOLF RESTORATION

Carnivore guild. We predict that coyotes will decline significantly following wolf restoration into YNP, based on observations in a wide variety of situations in North America (Seton 1929, Young and Goldman 1944, Munro 1947, Stenlund 1955, Mech 1966, Silver and Silver 1969, Mech 1970, Berg and Chesness 1978, Carbyn 1982, Chapter 6 this volume). We cannot predict how much coyotes will decline, but some insights are provided by historical observations from YNP and by studies of coyotes and wolves in Riding Mountain National Park. Schullery and Whittlesey (1996) concluded that coyotes were still common but less abundant in the 1890s when wolves were still extant in YNP. Paquet (1989) concluded that sympatric coyotes and wolves can coexist in situations like Riding Mountain National Park where the two canids can specialize on different-sized ungulates. Where both canids share a single major prey, such as white-tailed deer, competition between the two is more severe (Stenlund 1955, Silver and Silver 1969). We predict that as coyotes decline in YNP following wolf recovery, red foxes will increase (Harrison et al. 1989, Singer 1991b, Chapter 6 this volume).

Grizzlies and black bears occasionally fight and kill wolves and interact with them over carcasses. Wolves may also occasionally kill bears, but no population-level effects are predicted (Lime et al. 1993, Servheen and Knight 1993). We predict similar occasional agonistic or lethal interactions between wolves and mountain lions and between wolves and wolverines (Burkholder 1961, Murie 1963), but insufficient information is available to predict any population-level consequences.

Vegetation. We predict changes in vegetation abundance if elk and bison decrease to less than 50 percent of current populations under the scenario of high wolf density of Messier et al. (1995). Willow and aspen may increase. But large elk declines during the reductions of the 1960s did not result in any statistical increase in willow or aspen heights or any new recruitment of suppressed shrubs. Several decades of suppressed elk densities may be required before any vegetative changes occur. We predict that wolf densities will be moderate and that vegetation changes attributable to wolf restoration will be relatively minor.

Ungulate guild. If coyotes decline as a result of wolf recovery, coyotes will kill fewer elk calves. Thus wolf predation on elk might be compensatory with coyote predation. If coyotes decline following wolf restoration, we agree with Berger's (1991) prediction that pronghorns may increase on the Northern Range. We predict that wolves will not prey on pronghorns extensively because pronghorns winter near Gardiner, where wolves should not be common. The speedy pronghorns should not be particularly vulnerable to wolves in open grasslands when they migrate into wolf habitat during the summer months.

We predict that a population of large herbivores might increase following wolf restoration if one or all of the following criteria are met: (1) if the population is currently being suppressed by elk, that is, the primary prey item for wolves, (2) if any decrease in elk resulting from wolves then releases the large herbivore population, and (3) if the large herbivore is less vulnerable to wolves than elk are. For example, moose may increase following wolf restoration because they are less vulnerable than elk to wolves and may currently be suppressed by high densities of elk. Moose might reoccupy former range currently occupied by high densities of elk. Bighorn sheep and beaver may also increase because both species may currently be suppressed by elk and both are less vulnerable to wolves than elk (figure 8.9).

We predict little opportunity for any classic "predator pits" following wolf restoration, with the single exception of mule deer. Mule deer are more vul-

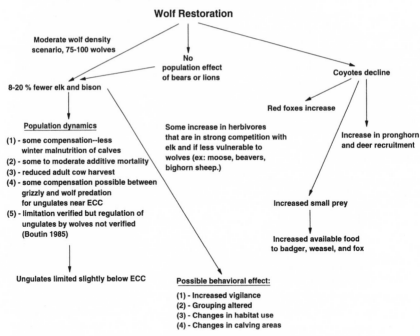

RETURN OF A KEYSTONE PREDATOR
Wolf Restoration

Moderate wolf density
scenario, 75-100 wolves

No
population effect
of bears or lions

Coyotes decline

8-20 % fewer elk and bison

Red foxes increase

Increase in pronghorn
and deer recruitment

Population dynamics

(1) - some compensation--less
winter malnutrition of calves
(2) - some to moderate additive mortality
(3) - reduced adult cow harvest
(4) - some compensation possible between
grizzly and wolf predation
for ungulates near ECC
(5) - limitation verified but regulation of
ungulates by wolves not verified
(Boutin 1985)

Some increase in herbivores
that are in strong competition with
elk and if less vulnerable to
wolves (ex: moose, beavers,
bighorn sheep.)

Increased small prey

Increased available food
to badger, weasel, and fox

Ungulates limited slightly below ECC

Possible behavioral effect:

(1) - Increased vigilance
(2) - Grouping altered
(3) - Changes in habitat use
(4) - Changes in calving areas

8.9. Possible changes to ungulates and other predators following return of a keystone
predator, the gray wolf, to the Northern Range of Yellowstone National Park.

nerable to wolf predation than elk, they are less abundant on Yellowstone's
northern winter range than elk (about nine elk per one mule deer), and wolf
densities are likely to be set primarily by the abundance of elk, not mule deer.
Thus we suggest that mule deer populations and their harvests by humans be
closely monitored following wolf restoration.

An Ecosystem Macrohypothesis, Conclusions, and Recommendations

Yellowstone National Park is a complex, multiequilibrial ecosystem. A sin-
gle-equilibrium or steady-state condition between vegetation, ungulates, and
large predators should not be expected. A good example was seen in the 1870s,
which have repeatedly been proposed as the baseline by which all future park
management and all future changes should be compared (Houston 1971, 1982,
Kay 1990, Kay and Wagner 1994, Wagner et al. 1995). But other researchers have
pointed out that that was an atypical period and thus a poor choice for a baseline. The 1870s followed a series of exceptionally large fires in the 1860s, the climate was relatively wetter and cooler, and there were more large floods, more

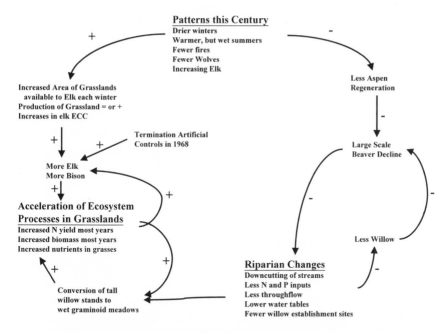

8.10. An ecosystem macrohypothesis for twentieth-century changes in elk, vegetation, climate, and wolves on Yellowstone's Northern Range.

wolves, more beaver, and fewer elk (figure 8.10, Balling et al. 1992, Meyer et al. 1992, Schullery and Whittlesey 1996). The more arid climate and lower frequency of fires this century—irrespective of the park's management policy for elk—predict significant changes in vegetation (Romme et al. 1995, Singer and Cates 1995). Similarly, the period of most intense artificial reductions of ungulates, 1948–67, when three species were reduced to one-fourth to one-fifth of their ECC, is often used as a baseline for comparisons (Kay 1990, Wagner et al. 1995), although it does not represent a suitable baseline with which to compare all later periods.

How does the restoration of a keystone predator compare quantitatively with other changes that have occurred in the ecosystem? Wolf restoration in YNP has been touted as the single biggest ecological change this century. We disagree. Although we recognize the important cascading effects that wolf restoration is likely to have on the ungulate and carnivore guilds, we propose that the changes to the climate and the fire frequency in this century have had much more profound effects on vegetation and the landscape. The fires of 1988, which altered entire landscapes and may result in 20–30 percent more elk and a large increase in bison, will probably have more significant effects than wolf restoration on ungulate populations. The influence of winter weather was larger than the fire influences on survival rates of elk (Coughenour and Singer 1996, Turner et al. 1994). The increase in usable winter range resulting from the purchase and easements of winter range outside YNP in the 1980s resulted in about twice the predicted increase in elk over any increase predicted from the fires of 1988 (Coughenour and Singer 1996). This expansion of the elk winter range is also likely a more important factor to elk populations than the restoration of wolves. Only under the high wolf density scenario of Messier et al. (1995) will wolf restoration have a larger effect on ungulates than the fires of 1988 or the expansion of the winter range.

Ungulates are the overwhelmingly abundant prey in the GYE, especially the grassland ungulates, elk and bison. Elk will remain the overwhelmingly abundant ungulate far into the future for two reasons: they will increase moderately as a result of the fires of 1988, and they will not decline disproportionately as a result of wolf restoration. Bison have increased more dramatically than any other ungulate over the past two decades. We also believe that bison will increase as a result of the fires, and because elk are already abundant in the area, bison, which are less vulnerable, will be influenced less by wolves. But bison populations are likely to be controlled by managers at or below ECC to reduce the potential of brucellosis transmission to other wild and domestic ungulates. Beaver and moose may increase in a few areas if elk decline substantially because of wolves. Few other changes in the relative abundance of the large herbivores are predicted in the study area in the near future.

Most populations of ungulates in the GYE are relatively large, their migrations are relatively undisturbed, and their populations are relatively secure. Nonetheless, a few areas of concern still exist. Some elk migrations have been abbreviated or altered (Jackson, northern Yellowstone, Gallatin). The Grand Teton bighorn sheep herd is currently isolated, precariously small, and declining. Similarly, the northern Yellowstone pronghorn population is isolated, declining, and it has lost habitat and migrations. Both populations are clearly threatened with possible extirpation in the near future.

We conclude that predators were not regulating elk populations before wolf restoration. Density dependence has a strong regulatory influence on the northern Yellowstone elk population. But predation, mostly by grizzlies and coyotes, does not appear to be density dependent (Singer et al. 1997, Boertje et al. 1989, Ballard 1992, Ballard et al. 1990). Typically, there is no feedback between bear and moose populations (Ballard 1992). This situation may change when wolves are restored to the ecosystem. Sinclair and Arcese (1995) concluded that in Africa, where the large predator fauna is intact, predation, along with density-dependent, intraspecific competition, jointly limited populations by removing animals that were in better condition than those that were starving.

We conclude that wolf predation is likely to compensate for winter malnutrition losses of elk following wolf restoration in YNP, but the precise level of this compensation cannot be predicted.

1. We recommend more intense monitoring of ungulate populations during wolf recovery. Harvest plans may need to curtail antlerless harvests in some ungulate herds. Wolf predation needs to be studied in detail. In particular, what wolf densities will be achieved? What kill rate will wolves achieve? What functional responses will occur? All of this information should be gathered to predict ungulate harvests that will not result in population declines. We suggest major updates for ungulate harvest plans every three to five years as new information accumulates.

2. More research is needed on a few declining, isolated, and threatened ungulate populations in the area.

3. Natural regulation management should continue, but should be closely monitored within YNP. The experiment is incomplete without wolf restoration (Peek 1980), and the success of the experiment needs to be evaluated with wolves in the system. Adaptive management of ungulates should occur, if necessary, only after the effects of wolves are observed.

4. The existence of density dependence in the regulation of the elk population is well established. But more work needs to be done on the strength of this density dependence and how tightly it may regulate elk and other ungulate populations. As Wolda (1989) notes, the real issue is no longer population regulation by density dependence, but the relative strength of density-dependent regulation and other disruptive forces in a population.

Literature Cited: Adams, L. G., F. J. Singer, and B. W. Dale. 1995. Caribou calf mortality in Denali National Park, Alaska. Journal of Wildlife Management 59:584–94.

Alt, K., and A. J. Foss. 1987. State wildlife survey and inventory. Project W-130-R-18, Job I-3(c), Montana Department of Fish, Wildlife, and Parks, Big Game Survey and Inventory, Helena.

Ballard, W. B. 1992. Bear predation on moose: A review. Alces Supplement 1:162–76.

Ballard, W. B., and S. D. Miller. 1990. Effects of reducing brown bear density on moose calf survival in southeastern Alaska. Alces 26:9–13.

Ballard, W. B., J. S. Whitman, and C. L. Gardner. 1987. Ecology of an exploited wolf population in south-central Alaska. Wildlife Monographs no. 98.

Balling, R. C., Jr., G. A. Meyer, and S. G. Wells. 1992. Climate change in Yellowstone National Park. Agricultural and Forest Meteorology 60:285–93.

Barmore, W. J. 1980. Population characteristics, distribution and habitat relationships of six ungulates in northern Yellowstone National Park. Final Report. Archives, Yellowstone National Park.

Bartholow, J. 1988. Pop-II system documentation. Fossil Creek Software, Fort Collins, Colorado.

Bartmann, R. M., G. C. White, and L. H. Carpenter. 1992. Compensatory mortality in a Colorado mule deer population. Wildlife Monographs 121:1–39.

Berg, W. E., and R. A. Chesness. 1978. Ecology of coyotes in northern Minnesota. Pp. 229–87 *in* M. Beckoff, ed., Coyotes: Biology, behavior, and management. Academic Press, New York.

Berger, J. 1991. Greater Yellowstone's ungulates: Myths and realities. Conservation Biology 5:353–63.

Bergerud, A. T. 1980. A review of the population dynamics of caribou and wild reindeer in North America. Pp 556–81 *in* E. Reimers, E. Gaara, and S. Skjenneberg, eds., Proceedings of the 2d International Reindeer-Caribou Symposium, Roros, Norway.

Bergerud, A. T., W. Wyett, and J. B. Snider. 1983. The role of wolf predation in limiting a moose population. Journal of Wildlife Management 47:977–88.

Biggs, T. 1995. The development and use of a Geographic Information Systems snow map for estimating equilibrium numbers of wolves in Yellowstone National Park. B.A. thesis, Princeton University, Princeton, New Jersey.

Blanchard, B. M., and R. R. Knight. 1990. Reactions of grizzly bears, *Ursus arctos horribilis,* to wildfire in Yellowstone National Park, Wyoming. Canadian Field-Naturalist 104:592–94.

Bodie, W. L., E. O. Garton, E. R. Taylor, and M. McCoy. 1995. A sightability model for bighorn sheep in canyon habitats. Journal of Wildlife Management 59:832–39.

Boertje, R. D., W. C. Gasaway, D. V. Grangaard, and D. G. Kellyhouse. 1989. Predation of moose and caribou by radio-collared grizzly bears in east-central Alaska. Canadian Journal of Zoology 66:2492–99.

Boutin, S. 1992. Predation and moose population dynamics: A critique. Journal of Wildlife Management 56:116–27.

Boyce, M. S. 1989. The Jackson elk herd: Intensive wildlife management in North America. Cambridge University Press, Cambridge.

———. 1991. Natural regulation or control? Pp. 183–208 *in* R. B. Keiter and M. S. Boyce, eds., The Greater Yellowstone Ecosystem: Redefining America's wilderness heritage. Yale University Press, New Haven.

———. 1993. Predicting the consequences of wolf recovery in Yellowstone National Park. Pp. 234–69 *in* R. S. Cook, ed., National Park Service Scientific Monograph no. 22. U.S.D.I., National Park Service, Denver.

Boyce, M. S., and E. H. Merrill. 1991. Effects of the 1988 fires on ungulates in Yellowstone National Park. Tall Timbers Fire Conference 29:121–32.

Burkholder, R. 1961. Observations concerning wolverine. Journal of Mammalogy 43:263–64.

Carbyn, L. N. 1974. Wolf predation and behavioral interactions with elk and other ungulates in an area of high prey diversity. Report. Canadian Wildlife Service, Edmonton, Alberta.

———. 1982. Coyote population fluctuations and spatial distribution in relation to wolf territories in Riding Mountain National Park, Manitoba. Canadian Field-Naturalist 96:176–83.

———. 1983. Wolf predation on elk in Riding Mountain National Park, Manitoba. Journal of Wildlife Management 47:963–76.

Carbyn, L. N., P. Paquet, and D. Melesko. 1987. Long-term ecological studies of wolves, coyotes and ungulates in Riding Mountain National Park. Mimeo. Canadian Wildlife Service, Edmonton.

Carbyn, L. N., and T. Trottier. 1987. Responses of bison on their calving grounds to predation by wolves in Wood Buffalo National Park. Canadian Journal of Zoology 65:2072–78.

———. 1988. Description of wolf attacks on bison calves in Wood Buffalo National Park. Arctic 41:297–302.

Cayot, L. J., J. Prukop, and D. R. Smith. 1979. Zootic climax vegetation and natural regulation. Wildlife Society Bulletin 7:162–69.

Chadde, S., and C. Kay. 1988. Willows and moose: A study of grazing pressure, Slough Creek exclosure, Montana, 1961–1986. Montana Forest Conservation Experiment Station Research Notes 24:1–5.

Chase, A. 1986. Playing God in Yellowstone. Atlantic Monthly Press, New York.

Clutton-Brock, T. H., F. E. Guiness, and S. D. Albon. 1982. Red deer: Behavior and biology of two sexes. University of Chicago Press, Chicago.

Cole, G. C. 1963. Range survey guide. Grand Teton National Park Natural History Association, Moose, Wyoming.

———. 1969. The elk of Grand Teton and southern Yellowstone national parks. Research Report GRTE-N-1. Moose, Wyoming.

———. 1971. An ecological rationale for the natural and artificial regulation of native ungulates in parks. Transactions of the North American Wildlife Conference 36:417–25.

Connell, J. H., and W. P. Sousa. 1983. On evidence needed to judge ecological stability or persistence. American Naturalist 121:789–824.

Connolly, G. E. 1981. Limiting factors and population regulation. Pp. 245–85 *in* O. C. Wallmo, ed., Mule deer and black-tailed deer of North America. University of Nebraska Press, Lincoln.

Consolo, S., and D. D. Hanson. 1993. Distribution of beaver in Yellowstone National Park, 1988–89. Pp. 38–48 *in* R. S. Cook, ed., Ecological issues on reintroducing wolves into Yellowstone National Park. National Park Service Scientific Monograph 93-22.

Consolo-Murphy, S., and R. B. Tatum. 1995. Distribution of beaver in Yellowstone National Park, 1994. Archives, Yellowstone National Park.

Coughenour, M. B. 1991. Biomass and nitrogen response to grazing of upland steppe on Yellowstone's northern winter range. Journal of Applied Ecology 28:71–82.

Coughenour, M. B., and F. J. Singer. 1995. Elk population processes in Yellowstone National Park under the policy of natural regulation. Ecological Applications 6:573–93.

———. 1996. Yellowstone elk population responses to fire: A comparison of landscape carrying capacity and spatial dynamic ecosystem modeling approaches. Pp. 169–80 *in* Ecological Implications of Fire in Greater Yellowstone. International Association of Wildland Fire, Fairfield, Washington.

Cowan, I. McT. 1947. The timber wolf in the Riding Mountain National Park of Canada. Canadian Journal of Research 25:139–74.

Craighead, J. J., G. Atwell, and B. W. O'Gara. 1972. Elk migrations in and near Yellowstone National Park. Wildlife Monographs no. 29.

DelGuidice, G. D., F. J. Singer, and U. S. Seal. 1991. Physiological assessment of winter nutritional deprivation in elk of Yellowstone National Park. Journal of Wildlife Management 55:653–64.

DelGiudice, G. D., F. J. Singer, U. S. Seal and G. Bowser. 1994. Physiological responses of Yellowstone bison to winter nutritional deprivation. Journal of Wildlife Management 58(1):24–34.

Dennis, B., and M. L. Taper. 1994. Density dependence in time series observations in natural populations. Ecological Monographs 64:205–24.

Despain, D. 1991. Yellowstone vegetation: Consequences of environment and history in a natural setting. Roberts Rinehart, Boulder, Colorado.

Dyskerius, E. J. 1949. Condition and management of rangeland based upon quantitative ecology. Journal of Range Management 2:104–15.

Eberhardt, L. E., L. L. Eberhardt, B. L. Tiller, and L. L. Caldwell. 1996. Growth of an isolated elk population. Journal of Wildlife Management 60:369–73.

Edwards, J. 1983. Diet shifts in moose due to predator avoidance. Oecologia 60:185–89.

Engstrom, D. R., C. Whitlock, S. C. Fritz, and H. E. Wright. 1991. Recent environmental change inferred from the sediments of small lakes in Yellowstone's Northern Range. Journal of Paleolimnology 5:139–74.

Farnes, P. 1996. A winter severity index for ungulates of Yellowstone National Park. *In* F. Singer, ed., Effects of grazing by wild ungulates in Yellowstone National Park. Technical Report NPS 96-01, NPS, Natural Resource Information Division, Denver.

Frank, D. A., R. S. Inouye, N. Huntley, G. W. Minshall, and J. E. Anderson. 1994. Biogeochemistry of a north-temperate grassland with native ungulates: Nitrogen dynamics in Yellowstone National Park. Biogeochemistry 26:163–88.

Frank, D. A., and S. J. McNaughton. 1992. The ecology of plants, large mammalian herbivores, and drought in Yellowstone National Park. Ecology 73:2043–58.

———. 1993. Evidence of promotion of aboveground grassland production by large herbivores in Yellowstone National Park. Oecologia 96:157–61.

Franzmann, A. W., and C. C. Schwartz. 1986. Black bear predation on moose calves in highly productive versus marginal moose habitats on the Kenai Peninsula, Alaska. Alces 22:139–54.

Franzmann, A. W., C. C. Schwartz, and R. O. Peterson. 1980. Moose calf mortality in summer on the Kenai Peninsula, Alaska. Journal of Wildlife Management 44:764–68.

French, S. P., and M. G. French. 1990. Predatory behavior of grizzly bears feeding on elk calves in Yellowstone National Park, 1986–88. International Conference on Bear Research and Management 8:335–41.

Fuller, T. K. 1989. Population dynamics of wolves in north-central Minnesota. Wildlife Monographs no. 105.

Garton, E. O., R. L. Crabtree, B. B. Ackerman, and G. Wright. 1990. The potential impact of a reintroduced wolf population on the northern Yellowstone elk population. Pp. 3-59 to 3-91 *in* Wolves for Yellowstone? A report to the United States Congress, vol. 2, research and analysis. National Park Service, Yellowstone National Park.

Gasaway, W. C., R. D. Boertje, D. V. Grangaard, D. G. Kelleyhouse, R. O. Stephenson, and D. G. Larsen. 1992. The role of predation in limiting moose at low densities in Alaska and Yukon and implications for conservation. Wildlife Monographs no. 20.

Geese, E. M., and S. Grothe. 1995. Analysis of coyote predation on deer and elk during winter in Yellowstone National Park, Wyoming. American Midland Naturalist 133:36–42.

Guiness, F. E., R. M. Gibson, and T. H. Clutton-Brock. 1978. Calving times of red deer on Rhum. Journal of Zoology 185:105–14.

Gunson, J. 1986. Wolves and elk in Alberta's Brazeau country. Bugle 87:29–33.

Haber, G. C. 1977. Socio-ecological dynamics of wolves in a subarctic-ecosystem. Ph.D. diss., University of British Columbia, Vancouver.

Haines, A. L. 1965. Osborne Russel's journal of a trapper. University of Nebraska Press, Lincoln.

Harrison, D. J., J. A. Bissonette, and J. A. Sherburne. 1989. Spatial relationship between coyotes and red foxes in eastern Maine. Journal of Wildlife Management 53:181–85.

Hatter I. W. 1982. Predator-ungulate relationships in second growth forests on Vancouver Island. IWIFR-5, Ministry of Environment and Forests, Victoria.

Hayden, J. A. 1984. Introduced mountain goats in the Snake River range, Idaho: Characteristics of vigorous population growth. Proceedings of the Biennial Symposium of the Northern Wild Sheep and Goat Council 4:94–119.

Hodgson, B. 1994. Buffalo: Back home on the range. National Geographic 186:65–89.

Houston, D. B. 1968. The Shiras moose in Jackson Hole, Wyoming. Grand Teton Natural History Association, Moose, Wyoming. Technical Bulletin 1:1–110.

———. 1971. Ecosystems of national parks. Science 172:648–51.

———. 1976. The status of research on ungulates in northern Yellowstone Park. Pp. 11–27 *in* Research in the parks. Transactions of the National Parks Centennial Symposium, National Park Service Symposium Series no. 1.

———. 1978. Elk as winter-spring food for carnivores in northern Yellowstone National Park. Journal of Applied Ecology 15:653–61.

———. 1982. The northern Yellowstone elk: Ecology and management. Macmillan, New York.

Hurley, K. P., L. Roop, and V. Stelter. 1989. Elk. Pp. 167–87, 254–74, 286–306, 348–89, 448–68, 487–95, 505–52, and 611–80 *in* Annual big game herd unit reports, 1988, District 2. Wyoming Game and Fish, Cody.

Jonas, R. J. 1955. A population and ecological study of the beaver *(Castor canadensis)* of Yellowstone National Park. M.S. thesis, University of Idaho, Moscow.

Kay, C. E. 1984. Aspen reproduction in the Yellowstone Park–Jackson Hole area and its relationship to the natural regulation of ungulates. Pp. 131–60 in G. W. Workman, ed., Western elk management: A symposium. Utah State University, Logan.

——. 1990. Yellowstone's northern elk herd: A critical evaluation of the "natural regulation" paradigm. Ph.D. diss., Utah State University, Logan.

——. 1994. Aboriginal overkill: The role of Native Americans in structuring western ecosystems. Human Nature 5:359–98.

Kay, C. E., and F. H. Wagner. 1994. Historical condition of woody vegetation on Yellowstone's Northern Range. Pp. 151–67 in D. G. Despain, ed., Plants and their environments. Technical Report no. 93, National Park Service, Denver.

Keating, K. A. 1985. The effects of temperature on bighorn population estimates in Yellowstone National Park. International Journal of Biometeorology 29:47–55.

Keating, K. A., L. R. Irby, and W. F. Kasworm. 1985. Mountain sheep food habits in the upper Yellowstone Valley. Journal of Wildlife Management 49:156–61.

Knapp, A. K., and T. R. Seastedt. 1986. Detritus accumulation limits productivity of tallgrass prairie. BioScience 36:662–68.

Knight, R. R., D. J. Mattson, and B. M. Blanchard. 1984. Movements and habitat use of the Yellowstone grizzly bear. Interagency Grizzly Bear Study Team Report, National Park Service, Bozeman.

Kuck, L., J. Nelson, and J. Turner, eds. 1989. Statewide surveys and inventories. Mule deer. Project W-170-R-13, Job 2. Idaho Department of Fish and Game, Boise.

Lahren, L. A. 1976. The Myers-Hindman site, Anthropologos Research International.

Laundré, J. W. 1990. The status, distribution, and management of mountain goats in the Greater Yellowstone Ecosystem. Report to National Park Service, Idaho State University, Pocatello.

Lemke, T. O., J. A. Mack, and D. B. Houston. 1998. Winter range expansion by the northern Yellowstone elk herd. Intermountain Journal of Sciences 4:1–9.

Leopold, A. S., S. A. Cain, C. M. Cottam, I. N. Gabrielson, and T. L. Kimball. 1963. Wildlife management in the national parks. Transactions of the North American Wildlife Conference 24:28–45.

LeResche, R. E., R. H. Bishop, and J. W. Coady. 1974. Distribution and habitats of moose in Alaska. Naturaliste Canadien 101:143–78.

Lime, D. W., B. A. Koth, and J. C. Vlaming. 1993. Effects of restoring wolves on Yellowstone area big game and grizzly bears: Opinions of scientists. Pp 306–26 in R. S. Cook, ed., Ecological issues on reintroducing wolves into Yellowstone National Park. National Park Service Scientific Monograph no. 22. National Park Service, Denver.

Lockman, D., G. Roby, and L. Wollrab. 1989. Big game. Pp. 1–20, 59–71, 214–63, 301–13, 354–77, 414–26, 445–81 in Annual big game herd unit reports, 1988. District 1. Wyoming Game and Fish, Cody.

Lonner, T. N., and P. Schladweiler. 1986. Elk population dynamics and breeding biology. Project Report W-120-R-17, Montana Department of Fish, Wildlife, and Parks, Helena.

Lovaas, A. L. 1970. People and the Gallatin elk herd. Montana Department of Fish and Game, Helena.

MacCracken, J. G., and L. A. Viereck. 1990. Browse regrowth and use by moose after fire in interior Alaska. Northwest Science 64:110–18.

Mack, J. A., and F. J. Singer. 1993a. Predicted effects of wolf predation on Northern Range elk using Pop-II models. Pp. 49–74 in R. Cook, ed., Ecological issues on reintroducing wolves into Yellowstone National Park. National Park Service Scientific Monograph no. 22. National Park Service, Denver.

——. 1993b. Population models for elk, mule deer, and moose on Yellowstone's northern winter range. Pp. 270–305 in R. Cook, ed., Ecological issues on reintroducing wolves into Yellowstone National Park. National Park Service Scientific Monograph Series no. 22. National Park Service, Denver.

Mattson, D. J., B. M. Blanchard, and R. R. Knight. 1991. Food habits of the Yellowstone grizzly bear, 1977–87. Canadian Journal of Zoology 69:1619–29.

May, R. M. 1977. Thresholds and breakpoints in ecosystems with a multiplicity of stable states. Nature 269:471–77.

McCullough, D. R. 1979. The George Reserve deer herd. University of Michigan Press, Ann Arbor.

McNaughton, S. J. 1979. Grazing as an optimization process: Grass-ungulate relationships in the Serengeti. American Naturalist 113:691–703.

———. 1988. Mineral nutrition and spatial concentrations of ungulates. Nature 34:343–45.

Meagher, M. 1973. The bison of Yellowstone National Park. National Park Service Scientific Monograph Series no. 1. National Park Service, Washington.

———. 1989a. Evaluation of boundary bison control for bison of Yellowstone National Park. Wildlife Society Bulletin 17:15–19.

———. 1989b. Range expansion by bison of Yellowstone National Park. Journal of Mammalogy 70:670–75.

Mech, L. D. 1966. The wolves of Isle Royale. U.S. National Park Service Fauna Series no. 7.

———. 1970. The wolf: Ecology and behavior of an endangered species. Doubleday, New York.

Messier, F. 1991. The significance of limiting and regulating factors on the demography of moose and white-tailed deer. Journal of Animal Ecology 60:377–93.

Messier, F., and M. Crete. 1985. Moose-wolf dynamics and the natural regulation of moose populations. Oecologia 65:503–12.

Messier, F., W. C. Gasaway, and R. O. Peterson. 1995. Wolf-ungulate interactions in the Northern Range of Yellowstone: Hypotheses, research priorities, and methodologies. A report to the National Biological Service, Fort Collins, Colorado.

Meyer, G. A., S. G. Wells, R. C. Balling Jr., and A. J. T. Jull. 1992. Responses of alluvial systems to fire and climate change in Yellowstone National Park. Nature 357:147–50.

Munro, J. A. 1947. Observations of birds and mammals in central British Columbia. British Columbia Provincial Museum Occasional Papers no. 6.

Murie, A. 1940. Ecology of coyotes in the Yellowstone. National Park Service Fauna Series no. 4.

———. 1963. A naturalist in Alaska. Doubleday, New York.

Murie, O. J. 1951. The elk of North America. Stackpole, Harrisburg, Pennsylvania.

Norland, J., F. J. Singer, and L. Mack. 1996. Effects of the fires of 1988 on elk habitats in Yellowstone National Park. Pp. 223–32 in J. M. Greenlee, ed., Ecological effects of the fires of 1988 in Yellowstone National Park. International Association of Wildland Fire, Fairfield, Washington.

Norris, P. 1877. Annual report of the superintendent, Yellowstone National Park. U.S. Government Printing Office, Washington.

O'Gara, B. W. 1968. A study of the reproductive cycle of the female pronghorn. Ph.D. diss., University of Montana, Missoula.

Paine, R. T. 1966. Food web complexity and species diversity. American Naturalist 100:65–75.

Paquet, D. C. 1989. Behavioral ecology of sympatric wolves and coyotes in Riding Mountain National Park, Manitoba. Ph.D. diss., University of Alberta, Edmonton.

Pastor, J., and R. J. Naiman. 1992. Selective foraging and ecosystem processes in boreal forests. American Naturalist 139:690–705.

Peck, S. V. 1972. The ecology of the Rocky Mountain goat in the Spanish Peaks area of southwestern Montana. M.S. thesis, Montana State University, Bozeman.

Peek, J. M. 1980. Natural regulation of ungulates. Wildlife Society Bulletin 8:217–27.

Pengelley, W. L. 1963. Thunder on the Yellowstone. Naturalist 14:18–25.

Peterson, R. O. 1977. Wolf ecology and prey relationships on Isle Royale. National Park Service Scientific Monograph no. 11.

Peterson, R. O., and D. L. Allen. 1974. Snow conditions as a parameter in moose-wolf relationships. Naturaliste Canadien 101:481–92.

Pimlott, D. H., J. A. Shannon, and G. B. Kolenosky. 1969. Ecology of the timber wolf in Algonquin Provincial Park. Ontario Department of Lands and Forests, Research Paper (Wildlife) no. 87.

Power, M. E., and L. S. Mills. 1995. The Keystone cop meets Hilo. Trends in Ecology and Evolution 10:182–84.

Ritchie, B. W. 1978. Ecology of moose in Fremont County, Idaho. Wildlife Bulletin no. 7, Idaho Department of Fish and Game, Boise.

Romme, W. H., M. G. Turner, L. L. Wallace, and J. S. Walker. 1995. Aspen, fire and elk in northern Yellowstone National Park. Ecology 76:2097–2104.

Rudd, W. J., A. L. Ward, and L. L. Irwin. 1983. Do split hunting seasons influence elk migrations from Yellowstone National Park? Wildlife Society Bulletin 11:328–31.

Samuel, M. D., E. O. Garton, M. W. Schlegel, and R. G. Carson. 1987. Visibility bias during aerial surveys of elk in north-central Idaho. Journal of Wildlife Management 51:622–30.

Sauer, J. R., and M. S. Boyce. 1983. Density dependence and survival of elk in northwestern Wyoming. Journal of Wildlife Management 47:31–37.

Schlegel, M. 1976. Factors affecting calf elk survival in north-central Idaho. Proceedings of the Annual Conference of Western Associations, State Game and Fish Commissions 56:342–55.

Schullery, P. 1989. Yellowstone fires: A preliminary report. Northwest Science 63:44–55.

Schullery, P., and L. Whittlesey. 1996. The documentary record of wolves and related wildlife species in the Greater Yellowstone prior to 1882. Pp. 63–76 *in* L. N. Carbyn, S. H. Fritts, and D. R. Seip, eds., Ecology and conservation of wolves in a changing world. Canadian Circumpolar Institute, University of Alberta, Edmonton.

Scott, M. D., and H. Geisser. 1996. Pronghorn migration and habitat use following the 1988 Yellowstone fires. Pp. 123–32 *in* J. M. Greenlee, ed., Ecological effects of the fires of 1988 in Yellowstone National Park. International Council on Wildland Fire, Fairfield, Washington.

Seton, E. T. 1929. Lives of game animals. Doubleday, Doran, and Co., New York.

Ser(veen, C., and R. R. Knight. 1993. Possible effects of a restored gray wolf population on grizzly bears in the Greater Yellowstone area. Pp. 28–37 *in* R. Cook, ed., Ecological issues on reintroducing wolves into Yellowstone National Park. National Park Service Scientific Monograph no. 22. National Park Service, Denver.

Servheen, C., R. Knight, D. Mattson, S. Mealy, D. Strickland, J. Varley, and J. Weaver. 1986. Report to the IGBC on the availability of food for grizzly bears in the Yellowstone ecosystem. Interagency Grizzly Bear Committee, Missoula.

Silver, H., and W. T. Silver. 1969. Growth and behavior of the coyote-like canid of northern New England. Wildlife Monographs no. 17.

Sinclair, A. R. E. 1989. Population regulation in animals. Pp. 197–241 *in* J. M. Cherrett, ed., Ecological concepts. Blackwell Scientific Publications, Oxford.

Sinclair, A. R. E., and P. Arcese. 1995. Population consequences of predation-sensitive foraging: The Serengeti wildebeest. Ecology 76:882–91.

Singer, F. J. 1987. Dynamics of caribou and wolves in Denali National Park. Pp. 117–57 *in* 4th Triennial Conference on Science in the National Parks, vol. 2, George Wright Society, Houghton, Michigan.

———. 1991a. Some predictions concerning a wolf recovery into Yellowstone National Park: How wolf recovery may affect park visitors, ungulates and other predators. Transactions of the North American Wildlife and Natural Resources Conference 56:567–83.

———. 1991b. The ungulate prey base for wolves in Yellowstone National Park. Pp. 323–48 *in* R. E. Keiter and M. S. Boyce, eds., The Greater Yellowstone ecosystem: Redefining America's wilderness heritage. Yale University Press, New Haven.

Singer, F. J., and R. G. Cates. 1995. Response to comment: Ungulate herbivory on willows on Yellowstone's northern winter range. Journal of Range Management 48:563–65.

Singer, F. J., and E. O. Garton. 1994. Super cub, elk sightability model. Pp. 47–49 *in* J. W. Unsworth, F. A. Leban, D. J. Leptich, E. O. Garton, and P. Zager, eds., Aerial survey: User's manual. Idaho Fish and Game Department, Boise.

Singer, F. J., and M. K. Harter. 1996. Comparative effects of elk herbivory and the fires of 1988 on grasslands in northern Yellowstone National Park. Ecological Applications 6:185–99.

Singer, F. J., A. Harting, K. S. Symonds, and M. B. Coughenour. 1997. Elk calf mortality in Yellowstone National Park: The evidence for density dependence and compensation. Journal of Wildlife Management 61:12–25.

Singer, F. J., and J. A. Mack. 1993. Potential ungulate prey for wolves. Pp. 75–117 *in* R. S. Cook, ed., Ecological issues on reintroducing wolves in Yellowstone National Park. National Park Service Scientific Monograph no. 22. National Park Service, Denver.

Singer, F. J., L. Mack, and R. G. Cates. 1994. Ungulate herbivory of willows on Yellowstone's northern winter range. Journal of Range Management 47:435–43.

Singer, F. J., and J. E. Norland. 1994. Niche relationships within a guild of ungulates following release from artificial controls. Canadian Journal of Zoology 72:1383–94.

Singer, F. J., and R. A. Renkin. 1995. Effects of browsing by native ungulates on the shrubs in big sagebrush communities in Yellowstone National Park. Great Basin Naturalist 55:201–12.

Singer, F. J., W. Schreier, J. Oppenheim, and E. O. Garton. 1989. Drought, fires, and large mammals. Bioscience 39:716–22.

Smith, B. L., and S. H. Anderson. 1996. Patterns of neonatal mortality of elk in northwest Wyoming. Canadian Journal of Zoology 74:1229–37.

Smith, B. L., and R. L. Robbins. 1994. Migrations and management of the Jackson elk herd. National Biological Survey Research Publication no. 199. National Biological Survey, Washington.

Steele, R., S. V. Cooper, D. V. Ondov, D. W. Roberts, and R. D. Pfister. 1983. Forest habitat types of eastern Idaho–western Wyoming. Forest Service, Gen. Tech. Sep. INT-44, Intermountain Forest and Range Experiment Station, Ogden, Utah.

Stenlund, M. H. 1955. A field study of the timber wolf *(Canis lupus)* on the Superior National Forest. Minnesota Division of Fish and Game, Department of Conservation Bulletin 4:1–55.

Stephens, P. W., and R. O. Peterson. 1984. Wolf avoidance strategies of moose. Holarctic Ecology 7:239–44.

Swenson, J. E. 1985. State wildlife survey and inventory. Project W-130-R-16, Job I-3(c), Montana Department of Fish, Wildlife and Parks, Big Game Survey and Inventory, Region 3, Helena.

Telfer, E. S., and J. P. Kelsall. 1984. Adaptation of some large North American mammals for survival in snow. Ecology 65:1828–34.

Trent, T., T. Parker, and J. Naderman. 1984. Pp. 41–46 *in* L. E. Oldenburg, ed., Statewide surveys and inventories. Project W-170-R-8, Job 8. Moose. Idaho Department of Fish and Game, Boise.

Turner, M. G., Y. Wu, L. L. Wallace, W. H. Romme, and A. Brenkert. 1994. Simulating winter interactions among ungulates, vegetation, and fire in northern Yellowstone National Park. Ecological Applications 4:472–96.

Tyers, D. B. 1996. Shiras moose winter habitat use in the upper Yellowstone River valley prior to and after the 1988 fires. Alces 31:35–43.

U.S.D.I., National Park Service. 1968. Senate hearings, control of elk populations in Yellowstone National Park. 90th Congress, 1st Session. U.S. Government Printing Office, Washington.

Vales, D. J., and J. M. Peek. 1993. Estimation of the potential interaction between hunter harvest and wolf predation on the Sand Creek, Idaho, and Gallatin, Montana, elk populations. Pp. 118–72 *in* National Park Service Scientific Monograph no. 22. National Park Service, Denver.

Wagner, F. H., R. Foresta, R. B. Gill, D. R. McCullough, M. R. Pelton, W. F. Porter, and H. Salwasser. 1995. Wildlife policies in the U.S. national parks. Island Press, Washington, D.C.

Walcheck, K. 1976. Montana wildlife 170 years ago. Montana Outdoors 7:15–30.

Warren, C. N. 1926. A study of beaver in the Yancey region of Yellowstone National Park. Roosevelt Wildlife Annual no. 1.

Wiens, J. A. 1977. On competition and variable environments. American Science 65:590–97.

Wolda, H. 1989. Density dependence tests, are they? Oecologia 95:581–91.

Wolff, J. O. 1978. Burning and browsing effects on willow growth in interior Alaska. Journal of Wildlife Management 42:135–40.

Yellowstone National Park. 1996. Final interim bison management plan. Yellowstone National Park, Mammoth, Wyoming.

——. 1997. Yellowstone's Northern Range: Complexity and change in a wildland ecosystem. Yellowstone National Park, Mammoth, Wyoming.

Young, S. P., and E. A. Goldman. 1944. The wolves of North America. Dover, New York.

Uinta ground squirrel (Franz J. Camenzind)

Chapter Nine

Small Prey of Carnivores in the Greater Yellowstone Ecosystem

Kurt A. Johnson and Robert L. Crabtree

Although carnivores of the Greater Yellowstone Ecosystem (GYE) prey on a variety of invertebrates (for example, insects, crayfish) and all five classes of vertebrates, small mammals are the most important component of the "small prey" consumed by carnivores in terms of numbers and biomass. This chapter integrates information from a review of the literature and results of a cooperative small mammal study on the Northern Range of Yellowstone National Park (YNP), hereafter called the Northern Range Small Mammal Study. This study was initiated by YNP in 1990 and has been continued since 1993 by Yellowstone Ecosystem Studies (YES). We summarize small-mammal habitat relations, community structure, and landscape patterns (including the effects of disturbance) in the GYE, and we discuss relevant carnivore food habits studies as they pertain to small mammals. We analyze the implications of current knowledge about small prey for carnivore conservation and make recommendations for small-mammal research.

The GYE covers twenty-four thousand square kilometers in northwestern Wyoming and adjacent areas of Idaho and Montana. Much of the GYE is protected in YNP and Grand Teton National Park (GTNP) and national forest wilderness areas (Beartooth, North Absaroka, South Absaroka, and Teton). Vegetation types of the GYE are described in detail by Houston (1982) and Despain (1991). A simplified habitat classification for small mammals would include alpine tundra, subalpine coniferous forest, aspen forest, moist montane coniferous forest, dry montane coniferous forest, sagebrush steppe, dry grassland, wet grassland–meadow, riparian shrub–forest, riparian sedge meadows, and special habitats (talus slopes, rocky canyons, and aquatic). Natural and manmade disturbances may modify these habitat types.

The Northern Range Small Mammal Study is totally within the boundaries of YNP. Trapping and data analysis methods are described in detail by Harter and Crabtree (1991). Results from 1992–93, published in Crabtree et al. (1997) and summarized in table 9.2, are used in this chapter. The Northern Range covers about 100,000 hectares, extending from Corwin Springs to Silver Gate, Montana (Crabtree et al. 1997). It is lower, warmer, and drier on average than other areas of YNP. The Northern Range Small Mammal Study includes six habitat types: (1) wet grassland (including riparian sedge meadow); (2) dry grassland (primarily Idaho fescue, *Festuca idahoensis*); (3) sagebrush steppe (primarily big sagebrush, *Artemisia tridentata*); (4) burned sagebrush steppe; (5) montane coniferous forest (primarily Douglas fir, *Pseudotsuga menziesii*); and (6) burned montane coniferous forest (Crabtree et al. 1997). The burned sagebrush steppe and burned montane coniferous forest habitats resulted from one of the most intensive and extensive disturbance events of recent history—the 1988 fires. Main trapping grids are located in (1) Little America Valley (sagebrush steppe and riparian) near Tower Junction, (2) the northeast entrance (spruce-fir forest), (3) Soda Butte (sagebrush steppe, riparian, and forest) in Lamar Valley, and (4) Rose Creek (mesic sedge) in Lamar Valley (Crabtree et al. 1997).

Nonmammalian Small Prey

GYE carnivores consume diverse nonmammalian small prey, including invertebrates, fish, amphibians, reptiles, and birds.

INVERTEBRATES

Invertebrate prey of carnivores in the GYE includes insects, particularly grasshoppers and beetles (Insecta), crayfish (Crustacea), and mollusks (Mollusca). Red foxes are known to eat insects (Clark and Stromberg 1987). Black bears eat insects, while grizzly bears eat ants and moths (Clark and Stromberg 1987). Raccoons are known to take crayfish, mollusks, and insects (Clark and Stromberg 1987, Streubel 1989). Ermine and long-tailed weasels eat insects, especially beetles and grasshoppers (Clark and Stromberg 1987). Mink take crayfish and other invertebrates (Clark and Stromberg 1987, Streubel 1989). Western spotted skunks eat beetles, crickets, grasshoppers, grubs, worms, and crayfish (Clark and Stromberg 1987). Martens and fishers eat insects (Clark and

Stromberg 1987). Insects were found in 32 percent of marten scats analyzed in a Northwest Territories study and in 19 percent of marten scats from western Montana (see studies cited in Buskirk and Ruggiero 1994). Arthropods were found in 37 percent of the fisher stomachs analyzed in a California study (Powell and Zielinski 1994).

FISH

Fish are the prey of carnivores specialized for an aquatic or semiaquatic existence, including mink (Clark and Stromberg 1987, Streubel 1989) and river otters (Clark and Stromberg 1987). Grizzly bears and wolverines also consume fish at certain times (Banci 1994). Trout (for example, cutthroat trout, *Salmo clarki*), whitefish (for example, mountain whitefish, *Prosopium williamsoni*), and suckers (for example, mountain sucker, *Catastomus platyrhynchus*) are the primary prey fish in the GYE.

HERPTILES

Herptile prey of carnivores in the GYE include frogs, toads, salamanders, lizards, and snakes. Coyotes are known to eat frogs and lizards (Clark and Stromberg 1987). Mink take frogs and snakes (Clark and Stromberg 1987, Streubel 1989). Western spotted skunks take frogs and lizards (Clark and Stromberg 1987). On occasion, fishers take snakes, frogs, and toads (Powell and Zielinski 1994).

BIRDS

Birds ranging in size from songbirds to blue grouse *(Dendragapus obscurus)* and white-tailed ptarmigan *(Lagopus leucurus)* are prey for carnivores in the GYE. Bird eggs and chicks are also eaten. Ermine eat songbirds (Clark and Stromberg 1987). Long-tailed weasels will take birds up to the size of grouse, as well as eggs and chicks (Clark and Stromberg 1987). Mink also eat birds (Clark and Stromberg 1987). Western spotted skunks eat bird eggs (Clark and Stromberg 1987). American martens eat birds and bird eggs; in eight studies cited in Buskirk and Ruggiero (1994), bird remains were present in 9–30 percent of the scats or gastrointestinal tracts analyzed. Fishers eat a wide array of birds, including gray jays *(Perisoreus canadensis)*, ruffed grouse *(Bonasa umbellus)*, ducks, and a variety of passerines (Powell and Zielinski 1994). Lynx eat grouse, especially when snowshoe hares *(Lepus americanus)* are not abundant (Koehler and Aubry 1994). Wolverines eat birds; ptarmigan are taken in the winter in the Yukon (Banci 1987 cited in Banci 1994), Alaska (Gardner 1985 cited in Banci 1994), and the Northwest Territories (Boles 1977 cited in Banci 1994).

Ecology of Small Mammals in the GYE

Small mammals are defined as noncarnivores ranging in adult weight from two grams (the smallest adult shrews, *Sorex nanus* and *Sorex hoyi,* the smallest mammals in the GYE) to thirty-five kilograms (the size of an adult beaver, *Castor canadensis*). Forty-five species of noncarnivorous small mammals from four orders and ten families are found in the GYE (Streubel 1989); common and scientific names and weight ranges are presented in table 9.1. This section discusses available information on species-habitat associations of small mam-

Table 9.1

Scientific names and adult body weights of small mammals of the Greater Yellowstone Ecosystem

Common name	Scientific name	Adult body weight
Dwarf shrew	*Sorex nanus*	0.002 – 0.003 kg
Pygmy shrew	*Sorex hoyi*	0.002 – 0.004 kg
Preble's shrew	*Sorex preblei*	0.003 – 0.006 kg
Masked shrew	*Sorex cinereus*	0.003 – 0.006 kg
Wandering shrew	*Sorex vagrans*	0.003 – 0.007 kg
Merriam's shrew	*Sorex merriami*	0.004 – 0.007 kg
Dusky shrew	*Sorex monticolus*	0.007 kg
Water shrew	*Sorex palustris*	0.012 – 0.018 kg
Red-backed vole	*Clethrionomys gapperi*	0.018 – 0.022 kg
Western jumping mouse	*Zapus princeps*	0.018 – 0.024 kg
Sage vole	*Lemmiscus curtatus*	0.017 – 0.035 kg
Deer mouse	*Peromyscus maniculatus*	0.020 – 0.035 kg
Heather vole	*Phenacomys intermedius*	0.025 – 0.040 kg
Least chipmunk	*Tamias minimus*	0.030 – 0.055 kg
Long-tailed vole	*Microtus longicaudus*	0.040 – 0.055 kg
Yellow pine chipmunk	*Tamias amoenus*	0.029 – 0.062 kg
Meadow vole	*Microtus pennsylvanicus*	0.030 – 0.065 kg
Uinta chipmunk	*Tamias umbrinus*	0.055 – 0.080 kg
Montane vole	*Microtus montanus*	0.034 – 0.090 kg
Richardson's water vole	*Microtus richardsoni*	0.12 kg
Northern flying squirrel	*Glaucomys sabrinus*	0.105 – 0.178 kg
Northern pocket gopher	*Thomomys talpoides*	0.075 – 0.180 kg
Pika	*Ochotona princeps*	0.150 – 0.175 kg
Red squirrel	*Tamiasciurus hudsonicus*	0.190 – 0.200 kg
Bushy-tailed woodrat	*Neotoma cinerea*	0.240 – 0.280 kg
Uinta ground squirrel	*Spermophilus armatus*	0.195 – 0.295 kg
Golden-mantled ground squirrel	*Spermophilus lateralis*	0.210 – 0.315 kg
Mountain cottontail	*Sylvilagus nuttallii*	0.630 – 0.870 kg
Desert cottontail	*Sylvilagus audubonii*	0.835 – 1.2 kg
White-tailed prairie dog	*Cynomys leucurus*	0.650 – 1.4 kg
Muskrat	*Ondatra zibethicus*	0.700 – 1.8 kg
Snowshoe hare	*Lepus americanus*	1.4 – 1.8 kg
Yellow-bellied marmot	*Marmota flaviventris*	1.6 – 5.0 kg
Whitetail jackrabbit	*Lepus townsendii*	3.0 – 6.0 kg
Porcupine	*Erethizon dorsatum*	3.5 – 18.0 kg
Beaver	*Castor canadensis*	10.0 – 35.0 kg

Note: Adult body weight ranges taken from Streubel 1989. Nine bat species are omitted because they are not important prey for carnivores.

mals, small-mammal community structure, landscape patterns and effects of disturbance on small-mammal communities, and ecosystem influences of small mammals in the GYE and similar ecosystems.

SPECIES-HABITAT ASSOCIATIONS

Insectivores. Insectivores (Order Insectivora) are represented in the GYE by eight shrew species (Streubel 1989). The dusky shrew *(Sorex monticolus)* and masked shrew *(S. cinereus)* are considered the most common species in the GYE (Streubel 1989). Preble's shrew *(S. preblei)* and dwarf shrew *(S. nanus)* are known from only one specimen each in the GYE, and Merriam's shrew *(S. merriami)* and pygmy shrew *(S. hoyi)* have yet to be documented in the GYE but are likely to occur there (Streubel 1989).

In a study of small mammals in GTNP, Clark (1973a) captured forty-four wandering shrews *(S. vagrans),* forty-two masked shrews, and nine water shrews *(S. palustris).* The masked shrew occurred over the widest range of plant communities but was most abundant in sedge meadows and shrub swamp. The wandering shrew was most abundant in lowland aspen. The water shrew was restricted to the shrub swamp community. In the Northern Range Small Mammal Study, shrews were trapped in all habitat types but were most common in mesic grasslands (Crabtree et al. 1997). Shrews were less common in burned sagebrush steppe and burned forest than in their nonburned counterparts (Crabtree et al. 1997).

Bats. Nine bat species (Order Chiroptera) are known in the GYE (Streubel 1989). Because they are not important prey for carnivores, bats are not discussed in this chapter, except to note that they are an important component of the ecosystem and that some species are becoming increasingly rare throughout their range.

Lagomorphs. Lagomorphs (Order Lagomorhpa) of the GYE include the pika *(Ochotona princeps),* mountain cottontail *(Sylvilagus nuttallii),* desert cottontail *(S. audubonii),* snowshoe hare, and whitetail jackrabbit *(L. townsendii).*

Pikas are almost always found in association with talus slopes that consist of fairly large rocks with patches of vegetation, often near forested areas (Streubel 1989). In the GYE pikas are found at nearly all elevations, from above timberline to lower elevations (Streubel 1989). Pikas were seen by Northern Range Small Mammal Study researchers but were not captured because trapping grids were not close enough to talus slopes.

Cottontails are not frequently seen on the Northern Range (Crabtree unpublished data) but are common in canyon habitats along the Yellowstone River from Hellroaring Creek to Crevice Creek (G. Felzien personal communication 1991). Snowshoe hares are relatively uncommon in lower elevation habitats of the Northern Range; sightings are typically associated with younger dense stands of Douglas fir and other forest types with a dense understory on north-facing slopes (Crabtree unpublished data). Whitetail jackrabbits are uncommon on the Northern Range. Extensive surveys conducted in the Lamar Valley and Blacktail Plateau during 1990 and 1991 resulted in only one sighting (Crabtree unpublished data). Whitetails are somewhat more common in the lower sagebrush habitats around the Gardiner and Mammoth areas.

Rodents. Rodents (Order Rodentia) are the most diverse order of small mammals in the GYE, with twenty-three species from six families (Streubel 1989).

The squirrel (Sciuridae) family is represented by nine species in the GYE: yellow-bellied marmot *(Marmota flaviventris),* two ground squirrel species *(Spermophilus* spp.), white-tailed prairie dog *(Cynomys leucurus),* three chipmunk species *(Tamias* spp.), red squirrel *(Tamiasciurus hudsonicus),* and northern flying squirrel *(Glaucomys sabrinus).*

Yellow-bellied marmots are found throughout the GYE, from low elevation valleys to alpine tundra (Streubel 1989). They are usually found in open, grassy areas, and almost always near rocks (Streubel 1989). In the Northern Range Small Mammal Study, marmots were seen by researchers in talus slopes or clusters of rocky boulders, but trapping grids were not close enough to suitable marmot habitat, nor were traps large enough to capture marmots.

Uinta ground squirrels *(Spermophilus armatus)* occur in disturbed or heavily grazed grasslands, sagebrush steppe, and mountain meadows up to 3,400 meters (Streubel 1989). Rieger (1996) studied Uinta ground squirrel reproduction in GTNP. He found that timing of reproduction affected litter size, offspring mass, and offspring survival. In the Northern Range Small Mammal Study, Uinta ground squirrels were trapped predominantly in burned sagebrush and mesic grasslands (Crabtree et al. 1997).

Golden-mantled ground squirrels *(S. lateralis)* occur in rocky areas along the edges of lowland and mountain meadows, forest openings, and alpine tundra (Streubel 1989). Golden-mantled ground squirrels were observed by Northern Range Small Mammal Study researchers but were not captured on trapping grids because the grids were too far from rocky terrain (Crabtree et al. 1997).

White-tailed prairie dogs occur along the eastern edge of the GYE and are not present in the Northern Range (Streubel 1989). Accordingly, they were not captured on any trapping grids.

The least chipmunk *(Tamias minimus)* favors open sagebrush steppe and other dry habitats including forest openings (Streubel 1989). Yellow-pine chipmunks *(T. amoenus)* occur in dry forests (such as lodgepole pine or Douglas fir) and forest openings (Streubel 1989). The Uinta chipmunk *(T. umbrinus)* seems to favor openings in dense spruce-fir forests at higher elevations (above 2,400 meters; Streubel 1989). In the Northern Range Small Mammal Study, chipmunks were trapped predominantly in forest and sagebrush (Crabtree et al. 1997). As all trapping grids were below 2,200 meters, no Uinta chipmunks were trapped.

Red squirrels are found in spruce, fir, and pine forests throughout the GYE; subadult red squirrels may also be found in aspen forests (Streubel 1989). Gurnell (1984) studied caching behavior, the size of cone caches, home range, and territoriality in red squirrels in a lodgepole pine stand on the east slope of the Colorado Front Range. Gurnell found that squirrel middens varied in size and construction, and that there was considerable individual variation in caching behavior and the number of cones cached in each midden. Sherburne and Bissonette (1993) found that marten use of subnivean access points in YNP was influenced significantly by the presence of red squirrel middens.

Reinhart and Mattson (1990) and Mattson et al. (1992) studied the interrelations of grizzly bears, red squirrels, and whitebark pine *(Pinus albicaulis)* in the Mount Washburn massif in north-central YNP and an area in the Gallatin National Forest near Cooke City, Montana. Indices of red squirrel activity and abundance were highest in the mesic and wet habitat types; pure whitebark pine stands were apparently not favorable habitat for red squirrels (Reinhart and Mattson 1990). Optimal red squirrel habitat in the whitebark pine zone consisted of stands with high tree species diversity, basal area, and environmental favorability (Reinhart and Mattson 1990).

Although northern flying squirrels are rarely seen, they occur in conifer forests throughout the GYE (Streubel 1989).

The gopher family (Geomyidae) is represented by one species in the GYE, the northern pocket gopher *(Thomomys talpoides)*. Hadly (1997) documented late-Holocene (that is, the past 3,200 years) ecological responses of pocket gophers to climate change in the sagebrush-grassland ecotone in YNP. Gophers increased in abundance during mesic intervals and declined in abundance during xeric intervals. Hadly (1997) and Hadly et al. (1998) measured genetic and morphologic variation over a late-Holocene sequence of pocket gophers from Lamar Cave in YNP. Their results indicate that the same subspecies of pocket gopher has occupied northern Yellowstone for at least 3,200 years.

This species occurs in a wide diversity of habitats in the GYE; its principal limiting factor appears to be topsoil depth (Streubel 1989). Youmans (1979) studied northern pocket gopher populations in relation to vegetation, soil texture, soil moisture, and snow melt phenology in Pelican Valley of YNP in 1977–78. He found that soil textures along the belt transects did not appear to influence pocket gopher numbers, but soil depths and soil temperatures may have done so. Soil moisture also limited pocket gopher distribution (Youmans 1979). The northern pocket gopher occurred in all six habitats trapped during the Northern Range Small Mammal Study, with greatest abundance in the dry grassland followed by the mesic grassland and forest (table 9.2). Captures and mounding rates of northern pocket gophers increased in all habitats between 1992 and 1993.

Beavers (Castoridae) appear to be rare on the Northern Range (Consolo and Hanson 1990) because of a lack of food (woody shrubs) in riparian areas (Streubel 1989).

The cricetid (Cricetidae) family is represented in the GYE by ten species from seven genera *(Peromyscus, Neotoma, Phenacomys, Clethrionomys, Microtus, Lemmiscus,* and *Ondatra.*

Forty-four deer mice *(Peromyscus maniculatus)* were captured during Clark's (1975) small-mammal community studies in GTNP. Ninety-eight percent of the deer mice were captured in two plant communities, lowland aspen and big sagebrush, both of which possessed overstory vegetation and were the driest habitats trapped (Clark 1975). In the Northern Range Small Mammal Study, the deer mouse was trapped in all six habitats; it was the most abundant small mammal in each habitat except mesic grassland in 1993 (Crabtree et al. 1997).

The bushy-tailed woodrat *(Neotoma cinerea)* is found throughout the GYE from lower elevations up to the top of the Beartooth Plateau (Streubel 1989). They den most frequently in rocky crevices, caves, and on cliff faces (Streubel 1989).

Little is known about the heather vole *(Phenacomys intermedius)* in the GYE, except that it probably occurs in a variety of habitats from lower sagebrush valleys up to treeline on the Beartooth Plateau (Streubel 1989). Negus (1950) found heather voles in sagebrush at 2,030 meters elevation in Jackson Hole and along a small stream at the edge of a spruce-fir forest at 2,100 meters.

Clark (1973b) studied local distributions and interspecies interactions among four species of microtines in GTNP: meadow vole *(Microtus pennsylvanicus),* montane vole *(M. montanus),* long-tailed vole *(M. longicaudus),* and redbacked vole *(Clethrionomys gapperi).* Six different plant community structures were identified; all six communities contained voles, but they varied considerably in numbers of species and individuals (Clark 1973b). Hodgson (1970) studied the ecological distribution of *M. montanus* and *M. pennsylvanicus* in an area in southwestern Montana where they were geographically sympatric.

In the Northern Range Small Mammal Study, meadow and montane voles were trapped in all habitats except forest (Crabtree et al. 1997). Long-tailed voles were present in mesic grasslands and burned forests and were always associated with nearby aspen trees (Crabtree et al. 1997). Richardson's or water voles *(M. richardsoni)* were trapped only in the mesic grassland associated with small streams (Crabtree et al. 1997).

Table 9.2

Mean number of individual small mammals caught per minigrid in Yellowstone's Northern Range, 1992–93

| | | Habitat types | | | | | |
| | | Grasslands | | | | | |
Species	Year	Mesic	Xeric	Sagebrush	Burned sagebrush	Forest	Burned forest
Deer mouse (Peromyscus maniculatus)	1992	1.10	6.25	6.07	3.06	3.52	5.13
	1993	2.63	3.50	3.67	3.25	4.56	4.83
Meadow/montane vole (Microtus spp.)	1992	7.00	.58	.93	.31	0	.13
	1993	0.11	1.33	0.50	0.08	0	0
Red-backed vole (Clethrionomys gapperi)	1992	0	0	0.07	0	3.95	1.5
	1993	0	0	0	0	0.78	0
Uinta ground squirrel (Spermophilus armatus)	1992	0	0.33	0.11	1.56	0	0
	1993	0	0.16	0	0	0	0
Shrews (Sorex spp.)	1992	0.95	0.17	0.53	0.06	0.38	0.69
	1993	0.89		0.67	0	0.56	0
Western jumping mouse (Zapus princeps)	1992	0.15		0	0	0	0.13
	1993	0		0	0.08	0	0
Water vole (Microtus richardsoni)	1992	0.13		0	0	0	0
	1993	0		0	0	0	0
Chipmunks (Tamias spp.)	1992			0.07	0	0.05	0.13
	1993			0.83	0.17	1.22	0.17
Long-tailed vole (Microtus longicaudus)	1992	0.08		0	0	0	0.06
	1993	0		0	0	0	0
Northern pocket gopher (Thomomys talpoides)							
Captures	1992	0.17	0.17	0.20	0.06	0.29	0.06
	1993	0.56	0.67	0.50	0.08	0.33	0.33
Mounding rates	1992	10.69	11.42	2.87	9.78	5.56	7.50
	1993	15.22	42.83	17.17	8.17	9.89	8.17
Long-tailed weasel (Mustela frenata)	1992	0.04	0	0	0	0	0
	1993	0	0	0	0	0	0
Short-tailed weasel (Mustela erminea)	1992	0.04	0	0	0	0	0
	1993	0.11	0	0	0	0	0

Source: Crabtree et al. 1997.
Note: Northern pocket gopher mounding rates refer to the mean number of new mounds counted per minigrid in each habitat.

The red-backed vole is a forest species most commonly associated with spruce-fir forests, though it also occurs in lodgepole pine and mixed aspen-Douglas fir forests (Streubel 1989). Nordyke and Buskirk (1991) studied red-backed voles in conifer forests of southeastern Wyoming in 1986-87. Red-backed voles were trapped in lodgepole pine, mature spruce-fir, and old-growth spruce-fir habitats but were most abundant and had best body condition in old-growth spruce-fir. In the Northern Range Small Mammal Study, red-backed voles were most abundant in the forest habitat, followed by the burned forest habitat (table 9.2, Crabtree et al. 1997). They were rare in the sagebrush habitat and were not trapped in mesic grassland, dry grassland, or burned sagebrush (Crabtree et al. 1997).

The sagebrush vole *(Lemmiscus curtatus)* prefers dry sagebrush communities (Streubel 1989). It has been reported from Jackson Hole but not from YNP (Streubel 1989). This species is apparently quite rare in the GYE (Streubel 1989).

The muskrat *(Ondatra zibethicus)* is widespread throughout the GYE, where it is found in lakes, beaver ponds, small streams, and marshes with standing water (Streubel 1989).

The family Zapodidae is represented by the western jumping mouse *(Zapus princeps)*. In the GYE this species is typically associated with lush streamside or marshy vegetation (Streubel 1989). Clark (1971) studied the activity patterns, habitat affinities, reproduction, and food habits of the western jumping mouse in GTNP. This species is active aboveground about three months each year. In the Northern Range Small Mammal Study, western jumping mice were present in low numbers in mesic grasslands, burned sagebrush, and burned forest habitats (Crabtree et al. 1997).

The family Erethizontidae is represented by the porcupine *(Erethizon dorsatum)*. Porcupines occur in a variety of habitats in the GYE, including forests, shrub communities, and sagebrush steppe (Streubel 1989).

SMALL-MAMMAL COMMUNITIES

Only a small number of studies have investigated small mammal communities in relatively undisturbed habitats in the GYE. Pattie and Verbeek (1967) studied small mammals in the alpine zone of the Beartooth Mountains northeast of YNP. They documented thirty-seven species in the alpine zone, which was composed of five community types—wet sedge meadow, moist meadow, dry meadow, fellfield, and krummholtz.

Northern Range Small Mammal Study. Northern Range Small Mammal Study trapping results for 1992 and 1993 are summarized in table 9.2. Seven taxa of noncarnivore small mammals were trapped in *mesic grasslands* in 1992, and four taxa were trapped in 1993. In 1992 meadow and montane voles were the most commonly trapped taxa (in terms of mean number of individuals caught per minigrid), with deer mice and shrews the other principal taxa. Northern pocket gopher mounding rates—an index of relative abundance—were high. Western jumping mice, water voles, and long-tailed voles were also captured. In 1993 the deer mouse was the most commonly trapped species. Meadow and montane vole numbers dropped precipitously from 1992. Northern pocket gopher mounding rates increased by 50 percent from 1992. Western jumping mice, water voles, and long-tailed voles were not captured at all.

Five taxa were trapped in *dry grasslands* in 1992 and four taxa in 1993. In 1992 the deer mouse was by far the most commonly trapped species. Meadow and montane voles, Uinta ground squirrels, and shrews were also captured. Northern pocket gopher mounding rates were high. In 1993 the deer mouse was still the most commonly trapped species, although numbers captured were one-half what they were in 1992. Meadow and montane voles were considerably more abundant than in 1992. Northern pocket gopher mounding rates were almost four times what they were in 1992. Uinta ground squirrel numbers dropped by one-half, while no shrews were captured.

Seven taxa were captured in *sagebrush steppe* in 1992 and five taxa in 1993. The deer mouse was the most abundant species captured in 1992. Meadow and montane voles, red-backed voles, Uinta ground squirrels, shrews, and chipmunks were also captured. Northern pocket gopher mounding rates were quite low. In 1993 the deer mouse was still the most abundant species, although numbers captured declined by 40 percent. Chipmunks and shrews increased in abundance, and northern pocket gopher mounding rates increased eightfold. Meadow and montane vole numbers declined, while red-backed voles and Uinta ground squirrels were not captured.

Five taxa were captured in the *forest* in both 1992 and 1993. Red-backed voles were the most abundant species in 1992, followed closely by the deer mouse. Shrews and chipmunks were also captured. Northern pocket gopher mounding rates were intermediate.

Temporal variability in small mammal communities. Small-mammal communities in the GYE can experience tremendous seasonal and yearly variability in species composition and numbers. Seasonal variability results primarily from life-history characteristics of species, such as reproductive strategy (for example, timing of breeding season) and overwintering strategy (for example, hibernation). Yearly variability results primarily from environmental influences on life-history characteristics of species, such as effects of winter weather on overwinter survival, and effects of spring weather on reproduction.

Northern Range Small Mammal Study data illustrate the variation in species composition and numbers that can occur between years as a result of environmental influences. The spring and early summer weather in 1992 was more or less normal in terms of temperature and precipitation, but 1993 had near-record precipitation and below-normal temperatures. Meadow and montane voles declined drastically in mesic grasslands between 1992 and 1993, probably because of flooding of their underground nests caused by the above-average precipitation in spring 1993. Pinter (1988) documented a significant negative correlation between precipitation during May and montane vole population dynamics in a nineteen-year study in Grand Teton National Park. Peak precipitation in May was correlated with a decline phase in the vole population cycle in the same year. Pinter hypothesized that spring weather may influence montane vole population dynamics in northwestern Wyoming by influencing survival and reproduction at the beginning of the breeding season. Other species—western jumping mouse, water vole, and long-tailed vole—also disappeared from mesic grasslands in 1993. Meadow and montane vole dispersal from the mesic grasslands to escape high water would also explain the increase in vole captures in the more upland, dry grassland habitats. Uinta ground squirrels disappeared from unburned and burned sagebrush steppe in 1993 and declined in the xeric grasslands. Red-backed voles declined drastically in the unburned and burned forest between 1992 and 1993. Conversely, chipmunk numbers and northern pocket gopher mounding rates increased in virtually every habitat between 1992 and 1993.

EFFECTS OF DISTURBANCE ON SMALL-MAMMAL COMMUNITIES

A landscape is the pattern of soils, landforms, water, and vegetation (or habitat) types that occur in a specified area during a specified period of time. As with ecosystems, landscapes vary from the simple to the complex. Landscape complexity can have a strong influence on small-mammal diversity in a given area, which, in turn, can influence the diversity of carnivores in that area. Thus a simple landscape (for example, an unbroken plain covered only with sagebrush steppe) would be expected to have lower diversity of small mammals and carnivores than a more complex landscape (for example, undulating hills of sagebrush steppe broken by small watercourses lined with riparian shrubs and north-facing slopes with pine forest patches).

Landscapes are created and maintained by environmental influences, biotic processes, and disturbances (Urban et al. 1987). Disturbances occur over a variety of spatial and temporal scales and intensity levels and can be both natural and human-caused. Climate change could be considered a long-term natural disturbance in the GYE. Short-term natural disturbances in the GYE include windthrows, avalanches, landslides, floods, pathogens, and fires. Human-caused disturbances in the GYE include timber harvest, grazing, mining, fire suppression, predator control or elimination, and introduction of exotic species.

Climate change. Hadly (1997) studied late-Holocene ecological responses of small mammals to climate change in the vicinity of Lamar Cave, northern YNP. She determined that pocket gophers, *Microtus* spp., deer mice, bushy-tailed woodrats, and Uinta ground squirrels all showed ecological responses to climate change, increasing or decreasing in abundance according to changes in their preferred habitat.

Timber harvest. Campbell and Clark (1980) studied the effects of clearcutting and selective logging on red-backed voles and deer mice in Bridger-Teton National Forest. Snap-trapping indicated that voles were most abundant on the unlogged and selectively logged mesic sites, while deer mice were more abundant on the xeric clearcuts. Species composition remained unchanged on selective cuts following logging, but changed from predominantly red-backed voles to predominantly deer mice (Campbell and Clark 1980).

Heath (1973) studied the species composition and density of small mammals on four-, six-, eight-, and ten-year-old clearcuts in a subalpine fir *(Abies lasiocarpa)* forest in Gallatin County in south-central Montana; uncut forest served as a control for each clearcut. Deer mice were most abundant in the four-year-old clearcut and decreased in abundance on older clearcuts and within the forest. Red-backed voles were more abundant in the forest and older clearcuts than in young clearcuts. Shrews were most abundant in the older clearcuts and less abundant both in the forest and in younger clearcuts.

Beauvais (1997) studied the effects of forest clearcutting on species richness of mammals in the Rocky Mountains. He found that habitat changes accompanying clearcutting allowed common habitat generalists to replace rarer, boreal-adapted species, and he concluded that persistent clearcutting of Rocky Mountain forests, especially on isolated mountain ranges, threatened to reduce mammal diversity at a regional scale. Hargis (1996) investigated the effects of forest fragmentation (due to natural openings and timber clearcuts) on marten and their prey in mature forest habitats in Utah. Capture rates of red-backed voles declined with increasing fragmentation, while deer mouse capture rates increased.

Wildfire. Wildfires can have a tremendous impact on vegetation (habitat) over a wide geographic area, with a concomitant impact on small mammal populations and communities. Romme (1982) studied the effects of fire on landscape diversity in subalpine forests of YNP. On the basis of fire scar analysis, he found evidence of fifteen fires since 1600 in his 73–square kilometer study watershed. Seven of those fires were considered major (greater than four hectares), destroying the existing forest and initiating secondary succession.

Romme (1982) concluded that the extensive subalpine plateaus of YNP appear to have a natural fire cycle of three hundred to four hundred years in which large areas burn during a short period, followed by a long, relatively fire-free period during which fuel loads develop.

Wood (1981) studied small-mammal communities after two fires in YNP. Intense fires in forest and shrub communities can significantly reduce small mammals, but grassland fires usually move fast and minimally harm mammals that live below ground. Alley and Moore (1989) also studied small-mammal communities in the first year following a wildfire in YNP—the Fan Creek fire of 1979. Small-mammal populations tend to recover rapidly following light fires, and even after intense fires they recover after a few years.

The wildfires of 1988 were perhaps the most important disturbance event of recent years in the GYE. The Northern Range Small Mammal Study investigated the effects of these fires on small-mammal communities in two habitats—sagebrush steppe and forest—on the Northern Range of YNP (table 9.2). In *burned sagebrush steppe* in 1992, deer mice were the most abundant species, although they were only half as abundant as in unburned sagebrush. Meadow and montane voles and shrews were present, although in much lower numbers than in unburned sagebrush. Chipmunks and red-backed voles were not present, although they occurred in low numbers in unburned sagebrush. Uinta ground squirrel numbers and northern pocket gopher mounding rates increased in burned sagebrush steppe compared to unburned sagebrush. In 1993 deer mice were equally abundant in unburned and burned sagebrush. Meadow and montane voles and shrews virtually disappeared from burned sagebrush steppe, whereas they were present in unburned sagebrush. (Shrew numbers actually increased over 1992.) Chipmunks appeared in the burned sagebrush, but their numbers were four times higher in unburned sagebrush. Northern pocket gopher mounding rates stayed static in 1993 in burned sagebrush, whereas they increased almost sixfold in unburned sagebrush.

In *burned forest* in 1992, deer mice were the most abundant species captured, and they were about one and one-half times more numerous than in unburned forest. Shrew numbers were also higher in the burned than unburned forest. Red-backed vole numbers were about one-third what they were in unburned forest.

ECOSYSTEM INFLUENCES OF SMALL MAMMALS

Small mammals are an important ecological component of the GYE. Not only are they a critical prey base for mammalian, avian, and reptilian predators, but they can also have a strong influence on ecosystem structure and function through their impacts to the abiotic environment and biotic community.

Important impacts to the abiotic environment include the soil-modifying activities of fossorial species (prairie dogs, ground squirrels, pocket gophers) and the stream modification caused by beavers. Important impacts to the biotic community include herbivory on grasses and shrubs by microtines and leporids, nutrient cycling by prairie dogs, ground squirrels, and northern pocket gophers, dispersal of mycorrhizal fungi by red-backed voles and northern pocket gophers, seed predation and dispersal by red-backed voles and deer mice, and competitive interactions.

Sirotnak (1998) studied the effects of vole *(Microtus)* herbivory on nitrogen dynamics in natural riparian meadows in northern YNP through the use of small-mammal exclosures. Immediately following a peak in vole numbers, more standing herbaceous litter accumulated in exclosures than in unexclosed plots, and litter inside exclosures had a higher *C:N* ratio than litter from control plots. Huntley (1995) also used exclosures to assess small-mammal effects on vegetation in a subalpine meadow in southwestern Colorado. In areas where pikas were excluded, total vegetation abundance and species richness increased. The increase in exclosures compared with control plots was greatest near talus and decreased with distance. Exclusion of pocket gophers or montane voles significantly increased vegetation abundance within exclosures compared to controls.

Small Mammals as Carnivore Prey

Virtually every species of small mammal in the GYE is preyed on by carnivores. Known and suspected small mammal–carnivore prey relations of the GYE are summarized in table 9.3. These relations are based on published studies of carnivore food habits in the GYE—for example, coyotes (Murie 1944, Weaver 1977, Gese et al. 1996), badgers (Minta 1990), martens (Fager 1991, Coffin, 1994, Kujala 1993), and other studies in similar habitats (studies cited in Banci 1994 for wolverine, Buskirk and Ruggiero 1994 for marten, Koehler and Aubry 1994 for lynx, and Powell 1994 for fisher). On the basis of food habits studies, seven species stand out as the most important prey for carnivores in the GYE: northern pocket gophers, voles, Uinta ground squirrels, deer mice, red-backed voles, snowshoe hares, and beavers.

CARNIVORE HABITAT AND FEEDING SPECIALIZATION

AND SMALL-MAMMAL PREY

For purposes of comparison, GYE carnivores can be classified by size, habitat specialization, and feeding specialization. Three body weight categories are used in this volume: (1) small carnivores (less than one kilogram); (2) mesocarnivores (one to fifteen kilograms), and (3) large carnivores (more than fifteen

Table 9.3

Known and suspected small mammal–carnivore prey relations in the Greater Yellowstone Ecosystem

Small mammals	Carnivores																		
	Short-tailed weasel/ermine	Long-tailed weasel	Western spotted skunk	Marten	Mink	Striped skunk	Fisher	Red fox	Raccoon	River otter	Badger	Lynx	Bobcat	Coyote	Wolverine	Gray wolf	Mountain lion	Black bear	Grizzly bear
Shrew spp.	+	+	+		+	+	+		+					+					
Red-backed vole	+	+	+	+		+	+	+	+			+	+	+	+				
Western jumping mouse	+	+	+			+		+			+		+	+					
Sage vole	+	+	+					+			+		+	+					
Deer mouse	+	+		+	+	+	+	+			+	+	+	+	+				
Heather vole	+	+	+	+	+	+	+	+	+		+	+	+	+	+				
Least chipmunk	+	+		+	+	+	+	+			+	+	+	+	+				
Long-tailed vole		+			+		+	+			+		+	+		+			
Yellow-pine chipmunk	+	+		+			+	+			+		+	+					
Meadow vole	+	+	+	+	+	+	+	+	+		+	+	+	+	+	+			
Uinta chipmunk	+	+		+	+		+	+			+		+	+		+			
Montane vole	+	+	+	+		+		+	+		+	+	+	+	+	+			
Water vole	+				+					+				+		+			
Northern flying squirrel		+		+			+							+					

Prey species																
Northern pocket gopher	+			+	+	+	+	+	+							
Pika						+	+						+			
Red squirrel					+	+	+					+	+		+	+
Bushy-tailed woodrat						+	+	+	+			+		+		+
Uinta ground squirrel	+		+	+	+	+	+		+							
Golden-mantled ground squirrel			+	+	+	+	+	+	+			+	+	+		+
Mountain cottontail			+	+		+	+		+			+				
Desert cottontail			+	+		+	+		+			+				
White-tailed prairie dog						+	+			+				+		
Muskrat						+	+				+				+	+
Snowshoe hare	+		+	+	+	+	+	+	+			+	+			
Yellow-bellied marmot			+	+	+	+	+	+					+			
White-tailed jackrabbit			+	+		+	+	+	+			+	+			
Porcupine			+	+	+	+	+									
Beaver			+	+	+	+	+			+	+					

Note: A + symbol indicates that the small mammal is known or suspected to be prey of indicated carnivore in the GYE.

kilograms). Buskirk (Chapter 7) has classified carnivores as habitat generalists or habitat specialists and feeding generalists or feeding specialists. This provides a 3-by-2-by-2 matrix of body size, habitat specialization, and feeding specialization that can be used for discussion of the importance of small-mammal prey and the importance of habitat disturbance.

Small and large carnivores. Small carnivores of the GYE (ermine, long-tailed weasel, striped skunk) tend to be both habitat and feeding generalists. Although their body size limits the size and type of prey they can capture, they still consume a variety of prey within those limits. Large carnivores of the GYE (gray wolf, mountain lion, black bear, grizzly bear) tend, because of their mobility and other factors (Chapter 7), to be habitat generalists. And although some large carnivores clearly emphasize one species or size of prey (for example, mountain lions and mule deer, wolves and elk), large carnivores are feeding generalists in that they can take a variety of prey if the need or opportunity arises.

Mesocarnivores. It is the mesocarnivores of the GYE that exhibit the greatest variation in habitat and feeding specialization. Buskirk (Chapter 7) discusses these in more detail.

1. Habitat generalist, feeding generalist. Habitat and feeding generalists are rather resilient to both seasonal and yearly fluctuations in prey abundance and to habitat and prey changes brought about by disturbance events. Normal seasonal or yearly fluctuations in prey abundance are met with functional responses—simple "prey switching" to more abundant species. Drastic seasonal or yearly changes in prey abundances are met with a numerical response—increases or decreases in reproductive rates and survival of young. Habitat disturbance events tend to have a relatively smaller impact on these mesocarnivores because they can utilize whatever residual or newly colonizing prey is available. Coyotes typify the category of mesocarnivores that use a variety of habitats and a variety of prey (e.g., Murie 1944).

2. Habitat generalist, feeding specialist. The badger is perhaps the best example of a mesocarnivore that exhibits great flexibility in habitats (ranging from sagebrush steppe to alpine tundra), but specializes in one form of prey—fossorial small mammals (for example, northern pocket gophers, Uinta ground squirrels). Because of their prey specialization, badgers will be influenced by seasonal and yearly fluctuations in prey abundance and by habitat and prey changes brought about by disturbance events. Seasonal reductions in food availability in winter, for example, result in lowered levels of activity. Minor yearly fluctuations in prey produce limited "prey switching"; major fluctuations result in a numerical response. Disturbance events that produce favorable habitats for fossorial prey—and

therefore increased prey populations—have a positive impact on badgers. Conversely, disturbance events that negatively affect habitat for prey populations will have a negative impact on badgers. The 1988 fires had a positive impact on Northern Range sagebrush steppe habitats for fossorial prey, as indicated by the higher numbers of Uinta ground squirrels trapped and northern pocket gopher mounds counted there compared with the counts on unburned sagebrush (Crabtree et al. 1997). Badgers almost certainly benefited from this increased prey base.

3. Habitat specialist, feeding generalist. American martens are considered a habitat specialist in that they are more or less restricted to conifer-dominated forests and, more specifically, to late-successional stands of mesic coniferous forest (Fager 1991, Kujala 1993, Buskirk and Ruggiero 1994, Coffin 1994). They are considered feeding generalists in that they consume a wide variety of small prey within their preferred habitat. Normal seasonal and yearly fluctuations in prey will result in functional and numerical responses by martens. Thus martens need forest, but not solely because of their prey. Intense forest disturbances would have a profound impact on marten populations through both changes in prey species or numbers and changes in habitat structure. The 1988 fires probably had a profound effect on American marten populations, especially in areas where the fires were most intense. In Hargis's (1996) investigation of the effects of forest fragmentation on marten and their prey in mature forests in Utah, martens were negatively correlated with increasing fragmentation (due to both natural openings and timber clearcuts). Fishers are also forest habitat specialists in the GYE.

4. Habitat specialist, feeding specialist. The lynx is both a habitat specialist and a feeding specialist. It occurs in mature coniferous forests and specializes on snowshoe hares and, to a lesser extent, grouse. That lynx are habitat and feeding specialists, at least in the northern part of their range, is indicated by the cyclic changes in their populations that are a consequence of the "snowshoe hare cycle." Large-scale fires in 1988 and the subsequent regrowth of vegetation could lead to an increase in snowshoe hare populations over the next five to thirty years. This may, in turn, provide an excellent food source for lynx and could lead to functional or numerical responses.

River otters are also both habitat and feeding specialists, although their principal prey is fish, not small mammals. Buskirk (Chapter 7) and others have noted that introduction of lake trout to Yellowstone Lake could have profound effects on other fish species that could, in turn, affect densities and distributions of fish-eating carnivores such as mink and river otters. Activities that degrade water quality and thereby reduce prey populations in rivers and streams where river otters occur will also have profound effects on otter populations.

WOLF RESTORATION AND SMALL-MAMMAL PREY

Although ungulates (mainly elk) are the predicted major prey for wolves reintroduced to the GYE (Singer 1990), the importance of small mammals to wolves and the impact of wolves on small-mammal populations should not be overlooked (Crabtree 1992). Although the majority of food biomass consumed by wolves is ungulate prey, small mammals may be important at certain times of the year and in certain locations.

A large majority of the studies reviewed in Crabtree (1992) indicate that nonungulate prey (primarily rodents and lagomorphs) constitutes less than 10 percent of the wolf diet on an annual basis. However, some studies have reported that a majority of the diet within a given season, in both biomass and number of individuals, was small mammals (Voigt et al. 1976, Theberge et al. 1978). Several studies demonstrate the importance of small mammals as a "buffer" prey during the summer, when ungulates are less available (Murie 1944, Cowan 1947, Mech 1970, Carbyn and Kingsley 1979, Fuller 1989). Many wolf food habits studies indicate significant seasonal variation in availability and amount consumed of various prey species. Besides seasonal prey switching, small-mammal prey supplement ungulate prey in wolf diets on an annual basis, especially when ungulate numbers are declining (Voigt et al. 1976).

Many wolf food habits studies note the importance of small-mammal prey during the pup-rearing season. Small-mammal prey consistently increases in the wolf diet during the spring and summer months. Wolves are energetically stressed during this period because they must feed dependent pups, yet their activity is centered at dens and rendezvous sites. Numerous studies have noted the relation between the location of wolf rendezvous sites and substantial exploitation of locally abundant small-mammal prey (e.g., Theberge and Cottrell 1977 for ground squirrels, Theberge et al. 1978 for beaver). Summer food availability and subsequent pup starvation have been singled out as a major limiting factor in wolf populations (Packard and Mech 1980). Peterson (1977) demonstrated that increased beaver abundance can result in increased pup survival. Small-mammal prey may serve also as a learning or "weaning" source for larger pups becoming independent from their parents. As a result, small-mammal prey—or the lack of it—could play a significant role in the wolf recovery attempt because successful restoration is measured by the survival of pups and their recruitment into the adult population.

Beavers, marmots, and, to a lesser extent, snowshoe hares, are fairly large prey "packages" compared with small rodents. When locally abundant, they can constitute a substantial portion of the wolf's seasonal diet (75 percent for beaver [Voigt et al. 1976] and 27 percent for marmots [Fox and Streveler 1986]).

Large-scale fires in 1988 and the subsequent regrowth of vegetation could lead to an increase in snowshoe hare populations over the next five to thirty years. This may, in turn, provide an excellent food source for wolves and medium-sized carnivores, especially lynx. In a recent study, Fuller (1989) calculated that snowshoe hares were killed and consumed by wolves as often as ungulate prey were, yet constituted only 2–3 percent of the total biomass consumed.

Wolf restoration is predicted to have an indirect impact on small mammal populations and communities through two processes: reduction in numbers of coyotes through interference competition with wolves (see Chapter 6), and vegetation changes resulting from increased predation on ungulates, primarily elk (see Chapters 8 and 10).

Coyotes are expected to decline significantly following wolf restoration in the GYE, based on observations in many North American studies (see studies cited in Chapter 8). Because coyotes are a major predator of a wide variety of small mammals (principally voles, pocket gophers, snowshoe hares, marmots, muskrats, and Uinta ground squirrels), their competitive displacement should result in population increases among these species that will be offset only partially by wolf predation. Prey increases should, in turn, result in functional and numerical responses among other mammalian predators of small mammals. On the Northern Range badgers, foxes, ermine, long-tailed weasels, and mink are predicted to respond functionally or numerically to increased small-mammal prey. In addition, because coyotes are dominant to certain other carnivores in the GYE, decreased coyote numbers are expected to result in an increased number of those other carnivores, especially red foxes (Singer 1991, Chapter 6 this volume).

Under the moderate wolf density scenario that they expect to occur, Singer and Mack (Chapter 8) have predicted that vegetation changes resulting from wolf restoration will be relatively minor, especially in comparison with changes resulting from the 1988 fires. Willow and aspen might increase over a period of years, with possible positive effects on beaver populations.

Recommendations for Small-Mammal Research and Management

Small mammals are important prey not only for the carnivores of the GYE, but for avian and reptilian predators as well. A large number of hawk and owl species depend heavily on small mammals, and snakes also eat them. This importance as a prey base for a variety of predators argues for long-term research and monitoring of small-mammal populations and communities in a number of habitats throughout the ecosystem. Continuation of the Northern Range Small Mammal Study is especially important in order to monitor the direct and indirect effects of wolf reintroduction on small mammals. Long-term studies should also be initiated in less-studied habitats and regions of the GYE, both within and outside protected areas. Such population and community studies should be conducted in concert with long-term studies of the entire predator guild in the area—mammals, birds, and perhaps even reptiles. Only in this way can competitive relations among predators be fully understood (see Chapter 12).

The most important small-mammal prey species in the GYE—northern pocket gophers, microtine voles, Uinta ground squirrels, deer mice, red-backed voles, snowshoe hares, and beavers—should be the focus of additional, in-depth research on population dynamics, habitat use, and ecosystem influences. Regular population monitoring of these species would also be valuable. Increased inventory and monitoring of beavers are especially important because they could serve as an alternate food for reintroduced wolves. Monitoring of snowshoe hare populations is needed because the species is vital to lynx populations. Finally, Uinta ground squirrels should be monitored because of their importance to an entire suite of predators in the GYE.

The role of landscape patterns in influencing small-mammal population dynamics, small-mammal communities, and predator-prey relations remains rather poorly understood in the GYE and deserves additional research. Two topics are of prime interest—the influence of highly patchy landscapes on small-mammal and carnivore biodiversity, and the influence of predation on small-mammal populations in small, isolated patches.

The effects of disturbance and habitat fragmentation on small mammals and their predator-prey relations is also rather poorly understood in the GYE. Disturbance, whether natural or human-caused, can have long-term effects on vegetation and thus on small mammals. Such long-term influences call for long-term studies of small mammals in disturbed areas of the GYE, with initial emphasis on the two most important disturbance factors in the region—wildfire and timber harvest. Grazing impacts by both wild ungulates and domestic livestock on small-mammal communities are also worthy of long-term investigation and hypothesis testing.

The importance of nonmammalian, vertebrate small prey to carnivores also appears to be rather poorly understood in the GYE. Reptiles, amphibians, and birds are preyed on by a wide variety of both mammalian and avian predators. Their significance as prey, as well as their potential role in buffering predation on small mammals, deserves further research. Routine population monitoring of selected indicator species of reptiles, amphibians, and birds would be a useful component of an ecosystem monitoring program in the GYE. Monitoring is especially crucial for amphibians because this group has been declining drastically throughout the western United States in recent years, and the loss of amphibian populations or species would be a blow to the biological diversity of the GYE.

With the reintroduction of the wolf, the GYE's predator-prey system is now much as it was before the advent of Euro-Americans. The GYE now presents the opportunity to study and monitor an intact predator-prey system at the ecosystem level over, we hope, decades. Such studies would be vitally important to our understanding of how the GYE works, which in turn might give us the information and insight needed to manage the system better.

Literature Cited: Alley, J. J., and R. E. Moore. 1989. The small mammal communities in the first year following the Fan Creek fire in 1979. Final report. Addendum to contract no. CX-1200-9-B035, Yellowstone National Park.

Banci, V. 1987. Ecology and behavior of wolverine in Yukon. M.S. thesis, Simon Fraser University, Burnaby, British Columbia.

——. 1994. Wolverine. Pp. 99–127 *in* L. F. Ruggiero, K. B. Aubry, S. W. Buskirk, L. J. Lyon, and W. J. Zielinski, eds., The scientific basis for conserving forest carnivores: American marten, fisher, lynx, and wolverine in the western United States. General Technical Report RM-254, USDA, Forest Service, Rocky Mountain Forest and Range Experiment Station, Fort Collins.

Beauvais, G. P. 1997. Mammals in fragmented forests in the Rocky Mountains: Community structure, habitat selection, and individual fitness. Ph.D. diss., University of Wyoming, Laramie.

Boles, B. K. 1977. Predation of wolves on wolverines. Canadian Field-Naturalist 91:68–69.

Buskirk, S. W., and L. F. Ruggiero. 1994. American marten. Pp. 7–37 *in* L. F. Ruggiero, K. B. Aubry, S. W. Buskirk, L. J. Lyon, and W. J. Zielinski, eds., The scientific basis for conserving forest carnivores: American marten, fisher, lynx, and wolverine in the western United States. General Technical Report RM-254, USDA, Forest Service, Rocky Mountain Forest and Range Experiment Station, Fort Collins.

Campbell, T. M., III, and T. W. Clark. 1980. Short-term effects of logging on red-backed voles and deer mice. Great Basin Naturalist 40:183–89.

Carbyn, L. N., and M. C. S. Kingsley. 1979. Summer food habits of wolves with emphasis on moose in Riding Mountain National Park. Proceedings of the North American Moose Conference Workshop 15:349–61.

Clark, T. W. 1971. Ecology of the western jumping mouse in Grand Teton National Park, Wyoming. Northwest Science 45:229–38.

——. 1973a. Distribution and reproduction of shrews in Grand Teton National Park, Wyoming. Northwest Science 47:128–31.

——. 1973b. Local distribution and interspecies interactions in microtines, Grand Teton National Park, Wyoming. Great Basin Naturalist 33:205–17.

——. 1975. Ecological notes on deer mice in Grand Teton National Park. Northwest Science 49:14–16.

Clark, T. W., and M. R. Stromberg. 1987. Mammals in Wyoming. University of Kansas Museum of Natural History, Public Education Series no. 10. University Press of Kansas, Lawrence.

Coffin, K. W. 1994. Population characteristics and winter habitat selection by pine martens in southwest Montana. M.S. thesis, Montana State University, Bozeman.

Consolo, S. L., and D. D. Hanson. 1990. Distribution of beaver in Yellowstone National Park,

1988–1989. Pp. 2-217 to 2-235 *in* Wolves for Yellowstone? A report to the United States Congress, vol. 2, research and analysis. National Park Service, Yellowstone National Park.

Cowan, I. McT. 1947. The timber wolf in the Rocky Mountain national parks of Canada. Canadian Journal of Research Section D 25:139–74.

Crabtree, B., M. Harter, P. Moorcroft. K. Johnson, and D. Despain. 1997. A landscape approach to inventory and monitoring of small mammal prey in northern Yellowstone. Final report to the Yellowstone Center for Resources, Yellowstone National Park.

Crabtree, R. L. 1992. A preliminary assessment of the non-ungulate mammal prey base for wolves in Yellowstone National Park. Pp. 5-122 to 5-138 *in* J. D. Varley and W. G. Brewster, eds., Wolves for Yellowstone? A report to the United States Congress, v. 4, research and analysis. National Park Service, Yellowstone National Park.

Despain, D. 1991. Yellowstone vegetation: Consequences of environment and history in a natural setting. Roberts Rinehart, Boulder, Colorado.

Fager, C. W. 1991. Harvest dynamics and winter habitat use of the pine marten in southwest Montana. M.S. thesis, Montana State University, Bozeman.

Fox, J. L., and G. P. Streveler. 1986. Wolf predation on mountain goats in southeastern Alaska. Journal of Mammalogy 67:192–95.

Fuller, T. K. 1989. Population dynamics of wolves in north-central Minnesota. Wildlife Monographs no. 105.

Gardner, C. L. 1985. The ecology of wolverines in southcentral Alaska. M.S. thesis, University of Alaska, Fairbanks.

Gese, E. M., R. L. Ruff, and R. L. Crabtree. 1996. Intrinsic and extrinsic factors influencing coyote predation of small mammals in Yellowstone National Park. Canadian Journal of Zoology 74:784–97.

Hadly, E. A. 1997. Evolutionary and ecological response of pocket gophers *(Thomomys talpoides)* to late-Holocene climatic change. Biological Journal of the Linnean Society 60:277–96.

Hadly, E. A., M. H. Kohn, J. A. Leonard, and R. K. Wayne. 1998. A genetic record of population isolation in pocket gophers during Holocene climatic change. Proceedings, National Academy of Sciences, USA, 95:6893–96.

Hargis, C. D. 1996. The influence of forest fragmentation and landscape pattern on American martens and their prey. Ph.D. diss., Utah State University, Logan.

Harter, M. K., and R. L. Crabtree. 1991. A landscape approach to inventory and monitoring of small mammals. Yellowstone National Park progress report. Yellowstone National Park.

Heath, M. L. 1973. Small mammal populations in clearcuts of various ages in south-central Montana. M.S. thesis, Montana State University, Bozeman.

Hodgson, J. R. 1970. Ecological distribution of *Microtus montanus* (Peale) and *Microtus pennsylvanicus* in an area of geographic sympatry in southwestern Montana. Ph.D. diss., Montana State University, Bozeman.

Houston, D. B. 1982. The northern Yellowstone elk. Macmillan, New York.

Huntley, N. J. 1985. The influence of herbivores on plant communities: Experimental studies of a subalpine meadow ecosystem. Ph.D. diss., University of Arizona, Tucson.

Koehler, G. M., and K. B. Aubry. 1994. Lynx. Pp. 74–98 *in* L. F. Ruggiero, K. B. Aubry, S. W. Buskirk, L. J. Lyon, and W. J. Zielinski, eds., The scientific basis for conserving forest carnivores: American marten, fisher, lynx, and wolverine in the western United States. General Technical Report RM-254, USDA, Forest Service, Rocky Mountain Forest and Range Experiment Station, Fort Collins.

Kujala, Q. J. 1993. Winter habitat selection and population status of pine marten in southwest Montana. M.S. thesis, Montana State University, Bozeman.

Mattson, D. J., B. M. Blanchard, and R. R. Knight. 1992. Yellowstone grizzly bear mortality, human habituation, and whitebark pine seed crops. Journal of Wildlife Management 56:432–42.

Mech, L. D. 1970. The wolf: Ecology and behavior of an endangered species. Doubleday, New York.

Minta, S. C. 1990. The badger, *Taxidea taxus* (Carnivora: Mustelidae): Spatial-temporal analysis, dimorphic territorial polygyny, population characteristics, and human influences on ecology. Ph.D. diss., University of California, Davis.

Murie, A. 1944. The ecology of the coyote in the Yellowstone. U.S. National Park Service Fauna Series no. 4.

Negus, N. C. 1950. Habitat adaptability of *Phenacomys* in Wyoming. Journal of Mammalogy 31:351.

Nordyke, K. A., and S. W. Buskirk. 1991. Southern red-backed vole, *Clethrionomys gapperi*, populations in relation to stand succession and old-growth character in the central Rocky Mountains. Canadian Field-Naturalist 105:330–34.

Packard, J. M., and L. D. Mech. 1980. Population regulation in wolves. Pp. 135–50 *in* M. N. Cohen, R. S. Malpass, and H. G. Klein, eds., Biosocial mechanisms of population regulation. Yale University Press, New Haven.

Pattie, D. L., and N. A. M. Verbeek 1967. Alpine mammals of the Beartooth Mountains. Northwest Science 41:110–17.

Peterson, R. O. 1977. Wolf ecology and prey relationships on Isle Royale. U.S. National Park Service Science Monograph Series no. 11.

Pinter, A. J. 1988. Multiannual fluctuations in precipitation and population dynamics of the montane vole *Microtus montanus*. Canadian Journal of Zoology 66:2128–32.

Powell, R. A., and W. J. Zielinski. 1994. Fisher. Pp. 38–73 *in* L. F. Ruggiero, K. B. Aubry, S. W. Buskirk, L. J. Lyon, and W. J. Zielinski, eds., The scientific basis for conserving forest carnivores: American marten, fisher, lynx, and wolverine in the western United States. General Technical Report RM-254, USDA, Forest Service, Rocky Mountain Forest and Range Experiment Station, Fort Collins.

Reinhart, D. P., and D. J. Mattson 1990. Red squirrels in the whitebark zone. International Conference on Bear Research and Management 8:343–50.

Rieger, J. F. 1996. Body size, litter size, timing of reproduction, and juvenile survival in the Uinta ground squirrel, *Spermophilus armatus*. Oecologia 107:463–68.

Romme, W. H. 1982. Fire and landscape diversity in subalpine forests of Yellowstone National Park, Wyoming, USA. Ecological Monographs 52:199–221.

Sherburne, S. S., and J. A. Bissonette. 1993. Squirrel middens influence marten *(Martes americana)* use of subnivean access points. American Midland Naturalist 129:204–7.

Singer, F. J. 1990. Some predictions concerning a wolf recovery into Yellowstone National Park: How wolf recovery may affect park visitors, ungulates, and other predators. Pp. 4-3 to 4-34 *in* Wolves for Yellowstone? A report to the United States Congress, vol. 2, Research and analysis. National Park Service, Yellowstone National Park.

———. 1991. Some predictions concerning a wolf recovery into Yellowstone National Park: How wolf recovery may affect park visitors, ungulates and other predators. Transactions of the North American Wildlife and Natural Resources Conference 56:567–83.

Sirotnak, J. M. 1998. The effects of voles *(Microtus)* on nitrogen dynamics in riparian ecosystems. Ph.D. diss., Idaho State University, Pocatello.

Streubel, D. 1989. Small mammals of the Yellowstone ecosystem. Roberts Rinehart, Boulder, Colorado.

Theberge, J. B., and T. J. Cottrell. 1977. Food habits of wolves in Kluane National Park. Arctic 30:189–91.

Theberge, J. B., S. M. Oosenbrug, and D. H. Pimlott. 1978. Site and seasonal variations in food of wolves, Algonquin Park, Ontario. Canadian Field-Naturalist 92:91–94.

Urban, D. L., R. V. O'Neill, and H. H. Schugart, Jr. 1987. Landscape ecology: A hierarchical perspective can help scientists understand spatial patterns. BioScience 37:119–27.

Voigt, D. R., G. B. Kolenosky, and D. H. Pimlott. 1976. Changes in summer foods of wolves in central Ontario. Journal of Wildlife Management 40:663–68.

Weaver, J. L. 1977. Coyote-food base relationships in Jackson Hole, Wyoming. M.S. thesis, Utah State University, Logan.

Youmans, C. C. 1979. Characteristics of pocket gopher populations in relation to selected environmental factors in Pelican Valley, Yellowstone National Park. M.S. thesis, Montana State University, Bozeman.

Wood, M. A. 1981. Small mammal communities after two recent fires in Yellowstone National Park. M.S. thesis, Montana State University, Bozeman.

Alpha male of the Chief Joseph pack of gray wolves in Yellowstone National Park (William Campbell)

Chapter Ten

Evaluating the Role of Carnivores in the Greater Yellowstone Ecosystem

Mark S. Boyce and Eric M. Anderson

Wolf recovery has restored the primary large predator in the Greater Yellowstone Ecosystem (GYE). This finally brings to closure a long-standing debate about the management of Yellowstone that has revolved around the need to cull elk in the park to replace natural predation (Pengelly 1963, Chase 1986, Despain et al. 1986, Kay 1990). We now have the opportunity to document the consequences that wolves have on ungulates and vegetation, particularly on Yellowstone's Northern Range, which hosts the park's largest concentration of wintering ungulates.

In general, it is widely known that predators can play a major role in shaping the structure and composition of ecosystems. For example, on Isle Royale National Park wolves appear to limit moose *(Alces alces)* populations, and moose in turn have a large effect on vegetation succession (McLaren and Peterson 1994). Evidence suggests a top-down controlled system on Isle Royale, where fluctuations in moose numbers are influenced by wolf numbers, while fluctuations in vegetation appear to have little consequence to the moose population.

Might we expect such a pattern in Yellowstone? Will wolves dominate the dynamics of the system such that ungulate populations and even the vegetation structure and composition are affected in a major way, or are we likely to see little effect attributable to wolves because ungulate numbers are largely regulated by a plant-herbivore interaction? Our experiences in Yellowstone lead us to believe that fluctuations in vegetation and overwinter mortality will continue to cause major perturbations to ungulate populations on the Northern Range, irrespective of the presence of wolves (Merrill and Boyce 1991, Coughenour and Singer 1996, Chapter 8 this volume). Ungulate numbers fluctuate substantially without wolves; we expect that one of the impacts of wolves will be to alter the magnitude of these fluctuations.

Our objective is to review ideas on top-down vs. bottom-up control of ecosystem dynamics and to consider possible implications for the dynamics and management of the GYE. We conclude that the distribution of stochastic variance in the system is likely to determine the apparent role that predators have in ungulate communities, and we conclude that understanding this variance will probably require experimental manipulations. This points to the great scientific opportunity for study that wolf recovery in Yellowstone provides as a manipulation of the system.

Do Predators Regulate Ungulate Numbers?

Some ecologists have argued that ungulate populations are regulated ultimately by "bottom-up" interactions with their food resources, and although predators may crop the "harvestable surplus" (Errington 1967), predators usually do not drive the dynamics of ungulate populations (Cole 1971, Despain et al. 1986). The alternative view is that predators substantially limit ungulate populations and fundamentally structure the dynamics of ungulate populations from the "top down" (McLaren and Peterson 1994).

Top-down vs. bottom-up control of population processes does not necessarily imply how a population is regulated. Population "regulation" implies a density-dependent influence (Sinclair 1989, Messier 1991, 1995). When population size becomes large, regulating forces cause population growth rates to decline, bringing the population closer to some equilibrium level. Conversely, if numbers are low, regulating mechanisms permit populations to increase. Although predators certainly can limit prey abundance, there is less evidence that predators necessarily operate in a density-dependent manner. Demonstrating population regulation via predation becomes particularly difficult in multispecies systems, because alternative prey or alternative predators can complicate the possibilities (Erlinge et al. 1984).

In general, when prey are more abundant, individual predators kill more of them. However, for any predator there exists an upper-threshold consumption rate—that is, a satiation level. Satiation operates in an inverse density-dependent fashion such that, as prey increase from moderate to high density, fewer prey will be killed per capita. Especially for mammalian carnivores, we may see density-dependent predation occurring at lower prey densities, where increases in prey density result in learning or the development of a search image. Thus the predator becomes progressively more effective as a predator, and prey consumption expands at an increasing rate (Maynard Smith 1974). The combination of density-dependent predation rate and satiation results in a logistic or Type III functional response (figure 10.1). Garton et al. (1990) present evidence that a Type III functional response occurs for wolves preying on moose.

Overview of the Northern Range Ecosystem

Since 1968 no harvest of elk has been permitted in Yellowstone National Park (Houston 1982) despite recommendations by a committee commissioned by the National Academy of Sciences that elk numbers on the Northern Range should be kept below five thousand (Leopold et al. 1963; see also Chapters 2 and 8 this volume). The ostensible justification for this "natural regulation experiment" was that ungulate populations were regulated by their food and habitat resources and not by predators (Cole 1971, Despain et al. 1986).

Under natural regulation, the northern Yellowstone elk herd increased to approximately twenty thousand animals, declined by approximately 7,500 during the winter of 1988–89 (Singer et al. 1989), and subsequently increased to more than twenty thousand, consistent with model predictions (Boyce and Merrill 1991). Some scientists believe that such high densities of elk will "damage" the vegetation (Wagner and Kay 1993), whereas others argue that this is a subjective interpretation and that herbivore consequences to vegetation involve natural interactions that can persist indefinitely (Despain et al. 1986, Boyce 1991). There appears to be no question that high ungulate densities will alter the vegetation on which they forage, but conclusions that effects on vegetation are "bad" invoke value judgments that are typically based on standards set for livestock use of rangelands (Dodd 1993).

Regardless, the Yellowstone ecosystem was altered by the elimination of wolves from the 1880s through the late 1920s, and thus wolf recovery should help resolve the controversy about whether ungulate populations must be culled to replace natural predation. Since they were released in 1995, wolves have killed elk in the park at a relatively high rate (see Chapter 5). Models have been constructed to predict the consequences of wolf recovery to ungulate populations in Yellowstone National Park (Garton et al. 1990, Vales and Peek 1990, Boyce 1992, 1993, 1995, Boyce and Gaillard 1992, Chapter 8 this volume), but these models were designed only to predict the interaction between wolves and ungulates and do not attempt to include plant-herbivore interactions.

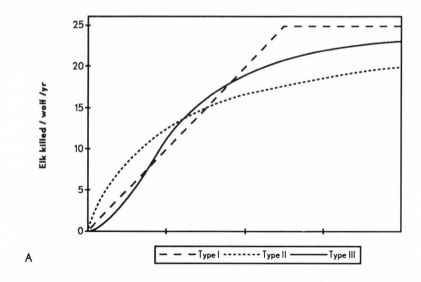

A

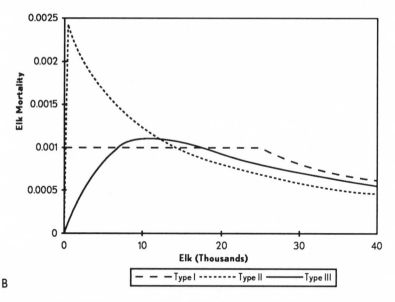

B

10.1. Type I, II, and III functional responses (Holling 1959) showing the number of prey killed per predator per year as a function of prey abundance (A), and expressed as the predation mortality rate on prey as a function of prey abundance (B). With a logistic (Type III) functional response at low to midsized populations, predation mortality increases with density.

A Model for Three Trophic Levels

We propose a simple three-trophic-level model for the Northern Range ecosystem. Ignoring most details of the food web, we deliberately focus on three components—vegetation, elk, and wolves. Only vegetation palatable to elk is considered; other herbivores are ignored because their relative abundances are low compared with the number of elk (Singer 1991); and to keep the model simple we assume that wolves are the primary predators on elk.

To frame an interactive three-trophic-level model we begin with the two-species Kolmogorov (1936) equations that Caughley (1976) adapted for a plant-herbivore model. We then complicate the system by adding logistic-functional-response predation by wolves and a density-dependent equation for the abundance of wolves. We assume that at time t the vegetation biomass, $V(t)$, elk numbers, $H(t)$, and wolves, $P(t)$, are governed by the following growth equations:

$$dV(t)/dt = bV(t)[1 - V(t)/K_r] - fH(t)\{1 - \exp[-a_1 V(t)]\} \tag{1}$$

$$dH(t)/dt = c_2 H(t)\{1 - \exp[-a_2 V(t)] - d_H H(t)\} - F(t)H(t)P(t) \tag{2}$$

$$dP(t)/dt = a_3 F(t)P(t) - (r_p/K_p)P(t)^2 - d_p P(t) \tag{3}$$

We define variables in table 10.1. Throughout, predators are assumed to kill prey according to a logistic functional response, $F(t)$, of the form:

$$F(t) = [A \cdot H(t)^2] / [1 + A \cdot T_h \cdot H(t)^2] \tag{4}$$

Approximate integration was used to derive the following set of difference equations for annual time increments, which were used for simulations:

Table 10.1

Definitions and values of variables used during simulation trials, per equations 1–7

Variable	Definition	Values used
V	Vegetation biomass	State variable, kg
H	Number of elk	State variable
P	Number of wolves	State variable
b	Potential growth of vegetation	1.4 (unitless)
K_r	Vegetation carrying capacity	30,000 kg
f	Maximum rate of food intake per elk	2.5 kg/elk/day
a_1	Herbivory efficiency with V small	1.5×10^{-5}
c_2	Rate of elk decline ameliorated by V	1.7
a_2	Elk demographic efficiency with small V	8×10^{-5}
d_H	Rate of elk population decline with small V	1.2
a_3	Wolf numerical response coefficient	0.075
r_p	Intrinsic rate of increase (wolf)	0.8
K_p	Wolf carrying capacity	100 wolves
d_p	Wolf death rate with low elk numbers	0.68
F	Functional response	$A \cdot H(t)^2/[1 + A \cdot T_h \cdot H(t)^2]$/wolf/day
A	Attack rate	1.2×10^{-6}
T_h	Prey handling time	0.04

Note: Estimation procedures are described in Boyce and Gaillard 1992.

$$V(t+1) = V(t) \cdot \exp\{b[1\text{-}V(t)/K_r] - [f \cdot H(t)/V(t)](1\text{-}\exp[-a_1V(t)])\} \qquad (5)$$

$$H(t+1) = H(t) \cdot \exp\{c_2(1\text{-}\exp[-a_2V(t)]\text{-}d_H) - F(t)P(t)\} \qquad (6)$$

$$P(t+1) = P(t) \cdot \exp\{a_3F(t) - (r_p/K_p)P(t) - d_p\} \qquad (7)$$

The model was parameterized using values similar to a more complex model developed to simulate the consequences of wolf recovery in Yellowstone (Boyce 1992, 1993, 1995, Boyce and Gaillard 1992).

We first performed deterministic recursive calculations to illustrate the magnitude of the influence of wolves on elk and vegetation. Without wolves, using the parameter values listed in table 10.1, the elk-vegetation interaction undergoes convergent oscillations to an equilibrium of about 21,000 elk and about 15,000 units of vegetation (figure 10.2). Adding wolves to the system (equations 5–7) has substantial consequences (figure 10.3). Although we still see convergent oscillations to equilibrium, we predict a 40 percent reduction in the number of elk to about 12,500 animals at equilibrium and a concomitant increase in vegetation to some 22,000 units associated with a population of about 120 wolves. Clearly, wolves have major consequences to the structure of this model ecosystem.

Next we added Gaussian stochastic variance to the system. First, we modeled a 20 percent coefficient of variation (cv) in vegetation carrying capacity,

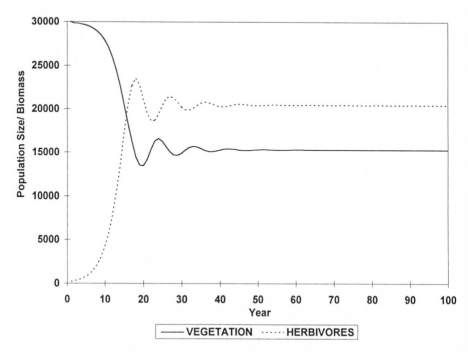

10.2. Dynamics of a deterministic vegetation-elk interaction showing convergent oscillations to equilibrium at about 21,000 elk and 15,000 units of vegetation.

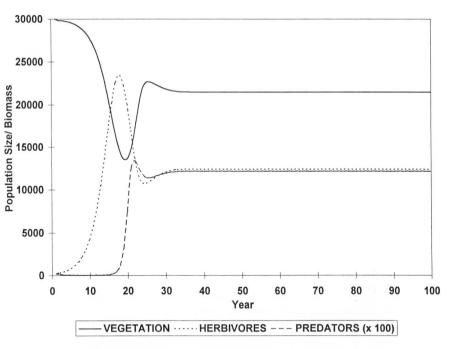

10.3. Dynamics of a deterministic three-species system to anticipate dynamics of the vegetation-elk-wolf interaction in Yellowstone National Park.

K_r, while leaving all other parameter variables exactly as in the deterministic case illustrated in figure 10.3. Our rationale was to simulate the dynamic growth conditions for the vegetation from year to year similar to the annual fluctuations in green herbaceous phytomass documented by Merrill et al. (1993). The resulting time series show high-frequency fluctuations in vegetation biomass, and lagged low-frequency oscillations in both elk and wolf numbers. Elk and wolf numbers do not vary nearly so much as the vegetation, and wolf numbers lag shortly behind those of elk (figure 10.4); average population sizes in these stochastic simulations are similar to those observed in the deterministic projections.

Patterns of population fluctuations can be characterized by the correlation between the population size this year and last year or two years ago. Such patterns in the structure of a time series are usefully summarized by plotting the autocorrelation function (ACF), which is the correlation of a time series with itself at a series of lags. For our simulation results we have characterized patterns of population fluctuations by plotting ACFs for first-order differences in $\log_e$ abundance for each of the three ecosystem components (figure 10.5). Predator population fluctuations have a high first-lag autocorrelation, indicative of the relatively smooth oscillations in numbers in contrast to the abrupt up-and-

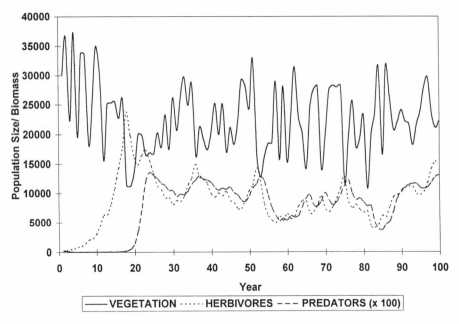

10.4. Dynamics of the three-species system plotted at figure 10.3, but with 20 percent stochastic variation in the carrying capacity for vegetation, K_r.

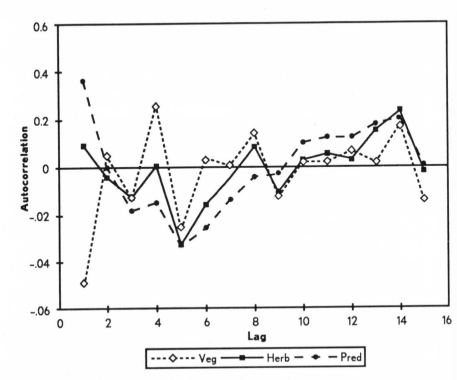

10.5. Autocorrelation functions for vegetation, elk, and wolves from the time series plotted in figure 10.4.

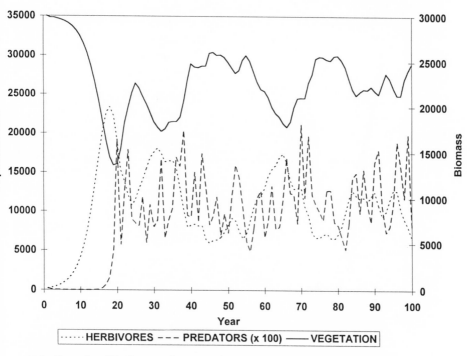

10.6. Dynamics of the three-species system plotted at figure 10.3, but with 20 percent stochastic variation in the functional response for predators.

down fluctuations in vegetation biomass reflected by the negative first-lag autocorrelation. No pronounced patterns are evident in the ACF for herbivores.

For the next set of simulations, we again set K_p equal to a constant but instead perturbed the functional response by adding normally distributed white noise to an average value of $F(t)$ that was calculated from equation 4. Again, variation surrounding the average $F(t)$ was set at 20 percent cv. This scenario might occur if wolf predation varied substantially from year to year as a result of variation in conditions for predation; for example, prey might be more vulnerable in years of deep snow. As we illustrate in the projected time series (figure 10.6) and corresponding ACF (figure 10.7), the pattern of dynamics is altered quite substantially by perturbing the system at the top rather than the bottom. We now see oscillations of much higher frequency in the predator population and comparatively smoother autocorrelated fluctuations in the vegetation.

We also studied a time series for elk and wolves projected from program WOLF5 (see Boyce 1992, 1995). Here we collapse the vegetation term into a simple carrying capacity term for the elk population, $K_{elk}(t)$, that varies stochastically to simulate the dynamics of winter range fluctuation as a consequence of annual variation in winter severity but is not influenced by the size of the elk population. Such a noninteractive system (Caughley and Lawton 1981) may be

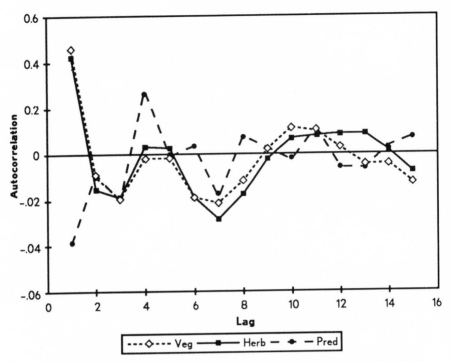

10.7. Autocorrelation functions for vegetation, elk, and wolves from the time series plotted in figure 10.6.

justified if we can assume that elk largely depend on the biomass of perennial herbaceous plants. We suspect that this model may be a reasonable approximation of elk herbivory on Yellowstone's Northern Range because the vegetation eaten by ungulates is mostly current annual growth of grasses and sedges (Houston 1982), although grazing may actually stimulate plant growth (Coughenour and Singer 1996). Grazing during winter does not kill the plant because the living portions are frozen in the ground, yielding a largely noninteractive plant-herbivore system (but see Frank and McNaughton 1992). Indeed, using data from Yellowstone for 1972–88, Merrill and Boyce (1991) were able to show a significant effect of vegetation on elk recruitment, survival, or population growth rates, but no effect could be demonstrated for elk numbers on vegetation the subsequent year. Thus, for simplicity, if we assume that elk numbers do not feed back to the vegetation biomass, then the available vegetation for elk varies according to normally distributed white noise in this model. The consequences to dynamics are most pronounced for the wolves, which assume greater periodicity than in our fully interactive three-species models (figure 10.8).

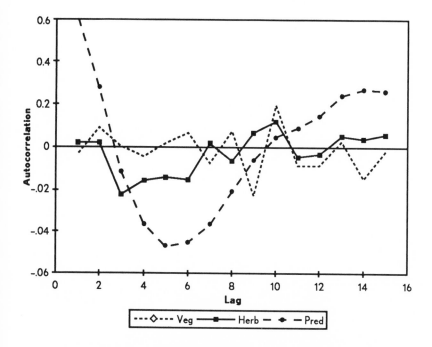

10.8.   Autocorrelation functions for elk and wolves from the time series generated by program WOLF5 (Boyce 1992), a model with a noninteractive vegetation-elk component where elk do not influence the biomass of vegetation.

Comparison with Field Data

We compared the dynamic patterns from simulations with time series of the moose and wolves on Isle Royale and of elk on the Northern Range of Yellowstone. An updated time series for wolves and moose on Isle Royale was obtained from Rolf Peterson (Michigan Technological University, Houghton) and for elk in Yellowstone from John Mack (Yellowstone National Park). The moose on Isle Royale have an ACF with a significant positive first-lag correlation and a subsequent pattern suggestive of periodicity (figure 10.9). This coincides with the pattern in the ACF for herbivores from the simulations with a stochastic functional response—that is, external perturbations to wolf numbers (figure 10.7). The wolf population on Isle Royale under the same model scenario ought to show a negative correlation at a low lag (figure 10.7). Although none of the autocorrelations are statistically significant, the pattern in the Isle Royale wolf ACF (figure 10.9) is likewise similar to the ACF for the predator in the stochastic-functional-response simulations (figure 10.7).

Analysis of the time series of elk on the Northern Range reveals no significant autocorrelations, but again the pattern in the ACF (figure 10.10) is in-

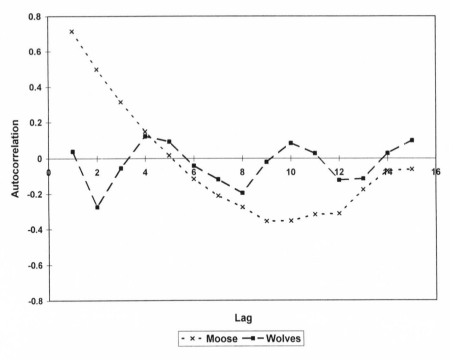

10.9. Autocorrelation functions for moose and wolves on Isle Royale National Park.

triguingly like that for herbivores in the simulations with stochastic vegetation, K_r (figure 10.5). For both the elk data ACF and the stochastic vegetation model ACF, the first lag is slightly positive but goes negative after several lags.

Ecological Implications

We focus our attention on the herbivores to evaluate the importance of top-down vs. bottom-up processes in determining ecosystem dynamics. On Isle Royale, McLaren and Peterson (1994) found that moose numbers appeared to be driven largely by fluctuations in wolf numbers, but there was a lack of evidence that vegetation influenced much of the variation in moose numbers. To place our simulation results in a similar context, we attempt to predict the per capita growth in herbivore numbers as a function of vegetation or predators. Such an approach is structurally similar to a key-factor analysis, where we estimate the variance in herbivore growth rates attributable to various ecological factors (Varley and Gradwell 1960); we also present the autoregressive models (see Royama 1992, Turchin and Taylor 1992). First, for the stochastic vegetation model, we find that we can predict 95 percent of the variance in the modeled herbivore dynamics by knowing the vegetation biomass in two time lags—that is, fitting the model:

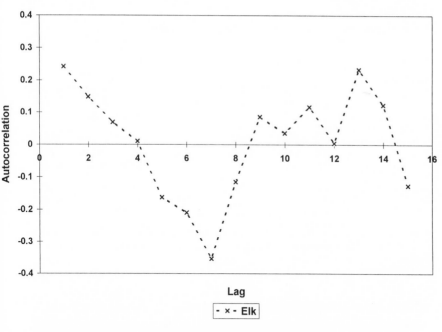

10.10. Autocorrelation functions for detrended counts of elk, 1970–96, from Yellowstone National Park, Wyoming.

$$\ln[H(t+1)/H(t)] = \beta_0 + \beta_1 \ln[V(t)] + \beta_2 \ln[V(t-1)] + \epsilon \qquad (8)$$

where $H(t)$ is the number of herbivores, $V(t)$ is vegetation biomass in year t, and ϵ is a random error term (see table 10.2). However, we could predict only 28 percent of the variance in herbivore dynamics with data on the number of predators in the first and second lags when we replace the $V(t)$s with $P(t)$s in the regression model:

$$\ln[H(t+1)/H(t)] = \beta_0 + \beta_1 \ln[P(t)] + \beta_2 \ln[P(t-1)] + \epsilon \qquad (9)$$

Thus, when stochastic perturbations influence the vegetation, by knowing vegetation biomass we can predict most of the variance in herbivore numbers; in contrast, knowledge of predator numbers does not help much in predicting ungulate numbers.

Next we performed the same regression analysis for the model with a stochastic functional response but found opposite patterns. Now we could predict only 21 percent of the variance in herbivore growth rates based on knowledge of vegetation in two lags, but we could account for 75 percent of this variance by knowing predator numbers.

Note that the predictability is asymmetrical. Although about 95 percent of the variance in elk numbers can be accounted for with vegetation when stochastic variation occurs in vegetation carrying capacity, only 75 percent of the variation in elk population growth rate can be accounted for by wolf numbers

Table 10.2

Regression analysis of herbivore per capita growth rate

Model	Predictor variables[a]	Constant	β_1	β_2	r^2	P
Stochastic						
vegetation	V(t), V(t-1)	−7.18	0.636	0.084	0.952	0.000
	H(t), H(t-1)	1.92	0.196	−0.408	0.332	0.000
	P(t), P(t-1)	0.532	−0.393	0.273	0.281	0.000
Stochastic func-						
tional response	V(t), V(t-1)	−5.39	−0.354	0.892	0.209	0.192
	H(t), H(t-1)	1.366	0.514	−0.622	0.455	0.000
	P(t), P(t-1)	1.43	−0.273	−0.033	0.748	0.000
Noninteractive	V(t), V(t-1)	0.029	0.08	−0.083	0.005	0.805
	H(t), H(t-1)	3.47	−0.132	−0.233	0.19	0.000
	P(t), P(t-1)	0.585	−0.47	0.337	0.193	0.000

Notes: Per capita growth rate measured as a function of 1n (vegetation biomass), 1n (herbivore numbers), and 1n (predator numbers). Sample size for each regression was one hundred years.
[a] 1n transformed.

when variation occurs in predation. The reason for this asymmetry probably relates to the form of the functional responses that link the pairs of trophic levels (DeAngelis 1992). The interaction between elk and wolves is governed by a Type III functional response, whereas the plant-elk interaction is driven by a Type II functional response. In general, a Type II functional response tends to amplify stochastic perturbations, whereas a Type III functional response has a dampening effect (Boyce 1992).

The magnitude of response in vegetation and herbivores attributable to predators is essentially the same, no matter where we perturb the system. All that we have done is to alter where noise enters the system. When noise enters at the bottom with stochastic vegetation, our regression analysis would lead us to conclude that predators are not having much effect relative to the vegetation. But when stochastic variation perturbs the system from the top, our analysis of the time series would lead us to believe that the system is governed by top-down processes. In terms of the variance structure of the system, such interpretations would be appropriate. But if one is attempting to understand the magnitude of consequence attributable to predators in the system, this sort of analysis could be seriously misleading.

Management Interpretations

The extent to which predators suppress their prey has been a contentious issue (Erlinge et al. 1984, Skogland 1991). If a system appears to be controlled from the top down, removal of predators would be an obvious way for managers to increase ungulate populations. Empirical evidence of the impact of predator control on ungulates has been mixed, with some apparent successes (Newsome et al. 1989, Gasaway et al. 1992), some failures (Miller and Ballard 1992), and a number of inconclusive results (Boutin 1992). The efficacy of wolf control in particular appears to be dependent on the presence of other predators (Messier 1994, Van Ballenberghe and Ballard 1994), the availability of alternate prey (Carbyn 1983), the impact of food competition on the prey species, degree of human impact on the system, and the mobility of the prey (Fryxell et al. 1988, Messier 1995).

The dichotomy of top-down vs. bottom-up regulation is overly simplistic. The interaction of top-down and bottom-up processes is what shapes population dynamics. We deliberately oversimplified our model so that we could understand it. In reality, variance enters the system from the top down as well as from the bottom up, and this variance is not independent; for example, bad years for elk are likely to be good years for wolves. Caughley's (1987) concept of "centripitality" suggests that density dependence of a population can be overwhelmed by wide fluctuations in the environment. Depending on the magnitude of those fluctuations, a population may appear to behave in a density-independent manner. Likewise, McLaren and Peterson (1994) suggest that extrinsic perturbations, such as fire and large windstorms, can override the usual top-down effects of wolves on the moose population on Isle Royale.

Many previous studies of the impact of predation on ungulates suffer from a number of limitations in drawing conclusions about the impact of predation on moose populations (Boutin 1992). Boutin argues that there is a lack of strong evidence to claim that predators limit or regulate moose populations because: (1) assumptions have not been tested, (2) alternate hypotheses have not been considered, and (3) experiments have not been conducted. The third factor, considered in context of the first two, offers the most valuable information.

As our analysis demonstrates, the source of variation in the system can shape conclusions about what factors are controlling the population. Because winter severity has such large influences in causing year-to-year variation in elk population size in Yellowstone before wolf recovery (Houston 1982, Merrill and Boyce 1991), we find it difficult to imagine that winter severity will not continue to be a major source of population fluctuations even in the face of wolf predation. Because winter severity influences the availability of forage and the extent of range used by elk, this will continue to be true now that wolves are in the system. We suspect that if McLaren and Peterson (1994) had attempted the same sort of analysis on Yellowstone's Northern Range as they did for wolves and moose on Isle Royale, they would have come to quite a different conclusion, because we believe that significant bottom-up influences on elk numbers would be detected. This does not imply that wolves will not be important components of the system, but studying the dynamics of ungulate numbers is more likely to reveal the sources of perturbations in the system rather than the strength of trophic-level interactions.

Several recent studies have shown the importance of predators in limiting ungulate populations (Gasaway et al. 1992, Van Ballenberghe and Ballard 1994, Messier 1994, 1995, Boertje et al. 1996). Some of this work has been done in order to justify wolf control programs to enhance moose hunting opportunities (Gasaway et al. 1992). We are concerned that the results of McLaren and Peterson (1994) might be generalized to imply that wolves dominate in the dynamics of ungulate populations, and that it therefore follows that wolf control efforts are generally necessary. Our results demonstrate that results of the sort presented by McLaren and Peterson (1994) are likely to reflect the sources of stochastic variance in the particular system rather than the importance of top-down control by wolves. Therefore we caution that inferences about the efficacy of predator control based on studies such as McLaren and Peterson (1994) may be inappropriate.

Being able to anticipate whether wolves will stabilize ungulate populations or cause increased variation in numbers would be of great interest to managers. Identifying trophic-level interactions using key-factor analysis cannot give insight on this question. Messier (1995) points out that *limiting factors* have a direct impact on the rate of population growth, mortality, or dispersal, without any relation to population density, while *regulating factors* are those density-dependent processes that keep populations within normal density ranges. Because ungulate densities must fall below the satiation thresholds for the predator functional response (figure 10.1) before we can discern whether or not predation is regulating, we would caution against implementing wolf culls prematurely. This might be difficult because current policy allows state agencies to begin predator control if they perceive that game herds are being jeopardized by wolf predation.

Although models can offer insights into how predator-prey relations are structured, experiments are essential to dissect the true processes driving ecosystem dynamics (see Chapter 12). The long-debated snowshoe hare population cycle is finally being elucidated by controlled experiments that suggest that neither a plant-herbivore model nor a predator-prey model is responsible for the cycles, but rather a synergistic, three-trophic-level interaction (Krebs et al. 1995). The release of wolves into Yellowstone National Park represents a unique and timely experiment that can yield important insights into the natural dynamics of the system. If carefully monitored, it will give insight into the population behavior of wolves and elk amidst the complexity of interactive effects in the broader ecological context of the park. In the 1920s no one had the foresight to monitor the consequences of wolf removal from the Yellowstone ecosystem. Now, seventy-five years later, we have another chance to assess the impact of wolves via reintroduction as opposed to removal. We should fully exploit the ecological insights offered by this opportunity.

Our objective is to encourage research on the consequences of wolf recovery in Yellowstone and to frame a conceptual context for this research. But an experiment of this scale is necessarily unreplicated. We submit that the only hope for reliable knowledge in the context of ecosystem management is through adaptive management protocols (Walters and Holling 1990, Chapter 12 this volume). As new insights are gained from monitoring, a shift in research emphases or direction may develop in the iterative process of feedback between research and management (Messier et al. 1995). The advantages of modeling for adaptive management are particularly powerful. The model provides predictions of system behavior under management manipulations, followed by monitoring to assess the validity of the predictions. Based on new information obtained through monitoring, adjustments in the model can be made, and new predictions formulated (Walters 1986).

In addition, only through such experimental manipulations (albeit without replication) can we dissect the magnitude of response associated with restoring wolves to the system. Yellowstone hosts nine elk herds (Singer 1991), but most are migratory and only two of these reside primarily in the park—the Northern Range herd and the Madison-Firehole herd. If predator control is used as a tool for management of the seven herds that migrate to winter ranges outside of the park, these herds may be replicates for an experimental design to evaluate alternative methods of wolf and ungulate management. In the spirit of ecosystem management, wolf recovery offers an outstanding opportunity to employ adaptive management approaches (Boyce and Haney 1997).

Literature Cited: Boertje, R. D., P. Valkenburg, and M. E. McNay. 1996. Increases in moose, caribou, and wolves following wolf control in Alaska. Journal of Wildlife Management 60:474–89.

Boutin, S. 1992. Predation and moose population dynamics: A critique. Journal of Wildlife Management 56:116–27

Boyce, M. S. 1991. Natural regulation or the control of nature? Pp. 183–208 *in* R. B. Keiter and M. S. Boyce, eds., The Greater Yellowstone Ecosystem: Redefining America's wilderness heritage. Yale University Press, New Haven.

———. 1992. Wolf recovery for Yellowstone National Park: A simulation model. Pp. 123–38 *in* D. R. McCullough and R. H. Barrett, eds., Wildlife 2001: Populations. Elsevier Applied Science, London.

———. 1993. Predicting the consequences of wolf recovery to ungulates in Yellowstone National Park. Pp. 234–69 *in* R. S. Cook, ed., Ecological issues on reintroducing wolves into Yellowstone National Park, Scientific Monograph NPS/NRYELL/NRSM-93/22, National Park Service. U.S. Government Printing Office, Washington.

———. 1995. Anticipating consequences of wolves in Yellowstone: Model validation. Pp. 199–

209 *in* L. N. Carbyn, S. H. Fritts, and D. R. Seip, eds., Ecology and conservation of wolves in a changing world. Canadian Circumpolar Institute, Occasional Publications no. 35.

Boyce, M. S., and J.-M. Gaillard. 1992. Wolf recovery in Yellowstone, Jackson Hole, and the North Fork of the Shoshone River: A computer simulation model. Pp. 4-71 to 4-115 *in* J. D. Varley and W. G. Brewster, eds., Wolves for Yellowstone? A report to the United States Congress, vol. 4, research and analysis. National Park Service, Yellowstone National Park.

Boyce, M. S., and A. W. Haney. 1997. Ecosystem management: Applications for sustainable forest and wildlife resources. Yale University Press, New Haven.

Boyce, M. S., and E. H. Merrill. 1991. Effects of the 1988 fires on ungulates in Yellowstone National Park. Proceedings of the Tall Timbers Fire Ecology Conference 17:121–32.

Carbyn, L. N. 1983. Wolf predation on elk in Riding Mountain National Park, Manitoba. Journal of Wildlife Management 47:963–76.

Caughley, G. 1976. Wildlife management and the dynamics of ungulate populations. Pp. 183–246 *in* T. H. Coaker, ed., Applied biology, vol. 1. Academic Press, London.

———. 1987. Ecological relationships. Pp. 159–87 *in* G. Caughley, N. Shepard, and J. Short, eds., Kangaroos: Their ecology and management in the sheep rangelands of Australia. Cambridge University Press, London.

Caughley, G., and J. H. Lawton. 1981. Plant-herbivore systems. Pp. 132–66 *in* R. M. May, ed., Theoretical ecology: Principles and applications. Blackwell Scientific Publications, Oxford.

Chase, A. 1986. Playing God in Yellowstone. Atlantic Monthly Press, Boston.

Cole, G. F. 1971. An ecological rationale for the natural or artificial regulation of native ungulates in parks. Transactions of the North American Wildlife and Natural Resources Conference 36:417–25.

Coughenour, M. B., and F. J. Singer. 1996. Elk population processes in Yellowstone National Park under the policy of natural regulation. Ecological Monographs 6:573–93.

DeAngelis, D. L. 1992. Dynamics of nutrient cycling and food webs. Chapman and Hall, London.

Despain, D., D. Houston, M. Meagher, and P. Schullery. 1986. Wildlife in transition: Man and nature on Yellowstone's Northern Range. Roberts Rinehart, Boulder, Colorado.

Dodd, J. 1993. Viewpoint: An appeal for riparian zone standards to be based on real world models. Rangelands 14(6):332.

Erlinge, S., G. Göransson, G. Högstedt, G. Jansson, O. Liberg, J. Loman, I. N. Nilsson, T. von Schantz, and M. Sylvén. 1984. Can vertebrate predators regulate their prey? American Naturalist 123:125–33.

Errington, P. L. 1967. Of predation and life. Iowa State University Press, Ames.

Frank, D. A., and S. J. McNaughton. 1992. The ecology of plants, large mammalian herbivores, and drought in Yellowstone National Park. Ecology 73:2043–58.

Fryxell, J. M., J. Greever, and A. R. E. Sinclair. 1988. Why are migratory ungulates so abundant? American Naturalist 131:781–98.

Garton, E. O., R. L. Crabtree, B. B. Ackerman, and G. Wright. 1990. The potential impact of a reintroduced wolf population on the northern Yellowstone elk herd. Pp. 3-59 to 3-91 *in* J. D. Varley and W. G. Brewster, eds., Wolves for Yellowstone? A report to the United States Congress, vol. 4, research and analysis. National Park Service, Yellowstone National Park.

Gasaway, W. C., R. D. Boertje, D. V. Grangaard, D. G. Kelleyhouse, R. O. Stephenson, and D. G. Larsen. 1992. The role of predation in limiting moose at low densities in Alaska and Yukon and implications for conservation. Wildlife Monographs no. 120.

Holling, C. S. 1959. The components of predation as revealed by a study of small mammal predation of the European pine sawfly. Canadian Entomology 91:293–320.

Houston, D. B. 1982. The northern Yellowstone elk herd. Macmillan, New York.

Kay, C. E. 1990. Yellowstone's northern elk herd: A critical evaluation of the "natural regulation" paradigm. Ph.D. diss., Utah State University, Logan.

Kolmogorov, A. N. 1936. Sulla teoria di Volterra della lotta per l'esistenza. Giornale Instituto Italiano Attuari 7:74–80.

Krebs, C. J., S. Boutin, R. Boonstra, A. R. E. Sinclair, J. N. M. Smith, M. R. T. Dale, K. Martin, and R. Turkington. 1995. Impacts of food and predation on the snowshoe hare cycle. Science 269:1112–15.

Leopold, A. S., S. A. Cain, C. M. Cottam, I. N. Gabrielson, and T. L. Kimball. 1963. Wildlife
management in the national parks. Transactions of the North American Wildlife and
Natural Resources Conference 28:28–45.

Maynard Smith, J. 1974. Models in ecology. Cambridge University Press, Cambridge.

McLaren, B. E., and R. O. Peterson. 1994. Wolves, moose, and tree rings on Isle Royale. Science
266:1555–58.

Merrill, E. H., and M. S. Boyce. 1991. Summer range and elk population dynamics in
Yellowstone National Park. Pp. 263–73 *in* R. B. Keiter and M. S. Boyce, eds., The Greater
Yellowstone Ecosystem: Redefining America's wilderness heritage. Yale University Press,
New Haven.

Merrill, E. H., M. K. Bramble-Brodahl, R. W. Marrs, and M. S. Boyce. 1993. Estimation of green
herbaceous phytomass from Landsat MSS data in Yellowstone National Park. Journal of
Range Management 46:151–56.

Messier, F. 1991. On the concepts of population limitation and population regulation as
applied to caribou demography. Pp. 260–77 *in* C. E. Butler and S. P. Mahoney, eds.,
Proceedings, 4th North American Caribou Worshop. St. John's, Newfoundland.

——. 1994. Ungulate population models with predation: A case study with the North
American moose. Ecology 75:478–88.

——. 1995. Trophic interactions in two northern wolf-ungulate systems. Wildlife Research
22:131–36.

Messier, F., W. C. Gasaway, and R. O. Peterson. 1995. Wolf-ungulate interactions in the
Northern Range of Yellowstone: Hypotheses, research priorities, and methodologies.
Report to USDA, National Biological Service, Yellowstone National Park.

Miller, S. D., and W. B. Ballard. 1992. Analysis of an effort to increase moose calf survivorship
by increased hunting of brown bears in south-central Alaska. Wildlife Society Bulletin
20:445–54.

Newsome, A. E., I. Parer, and P. C. Catling. 1989. Prolonged prey suppression by carnivores:
Predator-removal experiments. Oecologia 89:458–67.

Pengelly, W. L. 1963. Thunder on the Yellowstone. Naturalist 14:18–25.

Royama, T. 1992. Analytical population dynamics. Chapman and Hall, London.

Sinclair, A. R. E. 1989. The regulation of animal populations. Pp. 197–241 *in* J. M. Cherrett, ed.,
Ecological concepts. Blackwells, Oxford.

Singer, F. J. 1991. The ungulate prey base for wolves in Yellowstone National Park. Pp. 323–48
in R. B. Keiter and M. S. Boyce, eds., The Greater Yellowstone Ecosystem: Redefining
America's wilderness heritage. Yale University Press, New Haven.

Singer, F. J., W. Schreier, J. Oppenheim, and E. O. Garton. 1989. Drought, fires, and large
mammals. BioScience 39:716–22.

Skogland, T. 1991. What are the effects of predators on large ungulate populations? Oikos
61:401–11.

Turchin, P., and A. D. Taylor. 1992. Complex dynamics in ecological time series. Ecology
73:289–305.

Vales, D. J., and J. M. Peek. 1990. Estimates of the potential interactions between hunter
harvest and wolf predation on the Sand Creek, Idaho and Gallatin, Montana, elk
populations. Pp. 3-93 to 3-167 *in* J. D. Varley and W. G. Brewster, eds., Wolves for
Yellowstone? A report to the United States Congress, vol. 4, research and analysis. National
Park Service, Yellowstone National Park.

Van Ballenberghe, V., and W. B. Ballard. 1994. Limitation and regulation of moose
populations: The role of predation. Canadian Journal of Zoology 72:2071–77.

Varley, G. C., and G. R. Gradwell. 1960. Key factors in population studies. Journal of Animal
Ecology 29:399–401.

Wagner, F. H., and C. E. Kay. 1993. "Natural" or "healthy" ecosystems: Are U.S. national parks
providing them? Pp. 257–70 *in* M. J. McDonnell and S. T. A. Pickett, eds., Humans as
components of ecosystems. Springer-Verlag, New York.

Walters, C. J. 1986. Adaptive management of renewable resources. Macmillan, New York.

Walters, C. J., and C. S. Holling. 1990. Large-scale management experiments and learning by
doing. Ecology 71:2060–68.

Sedated adult female grizzly bear with cub (F. Lance Craighead)

Chapter Eleven

Genetic Considerations for Carnivore Conservation in the Greater Yellowstone Ecosystem

F. Lance Craighead, Michael E. Gilpin, and Ernest R. Vyse

With the recent reintroduction of the gray wolf, the Greater Yellowstone Ecosystem (GYE) has logically recovered the carnivore community that persisted in the Northern Rockies since the end of the Pleistocene. The species list of carnivores is the same as one that might have been written ten thousand or five thousand years ago. Having now restored the preexisting carnivore diversity, conservationists and managers face the problem of sustaining this system in a "healthy" manner. Ecological health can be examined within the varying time frames of ecological, community, and population dynamic processes. It can also be investigated over various spatial scales—landscape, ecosystem, community, or population. In this chapter we consider the question of conserving genetic biodiversity within the GYE's carnivore species. Genetics cannot be isolated from other ecological processes, but maintaining genetic diversity is the necessary foundation for resolving conservation issues at higher levels of ecological organization. While concentrating on population genetic processes, we point out connections to these other levels.

In this chapter, we discuss the relevant genetic theory, current applications of molecular tools to carnivore genetic research, and ecological and behavioral data with genetic implications for GYE's carnivores. We will explore genetic considerations for each species and place these in the context of management (e.g., Ralls and Ballou 1983). We will also touch on the implications of recent results for carnivore management in the GYE.

Genetics, Diversity, and Conservation of the Greater Yellowstone Ecosystem's Carnivores

The Carnivora occupy at least the third trophic level, and in general their body sizes are greater than those of their prey. In addition, they are warm-blooded and often must expend much energy to find sufficient prey over large home ranges or territories. Because of the reduced flow of energy to this trophic level, the thermodynamic constraints on its use, and the energetic requirements of relatively large-bodied animals, the maximum population sizes of carnivores tend to be small.

Table 11.1 shows very rough estimates of carnivore populations in the GYE, gathered from existing agency data or from the best guesses of field biologists. The GYE—a large, heterogeneous area of about 44,500 square kilometers—has not been systematically surveyed for most carnivores, and the accuracy of estimates for species varies greatly. Estimating total population size alone is difficult, and as we discuss below, social and reproductive behavior, degree of isolation, geographic features, population genetic history, and effects of human activities all work to reduce the genetic diversity within populations below that expected from a given census size.

For this discussion we will initially consider total numbers of adults (*N*) fewer than one thousand to constitute a small population, designated by an asterisk in table 11.1. The total population size (usually estimated as the number of adults) is not directly important in terms of genetics: an *N* of one thousand

Table 11.1

Body size and estimated number of Greater Yellowstone Ecosystem carnivores

Species	Body size	Estimated N (adults)
Grizzly bear (*Ursus arctos*)	large	<400*
Black bear (*Ursus americanus*)	large	<2,000
Gray wolf (*Canis lupus*)	large	<100*
Coyote (*Canis latrans*)	medium	<3,000
Red fox (*Vulpes vulpes*)	medium	<2,000
Mountain lion (*Felis concolor*)	large	<500*
Bobcat (*Felis rufus*)	medium	<500*
Lynx (*Felis lynx*)	medium	<500*
Wolverine (*Gulo gulo*)	large	<300*
Marten (*Martes americana*)	medium	<10,000
Fisher (*Martes pennanti*)	medium	<200*
River otter (*Lutra canadensis*)	medium	<500*
Badger (*Taxidea taxus*)	medium	<1,000*
Striped skunk (*Mephitis mephitis*)	medium	<2,000
Mink (*Mustela vison*)	medium	<3,000
Ermine (*Mustela erminea*)	small	<20,000
Long-tailed weasel (*Mustela frenata*)	small	<20,000
Raccoon (*Procyon lotor*)	small	<5,000

Note: * indicates a small population.

carnivores is well below Franklin's estimate of the genetic effective size needed to preserve adaptive potential (N_e of five hundred, or N of about 2,000, discussed below). The size of a population depends primarily on the amount of suitable habitat available (assuming human-caused mortality can be controlled). In the GYE, as a rule of thumb, the larger carnivores (bears, wolves, wolverines, and mountain lions) need much larger areas, and hence their populations are small. Smaller species such as otter, lynx, or fisher require less area, but their habitat (or prey) is restricted and thus their populations are also small.

In addition to small size, many of these populations in the GYE also show some effects of isolation, such as reduced heterozygosity. The movement of animals (and thus their genes) in and out of the ecosystem is restricted, and for many species movement between populations within the ecosystem is restricted. The central part of the GYE is a high-altitude, volcanic plateau with severe winters. The surrounding landscape is generally lower in elevation, supporting different habitat types and carrying the signature of human activity—a serious impediment to the movement of many carnivores.

Genetic and population genetic knowledge can enlighten and inform conservation decision making. We focus on the salient feature that, from the standpoint of many carnivore species, the GYE is an island ecosystem. The major management concern on islands is extinction, and in this chapter we explore the relation between genetics and extinction probabilities. Another genetic question is whether large enough island populations can be maintained to ensure long-term fitness and evolutionary potential (Soulé 1980). Such populations should be able to adjust to long-term environmental change. We feel that none of the GYE carnivore populations will ultimately be large enough to sustain sufficient levels of genetic diversity over millennia. Assuming that they will not remain that large, the next genetic question is whether the present population size has captured and can sustain sufficient genetic diversity to avoid short-term (decades or generations) loss of fitness through inbreeding and stochastic processes.

Genetic variation represents biodiversity within species (Cronin 1993). Alternate alleles represent options with which a species can respond to changes in the environment. Loss of variation limits the options and can increase risk of extinction. Fixation of alleles leaves a species with no genetic choice in subsequent generations unless mutation occurs. This may or may not be problematic, depending on whether or not the expression of a given allele is modified by other, perhaps more diverse alleles, and whether or not environmental changes have bearing on the trait in question. Thus loss of variation can make individuals more vulnerable to disease or climatic changes such as drought, increased cold, increased humidity, or increasing ultraviolet radiation. Effects on survival generally occur through inbreeding depression or susceptibility to disease.

EFFECTIVE POPULATION SIZE AND STOCHASTICITY

One of the consequences of small size and isolation is that genetic diversity can be reduced; genetic variation is related to population size (Frankham 1996). Deleterious genetic effects can hasten the extinction process through the short-term loss of fitness even in very small populations where the extinction threshold is within the range of demographic stochasticity. In populations that are large enough to exceed the range of demographic stochastic effects, genetic effects may act over time to reduce the population size enough that demographic effects become significant (Lacy 1997); this is the situation for most carnivores in the GYE because of the factors discussed in this section.

Effective population size, N_e. N_e is a measure of the rate of loss of heterozygosity (inbreeding N_e). It also refers to the rate at which neutral alleles are lost from a population or the rates of fluctuation in allele frequencies (variance N_e). N_e represents the size of an ideal, randomly mating, panmictic population, which would have the same rate of allelic loss as the actual population being studied (e.g., Franklin 1980). Given that allele frequencies do not remain stable from generation to generation, such allele frequencies drift in value over generations and may end up fixed at either 0.0 (lost) or 1.0. Accompanying such drift is loss of heterozygosity (H) and loss of alleles. Neutral genetic variation, or heterozygosity, is lost over time in a finite population by random genetic drift and by behavior that produces inbreeding. Inbreeding, however, does not cause loss of alleles.

Because large mammals like carnivores tend to show many characteristics that may reduce the amount of variation that is passed on from generation to generation, N_e may be substantially lower than N, the actual population size. Among these characteristics are unequal numbers of breeding males and females, fluctuations in population size, non-Poisson distribution in progeny number (unequal reproductive success), and geographic genetic structure in the population (nonrandom mating patterns). When examining populations of wild animals, though, it becomes difficult to estimate parameters for these characteristics.

The standard formula for heterozygosity is written as equation 1 (Wright 1969):

$$H_t = H_0[1 - 1/(2N_e)]_t \tag{1}$$

where H_0 = the amount of heterozygosity (H) at the beginning of a time period, H_t = the amount of heterozygosity (H) at the end of a time period, and t = the length of time in generations. The important point here is that N_e can be calculated from genetic data on heterozygosity. One can rearrange the terms in this equation to obtain equation 2:

$$N_e = [0.5]/[1 - (H_t/H_0)^{1/t}] \tag{2}$$

The problem with this approach is that it requires very accurate sampling of direct genetic data to determine the H levels, and H must be measured in at least two sequential generations; for many GYE carnivores this can easily be a decade. With current genetic techniques, however, particularly microsatellite analysis (Paetkau et al. 1997), accurate estimates of H can be obtained.

We also have other methods to estimate the ratio of N_e to the census N. In effect we can estimate N_e from a "snapshot" of the genetic data using demographic data. Different formulas, however, describe different factors that can reduce N_e; they are not alternatives but rather components that should be compounded to estimate the N_e resulting from multiple processes. Two important factors within a single generation are the ratio of males to females in the population and the variation in the sizes of families (Harris and Allendorf 1989, Hedrick and Gilpin 1996). One original formula (equation 3, Wright 1931) expressed N_e as a function of the number of successfully breeding females (N_f) and successfully breeding males (N_m):

$$N_e = 1/[1/(4N_m) + 1/(4N_f)] \text{ or } N_e = 4(N_m N_f)/(N_{m 1} N_f) \tag{3}$$

For an ideal population, the sex ratio is one-to-one and the variance in family size is Poisson distributed. Deviations from either of these assumptions in natural populations lower N_e relative to N. Lande and Barrowclough (1987) give formulas with which to calculate these discrepancies. Harris and Allendorf (1989) used a simulation model to estimate N_e by tracing the loss of heterozygosity through time and then comparing results with estimates produced by applying published formulas.

Over multiple generations, N_e is more sensitive to the harmonic mean of the population sizes. Thus one low population size per one hundred years is very important in setting the average N_e over the century. For example, if the grizzly bear population is one thousand for ninety years and ten for ten years, the average genetic effective size for the century is ninety-one.

In wild populations of carnivores, N_e is always lower than N. Frankham (1996) found that estimates of N_e/N varied according to fluctuation in population size, variance in family size, form of N used (adults, breeders, or total size), taxonomic group, and unequal sex ratio. Wilcox (1986) estimated the ratio of (N_e/N) to be 0.25 in mammals as a rule of thumb. Harris and Allendorf (1989) estimated a ratio of 0.24 to 0.32 for grizzly bears. Using only Wright's original equation and estimating the number of breeding males from paternity results of microsatellite analysis, two of the authors (Craighead 1994, Craighead and Vyse 1996) estimated an N_e/N ratio of 0.41 in an Arctic grizzly population.

Recent results from microsatellite analysis (Paetkau et al. 1997) demonstrate that the N_e to N ratio for grizzly bears may be only 0.037–0.12 (depending on the mutation rate). Whatever the actual value of N_e, it is clear that carnivores have effective population sizes that are much smaller than the actual numbers of animals censused.

Genetic stochasticity. Random (stochastic) events, or sampling effects, play a large role in the dynamics of small populations. As population size fluctuates near the extinction threshold, such factors as demographic stochasticity, genetic stochasticity, environmental stochasticity, and catastrophic events pose immediate risks (Shaffer 1978). Demographic stochasticity can rapidly reduce population size when N is below about twenty-five (Gilpin 1992). These four factors can be considered as extrinsic (environmental stochasticity and catastrophe) or intrinsic (demographic stochasticity and genetic deterioration) to a species (Hedrick 1996). Caughley (1994) coined the phrase "small population paradigm" for the approach focusing on effects of these four factors. He contrasted this to the "declining population paradigm," which focuses on deterministic factors such as habitat destruction, overkill, and fragmentation (Diamond 1989). Many researchers feel that deterministic factors such as human-caused mortality are the overriding cause of concern for many species such as grizzly bears (e.g., Primm 1996, Mattson et al. 1996, Peek et al. 1987). However, all these factors are interrelated in causing the decline of populations and eventual extinction, and genetic factors can be critical even in very small populations (Mills and Smouse 1994). The relative importance of each factor will vary in each situation.

Many of the larger carnivore populations may currently be at low enough levels that genetic effects could nudge them into the realm of demographic stochasticity in just a few generations, greatly increasing their risk of extinction. For medium-sized carnivores with larger population sizes, genetic factors can have long-term effects that act to reduce population sizes gradually, especially if movement between populations is restricted. For all carnivore species, a prudent approach is to conserve as much genetic variation as possible in order to avoid augmenting negative demographic effects, to allow an adaptive response to modified environmental conditions, and to stockpile against future evolutionary change (Soulé 1980).

GENETIC DIVERSITY IN TIME AND SPACE

The level of genetic diversity is controlled by five population genetic processes—mutation, migration, selection, drift, and inbreeding (e.g., Hedrick 1985)—which in turn are regulated by the size of the population, its biotic and abiotic environment, and its spatial structure. Thus the unique physiography of the GYE and the unique physiology of its inhabitants, communities, and systems impose limits and provide opportunities for genetic interplay. We are only beginning to understand how genetics are shaped by these processes in a spatially explicit setting, but the GYE provides a natural laboratory where some of these questions may be answered. Carnivore population genetics are affected by the basic processes outlined below, and carnivores themselves influence the genetics of other species.

Mutation. The processes of mutation and changes in allele frequency determine the amount of genetic variability within a closed population (that is, one with no immigration or emigration). Mutation is the inherent process whereby organisms change from one hereditary state to another (Suzuki et al. 1989). It can be in the form of a gene (or point) mutation, where one allelic form mutates to another, or a chromosome mutation, which refers to changes in chromosome structure or number involving segments of chromosomes, whole chromosomes, or even sets of chromosomes. Only mutations in reproductive (germline) cells can be transmitted to progeny. Different portions of the genome exhibit different mutation rates. Mitochondrial DNA (mtDNA) rates are estimated at 10 percent substitution of nucleotides per million years in mammals (Irwin et al. 1991). Microsatellite mutation rates are probably between 1×10^{-3} and 2×10^{-4} per generation (Amos et al. 1996, Paetkau et al. 1997). Most mutations are probably selectively neutral and do not affect the survival of the organism. Mutations are rare events and thus are less likely to occur in small populations.

Migration. Most carnivore populations are not closed systems, however, because animals disperse from their natal areas. Over time, unique alleles appear through mutation and become locally established in a population. If individuals carrying those alleles disperse into another population and breed successfully, those alleles can be introduced. The movement of alleles between populations (or subpopulations) is termed gene flow. At any point in time, different populations may exhibit different frequencies of each allele (except where fixed). The more frequent the interchange of individuals between populations, the greater the amount of gene flow and the closer the allele frequencies will be to each other.

Natural selection. Allele frequencies are also modified by selection. If environmental factors change enough so that one phenotype (and the alleles responsible) is selected for, or against, the frequencies of those alleles will increase or decrease. Over time, localized genetic differences arise as a population becomes more closely adapted to its immediate environment. There are thus two components that produce differences in allele frequencies between populations: random differences resulting from drift, and selected differences resulting from enhanced survival and reproduction conferred upon certain alleles.

If a population becomes totally isolated, or closed, it will diverge genetically from other populations over time because of the effects of drift and selection. At some point this will result in a locally adapted population that is sufficiently divergent (reproductively isolated) from its ancestral population that it can be considered a separate species (e.g., Mayr 1991).

Genetic drift and inbreeding. Changes in allele frequencies in small populations result primarily from random processes. Deviations from the expectations of randomness in population genetics data imply that some other process such as selection, nonrandom mating, or geographic separation may be at work. Alleles that are rare in a population (such as those that recently appeared by mutation) may not appear in the next generation because of random sorting, or genetic drift. Other alleles may increase in frequency. As the frequency of an allele increases, so does the probability of its persistence in the gene pool. Over the progression of generations, allele frequencies can drift as a result of sampling effects (drift) with the ultimate fixation of one allelic form at each locus. The effects of drift are increased in small populations.

LOSS OF GENETIC VARIABILITY

Inbreeding depression. In small populations, mate choice is limited and individuals become more closely related over time. Heterozygosity can be lost by inbreeding (the mating of closely related individuals, which is measured by Sewall Wright's F_{is}) or by random sorting of alleles in a small population (which is measured by divergence among populations, Wright's F_{st}). Loss of heterozygosity increases the chance that harmful recessive alleles will be expressed (Ralls et al. 1986). This expression of harmful alleles in the phenotype acts in a variety of ways to depress fitness by lowering fecundity and reducing survival rates. A concise discussion of the importance of genetic variation can be found in Lacy 1997.

The term *inbreeding depression* refers to a decrease in fitness, regardless of how heterozygosity is lost. The rate of inbreeding is generally measured by the reduction in heterozygosity (H) per generation. Heterozygosity has shown no correlation with fitness in several studies (Hedrick 1996), but lowering of H has been demonstrated to be positively correlated with reduced fitness in many other populations (Allendorf and Leary 1986, Ralls and Ballou 1983). The deleterious effects of inbreeding depression can be difficult to detect, particularly in small populations of mammals or birds. Inbreeding depression is generally expressed as reduced fertility or low juvenile survival (Ralls et al. 1979). Correlations between low heterozygosity and reduced juvenile survival in mammals have been demonstrated in inbred ungulates (Ballou and Ralls 1982), as well as in other species, including carnivores. Inbreeding has also been used to explain increased vulnerability of animals to disease (e.g., O'Brien and Evermann 1988).

Domestic animal breeding experience indicates that a 1 percent rate of inbreeding per generation is the maximum allowable to avoid the increased expression of deleterious alleles that have been exposed in homozygotes (Franklin 1980). The extent to which these generalizations are applicable to carnivores is not clear, but carnivores may tend to be less outbred than other mammals. Paetkau et al. (in press) found a reduction of heterozygosity of 15–20 percent over the past one hundred years in grizzly bears in the GYE, assuming that this population was once contiguous with and similar in variability to grizzlies in the Flathead River. This corresponds to a 1–4 percent rate of inbreeding.

However, reduced heterozygosity may or may not be immediately harmful in a given population. Thus elephant seals were able to pass through a population bottleneck of about one hundred individuals with a significant loss of allelic diversity and rebound to a current population of about 150,000. Hawaiian monk seals, on the other hand, after a similar bottleneck, continue to decline and now number about 1,300 (Kretzmann et al. 1997).

Thus species may differ in their sensitivity to inbreeding, depending on which alleles are lost. Inbreeding depression may not be manifested until a change in the environment provides an opportunity for other genetic options that have been lost through inbreeding. This may occur even in apparently highly successful species. Reduced genetic diversity therefore constitutes an increased risk of extinction to a population or a species: alleles that are lost in small populations might be important to survival now or at some time in the future under different conditions. No species of mammal has been shown to be unaffected by inbreeding, and genetic threats to viability will be expressed through their effects on and interactions with demographic and ecological processes (Lacy 1997).

Inbreeding in normally outbreeding species generally has deleterious effects. There is a possibility that harmful recessives can be purged from the population after a bottleneck, leaving the genome with less genetic load (accumulated deleterious genes) than before, but there is little evidence as yet (Lacy 1997). Though strongly deleterious recessive alleles may be removed in this fashion, overdominant or weakly deleterious recessives will remain (Ballou 1997, Hedrick 1994).

Genetic variation and spatial constraints. Conservation genetic research focuses on understanding the relation between genetics and the persistence of local populations, and this knowledge is used to sustain the fitness of small populations (Ralls and Ballou 1986). Lacy (1997) reviews evidence that lower genetic variation depresses individual fitness and mean fitness of populations. Frankham (1996) examines the relation between population size and genetic variation—the larger the population, the more variation. He also concludes that genetic variation within a species is related to "island" size or the amount of habitat a population can occupy.

Genetic variability is generally measured as the degree of heterozygosity. Although our models often concern single genetic loci, it is the genome-wide heterozygosity that is most useful for conservation analysis and conservation decision making. Heterozygosity is a relative concept. For small populations, we are interested in the level of heterozygosity relative to large, secure, mainland populations (if such exist) and in the internal spatial structure of heterozygosity within the conservation reserve.

Populations can originate with very low variability as a result of the founder effect. Two or more individuals may colonize unoccupied habitat, as when a species expands its range or when habitat has been vacated by a local extinction event. The total genetic diversity of the new population is low relative to other, larger populations. Without additional migration, the genetic diversity will remain low and will be further reduced by inbreeding. Species that were greatly reduced in number in the GYE, such as mountain lions and wolverines, have undergone genetic "bottlenecks" that probably lowered genetic diversity, but not as severely as a founding event.

Genetic variability is also influenced by population genetic structure, or subdivision: localized genetic differences within a population of interbreeding individuals. Differences in genotypes between populations, or between subgroups of populations, may reflect differential fitness values or local adaptation. Variation expressed as genetic structure can be viewed as a component of genetic diversity that has survival value for the population as a whole and is important for the conservation of the population or species as a whole.

Such differences can be caused by such nonrandom factors as assortative mating or from historical factors and drift. Gene flow, or the movement of alleles between component populations, acts to decrease genetic subdivision. The primary mechanism of gene flow in carnivores is dispersal of subadult animals. A subdivided population can theoretically maintain a greater amount of genetic variation than a single, large, panmictic population (Hedrick 1996, Hedrick and Gilpin 1996), as long as the subpopulations are stable.

GENETICS AND CARNIVORE CONSERVATION

To conserve carnivores we need to understand the genetic "health" of the populations in the GYE primarily through the measurement of heterozygosity and allele frequencies, and we need to understand the "structure" of this variation within and between population components. Overall, we need to examine the GYE as a conservation reserve, a refuge from extinction. In this section we shall discuss the genetic aspects relevant to species conservation—how genetic variation is measured, how much is enough, and how can we maintain that much.

Genetic tools and genetic markers. There are various molecules, markers, and methods that are applicable to the study of carnivore genetics. A genetic marker is an allele that can be marked or labeled and used as an experimental probe to keep track of an individual, tissue, cell, chromosome, or gene through the hereditary process. A genetic tool is a technique for manipulating DNA, DNA fragments, alleles, or proteins in order to identify and trace genetic markers through generations or populations.

There are three classes of molecules commonly studied: proteins, mitochondrial DNA, and nuclear DNA. Variation in these molecules at the level of species, populations, or individuals can be addressed using different analytical techniques. Because each focuses on a particular portion of the genome with varying degrees of resolution, different tools may produce different results when applied to the same genetic sample (for example, mtDNA reveals little about patterns of relatedness within family groups). This is important to keep in mind when interpreting results: methods appropriate to the spatial scale must be used.

The first tool is protein electrophoresis. Techniques of protein analysis focus on identifying and enumerating proteins, or allozymes, which are products of protein synthesis and indirectly reflect the underlying allelic variation. Protein electrophoresis is useful for comparing allele frequencies between populations or estimating genetic distances between species. Additional information relative to carnivore populations can be found in Goldman et al. (1989), Cronin (1993), and Hedrick (1996).

The second tool is DNA analysis. Techniques of DNA analysis focus on identifying and enumerating alleles (either directly or indirectly via "fingerprints"). The occurrence of a variety of alleles in a population (each individual has only two out of all alleles available) is important to produce a variety of genotypes and phenotypes, some of which may allow a population to survive and evolve under changing conditions. Factors that tend to reduce the amount of genetic variability in a population reduce its phenotypic options and thus reduce its chances for long-term survival. Analysis of the actual genetic material, DNA, allows quantification of the allele differentiation. These differences can be addressed by looking at multilocus groups or at single loci. Discussion of relevant DNA applications can be found in Cronin (1993), Talbot and Shields (1996), Paetkau and Strobeck (1994), Paetkau et al. (1995), and Kohn et al. (1997).

Rules of thumb for genetic management. An N of twenty-five is often used as a threshold for the effects of demographic stochasticity (Gilpin 1992). Both Franklin (1980) and Soulé (1980) estimated a minimum of fifty effective breeding individuals to maintain short-term fitness, and five hundred to maintain standing genetic variation. This "50\500 rule" has often been adopted as a guideline for conservation (Wilcox 1986), although there was little empirical basis for it. Subsequent analysis suggested a rule of thumb of five thousand to maintain standing genetic variation (Lande 1995).

A closed population will lose alleles through random events, or drift, while an open population may recover lost alleles through migration. One successful migrant individual per generation has been estimated as the minimum necessary to maintain an allele in a distant population (Wright 1969). Mills and Allendorf (1997) suggest that one migrant per generation may be an acceptable minimum, and ten migrants per generation is probably not too much to swamp localized differences in allele frequencies, but there are many mitigating factors in each population that make rules of thumb unreliable.

All of these rules or rough genetic guidelines were offered by their authors with cautionary caveats. They point out that every population has different effective population size, levels of variation, connectedness to other populations, and genetic history. In addition to deciding what population size is adequate for a desired genetic outcome, managers are faced with many other constraints. Often political, economic, or traditional considerations will interfere with optimal biological solutions. However, we are reaching a level of technological expertise that should make it possible to define the genetic health of a given population as a baseline for protecting genetic diversity. Rules of thumb are valuable tools for a first approximation, particularly when time is short and habitat is disappearing, but the techniques and approaches discussed below point the way toward measuring genetic diversity and prescribing more-precise spatially and numerically explicit solutions.

Genetics and reserve design. Shaffer (1987) introduced a systems approach to the study of extinction, distinguishing between deterministic and stochastic factors. Gilpin and Soulé (1986) further refined these ideas. The concerns for survival of small populations are: (1) deterministic extinction, which occurs when something essential (such as space, shelter, or food) is removed or when something irremediably negative is introduced, (2) environmental stochasticity, or random changes such as drought, global warming, or severe winters, (3) demographic stochasticity, or the chance variation in individual birth and death events, which has very large effects once a population becomes very small, (4) genetic deterioration, which can reduce fitness through increased genetic drift, inbreeding, and the subsequent loss of heterozygosity and genetic variance, and (5) catastrophes, or rare but widespread events that can extinguish a population or species, such as a huge meteorite, an epic flood, or the eruption of a large volcano. (The latter event formed the Yellowstone plateau and may have caused the extinction of several populations and perhaps even species.)

To reiterate, a conservation reserve can be considered a refuge from extinction. An adequate reserve will buffer a species from these five concerns for a certain length of time. A key concept in reserve design is the equilibrium theory of island biogeography (MacArthur and Wilson 1967, Harris 1984). This theory quantifies the number of species on an island as a function of the immigration rate, the extinction rate, the area of the island, and its distance from the mainland source of immigrants. The number of species on the island reaches an equilibrium, and although the number of species stays about the same, the composition of the fauna will change as species come and go. This theory is germane to our discussion because the GYE can be seen, from the viewpoint of many carnivore species, as an island of habitat.

To maintain carnivore populations over long time periods (tens or hundreds of generations), nature reserves need to be large enough to support an effective population size in which there is no significant loss of genetic diversity. Single large reserves, or groups of connected reserves, should ideally be large enough to contain subdivided populations (genetic structure) and therefore to conserve more genetic diversity within a species (Hedrick 1996, Hedrick and Gilpin 1996). A reserve the size of Greater Yellowstone may not be large enough to encompass and conserve genetic structure for large carnivores such as grizzly bears (Craighead et al. 1995), wolverines, or wolves but may capture this level of genetic diversity for other carnivores.

Frankel (1970) was probably the first to recommend genetic considerations in designing nature reserves. Schoenwald-Cox (1983) elaborated on the idea. Estimates of minimum population sizes and minimum area requirements have varied greatly, but as genetic techniques and understanding improve, these estimates are beginning to converge.

The GYE (figure 11.1) contains about 44,500 square kilometers of undisturbed roadless area, 17,800 square kilometers of which have no legal protection (Harting and Glick 1994). The Greater Yellowstone Grizzly Bear Recovery Zone covers an area of 24,605 square kilometers (USFWS 1993). Frankel and Soulé (1981) estimated that mountain lions would need about 13,000 square kilometers of habitat to survive, and wolves would need 39,000–78,000 square kilometers. Metzgar and Bader (1992) estimated that 129,500 square kilometers of habitat would be needed to maintain an N_e of five hundred grizzly bears. Paquet (unpublished, reported in Noss et al. 1996) estimated that wolves may require four times as much habitat as grizzly bears—518,000 to 1,295,000 square kilometers. Salwasser et al. (1987) estimated that the GYE may be the only place that could support 2,500 grizzly bears (an N_e of 625, according to Wilcox 1986). At the upper extreme, large mammals, with effective population sizes in the hundreds, may require 500,000 square kilometers of habitat for a 95 percent

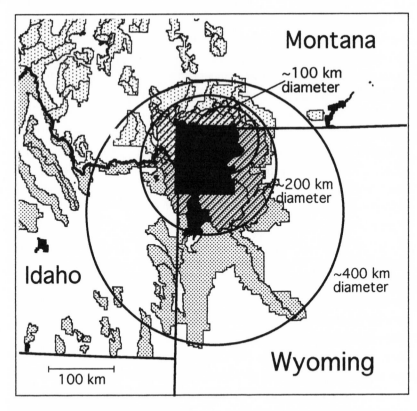

National Forest
GYE Grizzly Bear Recovery Zone
National Parks
State Boundaries
Dispersal Distances

11.1. The Greater Yellowstone Ecosystem, showing national park and forest lands, the grizzly bear recovery zone, and bear dispersal distances.

probability of persistence for one thousand years (Belovsky 1987, Shaffer 1987).

These estimates are, of course, inexact, but the trend is clear: very large reserves are needed for long-term persistence. Reserves of such size may be possible only in the Arctic. In Greater Yellowstone we must focus on lower certainty and explore other options for providing adequate habitat for carnivores. Recent theories of reserve design have focused on the sLOSS question—that is, should there be "single-large-or-several-small" reserves (Diamond 1976, Gilpin and Diamond 1988)—and on the provision of travel corridors between reserves (Noss et al. 1996, Beier and Noss 1998).

As far as large carnivores are concerned, the GYE is one large reserve. The core of this reserve is protected by law, but additional large, relatively road-less areas surround the core. Dispersal distances and the likelihood of gene flow between distant populations within and without the GYE are discussed below. At the very least, it seems crucial to maintain existing undisturbed habitat by appropriate conservation mechanisms (legal protection, conservation easements, land trades, and so on) and to maintain functional habitat connections (Noss et al. 1996) with other large reserve areas. Weaver et al. (1996) suggest that cougars and wolves may be maintained more successfully in a network of refugia spaced within dispersal distances (on a regional scale), while grizzly bears and wolverine may fare better in larger and more contiguous reserves.

Population Genetics of Greater Yellowstone Carnivores

From a management perspective, maintaining genetic diversity in carnivores requires populations large enough to contain historic levels of heterozygosity and to maintain historic levels of gene flow between subpopulations. In many cases, most notably the larger carnivores, these historic levels have already been lost, and we have no genetic baseline against which to compare current populations. We can only look at similar-sized populations in somewhat similar habitat to estimate the historic levels for the GYE. For the medium and smaller-sized carnivores, population size and genetic diversity have probably not been reduced critically from historic levels.

Population genetics is the study of gene (allele) behavior in populations. Research at this level is expanding rapidly with the advent of new tools. Below, for each family of Carnivora, we first discuss recent genetic research and its relevant findings. Since alleles are carried by individuals, dispersal and breeding behavior are important variables in understanding gene flow and genetic diversity. Next, for each family, we review behavioral and ecological studies that have bearing on the species and the environment of the GYE.

Dispersal patterns of subadults as they reach sexual maturity are an exceedingly important factor determining gene flow, population genetic structure, and ultimately genetic diversity. In mammals, the general pattern of dispersal seems to be that males are more mobile, while females remain nearer to their natal areas (Greenwood 1980). One reason may be that group-living mammals are primarily polygynous (Eisenberg 1981); subordinate males are excluded from breeding and forced to move to other groups to increase their reproductive potential. In monogamous species, both females and males are excluded from breeding by a dominant pair, and both sexes may need to disperse to reproduce (Rood 1987).

Table 11.2

Home range/territory size and dispersal distances of large and medium-sized carnivores

	Home range/territory size (km²)		Male dispersal distances (km)	
Species	Mean	Range	Mean	Maximum
Gray wolf	900	210–1,700	85	917
Coyote	5.6	–	48	81
Red fox	5.3	0.19–0.72	–	–
Grizzly bear, male	1,000	233–1,970	70	300
female	450	72–874		
Black bear, male	120	53–225	–	–
female	20	14–137		
Mountain lion, male	889	78–1,000	–	480
female	117	45–373		
Lynx, male	200	145–250	300	1,100
female	70	51–122		
Bobcat, male	122	21–200	109	–
female	43	18–59		
Wolverine, male	422	200–1,500	50	378
female	388	105–400		
Fisher, male	40	–	50	100
female	15	–		
Marten, male	5	–	27	–
female	3	–		
Badger, male	1.4	–	–	–
female	1.2	–		

Note: Data from across North America.

It is generally hypothesized that dispersing animals will occupy the first unoccupied habitat they encounter. However, there are instances where dispersers are tolerated within adult home ranges, and other instances where dispersers travel long distances even though suitable habitat appeared to be nearby.

Dispersal distances are summarized in table 11.2, with details and citations discussed below. These are estimates, extrapolated in some cases from the literature, to convey a rough sense of the spatial requirements of these species and the distances necessary for genetic subdivision and restriction of gene flow.

There are few data on genetic structure of carnivore populations in the GYE (or elsewhere, for that matter). Some inferences can be made by examining typical distances of home ranges and dispersal movements. If known centers of distribution are separated by distances greater than known dispersals or by barriers to movement, it is possible that they are isolated to some degree and have diverged genetically, creating a subdivided population or metapopulation.

At a regional scale (figure 11.2), broad forested dispersal routes exist between the GYE and the Northern Continental Divide Ecosystem (NCDE) and the Salmon-Selway Ecosystem (SSE). These areas are currently occupied by cougars

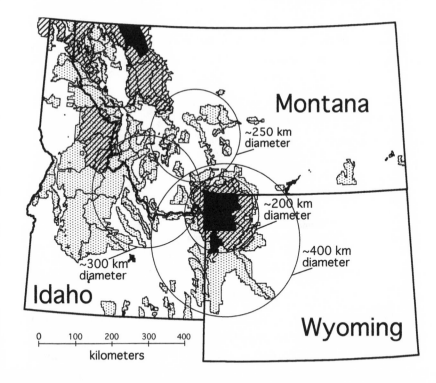

National Forest
Grizzly Bear Recovery Zones
National Parks
State Boundaries
Dispersal Distances

11.2. Grizzly bear recovery zones in Montana, Idaho, and Wyoming in relation to national park and forest lands in the Greater Yellowstone Ecosystem, the Northern Continental Divide Ecosystem, and the Salmon-Selway Ecosystem.

and have historically been occupied by grizzlies. Merriam (1922) reported grizzlies occupying the Tobacco Root, Gravelly, Crazy, Little Belt, Snowy, and Sapphire mountain ranges. Aune (personal communication) knew of grizzlies in the Big Belt Mountains in the early 1960s. Between 1976 and 1984 grizzlies were observed (or killed) in the Gravelly, Tobacco Root, Highland, Champion-Thunderbolt, and McDonald-Rogers mountain ranges (Picton 1986). Picton estimated that 3,600 square kilometers of habitat encompassing the Tobacco Roots, Snowcrests, and Gravellys could support about thirty-nine bears, while habitat farther north along the Continental Divide (3,250 square kilometers)

Table 11.3

Summary of carnivore genetic status in the Greater Yellowstone Ecosystem

Species	Genetic health	Genetic structure	Isolation
Wolves	*Normal–high H*	None	Moderate
Coyotes	*Normal H*	Little	None
Red foxes	Normal *H*	Moderate	Moderate
Grizzly bear	*Low H*	*None*	*High*
Black bear	Normal *H*	Little	None–little
Mountain lion	*Normal H*	None–little	None–little
Bobcat	Normal *H*	Little	None
Lynx	*Low – normal H*	Little	Moderate
Wolverine	Normal *H*	Little–none	Little–moderate
Fisher	Normal *H*	Little	Moderate
Marten	*Normal H*	Moderate	Little
Otter	Normal *H*	Moderate	Little
Badger	Normal *H*	Moderate	Moderate
Weasels	Normal *H*	Modetate	Moderate
Skunks	Normal *H*	Little	Little

Note: Items italicized are based upon some relevant genetic data.

could support about thirty-three. If managed with that goal in mind, these areas could support small populations of grizzly bears and wolves, as they already support other carnivores. Gene flow and population rescue could occur in the GYE for bears, wolves, cougars, and other species as long as dispersing animals could negotiate the filters and barriers (primarily human-caused) between islands of habitat.

Below we discuss the carnivore families in the GYE in light of current knowledge. Those species with normal to high heterozygosity (*H*), some degree of genetic structure within or between populations, and little to no isolation from other populations stand the best chance for long-term persistence. Table 11.3 summarizes our current state of knowledge (often no more than best guesses) of the major species.

CANIDAE: FOXES, COYOTES, AND WOLVES

The Canidae of the GYE consist of species with wide distribution. Although research here has focused mainly on coyotes and wolves, and mostly within the past decade, these species have been studied elsewhere and much of this research is directly applicable.

Current genetic research. Forbes and Boyd (1996, 1997) discussed microsatellite variation in naturally colonizing wolves in Glacier National Park, other Rocky Mountain populations, and reintroduced populations. They found high levels of genetic variation in multiple, unrelated founding wolves from Canada. They concluded that wolves in the United States and Canada in the

Rocky Mountains should be viewed as a single population and that maintaining gene flow throughout the region is necessary to retain this genetic variation. Wolf dispersal over long distances is likely if legal protection and public tolerance is adequate. Differentiation and hybridization of wolf and coyote populations has also been studied using microsatellite loci (Roy et al. 1994). Introgression of coyote mtDNA into gray wolf populations was demonstrated in Minnesota wolves (Lehman et al. 1991), demonstrating ancestral breeding between wolves and female coyotes. Twenty-six populations of gray wolves in North America were examined for mtDNA variability: eighteen genotypes were found, seven of which were derived from hybridization with coyotes (Wayne et al. 1992).

Little genetic research has been conducted with coyote populations. Results include the introgression studies discussed above (Lehman et al. 1991) and a mtDNA survey across North America in which coyotes showed no significant genetic differentiation, even between distant populations, and had six times as many genotypes as wolves (Lehman and Wayne 1991). This similarity is attributed to high rates of gene flow because of a rapid recent increase in coyote numbers and the expansion of their range (Wayne et al. 1992). Unhunted coyote populations in the GYE have established social behavior similar to wolves (Crabtree personal communication). Introgression between these species after wolf reintroduction is a distinct possibility.

Foxes have received little attention in the GYE. Red foxes from the Absaroka-Beartooth Plateau exhibit striking morphological differences from other areas (Crabtree personal communication), which may indicate genetic differences and population structure.

Patterns of dispersal. Coyotes and wolves tend to form social packs with a dominant breeding pair. Red foxes typically breed monogamously but also form polygynous groups of one male and two to five females (MacDonald 1981). Both sexes of juvenile wolves disperse and attempt to join other packs (Mech 1987); this is also the case in coyotes. Dispersal group size and timing of both wolves and coyotes appears to be influenced by local prey species and population density (Bekoff and Wells 1980). Where coyote population density is high, group size is larger (Andelt 1982). Wolves commonly travel fifty kilometers or more per day (Paquet personal communication), and males have been recorded traveling up to 917 kilometers (Fritts 1983). A young female dispersed 840 kilometers (Boyd et al. 1996). Wolf dispersal distance averaged over many studies is about eighty-five kilometers, with male dispersal only about eight kilometers greater than female (Weaver et al. 1996). High-volume roads like the Trans-Canada Highway near Banff are a barrier to wolves, but low-volume roads are crossed easily. Some wolves will use underpasses, but others will not (Paquet personal communication). Coyotes disperse in two modes in the GYE (Crabtree personal communication); either about fifty kilometers completely away from the natal area or into neighboring territories of other packs. The longest recorded dispersal distance for a coyote was 544 km (Carbyn and Paquet 1986).

Population sizes, home ranges, and inferences of genetic structure. Wolves were extirpated from Yellowstone Park (and the GYE) beginning in the 1870s by hide hunters poisoning elk carcasses, and a concerted effort to eliminate them was carried on from 1914 to 1926. The last two wolves killed by federal agents in the park were trapped in 1926, but a handful seem to have persisted into the 1930s, and one was killed in the GYE in 1943 (Varley personal communication). Thirty-one wolves were reintroduced into the GYE in 1995 and 1996 by the U.S. Fish and Wildlife Service and National Park Service under mandates of the Endangered Species Act. There were fifty-two known wolves by the winter of 1997 (Varley personal communication) and probably close to one hundred by late spring (Maughan personal communication). Wolves were captured in Alberta from presumably unrelated packs from different areas with the goal of introducing a genetically diverse founder population. It will be several generations before the population stabilizes, but distribution may eventually focus on two or three geographic areas, such as northern Yellowstone and geyser basins, Jackson Hole, and perhaps the Wind River Range foothills. A critical factor will be the availability of winter prey. Territory sizes of wolf packs range from 780 to 1,040 square kilometers in northwestern Montana. Lone wolves may range over 2,600 square kilometers (USFWS 1994). Wolves disperse long distances and gene flow occurs on a regional scale (Forbes and Boyd 1996, 1997). In light of this mobility, genetic subdivision probably occurs between the large core reserves but not within any of them.

Coyote populations are stable throughout Greater Yellowstone. High levels of gene flow discussed above should mitigate any genetic structure among areas within the ecosystem, but the lack of wintering habitat in the central Yellowstone Plateau may act as a partial barrier to dispersal between the northern and southern parts of the ecosystem. Coyote densities are probably highest in the Northern Range of Yellowstone Park where they average about 0.5 to 1.0 coyote per square mile. Territory size averages about 5.6 square kilometers (3.0 square miles, Crabtree personal communication). Throughout the entire ecosystem there are probably fewer than three thousand coyotes.

Red foxes appear to be stable in moderate numbers throughout the ecosystem. There are about two thousand in the entire GYE (Crabtree personal communication). Home-range sizes have been reported elsewhere as small as 0.19

to 0.72 square kilometers (MacDonald 1981) or as large as 5.3 square kilometers (Von Schantz 1981).

URSIDAE: BLACK AND GRIZZLY BEARS

Of the GYE's two bear species, the grizzly has been the subject of much more research, both ecological and genetic. Black bears, on the other hand, have been more or less taken for granted by the scientific community and even rough population estimates are uncertain.

Current genetic research. Allendorf and Knudsen have examined isozyme variation in grizzly bears and polar bears. Varying levels of polymorphism were found in different populations, with the least variation in grizzly bears from Kodiak Island (Knudsen personal communication). Cronin et al. (1991) found five mtDNA haplotypes in brown bears from sample locations from Alaska and Montana. Further mtDNA analysis by Waits et al. (1997) resolved four clades in North America that were probably formed before migration of the species from Asia. Three evolutionarily significant units are suggested; the GYE population is grouped with southern Alberta and British Columbia, Idaho, Montana, and Wyoming (Waits et al. 1997).

Simulations indicate that existing grizzly bear populations in the lower forty-eight states are not large enough to avoid detrimental loss of genetic variation in the short term (Harris 1986, Harris and Allendorf 1989). An initial small sample of sixteen grizzly bears from the northern Rockies was not found to differ significantly from Alaskan populations in heterozygosity, but five unique microsatellite alleles were found (Craighead 1994). However, analysis of a larger sample of 667 grizzlies, seventy-two of which were from Yellowstone, demonstrated a significantly lower H (55 percent) for Yellowstone grizzlies, compared with a high of 76 percent in the Kluane sample ($N = 50$) and a low of 26 percent in the Kodiak sample ($N = 34$). The main factor affecting levels of genetic diversity appears to be connectedness to larger populations (Paetkau et al. 1997). Estimates of N_e can be obtained from H within the limits of the accuracy with which mutation rates are known (assuming equilibrium for mutation, genetic drift, and migration). Thus estimates of N_e for the Yellowstone population are thirteen to sixty-five effective grizzly bears (Paetkau et al. 1997).

Zimmerman (1989) examined Restriction-Fragment Length Polymorphism variation in mtDNA among fifty-five black bears. Minor differences were found between three subspecies. Cronin et al. (1991) found six mtDNA haplotypes in black bears. Black bears ($N = 40$) examined were from Alaska, New Hampshire, Oregon, and Montana. Two genetically distinct groups of black bear haplotypes were found, and some types were found in only one location. Paetkau and

Strobeck (1994) examined genetic variation in black bears using four microsatellite loci. Black bears isolated on the island of Newfoundland were found to have relatively reduced levels of variation, analogous to brown bears on Kodiak.

Patterns of dispersal. Bear dispersal patterns follow the general mammalian pattern: females are more philopatric and often share a portion of their natal home range with the mother (Reynolds and Hechtel 1984). Males disperse farther, occasionally traveling hundreds of kilometers (Craighead 1994). In the GYE four dispersing males moved an average of seventy kilometers from their natal home range (Blanchard and Knight 1991). This pattern persists despite the fact that both brown and black bears exhibit characteristics of polygyny and polyandry. Male grizzly bears have been known to breed successfully with more than one female during a single season, and females have bred successfully with more than one male, producing litters with multiple paternity (Craighead et al. 1995). Factors other than breeding exclusion, therefore, also appear to be involved in dispersal behavior. Black bear dispersal is similar to that of grizzly bears, but the distances are smaller.

High-volume roads like the Trans-Canada Highway near Banff are a barrier to grizzly bears but not to black bears. Grizzlies avoid underpasses, but black bears use them (Woods and Munro 1996). The only effective barriers to grizzly dispersal (other than very large bodies of water) are products of human activities (Craighead and Vyse 1996). Although grizzlies are capable of long-distance dispersal, none of 460 radio-tagged grizzlies have traveled between the large core reserves in the northern Rockies within the past twenty-five years (Servheen personal communication, reported in Weaver et al. 1996). It is likely that barriers produced by human activities have impeded these longer movements, for grizzlies in contiguous habitat unaltered by humans often travel this far.

Population sizes, home ranges, and inferences of genetic structure. Recent estimates of grizzly numbers in the GYE include minimums of 133 and 245, an estimate of 390 using marked females, an estimate of 339 based on distinct families, a bootstrapped estimate of 344 with a 90 percent confidence interval of 280–610, and an interagency population review committee estimate of a minimum of 245 bears, of which sixty-seven are adult females (Eberhardt and Knight 1996). However, an estimate of N_e for the Yellowstone population is only thirteen to sixty-five effective grizzly bears (Paetkau et al. 1997). Female grizzly home ranges vary greatly in size (Craighead et al. 1995), ranging in the GYE from 233 square kilometers for males and seventy-three square kilometers for females (Craighead 1976) to 1,970 square kilometers for males and 874 square kilometers for females (Knight et al. 1984). In the Mission Mountains of Montana home ranges for males averaged 1,403 square kilometers (Servheen 1983). The average lifetime home range size is about 3,885 square kilometers (Mattson and Reid 1991), but female home ranges average around 450 square kilometers (Merrill personal communication).

In a study of microsatellite alleles in a grizzly population over 5,200 square kilometers in the western Brooks Range, Craighead et al. (1995) found no evidence of a Wahlund effect (deficiency of heterozygotes) and thus no indication of genetic structure. Other larger Arctic study areas also showed no evidence of genetic structure (Paetkau et al. 1997). The Greater Yellowstone Recovery Zone is almost five times as large as the western Brooks Range study area, and the GYE is about eight times as large, so that some genetic structure may once have existed between grizzlies at extreme ends of the ecosystem. Genetic structure is evident today on a larger metapopulation scale between the GYE and populations further north (Paetkau et al. 1997). Grizzlies, like wolves, could disperse between the large core reserves in a single season, but they typically travel shorter distances.

Black bear status and distribution has been given little attention in the GYE. In Montana about one thousand black bears are harvested annually (Montana Department of Fish, Wildlife and Parks 1994). Black bears are found throughout Yellowstone Park, and populations are presumed to be stable. One estimate of five hundred for Yellowstone National Park has been made by the U.S. National Park Service (1989). The status of the population is somewhat uncertain, however, for one estimate of illegal kills was forty black bears in 1991 and 1992. Black bears prefer more heavily forested areas than grizzlies and are more tolerant of human activities. In eastern Alberta, Young and Ruff (1982) found mean home-range size of males to be 119 square kilometers and 19.6 square kilometers for females. In Idaho Amstrup and Beecham (1976) reported home ranges from 109 to 115 square kilometers for males and seventeen to thirty square kilometers for females. In Montana male home ranges were 53 to 225 square kilometers and female home ranges were fourteen to 137 square kilometers (Kasworm and Thier 1991, Aune and Brannon 1987, Greer 1987). It is likely that slight genetic subdivision exists at the scale of the entire GYE. Because of black bears' cover and security requirements, immigration from distant populations is restricted to the same routes that grizzly bears use.

FELIDAE: BOBCATS, LYNX, AND MOUNTAIN LIONS

The Felidae have also received little attention in the GYE. Mountain lions have been studied only during the last decade as they have returned to the Northern Range following their virtual extirpation earlier in the century.

Current genetic research. Little genetic research has been done to date on the Felidae that inhabit the GYE. Microsatellite DNA analysis of mountain lions has revealed paternity of cubs and provides a baseline of allele frequency data and heterozygosity (Murphy personal communication). An ecological study of lynx in northwestern Montana is currently investigating lynx population genetics through the use of microsatellite DNA analysis of hairs collected from scent-rubbing posts (Weaver personal communication).

Patterns of dispersal. Felids are polygynous and polyandrous. Dispersal patterns in mountain lions, bobcats, and lynx also follow the general mammalian pattern, with male dispersers traveling farther. Young female mountain lions tend to remain near their maternal home ranges (Ross and Jalkotzy 1992), but males travel farther and have been known to move up to 480 kilometers (Logan et al. 1986). Female lynx offspring usually remain near their mothers' home ranges, while juvenile males tend to disperse (Koehler and Aubry 1994). When prey are scarce, even resident adult lynx may move long distances. Record distances include 1,100 kilometers and 700 kilometers in the Yukon (Ward and Krebs 1985) and 325 kilometers in western Montana (Brainerd 1985). Six male juvenile bobcats in British Columbia exhibited long-distance dispersal movements (a mean of 109 kilometers), while three juvenile females did not. Males actually traveled at least twice this distance before settling down (Apps 1996).

Population sizes, home ranges, and inferences of genetic structure. Mountain lion home ranges averaged forty-five square kilometers in Wyoming (Logan et al. 1986). Murphy et al. (1991) found lifetime home ranges of 117 square kilometers for females and 889 square kilometers for males in a northern Yellowstone study area. Winter home ranges were smaller. In central Montana Williams (1992) found female ranges of fifty-eight square kilometers and male ranges of 199 square kilometers. In Idaho, Seidensticker et al. (1973) found mean home ranges of 453 square kilometers for males and 268 square kilometers for females. Mountain lions were thought to have been completely eradicated by predator control programs in Yellowstone National Park: one hundred twenty-one had been removed by 1925 (Harting and Glick 1994). Since ecological studies began in the Northern Range in the 1980s, the population appears to be increasing. In Montana 424 lions were harvested in 1993, and fifty-four non-hunting mortalities were reported (Montana Department of Fish, Wildlife, and Parks 1995). Approximately 500 lions were harvested in 1998 (Williams personal communication). Mountain lion distribution is probably restricted by the availability of winter prey. This may act to isolate populations to some degree within the ecosystem, but genetic differences between areas would be ameliorated by gene flow. Cougars can easily disperse to distant areas within the ecosystem or to other ecosystems (all within three hundred kilometers) provided that human activities are not significant barriers to dispersal.

Lynx have been sighted in both Yellowstone and Grand Teton national parks and in the Centennial, Wind River, Gros Ventre, and Absaroka ranges (Harting and Glick 1994). The GYE is near the southern limit of their known range. Although records of lynx occur farther south, it is possible that these were not stable populations (Koehler and Aubry 1994). A population in the Wyoming Range, south of Jackson, is currently being studied by the Wyoming Department of Game and Fish. Lynx populations and their major prey, hares, do not cycle dramatically in the western United States as they do farther north. Populations of lynx in the northern Rockies are therefore relatively stable, perhaps because other predators (coyotes, bobcats, red foxes, hawks, and owls) and competitors help stabilize hare populations (Wolff 1982). Home-range sizes vary greatly with prey abundance but may range from one hundred to three hundred square kilometers in the GYE.

Male bobcats also have larger home ranges than females, according to most studies. In western Montana, Brainerd (1985) found males to average 122 square kilometers, while females averaged 43.1 square kilometers ($N = 17$). In British Columbia males averaged 138.5 square kilometers, while females averaged 55.7 square kilometers. The size of individual home ranges fluctuates with the prey base. When lagomorph numbers are low, bobcats must utilize larger areas. Habitat is chosen to minimize snow depth and energetic expenditure and to maximize hunting opportunities (Koehler and Hornocker 1989, Litvaitis et al. 1986). Koehler and Hornocker (1989) found that rocky terrain was a factor in resource partitioning between bobcats and coyotes, and it also provides security and escape habitat (Apps 1996). Bobcat numbers are probably sufficient for long-term stability in the GYE, but winter habitat and competition with other predators may be limiting and probably isolate bobcat populations somewhat into habitat surrounding the high central Yellowstone Plateau. These patterns of distribution probably contribute to genetic structure, but this has not yet been examined. Bobcats disperse on the order of one hundred kilometers and thus may exhibit some degree of genetic subdivision between extreme ends of the ecosystem. Some genetic structure almost certainly exists between ecosystems.

MUSTELIDAE: WEASELS, SKUNKS, BADGERS, MARTEN, FISHERS, OTTERS, AND WOLVERINES

Marten are the only mustelids in the GYE for which there is much genetic data available. Wolverines are low in number but disperse widely, which should maintain gene flow. Fishers are low in number and populations are widely separated, which should result in more genetic structure. Otters are low in number and restricted in distribution. The other species should have large enough populations over a large enough area to capture a degree of genetic structure.

Current genetic research. Mitton and Raphael (1990) examined allozyme variability in a marten population in the central Rocky Mountains. They found very high levels of mean multilocus heterozygosity (0.17) compared with carnivores in general (0.01, summarized by Kilpatrick et al. 1986). However, their sample size of ten was very small. Steven Fain at the National Fish and Wildlife Forensic Lab has developed multilocus DNA fingerprints of martens (Fain personal communication). Researchers at San Francisco State have examined D-loop sequences of mtDNA in California martens with much variation and in fishers with no variation (Zielinski personal communication).

Patterns of dispersal. Mustelids in general disperse as expected. Wolverine subadult females generally remain near their mothers, while males disperse when they reach sexual maturity at two years of age. Two dispersing males each traveled about two hundred kilometers and established home ranges. Two others dispersed and then returned to the natal area (Copeland 1996). A male wolverine in British Columbia dispersed sixty kilometers in ten days (Krebs personal communication). Wolverines do not seem to be deterred by such topographic features as mountains, lakes, and rivers (Hornocker and Hash 1981), but they avoid human developments, extensive settlements, and major access routes (Banci 1987, Ruggiero et al. 1994). In these respects they are similar to grizzly bears. One of the longest wolverine dispersals on record was 328 kilometers from south-central Alaska to the Yukon over about eight months (Gardner et al. 1986). Female fisher home ranges often overlap, and males disperse farther. One male traveled one hundred kilometers, crossing some fairly large open areas in Alberta. Fishers commonly cross one to two kilometers of open areas between woodlots at night (Badry personal communication). Two male fishers in Idaho dispersed twenty-six kilometers and forty-two kilometers (Buck et al. 1983). After reintroductions, fishers often move long distances, ranging from about forty to 163 kilometers from their release sites (summarized in Ruggiero et al. 1994).

Population sizes, home ranges, and inferences of genetic structure. Home-range sizes reported for wolverines averaged 1,522 square kilometers for males and 384 square kilometers for females in Idaho. Male home ranges overlapped about 16 percent, while females overlapped less than 8 percent. Male ranges often include two or three female home ranges (Copeland 1996). In northwestern Montana, Hornocker and Hash (1981) found a mean of 422 square kilometers for male home ranges and 388 square kilometers for females. Copeland reported an average density of one wolverine per 198 square kilometers of habitat. This translates to a maximum of 225 in the GYE, but the number may be much lower than that, because much of the area is not optimal habitat. Houston (1978) reported twenty-seven wolverine sightings around Yellowstone Park from 1970 to 1978, and Hoak et al. (1982) reported fifty sightings farther south in Wyoming since the 1950s. Given these low numbers and high mobility, it is unlikely that much historic genetic subdivision existed within the GYE, although human-caused fragmentation of habitat may currently be isolating some regions from others. Some genetic differences almost certainly exist between GYE populations and those in other ecosystems, but whether or not these are ecologically important remains to be seen.

A fisher habitat suitability model developed in Bristish Columbia assumes that 259 square kilometers of contiguous habitat are required for transplant attempts. Winter cover is the limiting factor, and road densities need to be very low (Apps 1996). Fishers avoid areas closer than two hundred to five hundred meters from roads and areas over one thousand meters from water (Badry personal communication). In general male fishers require about forty square kilometers of habitat and females fifteen square kilometers (Aubry personal communication). There appears to be a healthy population of fishers in the Bitterroot Mountains in Montana (Foresman personal communication) but very few in the GYE. Powell and Zielinski estimated that two thousand square kilometers of suitable habitat would maintain a population of fifty fishers in the Rocky Mountains (based on Jones 1991). Fisher populations in the GYE may be subdivided into two areas: the northeastern corner of the ecosystem in the Shoshone National Forest and the southwestern corner of Yellowstone Park and the Targhee National Forest (Crabtree personal communication). There are likely genetic differences between these population centers because of the distance apart and the small population sizes.

Marten male home-range sizes averaged about 0.8 square kilometers in Montana (Burnett 1981). Marten and weasel populations appear to be large enough in the GYE to be self-sustaining with adequate levels of genetic diversity, although little genetic work has been completed. There is probably important genetic subdivision across the ecosystem between distant areas such as the Wind River Range, the Beartooth Plateau, and the Madison Range.

Badger home ranges averaged 1.4 square kilometers for males and 1.2 square kilometers for females in southwestern Idaho (Messick and Hornocker 1981). In the GYE, where prey species are less dense, badger home ranges are probably much larger. Severe winters and dense snowpack, plus the lack of open habitat, probably limit badger numbers and distribution greatly in the central part of the ecosystem, the Yellowstone Plateau.

Otters prefer valley habitats over mountain lakes and streams, with use of large valleys predominant. They prefer open-water stream channels during winter (Melquist 1981). Otter populations appear stable along the main stem of major Montana rivers, such as the Jefferson, Madison, Gallatin, and Big Hole (Zackheim 1982). Extremely cold winter conditions with few open water stretches, coupled with the small size of many GYE headwater streams, probably limits the otter population in the central part of the ecosystem. The headwaters regions will likely be "rescued" by dispersal of individuals from source populations in the downstream reaches as long as those sections retain good otter habitat. Because otters are restricted primarily to riparian habitats, populations in the tributaries of the larger drainages may be genetically subdivided, much as fish stocks are, but much less distinctly.

In summary, it is likely that genetic subdivision of the GYE populations of all the medium- to small-sized species is present, particularly because habitat features such as large open areas like the larger valleys are barriers to the forest-dwelling carnivores. The larger lakes and rivers may hinder dispersal of most species during peak dispersal times in spring and fall. Upland areas, of course, act as barriers to riparian species such as mink and otters and perhaps even fishers.

GENETIC IMPLICATIONS FOR CARNIVORE MANAGEMENT

Species such as carnivores that require large areas of habitat can maintain only small populations on the Greater Yellowstone "island." It is estimated that the amount of habitat to support a grizzly bear, for example, ranges from 450 – 900 square kilometers. The number of breeding females fluctuates around fifty, but the effective breeding size appears to be in the range of thirteen to sixty-five (Paetkau et al. 1997).

Initial estimates suggest that the carrying capacity for wolves in Yellowstone Park (ten packs or one hundred animals) is close to being reached. The limiting factor is winter range for prey species and the tolerance of sympatric humans. Yellowstone Park will support only as many wolves as can survive through the winter. As wolves expand south, however, the Jackson Hole valley, with its wintering elk herd, should support many more. The first dispersing wolves observed ($N = 9$) reached this area in January 1999. With their great mobility, wolves in the GYE will soon be augmented by dispersers from the other core reserves.

Similarly, the population size of mountain lions is also limited by winter range and human attitudes: within the park this is essentially the Lamar-Gardiner valleys and the larger geyser basins. Prey species flourish at lower elevations around the perimeter of the ecosystem, but those areas are being populated by humans.

Wolverines can adapt to a variety of habitat types but are very sensitive to human disturbance. They prefer areas with year-round prey in undisturbed wilderness area (Banci 1994). Their denning habitat in the Rocky Mountains appears to be limited to high-elevation, rocky cirques (Koehler 1996). Denning habitat and the distribution of prey limit wolverine population size.

Lynx prefer boreal forest above about 1,900 meters in elevation, particularly early successional stages with high numbers of snowshoe hares and other prey (Koehler and Aubry 1994). Fishers prefer late-successional coniferous forests with high densities of snowshoe hare or porcupines (Powell and Zielinski 1994). The GYE is at the southern limit of the distribution of both lynx and fisher, but they are found in habitat connecting the GYE to the NCDE. Prey and habitat availability appear to limit fisher and lynx. River otter population sizes are constrained by the availability of riverine habitat, and otter populations thus will always remain small in the GYE.

It is the larger carnivore species for which genetic stochasticity is most critical: grizzly bear, wolf, wolverine, cougar, lynx, fisher, and otter are the species most at risk. If we can maintain populations of these species within the GYE, by maintaining all the habitat currently available, their population sizes will remain small enough for genetic effects to tip the balance that could lead to extinction. Inbreeding depression effects will always be critical for the species with small populations: the wolf, because of its high growth rate potential, will be affected least and will also benefit from a reintroduction scheme designed to increase genetic diversity.

Genetic diversity is an important concern for the long-term survival and adaptation of all carnivore species. Natural immigration by dispersing individuals from distant populations may serve to rescue declining populations and may represent selective forces and aspects of genetic diversity that human-managed translocation of captured animals ignores. The genetic future of these populations in the Greater Yellowstone, as well as that of other carnivores, is inextricably tied to the effects of human population growth and to human attitudes and values. Specifically, the more that human activities isolate the GYE from other populations (increase the distance to the "mainland") and alter the habitat within the ecosystem (reduce the size of the "island"), the greater the loss of genetic diversity and gene flow.

Summary and Conclusions

Biodiversity is much like the metaphorical Persian carpet from which Quammen (1996) launches his essay on islands and extinction: it is ancient, it is intertwined, and it is intricate. We have examined the genetic level of biodiversity in the GYE's carnivores, but genes empower and guide individuals on whom depend the fate of populations, and success and failure at individual and population levels feed back to alter gene frequencies. The connections run up and down the levels of biotic complexity, and it is only with our powers of scientific abstraction that we can think of them in isolation. The GYE is not a pristine Persian carpet of biodiversity, an isolated showcase for the ages. By using crude and blunt instruments of habitat alteration, inhibiting carnivore dispersal with ribbons of concrete and gravel, and replacing plant and animal communities with our own kind, modern humanity has torn the GYE from the broader fabric, leaving a tatter of unraveling edges. We must manage insightfully and purposefully to keep this tatter locally whole, locally a representation of the former whole, the colors strong and the details sharp.

The common thread that weaves through the fabric of life, from individual to individual, through populations and species and ancestral life forms, through years and decades and eons, is the genetic code—the abstract pattern book from which all is fashioned. It is a pattern book for species, and over time even the pattern can slowly change. Fundamental changes in the genetic code occur slowly, on an evolutionary time scale, as a response to long-term shifts of environmental factors. However, when the environmental conditions change suddenly, the genetic shift entails a rapid reduction in genetic variability. Further shifts in the environment can leave the species without the variation necessary to fashion adaptive responses.

Genetic diversity underlies the fitness of populations, including the short-term ability to adapt to local changes and the long-term potential to evolve. Each species has a basic genetic blueprint with essential features common to all individuals. Every individual blueprint is unique, however, because of individual variation, which occurs constantly through sexual reproduction and sorting of alleles. Genetic change occurs in every generation as the blueprint is fine-tuned to local environmental conditions through natural selection and through chance events that act to increase or decrease alleles in the population. Genetic change, particularly a reduction in genetic variability, is accelerated in small populations.

Rapid environmental changes are occurring now, reducing the numbers of wildlife populations and individuals. Worldwide habitat changes are reducing genetic diversity and reducing biodiversity. From the viewpoint of many species, human population growth can be considered a catastrophic event. The current catastrophe, however, has resulted from more than just an increase in our numbers. By altering our environment to achieve our own ends, we have drastically altered the environment for many other species as well. As a sentient and perhaps sapient species, we must be conscious of the effects of our actions. We know what we are doing, and we are responsible for the results. The current worldwide destruction of habitat and the wave of extinctions are perhaps avoidable. In many places on the planet we still have time to preserve habitats, populations, and species, and thus biodiversity. The Greater Yellowstone Ecosystem is one of these special places.

Carnivore populations are especially susceptible to reduced genetic diversity as a consequence of human activities. The GYE is one of the few remaining places with enough intact habitat to maintain stable carnivore populations and to provide for gene flow from one population to another. Grizzly bears, wolves, and wolverines probably need more habitat than is available in currently protected landscapes in the GYE to safeguard genetic diversity and provide a refuge from extinction. Lynx, fishers, and otters require less area, but their habitat and prey in the GYE are restricted, and thus their populations are small and may be genetically subdivided. To remain viable in the GYE, lynx and fishers need to maintain contact with populations and forested habitats farther north. Subpopulations of all carnivore species need to be identified better and their habitats protected. Baseline genetic research on the six species mentioned above is desperately needed. Additional research on other carnivore species is an important next step. Any carnivore that is handled by researchers should be sampled routinely for future genetic studies, including a blood sample if possible or at least a hair sample. Connectedness between subpopulations within the GYE should be a management priority. Human recreation and vehicle traffic should be managed with wildlife movements in mind.

The movement of large carnivores (and thus their genes) in and out of the GYE is restricted. One practical reserve design for these species is to maintain existing habitat connections with other wilderness ecosystems, specifically the Northern Continental Divide and Salmon-Selway ecosystems. The distance to both these large reserves is less than three hundred kilometers (figure 11.2), well within the dispersal range of the larger carnivore species. A network of reserves with connecting corridors is a workable solution. Minimally disturbed habitat sufficient to support small populations currently exists in intermediate locations. These areas are currently occupied by cougars and have historically been occupied by grizzly bears. Many of these areas also support all of the small and medium-sized carnivores.

Most of these islands of habitat, if managed with that goal in mind, could support one or more large carnivore family groups. Gene flow and population rescue could occur for bears, cougars, wolverines, and other species as long as dispersing animals could negotiate the human-altered landscape between islands of habitat. Such dispersal could occur in a single season, or it could take generations. Such obstacles to movement as human transportation corridors could be ameliorated in just a few places to greatly increase the probability of wildlife movement. The GYE could be managed as one population center of a regional metapopulation; gene flow within the metapopulation would probably increase genetic diversity within each of the component populations (Hedrick 1996, Hedrick and Gilpin 1996). As Doak and Mills (1994) point out, there are difficulties using broad theories such as metapopulation dynamics and island biogeography for conservation planning. One often cited caution is the allocation of scarce or otherwise valuable resources for unproven benefits. The attractiveness of a northern Rockies metapopulation scheme, then, is that most of the reserve connections are already in place and are still roadless and largely undisturbed. We can maintain them in this state and improve connectivity if we want. Thus implementation of reserve designs to maintain carnivores depends on the collective will of the human populations involved. The most significant variable affecting carnivore persistence may well be public opinion, which needs to be responsibly informed. It is our hope that this chapter and this book will help serve that need.

Acknowledgments: We would like to acknowledge the small circle of carnivore geneticists who have pioneered this rapidly expanding field of inquiry, particularly David Paetkau, Curtis Strobeck, Matt Cronin, Steven Fain, Sandra Talbot, Gerald Shields, Lizette Waits, Ryk Ward, Michael Kohn, Fred Allendorf, Cathy Knudsen, Pierre Taberlet, Jacques Bouvet, Robert Wayne, and Steven O'Brien. Equally important are the field biologists whose collections have made genetic analysis possible and whose work has provided the background for understanding the genetics of populations. We would like to thank L. Scott Mills for a critical review of an early draft, John Varley for reviewing draft sections of our review of the ecological literature, and Scott Creel and Robert Lacy for critical reviews of the manuscript. Thanks to the editors for their support and patience.

Literature Cited: Allendorf, F. W., and R. F. Leary. 1986. Heterozygosity and fitness in natural populations of animals. Pp. 57–76 *in* M. E. Soulé, ed., Conservation biology. Sinauer, Sunderland, Massachusetts.

Allendorf, F. W., and C. Servheen. 1986. Genetics and conservation of grizzly bears. Trends in Ecology and Evolution 1:88–89.

Amos, W., S. J. Sawcer, R. W. Feakes, and D. C. Rubinstein. 1996. Microsatellites show mutational bias and heterozygote instability. Nature Genetics 13:390–91.

Amstrup, S. C., and J. Beecham. 1976. Activity patterns of radio-collared black bears in Idaho. Journal of Wildlife Management 40:340–48.

Andelt, W. F. 1982. Behavioral ecology of coyotes on Welder Wildlife Refuge, South Texas. Ph.D. diss., Colorado State University, Fort Collins.

Apps, C. 1996. Bobcat *(Lynx rufus)* habitat selection and suitability assessment in southeast British Columbia. M.S. thesis, University of Calgary, Calgary.

Ballou, J. D. 1997. Ancestral inbreeding only minimally affects inbreeding depression in mammalian populations. Journal of Heredity 88:169–78.

Ballou, J. D., and K. Ralls. 1982. Inbreeding and juvenile mortality in small populations of ungulates: A detailed analysis. Biological Conservation 24:239–72.

Banci, V. 1987. Ecology and behavior of wolverine in the Yukon. M.S. thesis, Simon Fraser University, Burnaby, British Columbia.

Beier, P., and R. F. Noss. 1998. Do habitat corridors provide connectivity? Conservation Biology 12:1241–52.

Bekoff, M., and M. C. Wells, 1980. The social ecology of coyotes. Scientific American 242:130–48.

Belovsky, G. 1987. Extinction models and mammalian persistence. Pp. 35–58 *in* M. E. Soulé, ed., Viable populations for conservation. Cambridge University Press, Cambridge.

Blanchard, B. M., and R. R. Knight. 1991. Movements of Yellowstone grizzly bears. Biological Conservation 58:41–67.

Brainerd, S. M. 1985. Reproductive ecology of bobcats and lynx in western Montana. M.S. thesis, University of Montana, Missoula.

Buck, S., C. Mullis, and A. Mossman. 1983. Final report: Corral Bottom–Hayfork Bally fisher study. Unpublished report. USDA in cooperation with Humboldt State University, Arcata, California.

Caro, T. M., and M. K. Laurenson. 1994. Ecological and genetic factors in conservation: A cautionary tale. Science 263:485–86.

Caughley, G. 1994. Directions in conservation biology. Journal of Animal Ecology 63:215–44.

Copeland, J. 1996. Biology of the wolverine in central Idaho. M.S. thesis, University of Idaho, Moscow.

Craighead, F. C., Jr. 1976. Grizzly bear ranges and movement as determined by radiotracking. International Conference on Bear Research and Management 3:97–109.

Craighead, F. L. 1994. Conservation genetics of grizzly bears. Ph.D. diss., Montana State University, Bozeman.

Craighead, F. L., D. Paetkau, H. V. Reynolds, E. Vyse, and C. Strobeck. 1995. Microsatellite analysis of paternity and reproduction in Arctic grizzly bears. Journal of Heredity 86:255–61.

Craighead, F. L., and E. Vyse. 1996. Brown/grizzly bear metapopulations. Pp. 325–51 *in* D. McCullough, ed., Metapopulations and wildlife conservation management. Island Press, Washington, D.C.

Craighead, J. J., J. S. Sumner, and J. A. Mitchell. 1995. Pp. 303–5 in The grizzly bears of Yellowstone: Their ecology in the Yellowstone ecosystem, 1959–1992. Island Press, Washington, D.C.

Cronin, M. A. 1993. Mitochondrial DNA in wildlife taxonomy and conservation biology: Cautionary notes. Wildlife Society Bulletin 21:339–48.

Cronin, M. A., S. C. Amstrup, G. Garner, and E. R. Vyse. 1991. Intra- and interspecific mitochondrial DNA variation in North American bears (*Ursus*). Canadian Journal of Zoology 69:985–92.

Diamond, J. M. 1976. Island biogeography and conservation: Strategy and limitations. Science 193:1027–29.

———. 1989. Overview of recent extinctions. Pp. 37–41 *in* D. Western and M. Pearl, eds., Conservation for the twenty-first century. Oxford University Press, New York.

Doak, D. F., and L. S. Mills. 1994. A useful role for theory in conservation. Ecology 75:615–26.

Eberhardt, L. L., and R. R. Knight. 1996. How many grizzlies in Yellowstone? Journal of Wildlife Management 60:416–21.

Eisenberg. J. F. 1981. The mammalian radiations: An analysis of trends in evolution, adaptation, and behavior. University of Chicago Press, Chicago.

Forbes, S. H., and D. K. Boyd. 1996. Genetic variation of naturally colonizing wolves in the central Rocky Mountains. Conservation Biology 10:1082–90.

———. 1997. Genetic structure and migration in native and reintroduced Rocky Mountain wolf populations. Conservation Biology 11:1226–34.

Frankel, O. H. 1970. Variation: The essence of life. Sir William Macleay memorial lecture. Proceedings of the Linnean Society NSW 95:158–69.

Frankel, O. H., and M. E. Soulé. 1981. Conservation and evolution. Cambridge University Press, London.

Frankham, R. 1996. Relationship of genetic variation to population size in wildlife. Conservation Biology 10:1500–1508.

Franklin, I. A. 1980. Evolutionary change in small populations. Pp. 135–49 *in* M. E. Soulé and B. A. Wilcox, eds., Conservation biology: An evolutionary-ecological perspective. Sinauer Associates, Sunderland, Massachusetts.

Fritts, S. H. 1983. Record dispersal by a wolf from Minnesota. Journal of Mammalogy 64:166–67.

Gardner, C. L., W. B. Ballard, and R. H. Jessup. 1986. Long distance movement by an adult wolverine. Journal of Mammalogy 67:603.

Gilpin, M. 1992. Demographic stochasticity: A Markovian approach. Journal of Theoretical Biology 154:8–16.

Gilpin, M. E., and J. M. Diamond. 1988. A comment on Quinn and Hastings: Extinction in subdivided habitats. Conservation Biology 2:290–92.

Gilpin, M. E., and M. E. Soulé. 1986. Minimum viable populations: Processes of species extinction. Pp. 19–34 *in* M. E. Soulé, ed., Conservation biology: The science of scarcity and diversity. Sinauer Associates, Sunderland, Massachusetts.

Goldman, D. P., R. Giri, and S. J. O'Brien. 1989. Molecular genetic-distance estimates among the Ursidae as indicated by one- and two-dimensional protein electrophoresis. Evolution 43:282–95.

Greenwood, P. J. 1980. Mating systems, philopatry and dispersal in birds and mammals. Animal Behaviour 28:1140–62.

Greer, S. Q. 1987. Home range size, habitat use, and food habits of black bears in south-central Montana. M.S. thesis, Montana State University, Bozeman.

Halfpenny, J. C., and D. Thompson. 1966. Discovering Yellowstone's wolves. Yellowstone Center for Resources, Gardiner, Montana.

Harris, L. D. 1984. The fragmented forest: Island biogeography theory and the preservation of biotic diversity. University of Chicago Press, Chicago.

Harris, R. B., and F. W. Allendorf. 1989. Genetically effective population size of large mammals: An assessment of estimators. Conservation Biology 3:181–91.

Harting, A., and D. Glick. 1994. Sustaining Greater Yellowstone: A blueprint for the future. Greater Yellowstone Coalition, Bozeman, Montana.

Hedrick, P. W. 1985. Genetics of populations. Jones and Bartlett, Boston.

———. 1996. Genetics of metapopulations: Aspects of a comprehensive prospective. Pp. 29–51 *in* D. McCullough, ed., Metapopulations and wildlife conservation management. Island Press, Washington, D.C.

Hedrick, P. W., and M. Gilpin. 1996. Metapopulation genetics: Effective population size. Pp. 1–29 *in* I. Hanski and M. Gilpin, eds., Metapopulation dynamics: Ecology, genetics, and evolution. Academic Press, New York.

Hoak, J. H., J. L. Weaver, and T. W. Clark. 1982. Wolverine in western Wyoming. Northwest Science 56: 159–61.

Hornocker, M. G., and H. S. Hash. 1981. Ecology of the wolverine in northwestern Montana. Canadian Journal of Zoology 59:1286–1301.

Houston, D. B. 1978. Cougar and wolverine in Yellowstone National Park. Research Note no. 5. Yellowstone National Park, Mammoth.

Irwin, D. M., T. D. Kocher, and A. C. Wilson. 1991. Evolution of the cytochrome b gene of mammals. Journal of Molecular Evolution 32:128–44.

Jones, J. L. 1991. Habitat use of fisher in northcentral Idaho. M.S. thesis, University of Idaho, Moscow.

Kasworm, W. F., and T. J. Thier. 1991. Cabinet-Yaak ecosystem grizzly bear and black bear research. 1990 Progress Report. U.S. Fish and Wildlife Service, Missoula.

Knight, R. R., D. Mattson, and B. Blanchard. 1984. Movements and habitat use of the Yellowstone grizzly bear. USDI Interagency Grizzly Bear Study Team, Bozeman.

Koehler, G. M., and K. B. Aubry. 1994. Lynx. Pp. 74–94 *in* L. F. Ruggiero, K. B. Aubry, S. W. Buskirk, L. J. Lyon, W. J. Zielinski, eds., The scientific basis for conserving forest carnivores: American marten, fisher, lynx, and wolverine in the western United States. USDA Forest Service General Technical Report RM-254. Rocky Mountain Forest and Range Experiment Station, Fort Collins.

Koehler, G. M., and M. G. Hornocker. 1989. Influences of seasons on bobcats in Idaho. Journal of Wildlife Management 53:197–202.

Kohn, M., F. Knauer, A. Stoffela, W. Schroder, and S. Paabo. 1995. Conservation genetics of the European brown bear: A study using excremental PCR of nuclear and mitochondrial sequences. Molecular Ecology 4:95–103.

Kretzmann, M. B., W. G. Gilmartin, A. Meyer, G. P. Zegers, S. R. Fain, B. F. Taylor, and D. P. Costa. 1997. Low genetic variability in the Hawaiian monk seal. Conservation Biology 10:482–90.

Lacy, R. C., 1997. Importance of genetic variation to the viability of mammalian populations. Journal of Mammalogy 78(2):320–35.

Lande, R. 1995. Mutation and conservation. Conservation Biology 9:782–91.

Lande, R., and G. F. Barrowclough. 1987. Effective population size, genetic variation, and their use in population management. Pp. 87–123 *in* M. E. Soulé, ed., Viable populations for conservation. Cambridge University Press, Cambridge.

Lehman, N., A. Eisenhawer, K. Hansen, D. L. Mech, R. O. Peterson, and R. K. Wayne. 1991. Introgression of coyote mitochondrial DNA into sympatric North American gray wolf populations. Evolution 45:104–19.

Lehman, N., and R. K. Wayne. 1991. Analysis of coyote mitochondrial DNA genotype frequencies: Estimation of the effective number of alleles. Genetics 128:405–16.

Logan, K. A., L. L. Irwin, and R. Skinner. 1986. Characteristics of a hunted mountain lion population in Wyoming. Journal of Wildlife Management 50:648–54.

MacArthur, R. H., and E. O. Wilson. 1967. The theory of island biogeography. Princeton University Press, Princeton.

MacDonald, D. W. 1981. Resource dispersion and the social organization of the red fox *(Vulpes vulpes)*. Pp. 918–49 *in* J. A. Chapman and D. Ursley, eds., Proceedings of the Worldwide Furbearer Conference, Frostburg, Maryland.

Mattson, D. J., S. Herrero, R. G. Wright, and C. M. Pease. 1996. Science and management of Rocky Mountain grizzly bears. Conservation Biology 10:1013–25.

Mattson, D. J., and M. M. Reid. 1991. Conservation of the Yellowstone grizzly bear. Conservation Biology 5:364–72.

Mayr, E. 1991. One long argument: Charles Darwin and the genesis of modern evolutionary thought. Harvard University Press, Cambridge.

Mech, L. D. 1987. Age, season, distance, direction, and social aspects of wolf dispersal from a Minnesota pack. Pp. 55–74 *in* B. D. Chepko-Sade and Z. T. Halpin, eds., Mammalian dispersal patterns: The effects of social structure on population genetics. University of Chicago Press, Chicago.

Melquist, W. E. 1981. Ecological aspects of a river otter *(Lutra canadensis)* population in west-central Idaho. Ph.D. diss., University of Idaho, Moscow.

Merriam, C. H. 1922. Distribution of grizzly bears in U.S. Outdoor Life 50:405–6.

Messick, J. P., and M. G. Hornocker. 1981. Ecology of the badger in southwestern Idaho. Wildlife Monographs no. 76.

Metzgar, L. H., and M. Bader. 1992. Large mammal predators in the northern Rockies: Grizzly bears and their habitat. Northwest Environmental Journal 8:231–33.

Mills, L. S., and F. W. Allendorf. 1997. The one-migrant-per-generation rule in conservation and management. Conservation Biology 10:1509–18.

Mills, L. S., and P. E. Smouse. 1994. Demographic consequences of inbreeding in remnant populations. American Naturalist 144: 412–31.

Mitton, J. B., and M. G. Raphael. 1990. Genetic variation in the marten *(Martes americana)*, Journal of Mammalogy 71:195–97.

Montana Department of Fish, Wildlife, and Parks. 1994. Final EIS, Management of black bears in Montana. Montana Department of Fish, Wildlife, and Parks, Helena.

——. 1995. Draft EIS, Management of mountain lions in Montana. Montana Department of Fish, Wildlife, and Parks, Helena.

Mowat, G. 1995. Oral presentation to Western Forest Carnivore Committee meetings. Cranbrook, British Columbia.

Murphy, K. A., G. S. Felzien, and M. G. Hornocker. 1991. Ecology of the mountain lion *(Felis concolor missoulensis)* in the northern Yellowstone ecosystem. Cumulative Progress Report no. 4. Wildlife Research Institute, Moscow, Idaho.

Noss, R. F., H. B. Quigley, M. G. Hornocker, T. Merrill, and P. C. Paquet. 1996. Conservation biology and carnivore conservation in the Rocky Mountains. Conservation Biology 10:949–63.

O'Brien, S. J., and J. F. Evermann. 1988. Interactive influence of infectious disease and genetic diversity in natural populations. Trends in Ecology and Evolution 3:254–59.

Paetkau, D., W. Calvert, I. Stirling, and C. Strobeck. 1995. Microsatellite analysis of population structure in polar bears. Molecular Ecology 4:347–54.

Paetkau, D., and C. Strobeck. 1994. Microsatellite analysis of genetic variation in black bear populations. Molecular Ecology 3:489–95.

Paetkau, D., L. Waits, P. Clarkson, L. Craighead, E. Vyse, R. Ward, and C. Strobeck. 1997. Dramatic variation in genetic diversity across the range of North American brown bears. Conservation Biology 12:418–26.

Peek, J. M., M. R. Pelton, H. D. Picton, J. W. Schoen, and P. Zager. 1987. Grizzly bear conservation and management: A review. Wildlife Society Bulletin 15:160–69.

Picton, H. 1986. A possible link between Yellowstone and Glacier grizzly bear populations. Proceedings of the International Conference on Bear Research and Management 6:7–10.

Primm, S. A. 1996. A pragmatic approach to grizzly bear conservation. Conservation Biology 10:1026–35.

Quammen, D. 1996. The song of the dodo: Island biogeography in an age of extinctions. Scribner's, New York.

Ralls, K., and J. Ballou. 1983. Extinction: Lessons from zoos. Pp. 164–84 *in* C. M. Schonewald-Cox, S. M. Chambers, B. MacBryde, and L. Thomas. eds., Genetics and conservation: A reference for managing wild animal and plant populations. Benjamin-Cummings, Menlo Park, California.

——. 1986. Captive breeding programs for populations with a small number of founders. Trends in Ecology and Evolution 1(6):19–22.

Ralls, K., K. Brugger, and J. Ballou. 1979. Inbreeding and juvenile mortality in small populations of ungulates. Science 206:1101–3.

Ralls, K., P. H. Harvey, and A. M. Lyles. 1986. Inbreeding in natural populations of birds and mammals. Pp. 35–56 *in* M. E. Soulé, ed., Conservation biology: The science of scarcity and diversity. Sinauer Associates, Sunderland, Massachusetts.

Reynolds, H. V., and J. L. Hechtel. 1984. Structure, status, reproductive biology, movement, distribution, and habitat utilization of a grizzly bear population. Federal Aid to Wildlife Restoration Project, Final Report. Alaska Department of Fish and Game, Juneau.

Rood, J. P. 1987. Dispersal and intergroup transfer in the dwarf mongoose. Pp. 85–103 *in* B. D. Chepko-Sade and Z. T. Halpin, eds., Mammalian dispersal patterns: The effects of social structure on population genetics. University of Chicago Press, Chicago.

Ross, P., and M. Jalkotzy. 1992. Characteristics of a hunted population of cougars in southwestern Alberta. Journal of Wildlife Management 56:417–26.

Roy, M. S., E. Geffen, D. Smith, E. A. Ostrander, and R. K. L. Wayne. 1994. Patterns of differentiation and hybridization in North American wolflike canids, revealed by analysis of microsatellite loci. Molecular Biology and Evolution 11:553–70.

Ruggiero, L. F., K. B. Aubry, S. W. Buskirk, L. J. Lyon, and W. J. Zielinski, eds. 1994. The scientific basis for conserving forest carnivores: American marten, fisher, lynx, and wolverine in the western United States. USDA Forest Service General Technical Report RM-254. Rocky Mountain Forest and Range Experiment Station, Fort Collins.

Salwasser, H., C. Schonewald-Cox, and R. Baker. 1987. The role of interagency cooperation in managing for viable populations. Pp. 159–73 *in* M. E. Soulé, ed., Conservation biology: The science of scarcity and diversity. Sinauer Associates, Sunderland, Massachusetts.

Schoenwald-Cox, C. M. 1983. Guidelines to management: A beginning attempt. Pp. 414–45 *in* C. M. Schonewald-Cox, S. M. Chambers, B. MacBryde, and L. Thomas, eds., Genetics and conservation: A reference for managing wild animal and plant populations. Benjamin-Cummings, Menlo Park, California.

Seidensticker, J. C., M. G. Hornocker, W. V. Wiles, and J. P. Messick. 1973. Mountain lion social organization in the Idaho Primitive Area. Wildlife Monographs no. 35.

Servheen, C. 1983. Grizzly bear food habits, movements, and habitat selection in the Mission Mountains, Montana. Journal of Wildlife Management 47:1026–35.

Shaffer, M. 1978. Determining minimum viable population sizes: A case study of the grizzly bear *(Ursus arctos)*. Ph.D. diss., Duke University, Durham, North Carolina.

———. 1987. Minimum viable populations: Coping with uncertainty. Pp. 69–86 *in* M. E. Soulé, ed., Viable populations for conservation. Cambridge University Press, Cambridge.

Soulé, M. E. 1980. Thresholds for survival: Maintaining fitness and evolutionary potential. Pp. 151–69 *in* M. E. Soulé and B. A. Wilcox, eds., Conservation biology: An evolutionary-ecological perspective. Sinauer Associates, Sunderland, Massachusetts.

Suzuki, D. T., A. J. F. Griffiths, J. H. Miller, and R. C. Lewontin, eds. 1989. An introduction to genetic analysis. 4th ed. W. H. Freeman and Company, New York.

Talbot, S. L., and G. F. Shields. 1996. Phylogeography of brown bears *(Ursus arctos)* of Alaska and paraphyly within the Ursidae. Molecular Pylogenetics and Evolution 5:477–94.

U.S. Fish and Wildlife Service. 1993. Grizzly bear recovery plan. U.S. Fish and Wildlife Service, Missoula, Montana.

———. 1994. The reintroduction of gray wolves to Yellowstone National Park and central Idaho. U.S. Fish and Wildlife Service. Boise, Idaho.

U.S. National Park Service. 1989. Yellowstone National Park bear information book 1989. Yellowstone National Park, Mammoth.

Von Schantz, T. 1981. Female cooperation, male competition, and dispersal in red fox *(Vulpes vulpes)*. Oikos 37:63–68.

Waits, L. P., S. L. Talbot, R. H. Ward, and G. F. Shields. 1997. Mitochondrial DNA phylogeography of the North American brown bear and implications for conservation. Conservation Biology 12:408–17.

Ward, R. M. P., and C. J. Krebs. 1985. Behavioural responses of lynx to declining snowshoe hare abundance. Canadian Journal of Zoology 63:2817–24.

Wayne, R. K., N. Lehman, M. W. Allard, and R. L. Honeycut. 1992. Mitochondrial DNA variability of the gray wolf: Genetic consequences of population decline and habitat fragmentation. Conservation Biology 6: 559–69.

Weaver, J. L., P. C. Paquet, and L. R. Ruggiero. 1996. Resilience and conservation of large carnivores in the Rocky Mountains. Conservation Biology 10:964–76.

Wilcox, B. A. 1986. Extinction models and conservation. Trends in Ecology and Evolution 1:47–48.

Williams, J. S. 1992. Ecology of mountain lions in the Sun River area of northern Montana. M.S. thesis, Montana State University, Bozeman.

Wolff, J. O. 1982. Refugia, dispersal, predation, and geographic variation in snowshoe hare cycles. Pp. 441–49 *in* K. Myers and C. D. MacInnes, eds., Proceedings of the World Lagomorph Conference. University of Guelph, Guelph.

Woods, J., and R. H. Munro. 1996. Roads, rails, and the environment: Wildlife at the intersection in Canada's western mountains. *In* G. Evink, D. Ziegler, P. Garrett, and J. Berry, eds., Highways and movement of wildlife: Improving habitat connections and wildlife passageways across highway corridors. Proceedings of the Florida Department of Transportation/Federal Highway Administration Transportation-Related Wildlife Mortality Seminar, Orlando.

Wright, S. 1931. Evolution in mammalian populations. Genetics 16:97–159.

———. 1969. Evolution and the genetics of populations, vol. 2: The theory of gene frequencies. University of Chicago Press, Chicago.

Young, B. F., and R. L. Ruff. 1982. Population dynamics and movements of black bears in east-central Alberta. Journal of Wildlife Management 46:845–60.

Zackheim, H. 1982. Ecology and population status of the river otter in southwestern Montana. M.S. thesis, University of Montana, Missoula.

Zimmerman, E. G. 1989. Mitochondrial DNA analysis of North American black bears, *Ursus americanus*. Unpublished report, Department of Biological Sciences, University of North Texas, Denton.

Wolverine (Susan Morse)

Chapter Twelve

Carnivore Research and Conservation: Learning from History and Theory

Steven C. Minta, Peter M. Kareiva, and A. Peyton Curlee

The study of carnivores has contributed substantially to our understanding of natural history, ecological processes, and human impacts on ecosystems. The fact that we evolved with carnivores as predators and competitors has certainly shaped how we perceive them, how we manage them, and therefore why we study them. Carnivores have always had a deep grip on the human psyche and, historically, they were feared as predators, domesticated as pets, and intensively managed as pests or harvested species. In recent decades we have seen carnivores emerge as a rallying point for conservation biologists because they are viewed as an indicator of "wildness." In the current trend of ecosystem thinking, the scientific utility of carnivores is how they exemplify ecological concepts, such as metapopulation and landscape, and how they might play ecological roles that serve to simplify and represent the complexity of ecosystems.

Carnivores are now considered important indicators and arbiters of ecosystem responses to change because they are at the top of the food chain and so wide-ranging in their use of habitat. While we might be able to study plant or insect habitats at the scale of meters or kilometers, no one would dream of studying carnivore habitats at anything less than hundreds of hectares. In addition, viable carnivore populations are considered indicative of the integrity and health of an ecosystem (or landscape of connected ecosystems) because they are sensitive to human alterations and very susceptible to extinction in the face of landscape change.

Concern about large-scale changes in habitat distribution and their implications for how communities work has spawned the fields of landscape ecology and ecosystem management. Large carnivores have played a substantial role in catalyzing ecosystem thinking among managers and scientists. Yellowstone's grizzly bear, one of the most intensively studied carnivores in the world, served as an early and controversial icon of this emerging view of nature (Chapters 2 and 3). The eradication of wolves and their subsequent reintroduction into Yellowstone is a testimony to the shift in ecological values within the century (Chapters 2, 5, and 8).

Drawing from the work of many ecologists, especially the Craigheads (Craighead 1979, Craighead et al. 1995) and Sinclair (Sinclair and Norton Griffiths 1979, Sinclair and Arcese 1995), we build the theme of carnivore ecology as an archetype for understanding and monitoring ecosystem change, and we offer the Greater Yellowstone Ecosystem (GYE) as a prototype for understanding carnivores in other ecosystems. The ecosystem perspective as applied to terrestrial systems is a recent development in ecological theory and management (recent reviews in Gunderson et al. 1995, Hansson 1997, Boyce and Haney 1997, Daily 1997, Pickett et al. 1997). The Yellowstone and Serengeti ecosystems are two of the most notable, and carnivore studies have played a major role in their rise to preeminence (Keiter and Boyce 1991, Craighead et al. 1995, Sinclair and Arcese 1995, Clark et al. 1996). We believe that appreciating the contribution of carnivores to ecological theory—and vice versa, how theory has informed our understanding of carnivore ecology—requires an appreciation of history in this more than any other taxonomic group. The nature of carnivores has constrained our applications of the scientific method, and the nature of our relation to carnivores has dictated the questions we ask.

In this chapter, we examine the challenges inherent in studying carnivores and ecosystems, the methodological and conceptual advances that have been made through the study of carnivores, and the considerable gaps that remain in our understanding. This chapter differs from the preceding chapters in that it draws on an extensive body of literature and offers an overview of conservation science and conservation. We recommend several approaches for future research, and we examine how we might advance carnivore science, management, and policy for conservation gains. Protecting carnivores depends as much on our organizational and policy performance as on our understanding of their ecological requirements. Their future rests on a value choice by the public that would put carnivore conservation on an equal footing with other social and economic priorities. Let us first look at the limitations on gaining more certain ecological information about carnivores and at the history of research within the limitations imposed by these animals, as well as the self-imposed limitations of our scientific methodologies.

Inherent Limitations: Carnivores as Experimental Subjects

If carnivores are important to ecological theory and conservation science, certainly it is not for their experimental utility. They are terrible experimental subjects because they tend to be elusive, nocturnal, fast, solitary, and wideranging. They sense their environment and communicate primarily by using olfactory systems beyond human perception. Because they are difficult and even dangerous to handle and monitor, their abundance, sometimes even their presence, is difficult to determine with any great confidence. These traits restrict the questions we ask and the hypotheses we test. Laboratory and captivity studies are limited, except behavioral aspects of predation (e.g., Curio 1976, Leyhausen 1979, Bateson and Turner 1988) and zoo research (Ballou et al. 1995, Hedrick 1992, Kleiman et al. 1996, Miller et al. 1996). Furthermore, because there is tremendous taxonomic, morphological, ecological, and behavioral heterogeneity among carnivores, our knowledge varies from species to species, system to system, and, more frustrating to researchers, even from population to population. Carnivore research is expensive and yields a small ratio of data to dollar compared with other taxonomic groups.

Given the logistical and financial constraints, how must carnivore ecologists adjust their scientific methodology? And what approaches should be developed to overcome experimental limitations? Even in the best cases, there are gaping holes in our natural-history information. There is always the desire to fill those holes with what we expect from theory or from studies of other carnivores. How safe or advisable is it to do so? To what extent can we rely on our observations, deductions, and inferences about carnivore ecology when for most carnivores there is virtually no hope of ever doing well-replicated, manipulative experiments? Thus the benchmark of modern ecology and positivistic science is often beyond our grasp.

For carnivore ecology, we cannot point to "model" systems, similar to those developed for many types of insects, lizards, fish, birds, and intertidal organisms, which allow easy experimental control and manipulation as well as theoretical generality (e.g., Paine 1974, 1992, Huey et al. 1983, Taylor 1984, Feder and Lauder 1986, Konishi et al. 1989, Powers 1989). Moreover, we cannot find genuine examples of "surrogate" systems, the way agriculture has served plant ecology, or examples of "ecological microcosms" (e.g., Lawton 1995, Drake et al. 1996, Fraser and Keddy 1997, Gonzalez et al. 1998). The closest examples to surrogate systems might be the use of weasels for studies of predator control (King

1989, Reynolds and Tapper 1996) and the use of captive or observable social carnivores for studies of kin selection and social structure (reviews in Wilson 1975, Lott 1991, Soloman and French 1997). In addition, harvest data have provided insights into population dynamics, such as lynx-hare cycles and demography of furbearers (see Chapman and Pursley 1981, Novak et al. 1987, Sinclair et al. 1993). For carnivores, islands may be viewed as microcosms for understanding patterns in biogeography, community composition, food webs, and sociodemographic structure (Williamson 1981, Stamps and Buechner 1985, Heaney and Patterson 1986, Adler and Levins 1994, Alcover and McMinn 1994, Dayan and Simberloff 1998, Grant 1998, Terborgh et al. 1999).

In another sense of the word, the most informative "model" systems are the intensively studied systems, such as the highly observable carnivores of the Serengeti, the grizzly bears of Yellowstone, and the wolf-ungulate system of Minnesota (Mech 1995, 1996), which have allowed comparisons and contributed to ecological theory in general. Of course, this has been greatly facilitated by leaders who have brought together researchers and synthesized the outcomes of many separate research projects (e.g., Sinclair and Arcese 1995). In many ways, these systems have become the prototypical references shared by the disciplines within ecosystem science—experimental and theoretical ecologists, land and resource managers, and conservation biologists.

Many critics have been skeptical of our ability to use ecology to guide practical management, believing ecology to be a science of case studies and rough generalizations (Caughley 1981, Peters 1991, Shrader-Frechette and McCoy 1993). While this criticism appears especially salient to carnivore ecology, carnivores nonetheless offer us great opportunity because research and management have long been integrated. Worldwide in this century, we cannot help but be influenced by the legacy of gamekeeping, hunting, trapping, predator control, fur farming, and animal husbandry (especially in Europe and Russia). Hypothesis testing with carnivores as subject species has never been lucrative for academic grant writers. Nonetheless, many ecologists would argue that planning and regulation of—as well as much research based on—hunting, fur harvest, predator control, and endangered species have constituted a series of experiments, some trial-and-error and some not unlike adaptive management. Carnivore researchers have produced data and insights that have factored into intelligent management, although much of it has taken the form of expert opinion rather than experimentally derived inferences, and there has been insufficient integration of knowledge above the species level. Given our historical desire to manage carnivores, they actually share more heritage of "ecological theory" with ungulates and waterfowl than other taxonomic groups of predators. Thus the manifest link of carnivore ecology to predator-prey theory may appear tenuous, especially for species whose population dynamics are driven by human manipulations through harvest or control.

The bias toward management-driven research has been further compounded by the difficulties encountered in studying carnivores, which limit our ability to design experiments that unequivocally offer theoretical advances. If the role of theory has been constrained in driving experimental design and methodology, then we might expect a corresponding decrease in the degree to which theory has motivated carnivore research direction and shaped questions. Assume for a moment that the strongest carnivore science has been in the realms of evolution: paleobiology, morphology, taxonomy, systematics, and genetics, with overlap into macroecology and historical biogeography. It follows that carnivore ecologists have traditionally been trained to think about ecological problems with this intellectual backdrop. Indeed, Lidicker (1994) and others have noted that data-poor, experimentally weak mammalogy (particularly carnivore ecology) has always tended to be more "holistic" and dominated by evolutionary and natural-history perspectives. Also, the study of carnivores requires an appreciation of community interactions, at least of human-predator-prey connections, whether the knowledge is gained inductively or deductively. That knowledge is often shaped by extensive experiences accumulated during the long field studies that characterize research on carnivores.

As a result of this intellectual tradition, carnivore ecologists may express a proclivity for a wider and longer view of the species they study and a different, perhaps broader, sense of their study system. This sense of "nature" may lack the tidy rigor of analysis-of-variance experiments, but it still offers insight into nature and an understanding that is neither more nor less penetrating. Certainly at the larger scales of many carnivores it may enhance understanding and intuition of the big picture: ecosystems. Perhaps what has been perceived to be a hindrance by mainstream ecologists has turned out to be a blessing in disguise as carnivore ecologists essentially have been forced to deal with a more complex reality of multiple scales and multiple disciplines from very early on, which is a precursor of adaptive management. Unfortunately, communication and cross-fertilization among related disciplines is hampered by overly divergent views of nature and noncongruent beliefs about the appropriate, reliable methods of study. The evidence may become clearer as we review the benefits and disadvantages of our historical approach to carnivore research.

Trends in Ecological Research on Carnivores

Before the early decades of this century, most of our knowledge of carnivores was based on cultural lore, gamekeepers, hunters and trappers, and the writings of naturalists and explorers. The scientific literature, which consisted of descriptive accounts of species, their habitats, and diets, began to contain more studies of predators as pests (reviews in Errington 1967, King 1984, 1989, Putman 1989, Reynolds and Tapper 1996, Taylor and Dunstone 1996). By midcentury, more scientific accounts of carnivores were accumulating (Murie 1940, 1944, Errington 1946, Young and Goldman 1944, 1946, Young and Jackson 1951, Young 1958).

Since Darwin's time, the most rigorous carnivore science was concentrated on paleontology, biogeography, and evolution, because the robust skulls of ancient carnivores were well preserved in the fossil record (Simpson 1940, 1944, 1953, Kurten 1968, 1972, Radinsky 1988). For example, by midcentury, this fossil record allowed integration of paleontological knowledge with paradigms derived from advances in population genetics, which set the stage for the "new synthesis" (Zakrzewski and Lillegraven 1994).

Theoretical forays into community organization and life-history strategies were initiated by Rosenzweig (1966, 1968) and Schoener (1969). Ewer (1973) was the first to summarize and integrate the carnivore literature to date. She did not discuss "the population dynamics of predator-prey relationships" but concentrated on "adaptive significance and evolutionary history ... to unite anatomy and physiology with ecology and behavior; ... for the business of the killer is highly competitive as well as a highly skilled occupation. This perfection, however, is the product of a long evolutionary history and we cannot understand the carnivores of today without some appreciation of the process which brought them into being" (pp. xiii, 1).

By the early 1970s, the theoretical foundations had been concentrated in the disciplines of macroecology, paleoecology, evolutionary ecology, and behavioral ecology. Ewer's powerful and guiding synthesis set the stage and tone of future research: Gittleman (1989:viii) prefaced *Carnivore behavior, ecology, and evolution* with "I and all the contributors feel a deep gratitude to R. F. Ewer for her monumental volume *The Carnivores* (1973), which laid the foundation for modern carnivore studies. It is testimony to the long-lasting effect of her work that most of the contributors, though using very different methodologies and theoretical predictions, refer to *The Carnivores* for framing their questions." Nevertheless, intensive "study system" research began to change our perception of carnivores, with a background of empirical work coming from studies of coyotes (Murie 1940), coatis (Kauffman 1962), wolves (Murie 1944, Mech 1966, 1970), mink (summarized in Errington 1967), mountain lions (Hornocker 1970), grizzly bears (see Craighead et al. 1995), and tigers (Schaller 1967). The Serengeti came to the forefront as the premier carnivore system with landmark studies of lions (Schaller 1972) and hyenas (Kruuk 1972).

The Serengeti's abundant and large species of carnivores and ungulates inspired an appreciation for interspecific interactions, culminating in the first synthetic volume in which carnivores played center stage in an ecosystem (Sinclair and Norton Griffiths 1979). Although population regulation and predator-prey dynamics were coming into prominence (e.g., coyote-rabbit, snowshoe hare–lynx, wolf-ungulate), behavioral ecological studies still dominated during the 1970s. The main topics were adaptation and phylogeny, communication, development, foraging and energetics, body size and sexual dimorphism, mating and reproduction, spacing, and group living (e.g., see Kleiman and Eisenberg 1973, Fox 1975, Curio 1976, Bekoff 1978, Bertram 1978, Kruuk 1978, Leyhausen 1979, Sinclair and Norton Griffiths 1979, Frame and Frame 1981, Ralls and Harvey 1985, Gittleman 1989, 1996, Sandell 1990, Lott 1991, Eisenberg and Wolff 1994, Creel and Macdonald 1995, Sinclair and Arcese 1995, Solomon and French 1997). Large, easily observable, social carnivores were still the most studied species, and advances continued in the fields of life history and mating strategies and social organization. Fortuitously, the terrain and climate of African savannas allowed continuous observation and high observer mobility. If there was an overarching theme, it was probably resource partitioning in strict carnivores: existing experimental methods were useful, such as those for diet and habitat studies, and many theoretical issues could be integrated. The plentiful studies of economically important carnivores (pests and furbearers) were far more illuminating to wildlife biologists, particularly for field methods, than to mainstream experimental ecologists.

Smaller and elusive species, particularly those in closed habitats, required applications of tracking (mostly over snow), trapping-eartagging, and collection of carcasses and scats for vital rates and diet. These studies tended to focus on prey availability and diet, or habitat (prey species in habitats). Technical advancements began in the 1960s with telemetry (see White and Garrott 1990). Spatial analysis has had an enormous impact on North American and European studies, particularly in temperate to boreal forests. The nature of telemetry data is that we gain an average picture of home ranges, activity areas, and habitat selection at the midrange scale. More recent advances, including scat marking (with isotopes) and biochemical analysis of scat and hair (Pelton and Marcum 1977, Kruuk et al. 1980, Major et al. 1980, Putman 1984, Kohn and Wayne 1997), have had far less impact. During the 1970s many pioneering carnivore studies used telemetry and influenced methods, questions, and approaches in subsequent research. The Craigheads were the earliest to apply many technological advancements to the study of grizzly bears (Craighead 1979, Craighead et al. 1995). Telemetry became so influential because, in addition to increasing sampling rates, it could be used to take advantage of what carnivores offered as experimental subjects. Important carnivore features that could simplify or strengthen experimental designs became attractive to ecologists and modelers. The complementarity of carnivore traits and telemetry as an experimental tool can be summarized as:

- High trophic position associated with uniformly high energy prey units that are mobile and, on the average, spatially dispersed.
- Severe intra- and interspecific competition.
- Coevolution with prey, characterized by highly developed morphology, sensory and learning abilities, and tightly coupled predator-prey interactions.
- Sophisticated tracking and accounting for renewal rates of resources.
- Solitary behavior with strong inhibitory mechanisms for mutual avoidance of conspecifics.
- Profound life-history variation, but often partitioned into discrete sex and age categories.
- Carnivory itself, which tends to limit the potential spacing and mating systems.

In addition, telemetry helped overcome some of the experimental weaknesses imposed by carnivore characteristics, such as:

- Elusive, nocturnal, secretive, fast, and solitary behavior.
- Wide-ranging movements.
- Low abundances and population densities.
- Limited clues left for researchers to find and track individuals (allowing population analyses and assessment of demographically and socially distinct tactics of resource exploitation).

And telemetry offered advantages like:

- More and better data per unit dollar.
- Less need for laboratory and captivity studies (except behavioral aspects of predation).

Along with methodological advances came a shift in research focus. Early carnivore research empires were founded on generous funding for predator control in the 1970s (see Connolly 1978a, Wagner 1988). For example, with his students, Fred Knowlton (e.g., Knowlton 1972, Linhart and Knowlton 1975) was a leader on coyotes, the premier livestock pest of western North America, while Al Sargeant led the research direction in the Midwest, centered on waterfowl depredation by foxes, coyotes, and skunks (e.g., Sargeant 1972, Eberhardt and Sargeant 1977, Sargeant et al. 1984). For more general conceptual reasons, Maurice Hornocker and his numerous students tackled the major carnivore species, producing seminal papers and monographs on their diets, habitat requirements, and breeding systems (e.g., Hornocker 1970, Messick and Hornocker 1981, Melquist and Hornocker 1983). Many of these modern pioneers extended our knowledge of natural history while making inroads in refining models of habitat selection, resource partitioning, and predation. Methods and techniques were allowing carnivores to emerge as tractable experimental subjects for hypothesis testing: Powell modeled the fisher-porcupine system (1979a, 1980; review in Powell 1993), Mech (1970) and others focused on wolf-ungulates, Keith explored the lynx–snowshoe hare cycles (review in Keith 1990), and Bekoff led the way for understanding coyote sociality (review in Bekoff and Wells 1986). Similar trends throughout the world are exemplified by numerous studies of African carnivores, as we shall see. At the same time, the culmination of extensive lab and semicaptive studies was summarized in volumes by Curio (1976) and Leyhausen (1979).

It was also during the 1970s that funding sources shifted from carnivores as harvested species, pests, and competitors (of humans) to carnivores as threatened and endangered species that were indicative of habitat loss, degradation, and fragmentation (Frankel and Soulé 1981). The grizzly bear, wolf, and other large carnivores were becoming icons or "umbrella species" of the new age of conservation biology. Without a doubt, the International Biological Program (IBP) of the 1960s provided an underlying current of "the big picture" approach to understanding ecological systems, even though it was perceived by many to be a failed program during an economic period of big budgets. Still, the shift toward conservation and large-scale, system issues was undeniable. If the IBP program gave an impetus to community and systems approaches (for herbivores, but very little involving carnivores: Golley 1993, Mares and Cameron 1994), then legislation, at least in the United States, gave teeth to regulating agencies along the lines of the new thinking on biodiversity, habitat protection, and water quality. Consequently, we were forced to shift our thinking from idiosyncratic (population to population) to generalized models, and from expert opinion to inference-based and legally defendable empirical and modeling approaches. These trends were in full bloom by the 1980s. The National Environmental Policy Act (NEPA), with its multilevel dictum on assessing direct, indirect, and cumulative effects of any human action, forced the federal agencies to retool their entire formulation of understanding ecological systems at various scales. Agency and academic research was forever changed.

Parallel to shifts in funding and research motivation, a new era for carnivore ecology began in the 1980s based on theoretical advancements. Eisenberg (1981) provided a comprehensive and innovative approach to understanding mammalian phylogeny, biogeography, behavior, and ecology on a worldwide scale. His sweeping systems approach to evolution, morphology, and behavior is the foundation for all subsequent work: "To bring the ecology of the various species into perspective, I will attempt to define trends in habitat utilization, mating systems, and life history patterns" (p. xvi). The flurry of research during the previous decade was extremely helpful: species accounts and natural-history descriptions were up to date, including summary volumes (e.g., Burton 1979, Chapman and Pursley 1981, Chapman and Feldhamer 1982). More importantly, from here it would be the large data sets and long-term studies that would be most helpful in understanding carnivores in communities and ecological systems (Morris et al. 1989, various publications from the Fifth Theriological Congress, Australian Mammalogical special issue). In biogeography, the equilibrium model stimulated much research activity, including a substantial body focused on carnivores and other mammals because they are well suited for tests of biogeographic theory (Lomolino 1984, 1986, Brown 1986, Pagel et al. 1991). Conservation biogeography was continuing to gain momentum as it was applied to issues of reserve design, species endemism and rarity, refugia, faunal interchange, and insularization (reviews in Glenn 1990, Brown 1995, Lomolino and Channell 1995, Newmark 1995, 1996, Soulé and Terborgh 1999). However, very little literature examined entire communities or ecosystems, with the exception of the Serengeti (Sinclair and Norton Griffiths 1979).

Toward the end of the 1980s review articles and synthetic volumes included more sections and chapters on community and ecosystem approaches. Theory in carnivore ecology increasingly emphasized predation (adaptations and behavior, but more on population effects), spatial variation (as response to conspecifics and resources), and community ecology (Macdonald 1983, Harris 1987, Marti 1987, Marti et al. 1993, Jaksic and Simonetti 1987, Novak et al. 1987, Morris et al. 1989, Gittleman 1989, Sinclair 1989, Skogland 1991, L. Oksanen et al. 1992, T. Oksanen et al. 1992a,b, Macdonald 1993, Dunstone and Gorman 1993, Lidicker 1995, Pech et al. 1995, Smallwood and Schonewald 1996). Species interactions included predator-prey, with increasing emphasis on landscape ecology and interspecific interactions among carnivores, as well as a greater emphasis on the complex involvement of human disturbance and management (Lindström 1989, McCullough and Barrett 1992, Buskirk et al. 1994, Ruggiero et al. 1994, Andren 1995, Boman 1995, Lidicker 1995, Creel and Creel 1996, Jaksic et al. 1996, McCullough 1996, Oehler and Litvaitis 1996, Haight et al. 1998, Chapters 4 and 8 this volume). Carnivores began to play a major role in our understanding of population dynamics and predator-prey theory (Erlinge et al. 1983, 1984, 1988, Fowler 1987, Newsome et al. 1989, Boutin 1990, 1992, McCullough and Barrett 1992, Jaksic et al. 1993, Johnsingh 1993, Messier 1995, Norrdahl and Korpimaki 1995a,b, Karanth and Sunquist 1995, Krebs 1996, Wolff 1997, Korpimaki and Norrdahl 1998, Lindström et al. 1998). Rodent studies continued to reveal insights into predation (Boutin 1990, 1995, Hanski et al. 1991, 1993, Hanski and Henttonen 1996, Korpimaki et al. 1994, Stenseth and Lidicker 1992, Hansson 1989, 1995a, Boutin et al. 1995, Boonstra et al. 1998, Inchausti and Ginzburg 1998). The most famous time series in ecology, the ten-year wildlife cycle in North America, became clearer by partitioning the effects of predators and vegetation on the snowshoe hare (Sinclair et al. 1993, Boutin et al. 1995, Krebs et al. 1995, Stenseth et al. 1997, Boonstra et al. 1998, O'Donoghue et al. 1998a,b). Because mammalian cycles dominate the world's boreal systems (Danell et al. 1998), Pastor et al. (1998b) summarize the emerging view that we should focus not on the factors that control the mean but on the factors that control oscillations (see also Holling 1987, Gunderson et al. 1995, Holling and Meffe 1996, Ludwig et al. 1997). The big carnivores, particularly wolves, cats, and bears, symbolized conservation and ecosystem studies (Hummel and Pettigrew 1992, Macdonald 1993, Burrows et al. 1994, Carbyn et al. 1995, Caro and Durant 1995, Estes 1995, Paquet and Hackman 1995, Pelton and Van Manen 1996, Fuller and Kittredge 1996, Jedrzejewska et al. 1996, Bothma 1998, Soulé 1999).

Intensive studies of single carnivore species tend to emphasize their role in ecological systems, and many books and monographs appeared in the 1980s and 1990s (e.g., Hanby 1982, Bibikov 1985, Brown 1985, Kruuk 1989, 1995, Rabinowitz 1986, Sunquist and Sunquist 1988, Alvarez 1993, Bailey 1993, Dunstone 1993, Schaller 1993, Beecham and Rohlman 1994, Caro 1994, Fox 1984, Henry 1986, Macdonald 1987, Mills 1990, Craighead et al. 1995, Powell et al. 1997, Turner and Anton 1997, Maehr 1997). Such studies are invaluable because they offer a unique form of ecosystem perspective—that is, a tremendous amount of knowledge about a single carnivore species and its environmental interactions. The monographic texts are frequently written in a longer expository style that is understandable to ecologists from many disciplines, including theoretical ecologists who might use information and data for model building. Another source for the latest data and synthesis can be found in the edited books on carnivores that have appeared in the last decade (e.g., Gittleman 1989, 1996, Dunstone and Gorman 1993, Buskirk et al. 1994).

Eisenberg (1989:7) introduced *Carnivore behavior, ecology, and evolution* by stating:

There has been much discussion of attempts to preserve ecosystems intact. If this course is to be followed, full recognition must be made that top carnivores play an important role in structuring communities, and, ultimately, of ecosystems. The removal of a top carnivore from an ecosystem can have an impact on the relative abundance of herbivore species within a guild. In the absence of predation, usually one or two species come to dominate the community. The consequence of this is often a direct alteration of the herbaceous vegetation fed on by the herbivore guild or assemblage. Thus, the preservation of carnivores becomes an important consideration in the discipline of conservation biology.

However, none of the contributions in the book addressed these assertions, except in tangential ways, although the far-reaching work of Soulé and others was becoming quickly absorbed by carnivore ecologists worldwide (e.g., Soulé 1986, 1987, Soulé and Kohm 1989).

In the next major summary of carnivore ecology, Dunstone and Gorman (1993:v) began their preface to *Mammals as predators* by noting, "Predators have an important role to play in community ecology and their removal has frequently had dire consequences for habitat management. Thus, it is fundamental that we understand the process of predation and the role that predators play in maintaining ecosystem diversity." This was made evident in the stress on large felids, but the emphasis remained on viewing single carnivore species, with some limited set of human interactions, as the dominant force rearing from a backdrop called the ecosystem.

If edited compendiums serve to summarize state-of-the-art science and motivation for research, then the most recent additions are revealing (Sinclair and Arcese 1995, Gittleman 1996). Despite aspirations of an ecosystem approach, which can be very costly and go against the tradition of funding limits and cycles, carnivore biologists remain largely constrained to single-species studies, frequently with a conservation theme (habitat loss, poaching, or small-population dynamics). However, conservation biology is a very active field, and currently one of the hot topics is the seeming dichotomy of single-species vs. ecosystem approaches. Indeed, the Sinclair and Arcese (1995) volume attempts to synthesize many studies to give a picture of the Serengeti as a system with carnivores providing the major theme. In Gittleman's second volume on carnivores (1996), the introduction by Eisenberg in the first volume (1989) is replaced by an extremely provocative plea by Schaller to address the rising conservation problem of carnivores worldwide. One book reviewer (Woodroffe 1997) interprets Schaller's "marvelous no-holds-barred review" as the hallmark of the volume in which conservation is the underlying theme.

In addition to the shift away from harvest and pest-management emphases, the ever-declining budgets of the 1980s and 1990s have resulted in carnivore studies that are motivated far more by conservation and integrated management issues (for GYE: Brussard 1991, Keiter and Boyce 1991, Clark et al. 1996, Keiter and Locke 1996). Ironically, we have set our sights on the bigger target of ecosystems and we have many advanced tools in our arsenal, but we cannot afford much target practice. Moreover, many ecologists say that it is one thing to be guided by the increasing vision provided by community, landscape, and ecosystem theory, but that we should not confuse increased conceptualization with increased generation of precise, testable hypotheses that characterize a solid research program. Many carnivore ecologists would say that top predators are near the bull's-eye for gaining insights into ecological systems, making carnivore studies worthy investments despite expense and experimental constraints. But has carnivore ecology reached a state in which that argument convinces other ecologists, and would other ecologists agree that carnivore studies are adequately generalizable? Of course, disentangling issues of "importance" is impossible and always involves compromising perceived value with perceived utility. From a purely utilitarian-ecologist point of view, if we take away issues of conservation and public awareness and interest, we are left with superior subject species such as insects, fish, and birds—all of which then should attract the funding for their importance to advancing ecological theory. We must look at the record of contributions to ecological theory by carnivore ecology and, if ecosystem science is to be a useful conceptual and experimental construct, the potential contributions of carnivore research to ecosystem science.

How Carnivores Best Inform Ecological Concepts

Slobodkin (1972, 1992) critically examined ecology and evolution, observing that population dynamics, predator-prey theory, food webs, and ecological energetics are now almost independent theoretical regions, each distinct from the others and all distinct from most theoretical formulations of population genetics. He thought the main factors were different biological interests, different intellectual history and goals, and time-scale disparities. Ecosystem science has the potential to overcome this fragmenting trend by bringing distinct theoretical regions together so that we can understand the interaction of three components: abiotic (geochemical, weather), the evolutionary-organizational levels of inquiry (individual, population, community), and the species- or guild-centered environmental matrix of resources, habitats, landscapes, and other elements that we contend to be distinguishable and useful for understanding species as a fundamental unit within an ecosystem.

Concern about dwindling populations and government budgets has encouraged us to recast our efforts on the scale of ecosystems and landscapes rather than single species at single sites. Carnivores have especially been and will continue to be a catalyst for large-area and long-term ecological science. It was almost as if the study of carnivore ecology had been cast to fit the new mold of ecosystem thinking by virtue of a historically tight coupling of science and management, the ecological scales and processes integrated by carnivores, and the high profile of carnivore species readily associated with ecosystems. All in all, the experimental weaknesses of carnivore research might be overshadowed by these advantages, which are more likely to be appreciated by managers, policy makers, politicians, and the public.

Just as critics have found carnivore research lacking in empirical rigor, ecosystem management is similarly criticized, especially for ongoing definitional ambiguity and a lack of well-designed experimental tests (Boyce and Haney 1997, Rapport et al. 1998, Schindler 1998, Sit and Taylor 1998). However, one can argue that the goal of ecosystem management is uniformly accepted by ecologists and conservation biologists as sustainability of constituents and processes, and we all accept that this means attending to such things as connectedness, scale, context, dynamism in space and time (disturbance, pattern, and probabilism), uncertainty, adaptability, accountability, complexity, and functional redundancy as a safety valve for evolutionary change. Furthermore, new management challenges such as this require a mix of old and new methodologies and yardsticks. Positivisitic science is not well suited to complex management issues that are a blend of biological and human social factors. Sometimes it appears that criticism is a form of the traditional scorn for applied vs. "pure" science and not a question of how good the science is. This seems unfair when we consider that human needs and wishes have driven technological and theoretical advances in physics, aerodynamics, medicine, and other fields, and if it is acceptable in those disciplines, why not in ecology?

Criticisms aside, the strongest contributions of carnivore ecology involve the theoretical regions of predation and foraging theory (spatial variation), population dynamics, habitat selection (at several scales), and refugia (metapopulations and landscape ecology). The chapters in this book support this contention. At the experimental level, predator-prey theory is the region in which we would have expected carnivores to have played the greatest role. We therefore focus on this topic as the most telling and best-case example of the relation between carnivore ecology and ecological theory.

PREDATOR-PREY THEORY: A CRITICAL EXAMPLE
AND CRITICAL ASSESSMENT

Carnivore ecology suffers as a result of its theoretical naïveté. Carnivores are predators, and there is no hope of understanding carnivore ecology or management without a deep understanding of predator-prey theory. Unfortunately, even those studies of carnivore ecology that make use of some form of predator-prey modeling have generally lacked an appreciation of the richness of possibilities for predator-prey dynamics and hence have been myopic in their view of the carnivore world. We review predator-prey theory with an emphasis on its implications for carnivore ecology in the context of both conservation and wildlife management programs (see also Chapters 8 and 9). But we do not want the reader to dismiss this as ungrounded theory, so we begin with some key practical questions that surround the management of carnivores throughout the world. After this grounding in practical problems we quickly distill the most pertinent features of predator-prey theory. Our review of predator-prey theory sets the stage for a series of recommendations regarding research priorities that we believe must be addressed but that have often been neglected because so much carnivore ecology suffers a theory deficit. Many of our ideas overlap those of Boyce and Anderson in Chapter 10, so we direct the reader there.

With such a long history, we would expect that our intensive efforts at predator control would have yielded inferences about the relation between carnivores and their prey populations. The irony of past and ongoing carnivore control is that unambiguous documentation of elevated prey populations following carnivore suppression is hard to come by, particularly for large carnivores (see perspective in Boutin 1995, Erlinge et al. 1988, Wagner 1988, Newsome et al. 1989, Sinclair et al. 1990, Skogland 1991, Pech et al. 1992, Hatter and Janz 1994, Borralho 1995, Lawler and Morin 1995, Norrdahl and Korpimaki 1995a, Reynolds and Tapper 1996, Sinclair and Pech 1996, Klemola et al. 1997, National Research Council 1997, Ballard and Van Ballenberghe 1998, Greenwood et al. 1998, Korpimaki and Norrdahl 1998, Sinclair et al. 1998). Will we get more moose, more elk, more caribou, or more deer if we eliminate large carnivores, and how many more of these prey will we get as a result of predator control? This question is the central controversy that plagues many wildlife management programs throughout the world. Thus while it is a simple fact that predators eat prey, the effect of predators on prey population dynamics is an open question. This is not an easy question to answer because few populations are controlled simply by single factors, and because clear-cut experiments at large landscape scales are difficult to execute (see also Chapters 5 and 8).

Review of predator-prey theory in the context of carnivore conservation and wildlife management. Predator-prey theory is one of the oldest and richest branches of theoretical ecology, with a bewildering array of results. Instead of reviewing all this theory, the portions most pertinent to questions about predator control are extracted in the following section. First the theoretical result is highlighted, followed by its implications for carnivore management.

1. Predator-prey interactions can either oscillate wildly or persist as very stable interactions, with deviations from equilibrium densities rapidly di-

minishing through time (May 1976, 1981). The key components that stabilize predator-prey interactions are resource-limited population growth of the prey and the existence of a refuge from predation. But predator-prey systems need not be stable, and the "natural" circumstance might well entail extreme oscillations in prey numbers. If this is the case, then wildlife management programs aimed at producing constant prey numbers year after year for constant hunter satisfaction may be fighting "nature" and may consequently require an endless input of energy and money.

2. The removal of predators from a plant-herbivore-predator interaction can either stabilize or destabilize the herbivore population dynamics. Whether predator removal destabilizes herbivore populations depends critically on the time scale at which the carrying capacity for the herbivores changes relative to the time scale at which predator populations change (Crawley 1983). If predators operate on longer time scales than vegetation, then their presence in a food chain can be stabilizing. Biologically, this means that long-lived (relative to the speed at which one sees changes in the plant resources consumed by herbivores) and starvation-tolerant or generalist predators can be stabilizing. Crawley (1983) and Jordan et al. (1971) suggest that wolves on Isle Royale stabilize moose population dynamics, thus stabilizing their effect on vegetation. Apparently, in the absence of wolves the moose on Isle Royale are much more prone to large fluctuations in numbers than in the presence of wolves. What is less clear from the Isle Royale example is the implications of wolves for the long-term average number of moose.

3. There is a possibility of two alternative stable states in predator-prey systems, with a lower equilibrium corresponding to very low prey and predator populations and a higher equilibrium corresponding to high predator and prey populations (where prey are close to their carrying capacity). This idea received much attention in the 1970s from mathematical biologists, many of whom were inspired by so-called "catastrophe theory" (Hilborn 1975, Ludwig et al. 1978). Each equilibrium will "attract" neighboring population trajectories, so that a shift from one equilibrium to the other can be abrupt and can occur simply because densities of predators or prey are temporarily altered, without really imposing any sustained management. The "predator pit" alluded to by wildlife biologists is a verbal version of the lower stable equilibrium in these models. The key requirement for two different equilibria is the presence of a Type III functional response on the part of the predators—so that the per capita risk to predators actually increases with increasing prey density in the neighborhood of the lower equilibrium. Obviously, the presence of a predator pit would provide a strong argument

for predator control because by eliminating or reducing predators for a brief period, the "system" could be kicked into the higher equilibrium with more predators and more prey. In other words, if there are two equilibria in carnivore-prey systems, then predator control might result in "having our cake and eating it too"—that is, in yielding more prey and more carnivores. Obviously, if this theoretical idea were experimentally supported, it would have major policy implications.

4. If a regression analysis is used to ask what controls prey populations in a predator-prey system, the factor that explains the greatest proportion of the variance in prey population growth rates depends largely on where "noise" enters the system and not on what actually controls the dynamics (Chapter 10). By simulating predator-prey dynamics with environmental variation entered in different places (for example, in herbivore carrying capacity, or in the feeding rate of wolves as might be expected if weather influenced prey vulnerability), it has been shown that regression analyses of wolf and ungulate population change can be very misleading. The pattern of random environmental variation dictates the outcome of the regressions much more than does the actual linkage between predators and prey. This is a disturbing result if one is going to rely on regressions and correlations to test hypotheses about what "controls" caribou or moose populations.

Summary of key theoretical implications. Models of predator-prey interactions make it clear that the possibilities are so varied that management requires the identification of which dynamic scenario or model applies. To identify appropriate, perhaps site-specific, scenarios, we should build explicit dynamic models and ask how well those models are supported by the data. This will get us farther than drawing correlations between hypothetical driving variables, which are weak and unreliable for explanatory purposes. However, Boyce and Anderson make the excellent argument in Chapter 10 that even if we know the correct carnivore-herbivore-plant model, it will generally be impossible to infer what controls the population changes in nature simply by collecting census data. From our brief review of the theory of predator-prey systems, it should be obvious that there is a wide variety of plausible models for carnivores, and it is unlikely that we will ever adequately know the correct underlying models. This makes the situation much worse than even Boyce and Anderson (Chapter 10) imply: observing predator-prey populations is unlikely to yield understanding. What we need are experiments. In the next section we discuss the sorts of experiments that would give the greatest information.

Adaptive management and an experimental approach to carnivore ecology. It is silly to imagine doing well-replicated manipulative experiments with carnivores in the traditional analysis-of-variance paradigm. However, in particular game management units one can attempt to manage populations so that data are collected on carnivore and prey demography under the following suite of conditions: high carnivore density and low prey density, high carnivore density and medium prey density, high carnivore density and low prey density, low carnivore density and low prey density, and so forth. In others, wildlife managers should seek to produce combinations of carnivore and prey densities that correspond to the following two-way analysis-of-variance design: carnivore density (low, medium, and high levels) crossed with prey density (low, medium, and high levels). By tracking short-term population dynamics for this combination of initial conditions, it will be possible to build a reasonable model for carnivore-prey dynamics. The key here is that because the "experiment" cannot be replicated, the data may be more informative in conjunction with a model such as that described by Boyce and Anderson in Chapter 10.

A recent analysis of predator management (wolf control) in Alaska indicates how ineffectively carnivore researchers have taken advantage of experimental design (National Research Council 1997). Although more than eleven so-called large-scale experiments have been performed with wolves, none was sufficiently well designed to yield clear-cut inferences. The fact that studies are conducted at large scales is no excuse for inferior experimental design. Even though such large-scale experiments cannot be replicated, they can take advantage of meta-analysis and before-after-control-impact (BACI) analysis. We do not know of a single published carnivore study that has used BACI analysis. This is an embarrassment.

A second challenge, probably the biggest, is that carnivores and their prey do not interact in a simple pairwise fashion. The above experiment reduces a complex system to a pairwise interaction. In fact, rarely is there only one carnivore or one prey in an ecosystem. In the GYE, wolves and mountain lions (or foxes and coyotes) interact with many of the same species of prey. What are we to do about studying carnivore communities? We know so little about the functioning of these communities that we need to begin modestly. The first step would be for ecologists to be less specialized (for example, bear ecologists, wolf ecologists, elk ecologists) and to pay attention to other species. More definitively, we suggest that other species be added as potential explanatory variables to models after first doing the best possible with two-species equations. In addition, theoreticians need to simulate multispecies systems and ask whether, if first pairwise interactions are extracted and then other species are added, such an approach produces appropriate insights. It is amazing how little theoretical work has been done with plausible multispecies systems.

CARNIVORE STUDIES AND HUMAN IMPACTS: A LIABILITY AND A BOON

Conservation biologists typically view human threats strictly in terms of extinction or population declines. But there are subtle threats that warrant our attention. In particular, humans may so disrupt the social system of a species that the species is fundamentally transformed. Biologists are beginning to appreciate the demographic and social differences between exploited and unexploited carnivore populations. For example, Hornocker (1970), Kleiman (1977) and Kleiman and Brady (1978) stressed early on how easily human impact can modify carnivore ecology and social organization. Human modification in the form of fur harvest, predator control, and hunting acts as a nonrandom, noncompensatory form of mortality, alters other demographic processes, and may disrupt social organization.

Ironically, the main problem confounding our understanding of the relation between carnivore population and social organization and community characteristics is that demographic analyses are available primarily for species that are regularly harvested for fur, sport, or pest removal, thus providing the motive and the opportunity to collect large samples of normally scarce carnivores (see also Chapters 4–7). Furthermore, sampling carnivore populations is riddled with biases, whether using harvest or scientific methods. It is well known that age and sex classes exhibit differential susceptibility to trapping, hunting, and control techniques and that, if trapping biases exist, males are more often overrepresented in samples (e.g., Buskirk and Lindstedt 1989). Age biases are more variable; generally, juveniles are overrepresented. Nonetheless, because furbearer species have been trapped or hunted using similar techniques in most studies, limited comparisons among populations can be made in order to understand trends. Optimal reproductive effort should vary with changing survivorship schedules (Goodman 1979). Because exploitation by humans often has differential consequences to various age classes, we may expect human predation to influence reproductive patterns (Law 1979, Boyce 1981, Chapters 4 and 6 this volume).

In an early landmark review of furbearer population dynamics, Storm and Tzilkowski (1982) found that exploited coyote populations show an increase in litter size, an increase in the percentage of pregnant females, and a decrease in emigration rates. Harvested or controlled coyote populations may increase recruitment rates through compensatory reproduction by 30–100 percent (Voigt and Berg 1987) or more (Connolly 1978b). In red foxes Storm and Tzilkowski (1982) found that litter size was maximal in highly exploited populations. Many carnivore biologists have concluded that an inverse relation exists between population density and average litter size. Intensive hunting pressure on red foxes in Japan was responsible for: (1) a decrease in the survival rate of adult foxes, (2) shortened mean longevity, (3) an increased proportion of young foxes in the age distribution, and (4) a decrease in the male-to-female sex ratio (Yoneda and Maekawa 1982). Allen (1984) found indications in a North Dakota spring breeding population that the more adult males there were per female, the lower the overall ovulation rate and embryonic litter size and the higher the prenatal mortality. As has been shown in other studies, Allen demonstrated increased reproductive performance as a function of female age. In a British study, fox control operations increased productivity by reducing the proportion of nonbreeding vixens and the proportion of dog foxes in the cub and adult populations (Harris and Smith 1987).

Bunnell and Tait (1985) compared age-sex structures of thirty-nine bear populations (three species). Hunting is the major mortality factor in most bear populations, accounting for at least 50 percent. Males are more vulnerable to hunting than females (see also Powell et al. 1996, 1997, Beringer et al. 1998, Chapter 3 this volume). Estimation of mortality rates of subadults, in the order of 15–35 percent per year, appears higher than among unhunted adults. Changes in juvenile survival of black bears are attributed to the numbers of adult males rather than the total population (Kemp 1976). Other researchers have found that continual removal of individuals may produce a bottom-heavy age structure in exploited vs. unexploited populations (Zezulak and Schwab 1979, Debrot 1984, Lembeck 1986, Minta 1993, Zezulak 1998, Chapter 6 this volume).

Hornocker (e.g., Hornocker and Hash 1981, Hornocker et al. 1983, Hornocker and Bailey 1986, Chapter 4 this volume) has long believed that human exploitation of carnivores may be severe enough to keep the population in a "state of flux" by removing individuals before they could establish tenure. Mortality may not be excessive enough to reduce population size (because productivity is increased), but it may contribute to behavioral instability in the population. Territorialism may not operate simply because individuals do not have sufficient time; individuals are killed and replaced before they can establish themselves. Evidence further suggests that continual and long-term exploitation by humans alters behavioral characteristics through learning, natural selection, or both (M. Hornocker personal communication). Unexploited populations of mountain lions in Idaho showed a highly refined system of territorialism (Hornocker 1969, Seidensticker et al. 1973); individuals in populations that were exploited year after year were not territorial (Hornocker and Bailey 1986).

Coyotes offer the clearest example of the effects of exploitation on carnivore population and social organization. The coyote is exploited throughout its entire range, and all studies have reported solitary or pair home ranges. The only documented occurrences of pack-living coyotes have been in protected populations (Camenzind 1978, Bowen 1981, 1982, Andelt 1985, Bekoff and Wells 1986, Chapter 6 this volume). Regardless of prey size, pack living in coyotes probably requires long-lived individuals and low disruption or persecution by humans (Bowen 1981, 1982, Andelt 1985, Chapter 6 this volume).

The mixture of land uses and management regimes in the Greater Yellowstone Ecosystem offers numerous conditions for comparing exploited and unexploited carnivore populations across differently managed landscapes. The previous chapters provide some excellent examples, although the confounding variables of interspecific interactions are difficult to disentangle, particularly the disruptive influence of human behavior and landscape modification. We will return to this theme when we offer advice on revising research foci for improved models.

Where Do We Go Next with Carnivore Research?

If carnivore ecology is to be supported, the public and policy makers must see that conservation and management are improved as a result of carnivore studies. We must be pragmatic: ecological research on carnivores will be funded for its utility to management and conservation. We must frame the questions that will maximize the utility of carnivores to experimental ecology, ecosystem science, and management, and that will communicate the importance of research to other disciplines and, ultimately, to policy makers and the public.

Yellowstone carnivores offer a special opportunity because they reside in an ecosystem that is well defined and has been the subject of numerous ecosystem-level analyses (Greater Yellowstone Coordinating Committee 1987, 1990, 1991, Varley and Brewster 1990, Brussard 1991, Keiter and Boyce 1991, Clark et al. 1996). Thus the ecology of these carnivores need not be simply stories about this or that species but can be substantially related (in theory at least) to their role in the ecosystem. Exactly how to do that is, of course, the research challenge. Yellowstone is also special because it has not been seriously disturbed by humans (in comparison with other North American habitats) and thus offers some hope of yielding insight about carnivores with and without excessive intrusion. We also have good, relatively long-term data on many carnivores in the Greater Yellowstone Ecosystem.

WHAT SPATIAL AND TEMPORAL SCALES ARE USEFUL FOR MANAGERS?

Levin (1992) asserts that conservation biology manages to span the middle ground between population biology and ecosystem science and that the issue of scale is the fundamental conceptual problem in ecology. Ecologists are now aware that dynamics are rarely confined within a focal area and that factors outside a system may substantially affect (and even dominate) local patterns and dynamics. Local populations are linked closely with other populations through such spatially mediated interactions as source-sink and metapopulation dynamics, supply-side ecology, and source pool–dispersal effects. The identification of landscape ecology as a specific discipline is recognition of multihabitat dynamics, and Polis et al. (1997) argue for the integration of landscape perspectives with consumer-resource and food-web interactions. The theoretical development of landscapes within ecosystems, with the emphasis on multiple scales, may be a strong catalyst for improving experimental conservation science (Hansson and Angelstam 1991, Englund 1997, Keitt et al. 1997, Pickett et al. 1992, 1997, Tilman and Kareiva 1997, Hochberg and Van Baalen 1998, Ritchie 1998). Even though a landscape is arbitrary in size, in practice it will generally be chosen on the basis of human management goals (kilometers-wide scale), while ecosystems tend to be at the hundreds of kilometers–wide scale.

However, a great deal of ecological research focuses on single species or interactions between two species, usually on a time scale of a few years or less and on spatial scales that are often smaller than the characteristic distance over which an individual member of the species moves in its lifetime (Kareiva 1989, May 1994). The pressing concerns of conservation and management are on longer time scales and vastly greater spatial scales, particularly for the larger, showcase carnivores. In this regard, we must evaluate whether specific carnivore research is matched appropriately to the scale of carnivore management.

Because studies of carnivores are often instigated by agencies responsible for managing particular parcels of land or resources, the management impetus tends to drive the scale and scope of the biological research. For example, in lieu of studying grizzly bears, we may study grizzly bears in Yellowstone National Park. The danger of this approach lies in the fact that the dynamics of carnivores within particular management units can be strongly influenced by processes going on outside the unit (see Chapters 3 and 5). Thus it may be impossible to understand why grizzly bear populations in Yellowstone are suddenly declining without appreciating that there is a "sink" for these bears outside the park—a place to which grizzlies tend to move but not return from.

Although ecosystem science often must address processes at large scales and makes frequent reference to "landscape ecology" and polemics about scale, rarely is anything done beyond talking about the issues (or maybe exploring "scale" through simulation models). Similarly, conservation biology is clearly a large-scale enterprise with criticisms of studies done at the wrong scale (see Murphy 1989, Minta and Kareiva 1994, Schindler 1998). Carnivores may force us to do something real, rather than simply talking about these concerns. Carnivore population processes occur at scales that are moderately tractable and that often encompass large-scale ecosystem processes. We elaborate on the issue of scale in subsequent sections.

COMMUNITY INTERACTIONS: KEYSTONE AND INDICATOR SPECIES

There is an overwhelming body of evidence in the ecological literature demonstrating that predators can have a tremendous influence on community structure and dynamics. The food-web or interaction-web approach has produced an apparent dichotomy for community interactions: the bottom-up view concentrates on how resources influence higher trophic forms; the top-down view examines how the interactions between high-level consumers and their prey influence lower trophic forms (trophic cascade). However, both top-down and bottom-up regulation can operate concurrently in the same system (Krebs et al. 1995, Englund 1997, Terborgh et al. 1999, Chapter 8 this volume). Recent conceptual work has tried to synthesize the two views into one that examines how productivity and predation jointly affect trophic structure (Leibold et al. 1997, Polis et al. 1997). The prevalence of cascades subsequent to single-species perturbations suggests that strong interactions are relatively common in well-studied systems (Paine 1992, 1993). We expect that carnivores at the "top" of the food web will play a powerful role in our understanding the complexity of community interaction strengths because carnivores "track" systems without driving them. Furthermore, we are convinced that there is a predominance of cases demonstrating top-down regulation and strong interactions with herbivores and mesopredators (Terborgh et al. 1999). Therefore, as organisms whose influence on ecosystem function and diversity can be disproportionate to their numerical abundance, carnivores are likely keystone species in diverse terrestrial systems.

Estimating community interaction strengths has proven to be difficult (Laska and Wootton 1998), but direct evidence would help us better use carnivores as keystone, umbrella, and indicator species (Mills et al. 1993, Paine 1992, 1993, McKenney et al. 1994, Estes 1996, Terborgh et al. 1999). Nevertheless, managers have operated under weak but hopeful evidence for carnivores; for example, much has been made of the importance of carnivores as a focal point or umbrella for management of other species that share the same habitats or regions (e.g., Terborgh 1988, Soulé and Noss 1998, Soulé 1999). Also, integrative, multispecies management efforts use umbrella species whose requirements overlap with a broad array of other species at large scales (e.g., FEMAT 1993, Ruggiero et al. 1994, Witmer et al. 1999). Using carnivores in these ecological roles seems to be widely practiced, but many recent studies have weakened the utility and acceptability of "focal species" (keystone, indicator, umbrella, and flagship species). De Leo and Levin (1997) stress that in most cases it is groups of species, rather than individual species, that assume importance, forming "keystone groups" or "functional groups"—a generalization of the notion of keystone species. Functional groups (guilds) are a collection of species that perform the same functions and that, to some extent, may be substitutable and viewed as a unit (Schulze 1982, Solbrig 1994, Folke et al. 1996). Keystone groups, as opposed to a single species, might better accommodate the realistic complexity of species compositional turnover, successional mosaics, and disparate disturbance regimes (Tanner et al. 1994, Leibold 1996, Liebold et al. 1997, Khanina 1998, Woodroffe and Ginsberg 1998).

Simberloff (1998) concluded that the indicator species concept is problematic because there is no consensus on what the indicator is supposed to indicate and because it is difficult to know which is the best indicator species even when we agree on what it should indicate (see also Hurlbert 1997). The umbrella species (a species that needs such large tracts of habitat that saving it will automatically save many other species) seems like a better approach, although often whether many other species will really fall under the umbrella is a matter of faith rather than research (e.g., carnivores: Kerr 1997). Intensive management of an indicator or umbrella species (for example, by transplant or supplemental feeding) is a contradiction in terms because the rest of the community to be indicated or protected does not receive such treatment. A flagship species, normally a charismatic large vertebrate, is used to anchor a conservation campaign because it arouses public interest and sympathy, but a flagship need not be a good indicator or umbrella, and conservation of flagship species is often very expensive. Further, management regimes of two flagship species can conflict.

Some of the latest empirical evidence supports Simberloff's reasoning. Perhaps the notion of ecological integrity is so complex that its measure cannot be expressed through a single indicator but rather requires a set of indicators at different spatial, temporal, and hierarchical levels of ecosystem organization (De Leo and Levin 1997, Dufrene and Legendre 1997). In a tropical forest, Lawton et al. (1998) found that no one species group served as a good indicator taxon for changes in the species richness of other groups. Similarly, a lack of consistent patterns led Niemi et al. (1997) to conclude that improving monitoring and habitat classification techniques will prove more fruitful than focusing on a few "representative" species. Neither study included mammal species. On the other hand, in an exhaustive synthesis with new carnivore data from the Neotropics, Terborgh et al. (1999) conclude that top predators are often essential to the integrity of ecological communities. Top predators, including carnivores, can exert a strong top-down role in regulating prey populations—thereby stabilizing the trophic structure of terrestrial ecosystems. The authors claim that the absence of top predators appears to lead inexorably to ecosystem simplification accompanied by a rush of extinctions.

The recognition that some ecosystems have keystone species whose activities govern the well-being of many other species suggests an approach that may unite the best features of single-species and ecosystem management (Simberloff 1998, Soulé 1999). If we can identify keystone species and the mechanisms that cause them to have such wide-ranging impacts, we would almost certainly derive information on the functioning of the entire ecosystem that would be useful in its management. Some keystone species themselves may be appropriate targets for management, but even when they are not, our understanding of the ecosystem will be greatly increased. Keystone species may not be a panacea, however. We do not yet know how many ecosystems have keystone species, and the experiments that lead to their identification are often very difficult (Simberloff 1998).

One of the most widely discussed ideas in conservation biology is the notion that we can focus on a few indicator or keystone species, and, as long as we preserve viable populations of these species, we will also maintain intact ecosystems. Although there is virtually no evidence to support this hypothesis, it has strong appeal—it gives us a way to "preserve ecosystems." Carnivore species are a natural candidate for indicator species because the concept of keystone predators and trophic cascades has considerable experimental support. For example, the acceptance of carnivores as keystone species is at the heart of the most recent syntheses of regional and continental conservation strategies (Soulé and Noss 1998, Soulé and Terborgh 1999). Thus studies of Yellowstone carnivores could become a proving ground for research on focal species in general. Understanding more of the ecological background may help us choose carnivore species and design experiments for the better.

Keystone species: Few examples with a broadening search. The study of how predators affect prey populations includes classic predator-prey models about lethal direct effects and how these effects pass along the entire network of pathways in a community. The study of how predators alter prey behavior, life-history schedules, and habitat selection is termed risk effects, which includes such lethal indirect effects as reduced foraging time and such nonlethal indirect effects as habitat shift (Sih et al. 1985, Lima and Dill 1990, Werner 1992, Wootton 1994, Menge 1995). Conventional models of community dynamics do not account for indirect effects, particularly adaptive shifts in prey behavior and cascading trophic effects on herbivores and plants (top-down control). The challenge now is to derive an empirical understanding of the relative contribution of predator effects on community dynamics arising indirectly through the effects of predation risk and directly via predation events (Brown and Heske 1990, Abrams et al. 1996, Werner and Anholt 1996, Byers 1997, Schmitz et al. 1997). For example, recent work shows how local movement, combined with predator-prey dynamics, can lead to self-organized spatial structure that enables inherently unstable communities of interacting species to persist indefinitely. But the spatial extent of a system that is maintained this way typically is vastly greater than the area suggested by the scale of the local movement of individual organisms. At the very least, carnivores may help us gain an intuitive feeling for such systems and therefore help us understand and set the size of an ecosystem as a self-maintaining metacommunity (or reserve).

We have discussed how attempts at unraveling the manifold of predation effects are complex and difficult to elucidate (see also Jedrzejewska and Jedrzejewski 1989, Holt and Lawton 1994, Hansson 1995a, Krebs et al. 1995, Norrdahl and Korpimaki 1995a, Noss et al. 1996, Polis et al. 1996, 1997, Korpimaki and Norrdahl 1998, Chapter 6 this volume) and to model for carnivores (see also Holling 1992, McLaren and Peterson 1994, Post and Stenseth 1998, Chapter 10 this volume). For example, decades of predator control have led to the inference that the relative abundance of carnivore species has major effects on prey species (e.g., Garrettson et al. 1996, Côté and Sutherland 1997, McShea et al. 1997, Banks et al. 1998). More recently, the clearest empirical evidence comes either by comparing experimental sites or from studies in which a carnivore is removed by natural predation, resulting in changes in both the populations (or behavior) of other predators and the prey base. Soulé et al. (1988) demonstrated positive effects on prey populations when coyotes controlled the populations of smaller predators such as foxes and domestic cats. Thus the absence of coyotes may lead to higher levels of predation by a process of mesopredator release (see Chapter 6–8). Palomares et al. (1995, 1996) showed that a lynx may actually benefit its prey (rabbit) through intraguild predation on other smaller predators, such as mongooses, that share the prey. Mesopredator release may involve a keystone role for a carnivore species, but we do not have powerful evidence of a community-wide effect in these studies and others (Vickery et al. 1992, Sovada et al. 1995, Garrettson et al. 1996, Côté and Sutherland 1997, Greenwood et al. 1998, Rogers and Caro 1998), though good circumstantial evidence comes from canids (see Chapter 6). Although interspecific competition among carnivores is likely to be severe, we know little about its particulars (Moors 1984, Polis et al. 1989, Polis and Holt 1992, Creel and Creel 1996, Goszczynski 1997, Chapters 6 and 7 this volume).

The wolf–moose–balsam fir system on Isle Royale is the most notable example because it reinforces the value of good experimental design and especially choice of study system and subjects (McLaren and Peterson 1994, Pastor et al. 1998a, Peterson et al. 1998, Chapter 10 this volume). Isle Royale is a relatively closed study system with a relatively simplified animal community, which aided in demonstrating a wolf-induced trophic cascade—known strong community effects at many levels (main predator, main prey species, vegetation, soil microbes). The sea otter and American marten are examples of keystone and indicator species, respectively, but there appear to be few other carnivore candidates. For carnivores, Estes (1996) asks whether the seeming absence of these pivotal ecological roles is real or is simply a result of lack of identification (see also Noss et al. 1996, Terborgh et al. 1999). Carpenter and Cottingham (1997) note that ecosystem response to a given perturbation depends on only a fraction of the species pool, but the critical species are situation specific and can rarely be anticipated.

Historically, ecologists identified as keystone those species whose removal was expected to result in the disappearance of much of the assemblage (Mills et al. 1993). Such a restrictive criterion would rarely be met by a carnivore species; therefore, historically ecologists may have excluded carnivores as candidates. Recently, the criteria has broadened because the concept has successfully drawn our attention to differing interaction strengths in food webs and to the pattern in which only a few species have strong interactions that affect community composition, particularly in conservation biology (Bond 1993, Jones et al. 1994, Naeem et al. 1994, Tilman and Downing 1994, Holling et al. 1995, Folke et al. 1996, Colding and Folke 1997). Power et al. (1996) defined a keystone species less restrictively as one whose abundance is relatively low but whose effect on its community is relatively large. For practical purposes we may be able to identify such species without identifying and estimating the interactions involved. Estes (1996) suggests that previously overlooked examples of potential keystone carnivores involve the introduction of exotic predatory mammals (mongoose and domestic cat). Such cases may be useful in understanding the role of invasions and introductions and the importance of coevolution in structuring communities (Sinclair et al. 1998). Other evidence for carnivores as keystones is equivocal (Wright et al. 1994, Hofer and East 1995; but see Chapter 8 this volume). However, we agree with Terborgh et al. (1999) that the critical role of keystone species is gaining acceptance and that top predators appear to regulate many ecosystems, although large Pleistocene and Holocene carnivores were probably not effective regulators of the megaherbivores and the migratory ungulates. Due to the megafaunal collapse and more recent human persecution, both groups of species have been drastically reduced to their present-day distributions and numbers. Perhaps Terborgh et al. (1999) overstate that what remains nearly everywhere else are truncated mammal communities, which are regulated largely through top-down processes.

Clearly, the best conditions for detecting and demonstrating keystone roles have come or will come from experimental designs with the following characteristics:

- Confined or insular study systems.
- Controls in the form of manipulation (for example, adaptive management) or comparison (matching treatment sites with control sites within the study region).
- Simple systems with few trophic levels and disproportionately few species at higher levels; in other words, limited number of predators and prey species—ideally, two- to three-species systems in order to distinguish among alternative models.
- Tightly coupled systems in which interactions are few and strong.
- Well-studied systems (for example, Isle Royale) that have long-term data on the carnivore-herbivore-plant system over large areas.
- Dense populations (relative to human experimenters) or species that are easy to monitor.
- Little human modification, particularly involving habitat degradation, loss, and fragmentation, which increases confounding variation.

The Greater Yellowstone Ecosystem satisfies many of these conditions in that it is a peninsular, perhaps insular, region with a relatively depauperate fauna compared to the majority of tropical to temperate ecosystems (see also Caughley and Gunn 1993). Comparatively, the cumulative effects of human disturbance are low throughout enough of the region for contrasting the human role in ecosystems, which in itself is an interesting variable. The Yellowstone ecosystem is also very well studied, and wolves may emerge as a keystone species because of their ability to influence other predatory species (such as bears, coyotes, and foxes), the migratory patterns of ungulates, and populations of other possible keystone species (such as beaver and moose) (Hofer and East 1995, Estes 1996, Noss et al. 1996, Chapters 6 and 8 this volume).

A candidate keystone group for the Yellowstone system might be carnivores that potentially prey upon snowshoe hares. This group would include abundant predators such as coyotes, but it would also include the lynx and fisher, rare species in the region. Many ecologists have been puzzled by the scarcity of lagomorphs, particularly snowshoe hares, on the Yellowstone Plateau. Generally, lagomorphs are critical food sources during winter for most large predators. Lagomorphs may be replaced as prey sources by squirrels, voles, mice, and ungulates (primarily in the form of neonates and carrion) (Chapters 6–9). Much could be learned about potential keystones because the lynx–snowshoe hare cycle is one of the most studied predator-prey relations, and it has been the subject of the most rigorous experimental designs and sophisticated theoretical explorations involving carnivores. Coyote-rabbit cycles have also been the subject of intense scrutiny. Comparing ecosystems with and without important predator-prey combinations is valuable in itself (Sih et al. 1998) but also in devising conservation strategies that suggest protocols for an experimental management program aimed at conserving sensitive prey species. The task in the future is to determine how to change the vulnerability of the prey so that they can have a refuge at low numbers (Sinclair et al. 1998).

Snowshoe hares may be seen as simultaneously regulated from below and above, because food and predation together have a more than additive effect, which suggests that a three–trophic level interaction generates hare cycles (Krebs et al. 1995, Stenseth 1995, Stenseth et al. 1997). Thus Stenseth et al. (1997) argue that the classic food chain structure is inappropriate: the hare is influenced by many predators other than the lynx, and the lynx is influenced primarily by the snowshoe hare. Coyotes and lynx are the two most important mammalian predators of snowshoe hares throughout much of the boreal forest in North America. O'Donoghue et al. (1998a,b) found "generalist" coyotes and "specialist" lynx both fed mostly on hares during all winters, except during cyclic lows, when the main alternative prey of coyotes was voles (hunted in open cover), and lynx switched to hunting red squirrels (hunted primarily from ambush beds). Where snowshoe hares are the major prey, coyotes and lynx segregate along elevation and habitat gradients that are correlated with such snowpack characteristics as depth and crust firmness (Murray and Boutin 1991). Coyotes are more selective of snow conditions than lynx. Coyotes scavenge more often than lynx, but neither species seems to select habitats on the basis of carcass availability (Murray et al. 1994). Although lynx prefer dense forests over open habitats in Canada (e.g., Poole et al. 1996), lynx chase hares more frequently than coyotes in sparse spruce, whereas coyotes chase hares more often in dense spruce (Murray et al. 1995). For hares, the role of understory cover is ambiguous (Sinclair et al. 1988, Schmitz et al. 1992, Cox et al. 1997), as is the role of interspecific interactions with other herbivores (Boutin et al. 1995, Danell et al. 1998). Relevant insights on Yellowstone vegetation and cover can be found in Brussard (1991), Despain (1991), and Keiter and Boyce (1991).

Reviews of studies from the Palearctic region revealed that the share of hare species in the lynx diet positively correlated with latitude, whereas the share of ungulates was inversely related to latitude (Jedrzejewski et al. 1993, Sunde and Kvam 1997). In the southwest regions of the Palearctic, where lynx rely on ungulates, lynx numbers are more stable, but periodically they are more affected by man, with decadelong periods of near extermination

(Jedrzejewski et al. 1996). Knowledge of dispersal patterns in lynx is fragmentary, but mobility and vulnerability to human persecution of lynx populations are both high (Poole 1997, Sunde et al. 1998a). Refugia from trapping are believed to be important to support a long-term sustainable harvest of Canada lynx, but long-term studies in unharvested areas are lacking (Slough and Mowat 1996). Sunde et al. (1998b) conclude that lynx, even when suffering extensive, man-induced mortality, may tolerate high human activity within their range as long as sufficient stands of undisturbed, mature forest with dense horizontal cover are present.

Sustained low populations of lynx and hare seem possible. Coyote predation is the overwhelming determinant of hare survival and thus population trend in a highly fragmented habitat near the lynx's geographic limit in central Wisconsin (Keith et al. 1993). Moa et al. (1998) found that Norwegian lynx occupied much greater home ranges than reported in other lynx studies, which might be the result of a greater proportion of marginal habitats and a relatively smaller prey density in their study area. Islands and other sites along the periphery of a species' historic range represent critical refugia for many threatened species because of their relative isolation from central populations and from a suite of anthropogenic disturbances (Lomolino and Channel 1995). Increasingly, lynx and snowshoe hare are becoming part of enlightened management plans that integrate forestry and land use planning with an ecosystem perspective (e.g., Quade et al. 1996, Angelstam and Pettersson 1997, Travaini et al. 1997). One of the benefits of an ecosystem approach is that interesting interactions such as lynx-hare can be creatively explored at many levels while meeting other criteria or objectives.

Indicator species: In search of more than circumstantial evidence. A prominent idea in theoretical ecology is the possibility that there are certain "indicators" that foreshadow dramatic ecosystem change. The ideal indicator species gives a clear warning signal for major ecosystem deterioration long before chemical or physical changes are evident (see Noss 1990, Carpenter et al. 1993). Effective indicators are stenotopic or sensitive to environmental changes; have measurable responses, with low level of variability, to the environmental conditions of interest; and are year-round residents in the target area (Landres et al. 1988). Large body size and large area requirements are sometimes considered important selection criteria. Ehrlich (1996) argues how distributions of both species and population diversity are almost totally unexplored for temperate forests. Moreover, how well species diversity in different groups is correlated at a local scale is unknown; as a result it is unclear whether there are any "indicator groups" that can serve as surrogates for overall species diversity when designing specific conservation projects. Nonetheless, we should attempt to define those ecological circumstances when the indicator assumption is defensible, and we should incorporate uncertainty explicitly (Flather et al. 1997, Todd and Burgman 1998).

Carnivores are candidates for indicator status of habitat and landscape change, and we would expect them to offer the most illuminating case studies. Some ecologists have argued that carnivores are uniquely susceptible to extinction in the face of landscape change (Newmark 1986, 1987, 1995, 1996, Peterson 1988, Schonewald-Cox et al. 1991, Bixby 1991). Habitat fragmentation at spatial scales that might have little effect on most plant or herbivore species could profoundly affect carnivore populations (Schoener 1983, Lindstedt et al. 1986, Fowler 1987, Huntly 1991, Gilbert et al. 1998). Yet there are very few studies clearly relating landscape patterns to effects on carnivores (Sih et al. 1985, Hansson 1989, 1992, 1995b, Pagel et al. 1991, Powell 1994, Andren 1995, Maehr and Cox 1995, Mladenoff et al. 1995, Oehler and Litvaitis 1996, and reviews in Gittleman 1989, Buskirk et al. 1994, Ruggiero et al. 1994, Lidicker 1995, Hansson et al. 1995, Hansson 1997, Haight et al. 1998, Chapters 3 and 4 this volume).

The best case for a carnivore indicator species is the American marten, which requires coniferous forests, typically dominated by late successional stands of mesic conifers. Martens are among the most habitat-specific of North American mammals and are thus quite sensitive to habitat alteration (Harris 1984, Bissonette et al. 1989, Thompson 1991, McKenney et al. 1994, Bissonette and Broekhuizen 1995, Zielinski and Stauffer 1996).

Similar to the discussion of carnivores and keystones, several points emerge concerning the fact that managers will use indicator species whether or not academic ecologists are satisfied with the evidence. We need to strengthen the indirect evidence by shaping the way we monitor indicators. We could also use simulation models to determine whether a realistic monitoring program aimed at indicators could indeed foreshadow some imminent collapse of an ecosystem.

CARNIVORES AND THE FUNDAMENTAL CONCERNS
OF CONSERVATION SCIENCE

Population and community biology have been based largely on studies of abundant and widespread species, but most species are neither (Gaston 1994). Rarity is the state of having a low abundance or a small range size, broadly linked to either environmental variables or colonization abilities. The fossil record suggests that some species can be persistently rare. Conservation biology seems often to be largely concerned with rare species, where rarity takes the form of species restricted to limited regions or highly specialized habitats. However, organisms such as carnivores can also be quite rare in the sense of relatively low population densities (at least in comparison with herbivores). We need to know whether practical approaches or principles of conservation change fundamentally when dealing with rare carnivores that nonetheless can occupy a tremendous range of habitats at a wide range of densities (Dobson and Yu 1993, Prendergast et al. 1993, Thomas 1994, Tilman et al. 1997, Travaini et al. 1997).

For carnivore species with a variety of densities across their range, how are rarity and highly variable population densities related to center-boundary effects within the range (Hengeveld 1989)? How might the range of densities arise from interactions between local population dynamics, limited dispersal, and habitat geometry (Taylor 1990, Stenseth and Lidicker 1992, Brown et al. 1996, Fahrig 1997, Holt 1997)? What key demographic rates are changing?

To answer these questions, we must separate the effects of direct human influence. For example, intraspecific variation may be so great in part because many carnivore species can so easily alter their sociodemographic organization, in some cases showing an astonishing sensitivity to human persecution (Chapter 4). The best examples are drawn from extensive canid studies: compare Cavallini (1996) and Crabtree and Sheldon (Chapter 6) on canids with Balharry (1993) on martens.

Like any species, carnivores can disappear locally for a large variety of reasons. It is not particularly informative simply to recount and tally up all the horror tales of carnivore extinction. However, if we could develop a categorization of "causes for extinction," then it could become a context for attempting to make predictions. For instance, carnivores could dwindle because they were directly exploited or hunted, or they may diminish because their prey decline, or because small populations suffer genetic and demographic liabilities, and so on (Chapters 9 and 11). Identifying the hallmarks of these different threats to carnivores will sharpen the design of both our monitoring programs and our adaptive management experiments (Rapport et al. 1998, Schindler 1998, Sit and Taylor 1998, Soulé 1999). Monitoring of populations is politically attractive but ecologically banal unless it is coupled with experimental work to understand the mechanisms behind system changes (Krebs 1991). In particular, monitoring programs should be designed to enable us to distinguish among these different threats, and management experiments should be conducted in such a way as to expose which threats are most pressing for a carnivore species.

Comparative evaluation of factors that may affect carnivore distribution, abundance, and extinction is lacking (see Cutler 1991, 1994, Pagel et al. 1991, Schonewald-Cox et al. 1991, Holling 1992, Dobson and Yu 1993, Badgley et al. 1994, Cole et al. 1994, Young 1994, Alroy 1995, Ceballos and Brown 1995, Lomolino and Channel 1995, Newmark 1995, Rosenzweig 1995, Smallwood and Schonewald 1996, Weber and Rabinowitz 1996, Kerr and Packer 1997, McKinney 1997, Wolff 1997). We also need to spend more time evaluating the geographical scales that are most useful for applying species data (Murphy and Noon 1991, Maurer 1994, Brown 1995, Conroy and Noon 1996, Oehler and Litvaitis 1996, Blackburn and Gaston 1998).

Carnivore studies refine small-population and declining-population paradigms. In a highly synthetic paper, Caughley (1994) developed two main themes: the small-population paradigm that deals with the effect of smallness on the persistence of a population, and the declining-population paradigm that deals with the cause of smallness and its cure (Chapter 11). Although this division may be overly simplistic (Hedrick et al. 1996), Caughley suggests that processes relating to the problems of extinction of small populations (demographic and environmental stochasticity) are more attractive to researchers because they are more amenable to theoretical analysis and to valid generalizations. However, the declining-population paradigm is more relevant to most management and conservation problems, and we must codify the causes of decline and the antidotes. The declining-population paradigm is urgently in need of more theory; the small-population paradigm needs more practice. Caughley and others suggest less emphasis on elegant generalizations about minimum viable populations (MVPs) and more focus on the many factors that cause populations to decline in the first place.

What is the quality of our demographic information? Can we estimate populations for the long term—survival, reproduction, mortality sources, numbers, and density? Estimates of life-history parameters for most species either are unavailable, have unacceptable estimates of variance, are based on questionable approaches to sampling, or are based on a relatively short time interval (Dennis et al. 1991). Beissinger and Westphal (1998) believe that predictions from quantitative models are unreliable for endangered species due to poor quality of demographic data used in most applications, difficulties in estimating variance in demographic rates, and lack of information on dispersal (see also Diffendorfer 1998). Unreliable estimates also arise because stochastic models are difficult to validate, environmental trends and periodic fluctuations are rarely considered, the form of density dependence is frequently unknown but greatly affects model outcomes, and alternative model structures can result in very different predicted effects of management regimes. Thus reliable sources of demographic parameters are needed for linking population viability assessment (PVA) to habitat models that are spatially explicit (Boyce 1992, 1993, Possingham and Davis 1995, Lindenmayer and Possingham 1996, Linnell et al. 1997, Huxel and Hastings 1998, Kendall 1998, Lele et al. 1998, Reed et al. 1998, Kareiva 1999). Can we partition mortality sources as to cause and as to correlated human activities? Can we estimate additive mortality? What are the sensitivities and tolerance limits of carnivore species to natural vs. human-induced disturbance? Spatially explicit PVAs should help us probe these factors and identify the nature and form of critical threshold and compensatory variables (see Wikramanayake et al. 1998). Is it true that "current requirements for population viability analyses are fantasies: legislators ask for something that no scientist can provide" (Ludwig 1994, see also Rohlf 1991, Kareiva 1999)? Ludwig (1994) goes on to suggest the superior value and utility of presenting alternative models and consequences of alternative actions or scenarios (e.g., Swetnam et al. 1998).

INFORMATIVE COMPARISONS FROM PALEOHISTORY AND ACROSS PRESENT GEOGRAPHICAL REGIONS

Paleohistory and ecological processes. How much do we know about the pattern of paleoecological change in the Greater Yellowstone and other systems? Can study of past carnivore assemblages help us understand and predict the consequences of current extinction causes? A very long-term view of natural variation gives us perspective when viewing and evaluating the perturbation effects of European settlement. What factors ultimately determined the original range of the population and its average numbers? Rare events may often play a major role, which reinforces recent pleas for more dialogue between paleoecologists and resource managers and research ecologists (Willis 1993, Behrensmeyer et al. 1992, Erwin 1998). Although human-induced habitat and species disruptions may be qualitatively different from past events, pursuing the study of past extinctions and species interactions will surely give us some insight into our present snapshot of coevolved systems, particularly with regard to species invasions, extinctions, and the ecological roles of carnivores (Reed 1994, Jablonski and Sepkoski 1996, Palmqvist et al. 1996, McKinney 1997, Chiba 1998). The Yellowstone region is a "young" system with regard to changes from European colonization; to complete the picture we need to compare paleoecological trends with recent history of ecology and management.

The patterns of ecosystem change we see emerging from paleoecological analyses are not only relevant to the present crisis but also, to a chilling extent, remarkably similar (Burney 1993). It stands to reason that the past extinctions most applicable to the present situation are those relatively recent ones. The climate has made more than twenty shifts from glacial to interglacial conditions over the past 2.5 million years, and even within the past twelve thousand years landscapes have seen massive changes, chiefly in temperate and northern latitudes. The debate over the causes of the late-Pleistocene extinctions has centered on the Americas, where species turnover and habitat change were truly staggering, dwarfing even modern environmental catastrophes. No single factor is responsible; it is more likely that the primary lethal effect was a combination of factors, acting in a synergistic manner on a fauna unaccustomed to so many disruptions at once. These stresses included natural climate change, activities of human hunters, changes in fire regime and vegetative structure, and the arrival of exotic species (Behrensmeyer et al. 1992, Webb et al. 1995, Zimov et al. 1995a,b, Beck 1996).

Late-prehistoric extinctions constitute a series of cautionary tales for conservation biologists (Stuart 1991, Behrensmeyer et al. 1992, Burney 1993, Balmford 1996). Native faunas in most parts of the world today are only damaged fragments of what existed in the recent past. Most ecosystems were first disrupted by people not in recent centuries or decades but thousands of years ago. Most of the "communities" we seek to preserve are not highly integrated entities that have reached some long-standing equilibrium. Communities appear far more resilient to particular threats if they have faced similar challenges in the past (Balmford 1996). Modern community patterns emerged only in the last few thousand years, and many late Pleistocene communities do not have modern analogs (Graham et al. 1996). Places with long histories of human influence have already lost most of their human-sensitive species. Resulting biotas are primarily of two types: continental biotas with the largest herbivores and the large and medium-sized carnivores "skimmed off," and island biotas lacking many original components of all sizes and heavily invaded by cosmopolitan exotics. Most extinction in recent millennia is fundamentally different from earlier extinction because it is extinction without replacement, at least on human time scales. But studying past extinctions may lead us to some important conservation efforts, such as Owen-Smith's keystone herbivore hypothesis (Owen-Smith 1988, 1989). Large, slow-breeding herbivores play a role in shaping their own environments (such as habitat turnover, creating niches for smaller animals), and they are the first to go extinct, resulting in a "cascade" of extinctions of the middle-sized grazers and browsers (see also Young 1994, van de Koppel et al. 1997, Saether 1997). Despite large differences in phylogenetic composition, patterns of resource division among sympatric carnivores remain stable over many millions of years (Van Valkenburgh and Janis 1993, Andrews 1995, Van Valkenburgh 1995, Janis et al. 1998, Lambert and Holling 1998). In addition, species with naturally scarce, fluctuating, or isolated populations are less susceptible than other taxa to the demographic and genetic problems of reduced population size (Gaston 1994, Lawton and May 1995, Balmford 1996, Chapter 11). Comparing carnivores in relatively untouched ecosystems with human-impacted systems may be a key to understanding the concept of "filtering" and resulting variation in the susceptibility of biotas to current threats (Chapter 4).

We can also learn much from comparing the Americas, characterized by rapid, catastrophic extinctions, with Africa, the one place on Earth that seems to have virtually escaped the late-prehistoric extinction crisis. Hominids had much longer to coevolve with carnivores and herbivores in Africa compared with the Americas (Stuart 1991). Tool-using hominids, as carnivorous animals, would have been part of the various carnivore guilds present in Plio-Pleistocene Africa (Lewis 1997).

The adaptive history of most species is linked to natural disturbances. Many ecologists have found productive links between evolution, paleoecology, macroecology, and biogeography—links that help us understand the nature of species distribution, abundance, and persistence at scales from ecosystem to continent (Brown 1986, Delcourt and Delcourt 1988, Cutler 1991, Ceballos and Brown 1995, Lomolino and Channel 1995, Brown et al. 1996, Host et al. 1996, Byers 1997, Kerr and Packer 1997, Dayan and Simberloff 1998, M. Dobson 1998, Reed 1998). Detailed comparisons of past and present carnivore assemblages within the same systems, as well as comparisons across systems, reveal strong patterns in carnivore guilds as they relate to characteristics of prey species and ecosystems (Van Valkenburgh 1988, 1989, 1995, Van Valkenburgh and Janis 1993, Flessa and Kidwell 1995, Duckler and Van Valkenburgh 1998). Within the Holocene, historical analyses can tell us much about the effects of human disturbance and landscape alterations (e.g., Angelstam 1996 for Europe; Elias 1995, Hadly 1996 for Yellowstone). Resource agencies have shown considerable interest in evaluating how management actions might mimic natural disturbance regimes (for example, fire, forest harvest, and habitat changes). If we identify ecological processes that affect carnivore species, it would not only help our conceptualization of ecological processes and extinctions but also assist various methods of assessment.

Modern comparisons at the system level. Many have viewed the African Serengeti as a "sister ecosystem" to the Yellowstone, but the only rigorous comparisons come from paleontology (Van Valkenburgh and Janis 1993) and grassland ecology (Berger 1991, Frank et al. 1998). The Serengeti is the most intensively researched region in the world (Sinclair and Norton Griffiths 1979, Sinclair and Arcese 1995) and no doubt influenced the ecosystem approach in Yellowstone as it developed in the 1960s and 1970s. The centerpiece of the Serengeti is the extraordinarily complex predator-prey system. Comparing these and other intensively studied systems should be valuable (e.g., Angelstam et al. 1997). Other systems where carnivores have played a central role are the Bialowieza Primeval Forest in Poland (Jedrzejewski et al. 1989, 1993, 1995, 1996, Jedrzejewska et al. 1994, 1996, Okarma 1995, Okarma et al. 1995, 1997, 1998),

Chile (Jaksic and Simonetti 1987, Simonetti 1988, Jaksic 1989, Jaksic et al. 1993, 1996, Meserve et al. 1996), India (Johnsingh 1993, Karanth and Sunquist 1995, Sunquist and Sunquist 1988), Scotland (Kruuk 1978, 1989, 1995), and several regions in Scandinavia (e.g., Erlinge et al. 1988, Hansson 1992, 1997, Hanski et al. 1991, 1993, Korpimaki et al. 1994, Lindström 1989, Lindström et al. 1994, Pulliainen and Ollinmaki 1996, Oksanen and Henttonen 1996, Korpimaki and Norrdahl 1998).

The Yellowstone wolf reintroduction forced a successful comparison of Alaskan, Minnesotan, and Canadian systems in which wolf-prey systems were the focus of long-term research. More than an exercise in academic synthesis, this informed the models and shaped expert opinion at such a high level as to satisfy the scrutiny of the Endangered Species Act and most of the public.

METHODOLOGICAL IMPROVEMENTS

Telemetry has been the greatest, or at least the most influential, technological advance in the study of carnivores. However, even with telemetry studies, little is known about the movements and foraging of carnivores at the micro level (small scale: microhabitat use and foraging). Typically, telemetry studies sample animal locations every one to seven days, which effectively averages an animal's movement at midrange scale (activity area, home range) or sometimes at large scales (landscapes to ecosystems, as with wolves and grizzly bears). Therefore much of the literature reports space use and habitat (resource) selection—a midscale view of the animal's behavior and use of habitat, but not use of patches and use of habitat edge (e.g., Buskirk and Powell 1994, Powell 1994). Some studies have accomplished high sampling rates of locations by concentrating effort on a focal animal, which allows an assessment of selectivity at the micro scale, but results of these efforts are not commonly published (but see Johnson et al. 1998). When feasible, direct observation or snow-tracking is used for investigating micro-scale ecology.

Technological breakthroughs have opened up exciting and novel possibilities for research at large scales. Computer-intensive techniques allow innovative investigations of landscapes and ecosystems. Massive amounts of data can now be analyzed and modeled with sophisticated statistical and simulation tools. Integrated systems such as Geographic Information Systems (GIS) integrate sophisticated spatial databases and modeling methods with remote sensing and Global Positioning Systems (GPS), and in the past decade or two these technologies have released carnivore studies from some of the worst experimental constraints. Physical and vegetative data are available for regions of any size. Astonishing advances in sampling theory make it possible to extract the most from small samples. We can use simulation to optimize experimental designs and maximize the information from carnivore studies, particularly for estimating habitat selection and population characteristics. We are able to test for landscape effects on individual behavior and movement and on the distribution of carnivores at a variety of spatial scales. To date, few published studies have yielded significant biological insights, though many projects are promising (see Andren 1995, Brown and Litvaitis 1995, Maehr and Cox 1995, With and Crist 1995, Boone and Hunter 1996, Host et al. 1996, With 1997, Grunbaum 1998, Deangelis et al. 1998, Mladenoff and Sickley 1998, Haight et al. 1998, and see Chapter 3 this volume).

Our current ability to augment telemetry-based studies with information and analysis at multiple scales should be a boon to ecological investigations that inform management and conservation. Many of the fundamental issues are now much more approachable: what is the pattern and process of selection for resources and landscape elements, and how is individual behavior translated to population distribution and demographic structure? Telemetry studies have traditionally focused on the scale of home range and general habitat selection, but now we can connect this scale to selectivity at smaller and larger scales (figure 12.1). In doing so, we can refine our choice and estimation of pertinent variables and thus devise better experiments, all of which improves the models that rely on these underlying variables and mechanisms. It is these models that are eventually used for estimating ecosystem capacity, such as viability of populations and sensitivity to change. These models are also the foundation of understanding the ecological roles of carnivores as keystone or indicator species.

One line of promising and practical carnivore research will combine methods for studying resource selection with methods for studying spatial interaction (for example, movement models and GIS) (after Johnson 1980, Alldredge and Ratti 1992, Porter and Church 1987, Thomas and Taylor 1990, White and Garrott 1990, Aebischer et al. 1993, Manly et al. 1993, Arthur et al. 1996, Schumaker 1996, Conroy et al. 1997, Jackson 1997, Otis 1997, With et al. 1997, Duarte et al. 1998, Kernohan et al. 1998, Raphael et al. 1998, Turchin 1998, and see references cited above). Advances in computer-intensive methods offer the potential for analysis at several scales, ranging from landscape features down to the selection of particular habitat components, such as denning sites (Ruggiero et al. 1998). These exercises would naturally lead to habitat-based PVAS (Boyce 1992, 1993, Possingham and Davis 1995, Brooks 1997, Wikramanayake et al. 1998). One of the greatest concerns of managers is habitat loss and fragmentation and the resulting boundary or edge effects created by various sources of disruption. Vegetative and weather variables can be approximately matched to the scale of animal resource use (Holling 1992, Mysterud and Ims

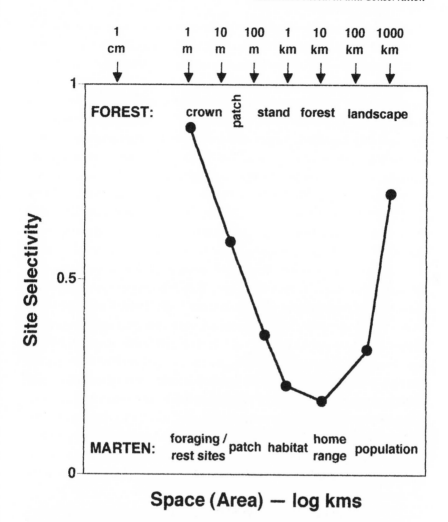

12.1. Hypothesized strength of site selection by a common carnivore (American marten, *Martes americana*) in the Greater Yellowstone Ecosystem occupying a forested system at different spatial scales. Strongest selectivity by marten has been estimated for foraging (rodent prey) and rest sites associated with coarse woody debris near the ground at the 1- to 100-meter scale, within patches of mature mixed-conifer habitats. Data are gathered by snowtracking and observation aided by telemetry. At habitat and home-range scales, to which telemetry data are typically limited, we know little about exact habitat selection and how finer-scaled sites are used because resources are "averaged" over space and time. Finally, at large spatial scales it is far easier to detect selection along a gradient of landscape types and habitat mosaics (for example, by monitoring marten distribution and abundance).

1998), with great care devoted to unraveling how carnivores use, and therefore define, habitat edge and environmental gradients (sensu Milne 1992, Hansen and di Castri 1992, Gosz 1993, Hansen et al. 1993).

Population estimation or monitoring over large regions is extraordinarily difficult (Nusser et al. 1998, Smallwood and Schonewald 1998), so we predict a nested approach of sampling intensities, from indices of presence or absence to absolute estimators for both sexes (e.g., Link et al. 1994, Wileyto et al. 1994, Zielinski and Kucera 1995, Gardner and Mangel 1996, Matlock et al. 1996, Miller et al. 1997, Bellehumeur and Legendre 1998, Bender et al. 1998, Mahon et al. 1998, Sibly and Smith 1998). Different estimators, based, say, on large-scale transect and fine-scale plots, can be used both independently and jointly as auxiliary data, and much progress has been made in this area during the past decade (Thompson et al. 1989, Becker 1991, Skalski 1991, 1994, Skalski and Robson 1992, T. Oksanen et al. 1992b, Van Sickle and Lindzey 1991, Thompson 1992, Buckland et al. 1993, Otis et al. 1993, Becker et al. 1998, Monkkonen and Aspi 1998, Swanson 1998). Snowtracking has been enhanced by GPS and telemetry for inferring individual behavior and for estimating population size and distribution. DNA analysis of scats and hair has recently become a burgeoning field, allowing cost-effective experimental designs for revealing and monitoring population characteristics (Foran et al. 1997a,b, Kohn and Wayne 1997, Haig 1998, Hughes 1998, Parker et al. 1998, Stander 1998). Similarly rapid advances have been made with estimating resource-selection probability functions (Agresti 1990, Pereira and Itami 1991, Trexler and Travis 1993). Additional methods can extend this estimation by including the amount of use, or changes in selection, related to factors such as weather, seasonal and annual differences, or subpopulation differences (Heisey 1985, Ludeke et al. 1990, Schoener and Adler 1991, Manly et al. 1993, McClean et al. 1998). Data resampling methods can yield a more realistic estimate of model classification and accuracy (Verbyla and Litvaitis 1989, Stoms et al. 1992), particularly for correcting errors that arise from aggregating fine-scale components (Rastetter et al. 1992).

The increased use of computer-intensive methods with advances in sampling theory also suggests greater cross-fertilization from other disciplines. For example, geostatistical approaches bring to ecology novel tools for the interpretation of spatial patterns of organisms, of the numerous environmental components with which they interact, and of the joint spatial dependence between organisms and their environment (Borcard et al. 1992, Rossi et al. 1992, Liebhold et al. 1993, Ver Hoef and Cressie 1993, Loehle and Li 1996). One of the most intractable problems of telemetry data involves dependence of observations (White and Garrott 1990, Rooney et al. 1998). Similarly, patches and gradients observed across geographic space exhibit spatial autocorrelation, which comes either from the physical forcing of environmental variables or from community processes. This presents a problem for statistical testing because autocorrelated data violate the assumption of independence of most standard statistical procedures. New statistical and mapping techniques explicitly introduce spatial structures into ecological models (e.g., Legendre 1993, Aspinall and Veitch 1993, Clark et al. 1993, Gustafson and Gardner 1996, Grunbaum 1998, Koenig and Knops 1998).

Spatially explicit population models are becoming increasingly useful tools for investigating scale-related questions in population and community ecology, especially the response of organisms to habitat change occurring at a variety of spatial scales, and this is particularly true of landscape models created with GIS (Dunning et al. 1995, Holt et al. 1995, Conroy et al. 1995, Schippers et al. 1996, Schumaker 1996, Meir and Kareiva 1998). These models purport to develop a unified approach to analyzing and modeling animal population density and distribution, resource distribution, habitat selection, physiography, weather variables, and landscape features (corridors, boundaries), all of which can be addressed only by considering the amount, geometry, and rate of change of habitat in a spatially explicit fashion at a variety of scales (Pulliam et al. 1992, McKelvey et al. 1993, Dunning et al. 1992, 1995, Boone and Hunter 1996). Spatially explicit population models integrate several scales; for example, movement models may be used to evaluate the interplay of landscape pattern and carnivore distribution. Some models can incorporate movement of animals between specific patches across the landscape, with movement rules allowing for both the temporary movement between patches (foraging, thermal refuges), or the more permanent movements of dispersing individuals. Movement rules can specify boundary effects, dispersal mortality, and use of corridors.

Modeling movement and dispersal is challenging because we know so little about the individual's perception of the landscape or the factors that influence the transition from one place to another (e.g., Turchin 1998). For example, biologists appreciate that olfaction is the primary sense that shapes a carnivore's mental map of the environment, but we have little comprehension of how to "observe" and measure its properties (e.g., Engen 1982, 1991, Macdonald 1985, Henry 1986, Smith et al. 1989, Gorman 1990, Stoddart 1991, Asa and Mech 1995, Eisthen 1997, Gese and Ruff 1997, Nams 1997, Rich and Hurst 1998). Olfaction provides a mechanism for sophisticated tracking of resources and for structured social interaction. Olfactory communication is highly developed, and scent glands allow long-lasting messages that operate in the sender's absence. Thus the temporal component of resource partitioning plays a large role (Minta 1992). Urine, feces, and glandular scents allow discrimination of conspecific age, sexual condition, social status, and individual identity. Even for less problematic animals that have been the subjects of optimal foraging models, most efforts have not addressed how an animal searches for food (Anderson et al. 1997). The implications of an olfactory-based carnivore are enormous for interpreting experiments and informing realistic models, particularly spatial models (e.g., Minta 1993, Lewis et al. 1997).

Considering our limitations, the promise of spatially explicit models may never be realized because we can never accurately give parameters to the models to make them very useful (Kareiva and Wennergren 1995, Ruckelshaus et al. 1997, Meir and Kareiva 1998). For example, in constructing a spatially explicit model, Moilanen and Hanski (1998) inferred that additional complexity beyond the effects of habitat patch area and isolation does not necessarily improve the predictive power of a metapopulation model. Given the shakiness of spatial models as a foundation for specific conservation recommendations, Wennergren et al. (1995) conclude they may be more useful as a tool for exploring the design of spatially structured monitoring schemes, so that management mistakes might be detected before they become irreversible. Yet understanding movement—the measurable expression of behavioral decisions about resources and conspecifics—provides a powerful basis for population and habitat models, and it links landscape ecology and food-web dynamics to metapopulation structure (e.g., Polis et al. 1996, 1997, Anholt and Werner 1998).

Whether we like it or not, we cannot give up on landscape models simply because they are difficult. Landscape models are a key to understanding ecosystems in an evolutionary context, and networks of corridors and reserves are essential to realistic and practical conservation strategies (Soulé 1996, Woodroffe and Ginsberg 1998, Soulé and Terborgh 1999). How then can we approach the problem of variation and scale in landscapes and ecosystems? In the following sections, we review and promote the integration of two long-standing approaches that have roots in carnivore biology—behavioral phenotypes (for pooling individual and life-history variation) and landscape ecology. We outline a basis for linking individual-based models to landscape models in a tractable and realistic manner. Rather than assuming that all individuals in a population are the same, or that average values of individuals in a population can be used to describe population phenomena, individual-based models have been developed by ecologists and evolutionary biologists who want to emphasize individual differences and local interactions. Traditionally, there have been two kinds of individual-based models: distribution models consider individual differences by lumping together individuals with common characters; configuration models track all individuals in a population to incorporate both individual differences and local interaction. For example, "behavioral minimalism" dictates a focus on only those few behavioral traits that are likely to be important to the question under study (Turchin 1991, 1996, 1998, Wiens et al. 1993, Lima and Zollner 1996). However, we may do better by exploiting our extensive knowledge base on carnivore behavior and social organization, particularly the patterns of variation observed among phenotypes, populations, and environments. In doing so we can account for variation, find a reduced set of parameters that are realistic, and construct less complicated models that produce interpretable outcomes.

BEHAVIORAL PHENOTYPES AND PHENOTYPIC PLASTICITY: THE POTENTIAL TO UNIFY MANY AREAS OF ECOLOGY

Ethology is traditionally the comparative study of behavioral phenotypes, while morphology and ontogeny address the comparative study of physical phenotypes. However, the behavioral phenotype has become an umbrella term for combined behavioral and physical variation—in the sense that categories are derived from lumping similar variants into types or categories for the purpose of linking individual- and population-level phenomena. For example, age and size are correlated with distinct reproductive tactics (such as alpha breeder, helper at the den, floater, disperser), and the frequency of each type varies between populations, depending on such factors as habitat saturation and level of human persecution (Chapters 4 and 6). Behavioral phenotypes help to explain patterns of life history and individual variation, and the resulting variability in social organization. The outcome of this variability at the population level is interpretable because we can make better sense of localized interactions (feedback) between resource distribution, behavior and movement, and population characteristics and structure. We propose that behavioral phenotypes might be an effective and measurable way of pooling individual variation so that individual behavior and movement can be mapped to higher-level population processes. If that is true, then we should expect to build more realistic landscape models without having to resort to extreme forms of spatially explicit or individual-based models, for which it is difficult to set parameters.

Individual variation is composed of differences between phenotypes and differences within the same phenotype at different ages (see reviews for mammals in Hayes and Jenkins 1997, Holmes and Sherry 1997). These sources of variation translate to higher-order outcomes in social and demographic structures, and it may be derived from genetic and behavioral variation in individuals (recent reviews in Bergstrom and Godfrey-Smith 1998, Chesser 1998, Coulson et al. 1998, F. Dobson 1998, Law and Dickman 1998, Wilson 1998). Two landmark papers appeared in 1977 that developed cohesive theories of behavioral phenotypes, one using marmots (Armitage 1977) and the other using canids (Bekoff 1977, based on Fox 1972). Although socially inherited behavior is understood best, other inherited environmental effects, such as maternal inheritance (apart from nuclear genes), are strongly implicated in mammals

(Krebs 1996, Boonstra et al. 1998, Inchausti and Ginzburg 1998). The resulting phenotypic variation influences survival and fecundity, particularly at the earliest stages of offspring life, a time when mortality and selection are often greatest (Rossiter 1996). Such variation has been characterized as a form of bet-hedging by increasing the likelihood of survival and reproduction in the face of environmental uncertainty (heterogeneity or transient environment inputs: Whitlock et al. 1995, Rossiter 1996, Bouskila et al. 1998, Scheiner 1998). For example, by producing young of varied phenotypes (for example, socially tolerant "stayers" vs. intolerant "dispersers"), a female increases the probability that over the long term some of her descendants will survive in varied and unpredictable social and ecological environments. The behavior of animals in mixed-age populations is strongly affected by the age-sex classes present and by kinship, which affects recruitment and dispersal (Armitage 1986, Smale et al. 1997). Individual behavioral phenotypes could be determined by varied social experiences during ontogeny (Bekoff 1977). Both genetic and developmental models agree that variation among individuals can be highly adaptive.

Omnivory and particularly carnivory are highly skilled occupations that tend to require a great deal of continuous, sophisticated learning throughout life. Carnivores also tend to be long-lived. Together, these characteristics are often associated with high individual variation and age-related behavior. The chapters in this book and the previous sections of this chapter have documented that many carnivore species exhibit a great deal of flexibility in social, demographic, and life-history patterns—both within and among populations. Of course, in contrast some carnivore species (for example, grizzly bear, marten) are very sensitive to environmental variation and are less flexible in their response to it, particularly human-induced disturbances, which are attributes that may make them potential keystone or indicator species. It is interesting to ask whether these among-species differences in "flexibility" lead to fundamental differences in vulnerability. The simple model is that inflexible species are vulnerable and flexible species (without the flexibility being an evolutionary response to relentless disturbance) are resilient.

Historically, the primary motivation for studying life histories was to demonstrate how differences in life-history patterns relate to ecological adaptation (for carnivores see Kleiman and Eisenberg 1973, Powell 1979b, Bekoff et al. 1981, 1984, Eisenberg 1986, Gittleman 1986, Boyce 1988, Bronson 1989, Ferguson et al. 1996, Geffen et al. 1996). In studies of carnivore life history, significant phylogenetic patterns are observed at familial levels, but ecological factors (at least in terms of diet) are surprisingly uninfluential (Gittleman 1993). However, most carnivore species exhibit a remarkable flexibility in diet and resilience to fluctuating food sources. Within a population, the outcome of individual responses is observed as changing schedules of birth and mortality.

In an exhaustive review, Bronson (1989) synthesized the complex variation observed in mammalian reproduction: of the many environmental factors that can influence a mammal's reproduction, food availability must be accorded the most important role. Levels of food restriction that allow ovulation seem to have little effect on pregnancy but have their most potent effects on the lactating female. Social cues can depress reproductive success in relation to crowding or social status and priming pheromones or to priming phenomena that involve other sensory modalities. The major advantages of such cuing systems would seem to be twofold: coordinating one's reproduction relative to existing social conditions and, related to this, enhancing synchrony in a population's reproductive effort under ecologically marginal conditions. Actions and interactions of environmental factors can vary greatly with the stage of the mammal's life cycle during which they are perceived. Thus a mammal's reproductive potential during adulthood may be modulated by a variety of earlier experiences, starting even in the uterus. The species concept that pervades our conceptualization of reproduction really is not valuable when applied in its traditionally rigid manner. Reproductive success in different environments leads to different strategies (see also reviews in Lott 1991, Rossiter 1996, Concannon 1997, Chesser 1998, Crews 1998, Asa and Valdespino 1998). The capacity of inherited environmental effects to influence population dynamics, with the potential to cause cycles or destabilization (either noncyclic outbreak or extinction), is supported by theoretical and empirical work on mammals (Rossiter 1996).

Life histories in mammals are also highly compensatory in balancing the effects of mortality. Hunting pressure accounts for 80–90 percent of the mortality rates in carnivores as a whole. However, variability in carnivore life histories may "compensate" for sources of juvenile and adult mortality (Gittleman 1993). Understanding the implications of life-history evolution and variation is critical to using life-history and demographic data in population, community, and habitat models (Holmes and Sherry 1997).

Having failed to produce a robust, predictive theory of population and community dynamics in the 1960s and 1970s (Bowers 1997), ecologists were forced to consider, among other things, the importance of large-scale habitat heterogeneity. At the same time, the importance of individual variation at small scales was becoming more inescapable. Ironically, the impressive progress made possible by the simplifying assumptions of evolutionary genetics was matched by the inability of evolutionary genetics to explain phenotypic design (Stearns 1989, 1992). People were well aware of that deficiency by the 1970s and early 1980s, when it was widely suggested that the role of development in evolution and the new phenotypic theories—optimality theory, quantitative genetics, and game theory—would provide the solution. Phenotypic plasticity and behavioral phenotypes became a potential rallying point for unifying many aspects of population ecology and genetics, life-history theory, developmental biology, and the study of variation in social and demographic organization. Conceptual progress continues to be made by key thinkers across many disciplines (Berger 1979, Caro and Bateson 1986, Clark and Ehlinger 1987, Chepko-Sade and Halpin 1987, Holt 1987, Parsons 1987, Bruton 1989, Geist 1989, Stearns 1989, Hazel et al. 1990, Chesson and Rosenzweig 1991, Werner 1992, Finch and Rose 1995, Rollo 1995, Via et al. 1995, Fryxell 1997, Komers 1997, Dewitt et al. 1998, Simons and Johnston 1997, Tienderen 1997, Lima 1998). Modeling progress has been limited (Gurney and Nisbet 1979, Real 1980, Anderson 1983, Caswell 1983, Via and Lande 1985, Buss 1988, Yoshimura and Clark 1991, Goldwasser et al. 1994, De Jong 1995, Sibly 1995, Behera and Nanjundiah 1996, McNamara and Houston 1996, Anholt 1997, Van Tienderen 1997, Pfister 1998, Anholt and Werner 1998).

Within a population, behavioral phenotypes may be produced by inherited differences or a fluctuating environment, or both interacting during ontogeny as a response to a changing social environment. In addition, carnivores are typically highly intelligent animals with tremendous capacity for learning and even the possibility of developing different "hunting cultures." For example, coyotes may learn to obtain their food from garbage dumps or domestic pets or by stalking. Yellowstone grizzlies have been categorized as "dump" bears and "wilderness" bears. Patterns of human activity and land management determine which of these behavioral phenotypes will prevail. One could have two different landscapes with identical numbers of carnivores, but dramatically different hunting phenotypes. The consequences of these behavioral differences would be profound for the impact of carnivores on the ecosystem they inhabit.

For modeling purposes, it is difficult to incorporate environmental modulation of phenotypic expression (during epigenesis and ontogeny) because it merges genetic and ecological change, which supposedly operate at two different time scales, at least in conventional models. For example, in resource partitioning, within-phenotype is the ecological time frame, while between-phenotype is the evolutionary time frame. Between-phenotype variation would occur or manifest itself between sex and age classes (e.g., Smale et al. 1997), but only if it realistically accounts for phenotypic differences, which is not the case for strong phenotypic variation that is sensitive and responsive to immediate environmental variation.

It is crucial for ecologists not to dismiss the existence and development of behavioral phenotype just because it is hard to treat statistically or incorporate into models (Anholt 1997). For carnivores, the explanatory power is huge. Most carnivore ecologists have a strong intuition of it and how it manifests itself in intraspecific variation within a population and dramatic differences among populations (e.g., Hayes and Jenkins 1997, Holmes and Sherry 1997, Lacy 1997). Theoretical ecologists would do a great service to these ideas by working with carnivore biologists in formulating specific and realistic models over general models that involve carnivores (e.g., Judson 1994, Lima and Zollner 1996, Dieckmann et al. 1995, Dieckmann 1997). Although the muddied relation of phenotype, genotype, and environmental variation concerns many ecologists, it is crucial for conservation biologists to find patterns of variation and plasticity even before we understand all of the mechanisms involved.

THE CHALLENGE OF LINKING INDIVIDUAL BEHAVIOR WITH PROCESSES

Conceptual landscape models are weak and overly generalized, attempting to describe many ecological scales, levels of inquiry, and sources of variation, even in light of new and powerful approaches to resolving variation (heterogeneity) and interactions. Explicit landscape models make predictions and test hypotheses but are too specific and difficult to establish parameters for. The problem remains of setting parameters for models—how to derive empirically based estimates and variances that are useful, meaningful, and interpretable to ecologists. Ecologists of all types have found increasing unity under the rubrics of landscape ecology, conservation biology, and ecosystem science. Within these superdisciplines, societal pressures have encouraged communication among disciplines. If this has indeed occurred, then we should be able to review a progression during the past decade of sound, rigorous field studies that validate and extend models that, in turn, inform and tune experimental designs for the purpose of refining model parameters. To a large degree Bowers (1997) concludes that this is not the case for mammalian landscape ecology, despite the optimistic overview by Lidicker (1995), which touted mammals, particularly rodents, as model organisms for testing various tenets of landscape ecology.

The conceptual elaboration of ecological themes and scales has been successful: the collective outcome of individual movements among patches within a landscape translates to spatially subdivided populations (metapopulation). However, if landscape ecology has taught us anything, it is that it is relatively easy to describe the spatial template we call a landscape, but less so to map processes onto that template (Krohne 1997, Pastor et al. 1997, Songer et al. 1997, Moilanen et al. 1998). The former suffers from human convenience or perception; the latter requires a subjective, species- and question-specific consideration of process. The central issue is scaling features of the landscape to fit the organism and the question being asked, and scaling our studies to provide information on pattern-process relations that are specific to one scale but sufficiently flexible to address processes over a range of scales (Bowers 1997). There is no single scale at which ecological phenomena *should* be studied.

A weakness of landscape ecology is that in order treat large-scale phenomena, we average out small-scale variability to focus on landscape patterns of populations, community, and ecosystems. By doing so, we have largely taken individual responses out of the domain of landscape ecology, but it is the individual that chooses habitats and patches (Bowers and Matter 1997). Movement patterns define the patch scales for a model. An emphasis on individual behavior allows us to examine ecological phenomena over a hierarchy of spatial scales and look to higher scales for significance but to lower scales for mechanism (Levin 1992, Wiens et al. 1993, Bowers and Matter 1997). Part of the attraction of carnivores for modelers is that many species match the scale of human perception, and many species are coupled to mammalian prey operating at similar scales and perceiving habitat similarly to the predators. These landscape scales are also relevant to resource and land managers, and human-induced habitat loss and fragmentation are apparent at these scales. For many purposes, the abundance and distribution of herbivore prey can define carnivore patches through either empirical data or correlations on prey-habitat relations. However convenient this might be for realistically simplifying some of the heterogeneities of physical and vegetative-based resource distribution, the issue of variability remains highly problematic. It is heterogeneity of movement modes by a carnivore that ties spatial patterns of resources to life-history requisites (foraging and breeding) and population dynamics (dispersal). Unfortunately, habitat selection and colonization can be confounding processes that impede the development of mechanistic and predictive models of landscapes and metapopulations (Rosenzweig 1991, Wiens et al. 1993). We must improve how we link individual-level variation, a rich source of explanatory power, with population-level processes, neither overgeneralizing nor becoming lost in a sea of detail (Hayes and Jenkins 1997). While we wait for more-detailed studies of individual behavior and movement, we can continue to simplify the variation we currently know so that it can be incorporated into models.

So in theory one can model an animal's behavior as it moves through a landscape and can predict population processes, but no one has done it yet. Between the extremes of conceptual and explicit landscape models, we can specify fewer parameters by finding the "lumps" in the most critical sources of variation for a reduced number of patch types and types of individual (besides age and sex subpopulations). Historically, population models gained realism by incorporating variation in carrying capacity, both spatially and temporally. In this context, animal movement has been investigated in models that varied spatial scales, interspecific interactions, age-sex structures, or behavioral strategies (such as territoriality and dominance). It is the latter two forms of variation that have received least attention yet show a great deal of promise for landscape models using carnivores. In an earlier section we espoused the relevance of behavioral phenotypes to understanding carnivores in ecosystems. Similarly, realistic model simplification can be derived from the correspondence of distinct behavioral strategies to age- and sex-specific life-history stages (Smale et al. 1997).

Life-history variation unifies what we know about individual- and population-level phenomena (in the context of patterns of phenotypic expression modified by all feedback), which helps us specify pooled types of individuals that represent different strategies for exploiting patches and moving between them for foraging, breeding, or dispersal. The variation we see in carnivore phenotypes is lumpy, and animals can be pooled into such sociodemographic categories—behavioral phenotypes—as resident territorial breeders, resident drifters, nomadic roamers, dispersers, and so on (for example, Chapter 6). That is, because of a convincing covariation, we can effectively combine life-history variation (such as flexible ontogenies subject to environmental and social feedback) with population characteristics (such as age-sex structure) that affect resource partitioning and movement (social organization; see Southwood 1977, 1988, 1996, Holm 1988, Bruton 1989). Thus numerous combinations of behaviors, developmental patterns, and reproductive tactics constitute a reduced set of behavioral phenotypes. We exploit rich data sources—individual behavior and movement, age and sex ratios, social structure—that combine many sources of variation into realistic and understandable life-history variants and that produce interpretable outcomes of resource partitioning, reproduction, population characteristics, and dispersal patterns. The linking of individual- and population-level processes can be mapped to the landscape pattern of disturbance, patch quality, and ultimately metapopulation structure in the landscape.

Carnivore ecologists, particularly wildlife researchers and managers, have always known about these ideas. Early on, mammalian pest and disease outbreaks predisposed thinking about populations in landscapes (e.g., Naumov 1948, Howard 1960, Lidicker 1962, Krebs et al. 1969, Anderson 1970). Although the best experimental work used small mammals, carnivores and big-game species have long been studied and managed as if they occupied landscapes subdivided into harvest units, zones, or refugia (e.g., Leopold 1933, Connolly

1978a, Jewell and Holt 1981, Banci 1989, Alaska Department of Fish and Game 1992, Powell 1993, Caughley and Sinclair 1994, Ruggiero et al. 1994, Borralho 1995, Slough and Mowat 1996, Taylor and Dunstone 1996, Angelstam et al. 1997, Powell et al. 1997, Durant 1998, and see preceding chapters). Any disjunct spatial units are naturally thought to be linked (or not) by corridors, which have served as a unifying topic for research, wildlife management, land planning, and conservation activism (Soulé 1986, 1987, Beier 1993, McEuen 1993, Mech 1995, 1996, Yahner et al. 1995, McCullough 1996, Newmark 1996, Gustafsson and Hansson 1997, Gonzalez et al. 1998, Witmer et al. 1998, Soulé and Terborgh 1999, and see preceding chapters).

The tremendous variation observed within and among studies by carnivore biologists encouraged the integration of behavior and population variability with habitat and landscape considerations (e.g., Lockie 1966, Hornocker 1970, Hornocker and Bailey 1986, Bekoff 1978, Powell 1979b, Eisenberg 1981, Banci 1989). Some species stood out as extreme forms of adaptability, particularly the highly generalized canids. For example, disturbance in the form of human persecution and habitat alteration causes a population flux that is measurable as a change in behavioral phenotypes. Coyotes and foxes offer the most extreme cases of flexibility: populations exhibit an extraordinary range of population structures, reproductive tactics, and social organizations (indeed, advantages to sociality, such as cooperative hunting, may have driven development of the anomalies of the reproductive system: Asa and Valdespino 1998). These range from bottom-heavy age structures of solitary animals under intense persecution to populations of wolf pack–like extended families of long-lived animals occupying protected and saturated habitats (e.g., Andelt 1985, Bowen 1981, 1982, Bekoff and Wells 1986, Moehlman 1989, Cavallini 1996, Geffen et al. 1996, Chapter 6 this volume).

Conclusions

While the ecosystem concept is appealing because it challenges us to integrate many ecological processes over large scales, beyond that obvious admonishment, it remains to be seen what practical or predictive value it has. We are left, then, with a daunting research challenge, but our methods and conceptualizations are better suited than ever to the research and management tasks we have set for ourselves. While we have traditionally paid poor attention to the match between experimental scale and ecological reality or management application, we now have tools that can help us incorporate multiple levels of data and variation. Carnivores have helped us appreciate this need and will continue to test our abilities. Similarly, what can be interpreted as weaknesses of past carnivore research—such as the potential bias of human impacts because much of our data comes from managed systems—can be used as an opportunity to meet new societal concerns about how to mitigate deleterious human actions.

Given their drawbacks as convenient study organisms, what are the benefits of studying carnivores? First of all, for their own sake, carnivores warrant attention. But more important, they challenge conservation biology. It is reasonable to argue that if we can preserve carnivores and narrowly distributed endemic species, we can preserve ecosystems. The question is, how do we modify existing conservation approaches to apply to carnivores?

First, in lieu of models based on population dynamics (which require that sampling error is small relative to environmental variation), carnivore management will require more attention to patterns of habitat and landscape use. Rather than tracking the "ups and downs" of carnivore populations, it makes more sense to focus on the condition and reproductive performance of individual animals as a function of how they use habitats. In this manner it should be possible to relate human perturbations to changes in ecosystems resulting from human activities. Thus carnivores will be the ideal case in which to proceed from individual behavior and demography up to ecosystem attributes, skipping the population level of analysis.

A second approach that needs more attention with carnivores is the study of interactions between carnivores and other species. Much of North American conservation biology proceeds one species at time, with PVA being the manifestation of this single-species dogma. But carnivores clearly interact strongly with prey and with other carnivore species. It is impossible to assess the fate and impact of reintroduced wolves in Yellowstone without also considering the density and distribution of elk as well as competing carnivores.

Finally, carnivore models have to build in interactions with humans explicitly. This means the behavior of humans, and the behavioral interactions between humans and carnivores, demand explicit modeling. This is especially obvious with studies of wolf reintroductions or wolf control. Wolf reintroductions may fail because of human behavior, as opposed to any ecosystem inadequacy. Similarly, the payoff from carnivore control depends on the hunting behavior of humans. While most conservation models include the effects of humans, this is quite different from including human behavior itself. Carnivore conservation should meld carnivore and human behavior together for long-range planning.

In summary, carnivore conservation will be distinguished by three unique foci: proceeding from individual behavior and demography to ecosystem levels, emphasizing interspecific interactions, and explicitly considering human behavior (not just human effects). The advances made along these three fronts will not be restricted to carnivores alone—indeed, for many species we need to pay more attention to these approaches, in lieu of classical PVAs, habitat association regressions, and abstract landscape ecology.

From this review and the preceding chapters we can see how many carnivore species are predisposed as focal points for ecosystem studies. This is not only because they are responsive and sensitive to ecological change but because the effects are manifested as observable forms of distinguishable variation at the individual and population level, and as strong community interactions. In a sense, the historical coupling of carnivore research and management, which involves large scales and direct human intervention, also cultivates collaborations with academic researchers and predisposes many carnivores as focal species for adaptive management, particularly in forested systems (Boyce and Haney 1997, Perry 1998). This is despite academic reward systems that are biased in favor of narrow disciplinary research and prevailing selection against research on complex, variable ecosystems that require long-term study and interdisciplinary collaboration (Gunderson et al. 1995, Carpenter 1998, Rapport et al. 1998). Many people believe the age of adaptive management is waiting for changes in institutional structures that will facilitate large-scale, long-term learning by doing. We believe carnivore study and management has always inspired productive and large thinking about ecological systems. For example, Soulé and colleagues have identified rewilding—the scientific argument for restoring big wilderness based on the regulatory roles of large predators—as the most recent and compelling strategy in the modern conservation movement (e.g., Soulé and Noss 1998, Soulé 1999). Rewilding extends the previous historical themes of biological conservation, ecosystem representation, and island biogeography by recognizing three fundamental features: core reserves, connectivity (corridors), and carnivores as keystone species.

Carnivores will continue to be tricky experimental subjects, but the greatest obstacles may reside more in our organizational and sociopolitical systems. As we send this book to press, the fate of Yellowstone's reintroduced wolves is being considered in court. Livestock interests and conservationists have sued over the reintroduction, and a judge has called for removal of all wolves, though any action has been suspended pending appeals. Clearly, carnivore conservation rests on both reliable scientific information and informed public consent. Today's managers must navigate troubled political waters, and all of us concerned about the conservation of biological diversity have to redouble our efforts (Clark et al. 1996, Clark 1998, Clark et al. in press a,b). It is instructive to keep in mind that what we ecologists take for granted as ground truth and of the highest value to society is not held so dear by others. As Flowers (1998:152) has pointed out, "Even as the twenty-first century nears, the United

States is the only developed country where a great many people who consider themselves educated dismiss Darwinian thought."

Literature Cited: Abrams, P. A., B. A. Menge, G. G. Mittelbach, D. Spiller, and P. Yodzis. 1996. The role of indirect effects in food webs. Pp. 371–95 *in* G. A. Polis and K. O. Winemiller, eds., Food webs: Integration of patterns and dynamics. Chapman and Hall, New York.

Adler, G. H., and R. Levins. 1994. The island syndrome in rodent populations. Quarterly Review of Biology 69(4):473–90.

Aebischer, N. J., P. A. Robertson, and R. E. Kenward. 1993. Compositional analysis of habitat use from animal radio-tracking data. Ecology 74(5):1313–25.

Agresti, A. 1990. Categorical data analysis. John Wiley and Sons, New York.

Alaska Department of Fish and Game. 1992. Strategic wolf management plan for Alaska. Alaska's Wolves. Alaska's Wildlife (Jan.–Feb. suppl.):S7–S14.

Alcover, J. A., and M. McMinn. 1994. Predators of vertebrates on islands. BioScience 44(1):12–18.

Alldredge, J. R., and J. T. Ratti. 1992. Further comparison of some statistical techniques for analysis of resource selection. Journal of Wildlife Management 56:1–9.

Allen, S. H. 1984. Some aspects of reproductive performance in female red fox in North Dakota. Journal of Mammalogy 65:246–55.

Alroy, J. 1995. Does climate or competition control mammalian diversity? Journal of Vertebrate Paleontology 15(3 suppl.):16A.

Alvarez, K. 1993. Twilight of the panther: Biology, bureaucracy, and failure in an endangered species program. Myakka River Publishing, Sarasota, Florida.

Andelt, W. F. 1985. Behavioral ecology of coyotes in south Texas. Wildlife Monographs 94:1–45.

Anderson, J. P., D. W. Stephens, and S. R. Dunbar. 1997. Saltatory search: A theoretical analysis. Behavioral Ecology 8(3):307–17.

Anderson, P. K. 1970. Ecological structure and gene flow in small mammals. Symposia of the Zoological Society of London 26:299–325.

Anderson, W. W. 1983. Achieving synthesis in population biology. Pp. 189–204 *in* C. R. King and P. S. Dawson, eds., Population biology: Retrospect and prospect. Columbia University Press, New York.

Andren, H. 1995. Effects of landscape composition on predation rates at habitat edges. Pp. 225–55 *in* L. Hansson, L. Fahrig, and G. Merriam, eds., IALE Studies in Landscape Ecology, vol. 2. Mosaic landscapes and ecological processes. Chapman and Hall, New York.

Andrews, P. 1995. Mammals as paleoecological indicators. Acta Zoologica Cracoviensia 38(1):59–72.

Angelstam, P. K. 1996. The ghost of forest past–natural disturbance regimes as a basis for reconstruction of biologically diverse forests in Europe. Pp. 287–337 *in* R. M. DeGraaf and R. I. Miller, eds., Conservation of faunal diversity in forested landscapes. Chapman and Hall, New York.

Angelstam, P. K., V. M. Anufriev, L. Balciauskas, A. K. Blagovidov, S. O. Borgegard, S. J. Hodge, P. Majewski, S. V. Ponomarenko, E. A. Shvarts, A. A. Tishkov, L. Tomialojc, and T. Weslowski. 1997. Biodiversity and sustainable forestry in European forests: How East and West can learn from each other. Wildlife Society Bulletin 25(1):38–48.

Angelstam, P. K., and B. Pettersson. 1997. Principles of present Swedish forest biodiversity management. Pp. 191–203 *in* L. Hansson, ed., Boreal ecosystems and landscapes: Structures, processes and conservation of biodiversity. Munksgaard, Copenhagen.

Anholt, B. R. 1997. How should we test for the role of behaviour in population dynamics? Evolutionary Ecology 11(6):633–40.

Anholt, B. R., and E. E. Werner. 1998. Predictable changes in predation mortality as a consequence of changes in food availability and predation risk. Evolutionary Ecology 12(6):729–38.

Armitage, K. B. 1977. Social variety in the yellow-bellied marmot: A population-behavioural system. Animal Behavior 25:585–93.

——. 1986. Individuality, social behavior, and reproductive success in yellow-bellied marmots. Ecology 67:1186–93.

Arthur, S. M., B. F. J. Manly, L. L. McDonald, and G. W. Garner. 1996. Assessing habitat selection when availability changes. Ecology 77(1):215–27.

Asa, C. S., and L. D. Mech. 1995. A review of the sensory organs in wolves and their importance to life history. Pp. 287–91 *in* L. N. Carbyn, S. H. Fritts, and D. R. Seip, eds., Ecology and conservation of wolves in a changing world. Canadian Circumpolar Institute, University of Alberta, Edmonton.

Asa, C. S., and C. Valdespino. 1998. Canid reproductive biology: An integration of proximate mechanisms and ultimate causes. American Zoologist 38(1):251–59.

Aspinall, R., and N. Veitch. 1993. Habitat mapping from satellite imagery and wildlife data using a Bayesian modeling procedure in a GIS. Photogrammetric Engineering and Remote Sensing 59:537–43.

Badgley, C., D. Fox, and L. Farber. 1994. Ecological structure of North American mammalian faunas in relation to climate and geography. Journal of Vertebrate Paleontology 14(3 suppl.):15A.

Bailey, T. N. 1993. The African leopard: Ecology and behavior of a solitary felid. Columbia University Press, New York.

Balharrry, D. 1993. Social organization in martens: An inflexible system? Pp. 321–45 *in* N. Dunstone and M. L. Gorman, eds., Mammals as predators. Symposia Zoological Society of London, no. 65. Clarendon Press, Oxford.

Ballard, W. B., and V. Van Ballenberghe. 1998. Predator-prey relationships. Pp. 247–73 *in* A. W. Franzmann and C. C. Schwartz, eds., Ecology and management of the North American moose. Smithsonian Institution Press, Washington.

Ballou, J. D., M. Gilpin, and T. J. Foose, eds. 1995. Population management for survival and recovery: Analytical methods and strategies in small population conservation. Columbia University Press, New York.

Balmford, A. 1996. Extinction filters and current resilience: The significance of past selection pressures for conservation biology. Trends in Ecology and Evolution 11(5):193–96.

Banci, V. A. 1989. A fisher management strategy for British Columbia. Wildlife Bulletin no. B-63, ISSN 0829-9560. British Columbia Ministry of Environment, Victoria, Canada.

Banks, P. B., C. R. Dickman, and A. E. Newsome. 1998. Ecological costs of feral predator control: Foxes and rabbits. Journal of Wildlife Management 62(2):766–72.

Bateson, P. P. G., and D. C. Turner, eds. 1988. The domestic cat: The biology of its behaviour. Cambridge University Press, Cambridge.

Beck, M. W. 1996. On discerning the cause of late Pleistocene megafaunal extinctions. Paleobiology 22(1):91–103.

Becker, E. F. 1991. A terrestrial furbearer estimator based on probability sampling. Journal of Wildlife Management 55:730–37.

Becker, E. F., M. A. Spindler, and T. O. Osborne. 1998. A population estimator based on network sampling of tracks in the snow. Journal of Wildlife Management 62(3):968–77.

Beecham, J. J., and J. Rohlman. 1994. A shadow in the forest: Idaho's black bear. Idaho Department of Fish and Game, Boise, and University of Idaho Press, Moscow.

Behera, N., and V. Nanjundiah. 1996. The consequences of phenotypic plasticity in cyclically varying environments: A genetic algorithm study. Journal of Theoretical Biology 178(2):135–44.

Behrensmeyer, A. K., J. D. Damuth, W. A. DiMichele, R. Potts, H.-D. Sues, and S. L. Wing, eds. 1992. Terrestrial ecosystems through time: Evolutionary paleoecology of terrestrial plants and animals. University of Chicago Press, Chicago.

Beier, P. 1993. Determining minimum habitat areas and habitat corridors for cougars. Conservation Biology 7(1):94–108.

Beissinger, S. R., and M. I. Westphal. 1998. On the use of demographic models of population viability in endangered species management. Journal of Wildlife Management 62(3):821–41.

Bekoff, M. 1977. Mammalian dispersal and the ontogeny of individual behavioral phenotypes. American Naturalist 111:715–32.

———, ed. 1978. Coyotes: Biology, behavior, and management. Academic Press, New York.

Bekoff, M., T. J. Daniels, and J. L. Gittleman. 1984. Life history patterns and the comparative social ecology of carnivores. Annual Review of Ecology and Systematics 15:191–232.

Bekoff, M., J. Diamond, and J. B. Mitton. 1981. Life history patterns and sociality in canids: Body size, reproduction, and behavior. Oecologia 50:386–90.

Bekoff, M., and M. C. Wells. 1986. Social ecology and behavior of coyotes. Advances in the Study of Behaviour 16:251–338.

Belew, R. K., and M. Mitchell, eds. 1996. Adaptive individuals in evolving populations: Models and algorithms. Proceedings of the Santa Fe Institute workshop. Plastic Individuals in Evolving Populations: Models and Algorithms, July 11–15, 1993, Santa Fe, New Mexico. Addison-Wesley, Reading, Massachusetts.

Bellehumeur, C., and P. Legendre. 1998. Multiscale sources of variation in ecological variables: Modeling spatial dispersion, elaborating sampling designs. Landscape Ecology 13(1):15–25.

Bender, D. J., T. A. Contreras, and L. Fahrig. 1998. Habitat loss and population decline: A meta-analysis of the patch size effect. Ecology 79(2):517–33.

Berger, J. 1979. Social ontogeny and behavioural diversity: Consequences for bighorn sheep, *Ovis canadensis,* inhabiting desert and mountain environments. Journal of Zoology 118:252–66.

———. 1991. Greater Yellowstone's native ungulates: Myths and realities. Conservation Biology 5:353–63.

Bergstrom, C. T., and P. Godfrey-Smith. 1998. On the evolution of behavioral heterogeneity in individuals and populations. Biology and Philosophy 13(2):205–31.

Beringer, J., S. G. Seibert, S. Reagan, A. J. Brody, M. R. Pelton, and L. D. Vangilder. 1998. The influence of a small sanctuary on survival rates of black bears in North Carolina. Journal of Wildlife Management 62(2):727–34.

Bertram, B. C. R. 1978. Pride of lions. Charles Scribner's Sons, New York.

Bibikov, D. I., ed. 1985. Volk (The Wolf). Izdatelstvo Nauka, Moscow (in Russian).

Bissonette, J. A., and S. Broekhuizen. 1995. *Martes* populations as indicators of habitat spatial patterns: The need for a multiscale approach. Pp. 95–121 *in* W. Z. Lidicker, ed., Landscape approaches in mammalian ecology and conservation. University of Minnesota Press, Minneapolis.

Bissonette, J. A., R. J. Fredrickson, and B. J. Tucker. 1989. American marten: A case for landscape-level management. Transactions of the North American Wildlife and Natural Resources Conference 54:89–100.

Bixby, K. 1991. Predator conservation. Pp. 199–213 *in* K. A. Kohm, ed., Balancing on the brink of extinction: The Endangered Species Act and lessons for the future. Island Press, Washington, D.C.

Blackburn, T. M., and K. J. Gaston. 1998. Some methodological issues in macroecology. American Naturalist 151(1):68–83.

Boman, M. 1995. Estimating costs and genetic benefits of various sizes of predator populations: The case of bear, wolf, wolverine, and lynx in Sweden. Journal of Environmental Management 43(4):349–57.

Bond, W. J. 1993. Keystone species. Pp. 237–53 *in* E. D. Schulze and H. A. Mooney, eds., Biodiversity and ecosystem function. Springer-Verlag, Berlin.

Boone, R. B., and M. L. Hunter, Jr. 1996. Using diffusion models to simulate the effects of land use on grizzly bear dispersal in the Rocky Mountains. Landscape Ecology 11(1):51–64.

Boonstra, R., C. J. Krebs, and N. C. Stenseth. 1998. Population cycles in small mammals: The problem of explaining the low phase. Ecology 79(5):1479–88.

Borcard, D., P. Legendre, and P. Drapeau. 1992. Partialling out the spatial component of ecological variation. Ecology 73:1045–55.

Borralho, R. 1995. Predation, hunting, and conservation. Revista de Ciencias Agrarias 18(2):35–46.

Bothma, J. P. 1998. Carnivore ecology in arid lands. Springer-Verlag, New York.

Bouskila, A., M. E. Robinson, B. D. Roitberg, and B. Tenhumberg. 1998. Life-history decisions under predation risk: Importance of a game perspective. Evolutionary Ecology 12(6):701–15.

Boutin, S. A. 1990. Food supplementation experiments with terrestrial vertebrates: Patterns, problems, and the future. Canadian Journal of Zoology 68:203–20.

———. 1992. Predation and moose population dynamics: A critique. Journal of Wildlife Management 56(1):116–27

———. 1995. Testing predator-prey theory by studying fluctuating populations of small mammals. Wildlife Research 22(1):89–100.

Boutin, S. A., C. J. Krebs, R. Boonstra, M. R. T. Dale, S. J. Hannon, K. Martin, A. R. E. Sinclair, J. N. M. Smith, R. Turkington, M. Blower, A. Byrom, F. I. Doyle, C. Doyle, D. Hik, L. Hofer, A. Hubbs, T. Karels, D. L. Murray, V. Nams, M. O'Donoghue, C. Rohner, and S. Schweiger. 1995. Population changes of the vertebrate community during a snowshoe hare cycle in Canada's boreal forest. Oikos 74(1):69–80.

Bowen, W. D. 1981. Variation in coyote social organization: The influence of prey size. Canadian Journal of Zoology 59:639–52.

———. 1982. Home range and spatial organization of coyotes in Jasper National Park, Alberta. Journal of Wildlife Management 46:201–16.

Bowers, M. A. 1997. Mammalian landscape ecology. Journal of Mammalogy 78(4):997–98.

Bowers, M. A., and S. F. Matter. 1997. Landscape ecology of mammals: Relationship between density and patch size. Journal of Mammalogy 78(4):999–1013.

Boyce, M. S. 1981. Beaver life-history responses to exploitation. Journal of Applied Ecology 18:749–53.

———. 1992. Population viability analysis. Annual Review of Ecology and Systematics 23:481–506.

———. 1993. Population viability analysis: Adaptive management for threatened and endangered species. Transactions of the North American Wildlife and Natural Resources Conference 58:520–27.

———, ed. 1988. Evolution of life histories of mammals: Theory and pattern. Yale University Press, New Haven.

Boyce, M. S., and A. W. Haney, eds. 1997. Ecosystem management: Applications for sustainable forest and wildlife resources. Yale University Press, New Haven.

Bronson, F. H. 1989. Mammalian reproductive biology. University of Chicago Press, Chicago.

Brooks, R. P. 1997. Improving habitat suitability index models. Wildlife Society Bulletin 25(1):163–67.

Brown, A. L., and J. A. Litvaitis. 1995. Habitat features associated with predation of New England cottontails: What scale is appropriate? Canadian Journal of Zoology 73(6):1005–11

Brown, D. E. 1985. The grizzly in the Southwest: Documentary of an extinction. University of Oklahoma Press, Norman.

Brown, J. H. 1986. Two decades of interaction between the MacArthur-Wilson model and the complexities of mammalian distributions. Biological Journal of the Linnean Society 28:231–51.

———. 1995. Macroecology. University of Chicago Press, Chicago.

Brown, J. H., and E. J. Heske. 1990. Control of a desert-grassland transition by a keystone rodent guild. Science 250:1705–7.

Brown, J. H., G. C. Stevens, and D. M. Kaufman. 1996. The geographic range: Size, shape, boundaries, and internal structure. Annual Review of Ecology and Systematics 27:597–623.

Brussard, P. F., ed. 1991. The Greater Yellowstone Ecosystem. Conservation Biology 5(3 suppl.):335-422.

Bruton, M. N., ed. 1989. Alternative life-history styles of animals. Kluwer Academic Publishers, Boston.

Buckland, S. T., D. R. Anderson, K. P. Burnham, and J. L. Laake. 1993. Distance sampling: Estimating abundance of biological populations. Chapman and Hall, London.

Bunnell, F. L., and D. E. N. Tait. 1985. Mortality rates of North American bears. Arctic 38:316–23.

Burney, D. A. 1993. Recent animal extinctions: Recipes for disaster. American Scientist 81(6):530–41.

Burrows, R., H. Hofer, and M. L. East. 1994. Demography, extinction, and intervention in a small population: The case of the Serengeti wild dogs. Proceedings of the Royal Society of London Series B Biological Sciences 256(1347):281–92.

Burton, R. 1979. Carnivores of Europe. Batsford, London.

Buskirk, S. W., and S. L. Lindstedt. 1989. Sex biases in trapped samples of Mustelidae. Journal of Mammalogy 70(1):88–97.

Buskirk, S. W., A. S. Harestad, M. G. Raphael, and R. A. Powell, eds. 1994. Martens, sables, and fishers: Biology and conservation. Cornell University Press, Ithaca.

Buskirk, S. W., and R. A. Powell. 1994. Habitat ecology of fishers and American martens. Pp. 283–96 *in* S. W. Buskirk, A. S. Harestad, M. G. Raphael, and R. A. Powell, eds., Martens, sables, and fishers: Biology and conservation. Cornell University Press, Ithaca.

Buss, L. W. 1988. The evolution of individuality. Princeton University Press, Princeton.

Byers, J. A. 1997. American pronghorn: Social adaptations and the ghosts of predators past. University of Chicago Press, Chicago.

Camenzind, F. J. 1978. Behavioral ecology of coyotes on the National Elk Refuge, Jackson, Wyoming. Pp. 267–96 *in* M. Bekoff, ed., Coyotes: Biology, behavior, and management. Academic Press, New York.

Carbyn, L. N., S. H. Fritts, and D. R. Seip, eds. 1995. Ecology and conservation of wolves in a changing world. Canadian Circumpolar Institute, University of Alberta, Edmonton.

Caro, T. M. 1994. Cheetahs of the Serengeti Plains: Group living in an asocial species. University of Chicago Press, Chicago.

Caro, T. M., and P. Bateson. 1986. Organization and ontogeny of alternative tactics. Animal Behavior 34:1483–99.

Caro, T. M., and S. M. Durant. 1995. The importance of behavioral ecology for conservation biology: Examples from Serengeti carnivores. Pp. 451–72 *in* A. R. E. Sinclair and P. Arcese, eds., Serengeti II: Dynamics, management, and conservation of an ecosystem. University of Chicago Press, Chicago.

Carpenter, S. R. 1998. Keystone species and academic-agency collaboration. Conservation Ecology [online] 2(1):R2. http://www.consecol.org/vol2/iss1/resp2

Carpenter, S. R., and K. L. Cottingham. 1997. Resilience and restoration of lakes. Conservation Ecology [online] 1(1):2. http://www.consecol.org/vol1/iss1/art1

Carpenter, S. R., T. M. Frost, J. F. Kitchell, and T. K. Kratz. 1993. Species dynamics and global environmental change: A perspective from ecosystem experiments. Pp. 267–79 *in* P. M. Kareiva, J. G. Kingsolver, and R. B. Huey, eds., Biotic interactions and global change. Sinauer Associates, Sunderland, Massachusetts.

Caswell, H. 1983. Phenotypic plasticity in life-history traits: Demographic effects and evolutionary consequences. American Zoologist 23:35–46.

Caughley, G. 1981. Overpopulation. Pp. 7–19 *in* P. A. Jewell, S. Holt, and D. Hart, eds., Problems in management of locally abundant wild mammals. Academic Press, New York.

———. 1994. Directions in conservation biology. Journal of Animal Ecology 63(2):215–44.

Caughley, G., and A. Gunn. 1993. Dynamics of large herbivores in deserts: Kangaroos and caribou. Oikos 67(1):47–55.

Caughley, G., and A. R. E. Sinclair. 1994. Wildlife ecology and management. Blackwell Science, New York.

Cavallini, P. 1996. Variation in the social system of the red fox. Ethology Ecology and Evolution 8(4):323–42.

Ceballos, G., and J. H. Brown. 1995. Global patterns of mammalian diversity, endemism, and endangerment. Conservation Biology 9(3):559–68.

Chapman, J. A., and G. A. Feldhamer, eds. 1982. Wild mammals of North America. Johns Hopkins University Press, Baltimore.

Chapman, J. A., and D. Pursley, eds. 1981. Worldwide furbearer conference proceedings. 3 vols. International Association of Fish and Wildlife Agencies, Washington.

Chepko-Sade, B. D., and Z. T. Halpin, eds. 1987. Mammalian dispersal patterns: The effects of social structure on population genetics. University of Chicago Press, Chicago.

Chesser, R. K. 1998. Relativity of behavioral interactions in socially structured populations. Journal of Mammalogy 79(3):713–24.

Chesson, P., and M. Rosenzweig. 1991. Behavior heterogeneity and the dynamics of interacting species. Ecology 72(4):1187–95.

Chiba, S. 1998. A mathematical model for long-term patterns of evolution: Effects of environmental stability and instability on macroevolutionary patterns and mass extinctions. Paleobiology 24(3):336–48.

Clark, A. B., and T. J. Ehlinger. 1987. Pattern and adaptation in individual behavioral differences. Perspectives in Ethology 7:1–47.

Clark, J. D., J. E. Dunn, and K. G. Smith. 1993. A multivariate model of female black bear habitat use for a geographic information system. Journal of Wildlife Management 57(3):519–26.

Clark, T. W. 1998. Interdisciplinary problem-solving: Next steps in the Greater Yellowstone Ecosystem. The theory and practice of interdisciplinary work, June 11–13, Foundation for Strategic Environmental Research (MISTRA) and Council for Planning and Co-ordination of Research (FRN), Stockholm, Sweden.

Clark, T. W., P. C. Paquet, and A. P. Curlee, eds. 1996. Large carnivore conservation in the Rocky Mountains of the United States and Canada. Conservation Biology (Special Issue) 10(4):936–1058.

Clark, T. W., R. P. Reading, D. Mattson, and B. J. Miller. In press (a). An interdisciplinary framework for effective carnivore conservation. Zoological Society of London, Cambridge University Press, Cambridge.

Clark, T. W., R. P. Reading, and B. J. Miller. In press (b). Resolving endangered species conflicts using an interdisciplinary "informational" approach. *In* R. P. Reading and B. J. Miller, eds., Endangered animals: Conflicting issues. Greenwood Publishing Group, Westport, Connecticut.

Colding, J., and C. Folke. 1997. The relations among threatened species, their protection, and taboos. Conservation Ecology [online] 1(1):6. http://www.consecol.org/vol1/iss1/art6

Cole, F. R., D. M. Reeder, and D. E. Wilson. 1994. A synopsis of distribution patterns and the conservation of mammal species. Journal of Mammalogy 75(2):266–76.

Concannon, P. W., ed. 1997. Reproduction in dogs, cats, and exotic carnivores. Journal of Reproduction and Fertility 51(suppl.):1–372.

Connolly, G. E. 1978a. Predators and predator control. Pp. 369–94 *in* J. L. Schmidt and D. L. Gilbert, eds., Big game of North America. Stackpole Books, Harrisburg, Pennsylvania.

———. 1978b. Predator control and coyote populations: A review of simulation models. Pp. 327–45 *in* M. Bekoff, ed., Coyotes: Biology, behavior, and management. Academic Press, New York.

Conroy, M. J., Y. Cohen, F. C. James, Y. G. Matsinos, and B. A. Maurer. 1995. Parameter estimation, reliability, and model improvement for spatially explicit models of animal populations. Ecological Applications 5:17–19.

Conroy, M. J., J. D. Nichols, and E. R. Asanza. 1997. Contemporary quantitative methods to understand and manage animal populations and communities. Interciencia 22(5):247–58.

Conroy, M. J., and B. R. Noon. 1996. Mapping of species richness for conservation of biological diversity: Conceptual and methodological issues. Ecological Applications 6(3):763–73.

Côté, I. M., and W. J. Sutherland. 1997. The effectiveness of removing predators to protect bird populations. Conservation Biology 11:395–405.

Coulson, T. N., S. D. Albon, J. M. Pemberton, J. Slate, F. E. Guiness, and T. H. Clutton-Brock. 1998. Genotype by environment interactions in winter survival in red deer. Journal of Animal Ecology 67(3):434–45.

Cox, E. W., R. A. Garrott, and J. R. Cary. 1997. Effect of supplemental cover on survival of snowshoe hares and cottontail rabbits in patchy habitat. Canadian Journal of Zoology 75(9):1357–63.

Craighead, F. C. 1979. Track of the grizzly. Sierra Club Books, San Francisco.

Craighead, J. J., J. S. Sumner, and J. A. Mitchell. 1995. The grizzly bears of Yellowstone: Their ecology in the Yellowstone ecosystem, 1959–1992. Island Press, Washington, D.C.

Crawley, M. J. 1983. Herbivory: The dynamics of animal-plant interactions. Studies in Ecology, vol. 10. University of California Press, Berkeley.

Creel, S. R., and N. M. Creel. 1996. Limitation of African wild dogs by competition with larger carnivores. Conservation Biology 10(2):526–38.

Creel, S. R., and D. Macdonald. 1995. Sociality, group size, and reproductive suppression among carnivores. Advances in the Study of Behavior 24:203–57.

Crews, D. 1998. On the organization of individual differences in sexual behavior. American Zoologist 38(1):118–32.

Curio, E. 1976. The ethology of predation. Springer-Verlag, New York.

Cutler, A. H. 1991. Nested faunas and extinction in fragmented habitats. Conservation Biology 5(4):496–505.

———. 1994. Nested biotas and biological conservation: Metrics, mechanisms, and meaning of nestedness. Landscape and Urban Planning 28(1):73–82.

Daily, G., ed. 1997. Nature's services: Societal dependence on natural ecosystems. Island Press, Washington, D.C.

Danell, K., T. Willebrand, and L. Baskin. 1998. Mammalian herbivores in the boreal forests: Their numerical fluctuations and use by man. Conservation Ecology [online] 2(2):9. http://www.consecol.org/vol2/iss2/art9

Dayan, T., and D. Simberloff. 1998. Size patterns among competitors: Ecological character displacement and character release in mammals, with special reference to island populations. Mammal Review 28(3):99–124.

Deangelis, D. L., L. J. Gross, M. A. Huston, W. F. Wolff, D. M. Fleming, E. J. Comiskey, and S. M. Sylvester. 1998. Landscape modeling for Everglades ecosystem restoration. Ecosystems 1(1):64–75.

Debrot, S. 1984. Dynamique du renouvellement et structure d'age d'une population d'hermines *(Mustela erminea)*. Review of Ecology (Terre Vie) 39:77–88.

De Jong, G. 1995. Phenotypic plasticity as a product of selection in a variable environment. American Naturalist 145(4):493–512.

Delcourt, H. R., and P. A. Delcourt. 1988. Quaternary landscape ecology: Relevant scales in space and time. Landscape Ecology 2(1):23–44.

De Leo, G. A., and S. Levin. 1997. The multifaceted aspects of ecosystem integrity. Conservation Ecology [online] 1(1):3. http://www.consecol.org/vol1/iss1/art3

Dennis, B., P. L. Munholland, and J. M. Scott. 1991. Estimation of growth and extinction parameters for endangered species. Ecological Monographs 61(2):115–43.

Despain, D. G. 1991. Yellowstone vegetation: Consequences of environment and history in a natural setting. Roberts Rinehart, Boulder, Colorado.

Dewitt, T. J., A. Sih, and D. S. Wilson. 1998. Costs and limits of phenotypic plasticity. Trends in Ecology and Evolution 13(2):77–81.

Dieckmann, U. 1997. Can adaptive dynamics invade? Trends in Ecology and Evolution 12(4):128–31.

Dieckmann, U., P. Marrow, and R. Law. 1995. Evolutionary cycling in predator-prey interactions: Population dynamics and the Red Queen. Journal of Theoretical Biology 176(1):91–102.

Diffendorfer, J. 1998. Testing models of source-sink dynamics and balanced dispersal. Oikos 81(3):417–33.

Dobson, F. S. 1998. Social structure and gene dynamics in mammals. Journal of Mammalogy 79(3):667–70.

Dobson, F. S., and J. Yu. 1993. Rarity in neotropical forest mammals revisited. Conservation Biology 7(3):586–91.

Dobson, M. 1998. Mammal distributions in the western Mediterranean: The role of human intervention. Mammal Review 28(2):77–88.

Drake, J. A., G. R. Huxel, and C. L. Hewitt. 1996. Microcosms as models for generating and testing community theory. Ecology 77(3):670–77.

Duarte, L. C., J. L. Boldrini, and S. F. D. Reis. 1998. Scaling phenomena and ecological interactions in space: Cutting to the core. Trends in Ecology and Evolution 13(5):176–77.

Duckler, G. L., and B. Van Valkenburgh. 1998. Exploring the health of Late Pleistocene mammals: The use of Harris lines. Journal of Vertebrate Paleontology 18(1):180–88.

Dufrene, M., and P. Legendre. 1997. Species assemblages and indicator species: The need for a flexible asymmetrical approach. Ecological Monographs 67(3):345–66.

Dunning, J. B., B. J. Danielson, B. R. Noon, T. L. Root, R. L. Lamberson, and E. E. Stevens. 1995. Spatially explicit population models: Current forms and future uses. Ecological Applications 5:3–11.

Dunning, J. B., B. J. Danielson, and H. R. Pulliam. 1992. Ecological processes that affect populations in complex landscapes. Oikos 65:169–75.

Dunstone, N. 1993. The mink. T. and A. D. Poyser, London.

Dunstone, N., and M. L. Gorman, eds. 1993. Mammals as predators. Symposia Zoological Society of London no. 65. Clarendon Press, Oxford.

Durant, S. M. 1998. Competition refuges and coexistence: An example from Serengeti carnivores. Journal of Animal Ecology 67(3):370–86.

Eberhardt, L. E., and A. B. Sargeant. 1977. Mink predation on prairie marshes during the waterfowl breeding season. Pp. 33–43 *in* R. L. Phillips and C. Jonkel, eds., Proceedings of the 1975 Predator Symposium. University of Montana, Missoula.

Ehrlich, P. R. 1996. Conservation in temperate forests: What do we need to know and do? Forest Ecology and Management 85(1–3):9–19.

Eisenberg, J. F. 1981. The mammalian radiations: An analysis of trends in evolution, adaptation, and behavior. University of Chicago Press, Chicago.

———. 1986. Life history strategies of the felidae: Variations on a common theme. Pp. 293–303 *in* S. D. Miller and D. D. Everett, eds., Cats of the world: Biology, conservation, and management. National Wildlife Federation, Washington.

———. 1989. An introduction to the Carnivora. Pp. 1–9 *in* J. L. Gittleman, ed., Carnivore behavior, ecology, and evolution. Cornell University Press, Ithaca.

Eisenberg, J. F., and J. O. Wolff. 1994. Behavior. Pp. 398–420 *in* E. C. Birney and J. R. Choate, eds., Seventy-five years of mammalogy (1919–1994). Special Publication no. 11, American Society of Mammalogists.

Eisthen, H. L. 1997. Evolution of vertebrate olfactory systems. Brain Behavior and Evolution 50(4):222–33.

Elias, S. A. 1995. Ice-age history of national parks in the Rocky Mountains. Smithsonian Institution Press, Washington.

Engen, T. 1982. The perception of odors. Academic Press, New York.

———. 1991. Odor sensation and memory. Praeger Publishers, New York.

Englund, G. 1997. Importance of spatial scale and prey movements in predator caging experiments. Ecology 78(8):2316–25.

Erlinge, S., G. Göransson, L. Hansson, G. Högstedt, O. Liberg, I. N. Nilsson, T. Nilsson, T. Von Schantz, and M. Sylvén. 1983. Predation as a regulating factor on small rodent populations in southern Sweden. Oikos 40: 36–52.

Erlinge, S., G. Göransson, G. Högstedt, G. Jansson, O. Liberg, J. Loman, I. N. Nilsson, T. Von Schantz, and M. Sylvén. 1984. Can vertebrate predators regulate their prey? American Naturalist 123:125–33.

———. 1988. More thoughts on vertebrate predator regulation of prey. American Naturalist 132:148–54.

Errington, P. L. 1946. Predation and vertebrate populations. Quarterly Review of Biology 21:44–177.

———. 1967. Of predation and life. Iowa State University Press, Ames.

Erwin, D. H. 1998. The end and the beginning: Recoveries from mass extinctions. Trends in Ecology and Evolution 13(9):344–49.

Estes, J. A. 1995. Top-level carnivores and ecosystem effects: Questions and approaches. Pp. 151–58 *in* C. G. Jones and J. H. Lawton, eds., Linking species and ecosystems. Chapman and Hall, New York.

———. 1996. Predators and ecosystems management. Wildlife Society Bulletin 24(3):390–96.

Ewer, R. F. 1973. The carnivores. Cornell University Press, Ithaca.

Fahrig, L. 1997. Relative effects of habitat loss and fragmentation on population extinction. Journal of Wildlife Management 61(3):603–10.

Feder, M. E., and G. V. Lauder, eds. 1986. Predator-prey relationships: Perspectives and approaches from the study of lower vertebrates. University of Chicago Press, Chicago.

FEMAT. 1993. Forest ecosystem management: An ecological, economic, and social assessment. Report of the Forest Ecosystem Management Assessment Team. J. W. Thomas, team leader. United States Department of Agriculture Forest Service, United States Department of the Interior Fish and Wildlife Service, National Park Service, Bureau of Land Management, Environmental Protection Agency, and U.S. Department of Commerce, NOAA and NMFS. July 1993.

Ferguson, S. H., J. A. Virgl, and S. Lariviere. 1996. Evolution of delayed implantation and associated grade shift in life history traits of North American carnivores. Ecoscience 3(1):7–17.

Finch, C. E., and M. R. Rose. 1995. Hormones and the physiological architecture of life history evolution. Quarterly Review of Biology 70(1):1–52.

Flather, C. H., K. R. Wilson, D. J. Dean, and W. C. McComb. 1997. Identifying gaps in conservation networks: Of indicators and uncertainty in geographic-based analyses. Ecological Applications 7(2):531–42.

Flessa, K. W., and S. M. Kidwell. 1995. The quality of the fossil record: Populations, species, and communities. Annual Review of Ecology and Systematics 26:269–99.

Flowers, C. 1998. A science odyssey: 100 years of discovery. William Morrow and Company, New York.

Folke, C., C. S. Holling, and C. Perrings. 1996. Biological diversity, ecosystems, and the human scale. Ecological Applications 6(4):1018–24.

Foran, D. R., K. R. Crooks, and S. C. Minta. 1997a. Species identification from scat: An unambiguous genetic method. Wildlife Society Bulletin 25(4):835–39.

Foran, D. R., S. C. Minta, and K. S. Heinemeyer. 1997b. DNA-based analysis of hair to identify species and individuals for population research and monitoring. Wildlife Society Bulletin 25(4):840–47.

Fowler, C. W. 1987. A review of density dependence in populations of large mammals. Pp. 401–41 in H. H. Genoways, ed., Current mammalogy, vol. 1. Plenum Publishing, New York.

Fox, M. W. 1972. Socio-ecological implications of individual differences in wolf litters: A developmental and evolutionary perspective. Behaviour 41:299–313.

———. 1984. The whistling hunters: Field studies of the Asiatic wild dog *(Cuon alpinus)*. State University of New York Press, Albany.

———, ed. 1975. The wild canids: Their systematics, behavioral ecology, and evolution. Van Nostrand Reinhold, New York.

Frame, G., and L. Frame. 1981. Swift and enduring: Cheetahs and wild dogs of the Serengeti. E. P. Dutton, New York.

Frank, D. A., S. J. McNaughton, and B. F. Tracy. 1998. The ecology of the earth's grazing ecosystems: Profound functional similarities exist between the Serengeti and Yellowstone. BioScience 48(7):514–21.

Frankel, O. H., and M. E. Soulé. 1981. Conservation and evolution. Cambridge University Press, Cambridge.

Fraser, L. H., and P. Keddy. 1997. The role of experimental microcosms in ecological research. Trends in Ecology and Evolution 12(12):478–81.

Fryxell, J. M. 1997. Evolutionary dynamics of habitat use. Evolutionary Ecology 11(6):687–701.

Fuller, T. K., and D. B. Kittredge, Jr. 1996. Conservation of large forest carnivores. Pp. 137–66 in R. M. DeGraaf and R. I. Miller, eds., Conservation of faunal diversity in forested landscapes. Chapman and Hall, New York.

Gardner, S. N., and M. Mangel. 1996. Mark-resight population estimation with imperfect observations. Ecology 77(3):880-84.

Garrettson, P. R., F. C. Rohwer, J. M. Zimmer, B. J. Mense, and N. Dion. 1996. Effects of mammalian predator removal on waterfowl and non-game birds in North Dakota.

Transactions of the North American Wildlife and Natural Resources Conference 61:94–101

Gaston, K. J. 1994. Rarity. Chapman and Hall, New York.

Geffen, E., M. E. Gompper, J. L. Gittleman, H. K. Luh, D. W. MacDonald, and R. K. Wayne. 1996. Size, life-history traits, and social organization in the Canidae: A reevaluation. American Naturalist 147(1):140–60.

Geist. V. 1989. Environmentally guided phenotype plasticity in mammals and some of its consequences to theoretical and applied biology. Pp. 153–76 *in* M. N. Bruton, ed., Alternative life-history styles of animals. Kluwer Academic Publishers, Dordrecht.

Gese, E. M., and R. L. Ruff. 1997. Scent-marking by coyotes, *Canis latrans:* The influence of social and ecological factors. Animal Behaviour 54(5):1155–66.

Gilbert, F., A. Gonzalez, and I. Evans-Freke. 1998. Corridors maintain species richness in the fragmented landscapes of a microecosystem. Proceedings of the Royal Society of London Series B. Biological Sciences 265(1396):577–82.

Gittleman, J. L. 1986. Carnivore life history patterns: Allometric, phylogenetic, and ecological associations. American Naturalist 127:744–71.

———. 1993. Carnivore life histories: A re-analysis in light of new models. Pp. 65–86 *in* N. Dunstone and M. L. Gorman, eds., Mammals as predators. Symposia Zoological Society of London no. 65. Clarendon Press, Oxford.

———, ed. 1989. Carnivore behavior, ecology, and evolution. Cornell University Press, Ithaca.

———, ed. 1996. Carnivore behavior, ecology, and evolution, vol. 2. Cornell University Press, Ithaca.

Glenn, S. M. 1990. Regional analysis of mammal distributions among Canadian parks: Implications for park planning. Canadian Journal of Zoology 68(12):2457–64.

Goldwasser, L., J. Cook, and E. D. Silverman. 1994. The effects of variability of metapopulation dynamics and rates of invasion. Ecology 75(1):40–47.

Golley, F. B. 1993. A history of the ecosystem concept in ecology: More than the sum of the parts. Yale University Press, New Haven.

Gonzalez, A., J. H. Lawton, F. S. Gilbert, T. M. Blackburn, and I. Evans-Freke. 1998. Metapopulation dynamics, abundance, and distribution in a microecosystem. Science 281:2045–47.

Goodman, D. 1979. Regulating reproductive effort in a changing environment. American Naturalist 113:735–748.

Gorman, M. L. 1990. Scent marking strategies in mammals. Revue Suisse de Zoologie 97(1):3–30.

Gosz, J. R. 1993. Ecotone hierarchies. Ecological Applications 3:369–76.

Goszczynski, J. 1997. From hostility to cooperation: The interactions among predators. Wiadomosci Ekologiczne 43(2):117–138.

Graham, R. W., E. L. Lundelius, Jr., M. A. Graham, E. K. Schroeder, R. S. Toomey III, E. Anderson, A. D. Barnosky, J. A. Burns, C. S. Churcher, D. K. Grayson, R. D. Guthrie, C. R. Harington, G. T. Jefferson, L. D. Martin, H. G. McDonald, R. E. Morlan, H. A. Semken, Jr., S. D. Webb, L. Werdelin, and M. C. Wilson. 1996. Spatial response of mammals to late quaternary environmental fluctuations. Science 272:1601–6.

Grant, P. R., ed. 1998. Evolution on islands. Oxford University Press, New York.

Greater Yellowstone Coordinating Committee. 1987. The Greater Yellowstone Area: An aggregation of national park and national forest management plans. U.S. Department of Agriculture Forest Service and U.S. Department of the Interior National Park Service. Distributed by the Yellowstone Association, Yellowstone National Park.

———. 1990. Vision for the future: A framework for coordination in the Greater Yellowstone area. U.S. Department of Agriculture Forest Service and U.S. Department of the Interior National Park Service. Greater Yellowstone Coordinating Committee, Billings, Montana.

———. 1991. A framework for coordination of national parks and national forests in the Greater Yellowstone Area. U.S. Department of Agriculture Forest Service and U.S. Department of the Interior National Park Service. Greater Yellowstone Coordinating Committee, Billings, Montana.

Greenwood, R. J., D. G. Pietruszewski, and R. D. Crawford. 1998. Effects of food supplementation on depredation of duck nests in upland habitat. Wildlife Society Bulletin 26(2):219–26.

Grunbaum, D. 1998. Using spatially explicit models to characterize foraging performance in heterogeneous landscapes. American Naturalist 151(2):97–115.

Gunderson, L. H., C. S. Holling, and S. S. Light., eds. 1995. Barriers and bridges to the renewal of ecosystems and institutions. Columbia University Press, New York.

Gurney, W. S. C., and R. M. Nisbet. 1979. Ecological stability and social hierarchy. Theoretical Population Biology 16:48–80.

Gustafson, E. J., and R. H. Gardner. 1996. The effect of landscape heterogeneity on the probability of patch colonization. Ecology 77(1):94–107.

Gustafsson, L., and L. Hansson. 1997. Corridors as a conservation tool. Pp. 182–90 *in* L. Hansson, ed., Boreal ecosystems and landscapes: Structures, processes and conservation of biodiversity. Munksgaard, Copenhagen.

Hadly, E. A. 1996. Influence of late Holocene climate on northern Rocky Mountain mammals. Quaternary Research 46(3):298–310.

Haig, S. M. 1998. Molecular contributions to conservation. Ecology 79(2):413–25.

Haight, R. G., D. J. Mladenoff, and A. P. Wydeven. 1998. Modeling disjunct gray wolf populations in semi-wild landscapes. Conservation Biology 12(4):879–88.

Hanby, J. 1982. Lions share: The story of a Serengeti pride. Houghton Mifflin, Boston.

Hansen, A. J., and F. di Castri, eds. 1992. Landscape boundaries: Consequences for biotic diversity and ecological flows. SCOPE Book Series, Ecological Studies, vol. 92. Springer-Verlag, New York.

Hansen, A. J., S. L. Garman, B. Marks, and D. L. Urban. 1993. An approach for managing vertebrate diversity across multiple-use landscapes. Ecological Applications 3:481–96.

Hanski, I., L. Hansson, and H. Henttonen. 1991. Specialist predators, generalist predators, and the microtine rodent cycle. Journal of Animal Ecology 60(1):353–68.

Hanski, I., and H. Henttonen. 1996. Predation on competing rodent species: A simple explanation of complex patterns. Journal of Animal Ecology 65(2):220–32.

Hanski, I., P. Tuchin, E. Korpimaki, and H. Henttonen. 1993. Population oscillations of boreal rodents regulation by mustelid predators lead to chaos. Nature 364(6434):232–35.

Hansson, L. 1989. Predation in heterogeneous landscapes: How to evaluate total impact? Oikos 54:117–19.

——. 1995a. Is the indirect predator effect a special case of generalized reactions to density-related disturbances in cyclic rodent populations? Annales Zoologici Fennici 32(1): 159–62.

——. 1995b. Development and application of landscape approaches in mammalian ecology. Pp. 20–39 *in* W. Z. Lidicker, ed., Landscape approaches in mammalian ecology and conservation. University of Minnesota Press, Minneapolis.

——, ed. 1992. The ecological principles of nature conservation: Applications in temperate and boreal environments. Elsevier Science Publishers, New York.

——, ed. 1997. Boreal ecosystems and landscapes: Structures, processes and conservation of biodiversity. Munksgaard International Publishers, Malden, Massachusetts.

Hansson, L., and P. Angelstam. 1991. Landscape ecology as a theoretical basis for nature conservation. Landscape Ecology 5:191–201.

Hansson, L., L. Fahrig, and G. Merriam, eds. 1995. IALE Studies in Landscape Ecology, vol. 2. Mosaic landscapes and ecological processes. Chapman and Hall, New York.

Harris, L. D. 1984. The fragmented forest: Island biogeography theory and the preservation of biotic diversity. University of Chicago Press, Chicago.

Harris, S., ed. 1987. Mammal population studies. Symposia Zoological Society of London, no. 58. Clarendon Press, New York.

Harris, S., and G. C. Smith. 1987. Demography of two urban fox *(Vulpes vulpes)* populations. Journal of Applied Ecology 24:75–86.

Hatter, I. W., and D. W. Janz. 1994. Apparent demographic changes in the black-tailed deer associated with wolf control on northern Vancouver Island. Canadian Journal of Zoology 72(5):878–84.

Hayes, J. P., and S. H. Jenkins. 1997. Individual variation in mammals. Journal of Mammalogy 78(2):274–93.

Hazel, W. N., R. Smock, and M. D. Johnson. 1990. A polygenic model for the evolution and maintenance of conditional strategies. Proceedings of the Royal Society of London Series B Biological Sciences 242(1305):181–88.

Heaney, L. R., and B. D. Patterson, eds. 1986. Island biogeography of mammals. Academic Press, New York.

Hedrick, P. W. 1992. Genetic conservation in captive populations and endangered species. Pp. 45–68 *in* S. K. Jain and L. W. Botsford, eds., Applied population biology. Kluwer Academic Publishers, Norwell, Massachusetts.

Hedrick, P. W., R. C. Lacy, F. W. Allendorf, and M. E. Soulé. 1996. Directions in conservation biology: Comments on Caughley. Conservation Biology 10(5):1312–20.

Heisey, D. M. 1985. Analyzing selection experiments with log-linear models. Ecology 66:1744–48.

Hengeveld, R. 1989. Dynamic biogeography. Cambridge University Press, Cambridge.

Henry, J. D. 1986. Red fox: The catlike canine. Smithsonian Institution Press, Washington.

Hilborn, R. 1975. The effect of spatial heterogeneity on the persistence of predator-prey interactions. Theoretical Population Biology 8:346–55.

Hochberg, M. E., and M. Van Baalen. 1998. Antagonistic coevolution over productivity gradients. American Naturalist 152(4):620–34.

Hofer, H., and M. East. 1995. Population dynamics, population size, and the commuting system of Serengeti spotted hyenas. Pp. 332–63 *in* A. R. E. Sinclair and P. Arcese, eds., Serengeti II: Dynamics, management, and conservation of an ecosystem. University of Chicago Press, Chicago.

Holling, C. S. 1987. Simplifying the complex: The paradigms of ecological function and structure. European Journal of Operational Research 30:139–46.

———. 1992. Cross-scale morphology, geometry, and dynamics of ecosystems. Ecological Monographs 62(4):447–502.

Holling, C. S., and G. K. Meffe. 1996. Command and control and the pathology of natural resource management. Conservation Biology 10(2):328–37.

Holling, C. S., D. W. Schindler, B. W. Walker, and J. Roughgarden. 1995. Biodiversity in the functioning of ecosystems: An ecological synthesis. Pp. 44–48 *in* C. Perrings, K-G. Mäler, C. Folke, C. S. Holling, and B. O. Jansson, eds., Biodiversity loss: Economic and ecological issues. Cambridge University Press, Cambridge.

Holm, E. 1988. Environmental restraints and life strategies: A habitat templet matrix. Oecologia 75:141–45.

Holmes, D. J., and D. Sherry. 1997. Selected approaches to using individual variation for understanding mammalian life-history evolution. Journal of Mammalogy 78(2):311–19.

Holt, R. D. 1987. Prey communities in patchy environments. Oikos 50:276–90.

———. 1997. On the evolutionary stability of sink populations. Evolutionary Ecology 11(6):723–31.

Holt, R. D., and J. H. Lawton. 1994. The ecological consequences of shared natural enemies. Annual Review of Ecology and Systematics 25:495–520.

Holt, R. D., S. W. Pacala, T. W. Smith, and J. Liu. 1995. Linking contemporary vegetation models with spatially explicit animal population models. Ecological Applications 5(1):20–27.

Hornocker, M. G. 1969. Winter territoriality in mountain lions. Journal of Wildlife Management 33:457–64.

———. 1970. An analysis of mountain lion predation upon mule deer and elk in the Idaho Primitive Area. Wildlife Monographs 21:1–39.

Hornocker, M. G., and T. Bailey. 1986. Natural regulation in three species of felids. Pp. 211–20 *in* S. D. Miller and D. D. Everett, eds., Cats of the world: Biology, conservation, and management. National Wildlife Federation, Washington.

Hornocker, M. G., and H. S. Hash. 1981. Ecology of the wolverine in northwestern Montana. Canadian Journal of Zoology 59:1286–1301.

Hornocker, M. G., J. P. Messick, and W. E. Melquist. 1983. Spatial strategies in three species of Mustelidae. Acta Zoologica Fennici 174:185–88.

Host, G. E., P. L. Polzer, D. J. Mladenoff, M. A. White, and T. R. Crow. 1996. A quantitative approach to developing regional ecosystem classifications. Ecological Applications 6(2):608–18.

Howard, W. E. 1960. Innate and environmental dispersal of individual vertebrates. American Midland Naturalist 63: 152–61.

Huey, R. B., E. R. Pianka, and T. W. Schoener, eds. 1983. Lizard ecology: Studies of a model organism. Harvard University Press, Cambridge.

Hughes, C. 1998. Integrating molecular techniques with field methods in studies of social behavior: A revolution results. Ecology 79(2):383–99.

Hummel, M., and S. Pettigrew. 1992. Wild hunters: Predators in peril. Roberts Rinehart, Niwot, Colorado.

Huntly, N. 1991. Herbivores and the dynamics of communities and ecosystems. Annual Review of Ecology and Systematics 22:477–503.

Hurlbert, S. H. 1997. Functional importance vs keystoneness: Reformulating some questions in theoretical biocenology. Australian Journal of Ecology 22(4):369–82.

Huxel, G. R., and A. Hastings. 1998. Population size dependence, competitive coexistence and habitat destruction. Journal of Animal Ecology 67(3):446–53.

Inchausti, P., and L. R. Ginzburg. 1998. Small mammals' cycles in northern Europe: Patterns and evidence for a maternal effect hypothesis. Journal of Animal Ecology 67(2):180–94.

Jablonski, D., and J. J. Sepkoski, Jr. 1996. Paleobiology, community ecology, and scales of ecological pattern. Ecology 77(5):1367–78.

Jackson, D. A. 1997. Compositional data in community ecology: The paradigm or peril of proportions? Ecology 78(3):929–40.

Jaksic, F. M. 1989. What do carnivorous predators cue in on: size or abundance of mammalian prey? A crucial test in California, Chile, and Spain. Revista Chilena de Historia Natural 62(2):237–50.

Jaksic, F. M., P. Feinsinger, and J. E. Jimenez. 1993. A long-term study on the dynamics of guild structure among predatory vertebrates at a semi-arid neotropical site. Oikos 67(1):87–96.

———. 1996. Ecological redundancy and long-term dynamics of vertebrate predators in semiarid Chile. Conservation Biology 10(1):252–62.

Jaksic, F. M., and J. A. Simonetti. 1987. Predator-prey relationships among terrestrial vertebrates: An exhaustive review of studies conducted in southern South America. Revista Chilena de Historia Natural 60(2):221–44.

Janis, C. M., K. M. Scott, and L. L. Jacobs. 1998. Evolution of Tertiary mammals of North America, vol. 1, Terrestrial carnivores, ungulates, and ungulatelike mammals. Cambridge University Press, New York.

Jedrzejewska, B., and W. Jedrzejewski. 1989. Evasive response of prey and its effect on predator-prey relationships. Wiadomosci Ekologiczne 35(1):3–22.

Jedrzejewska, B., W. Jedrzejewski, A. N. Bunevich, L. Milkowski, and H. Okarama. 1996. Population dynamics of wolves *(Canis lupus)* in Bialowieza Primeval Forest (Poland and Belarus) in relation to hunting by humans, 1847–1993. Mammal Review 26(2–3):103–26.

Jedrzejewska, B., H. Okarma, W. Jedrzejewski, and L. Milkowski. 1994. Effects of exploitation and protection on forest structure, ungulate density and wolf predation in Bialowieza Primeval Forest, Poland. Journal of Applied Ecology 31(4):664–76.

Jedrzejewski, W., B. Jedrzejewska, H. Okarma, K. Schmidt, A. N. Bunevich, and L. Milkowski. 1996. Population dynamics (1869–1994), demography, and home ranges of the lynx in Bialowieza Primeval Forest (Poland and Belarus). Ecography 19(2):122–38.

Jedrzejewski, W., B. Jedrzejewska, and A. Szymura. 1989. Food niche overlaps in a winter community of predators in the Bialowieza Primeval Forest, Poland. Acta Theriologica 34(29–43):487–96.

———. 1995. Weasel population response, home range, and predation on rodents in a deciduous forest in Poland. Ecology 76(1):179–95.

Jedrzejewski, W., K. Schmidt, L. Milkowski, B. Jedrzejewska, and H. Okarma. 1993. Foraging by lynx and its role in ungulate mortality: The local (Bialowieza Forest) and the Palearctic viewpoints. Acta Theriologica 38(4):385–403.

Jewell, P. A., and S. Holt., eds. 1981. Problems in management of locally abundant wild mammals. Academic Press, New York.

Johnsingh, A. J. T. 1993. Prey selection in three large sympatric carnivores in Bandipur. Mammalia 56(4):517–26.

Johnson, B. K., A. A. Ager, S. L. Findholt, M. J. Wisdom, D. B. Marx, J. W. Kern, and L. D. Bryant. 1998. Mitigating spatial differences in observation rate of automated telemetry systems. Journal of Wildlife Management 62(3):958–67.

Johnson, D. H. 1980. The comparison of usage and availability measurements for evaluating resource preference. Ecology 61:65–71.

Jones, C. G., J. H. Lawton, and M. Shachak. 1994. Organisms as ecosystem engineers. Oikos 69(3):373–86.

Jordan, P. A., D. B. Botkin, and M. L. Wolfe. 1971. Biomass dynamics in a moose population. Ecology 52(1):147–52.

Judson, O. P. 1994. The rise of the individual-based model in ecology. Trends in Ecology and Evolution 9(1):9–14.

Karanth, K. U., and M. E. Sunquist. 1995. Prey selection by tiger, leopard and dhole in tropical forests. Journal of Animal Ecology 64(4):439–50.

Kareiva, P. M. 1989. Renewing the dialogue between theory and experiments in population ecology. Pp. 68–88 *in* J. Roughgarden, R. M. May, and S. A. Levin, eds., Perspectives in ecological theory. Princeton University Press, Princeton.

——, ed. 1999. Using science in Habitat Conservation Plans. National Center for Ecological Analysis and Synthesis, Santa Barbara, and American Institute of Biological Sciences, Washington.

Kareiva, P. M., and U. Wennergren. 1995. Connecting landscape patterns to ecosystem and population processes. Nature 373(6512):299–302.

Kaufmann, J. H. 1962. Ecology and social behavior of the coati, *Nasua narica,* on Barro Colorado Island, Panama. University of California Press, Berkeley.

Keiter, R. B., and M. S. Boyce, eds. 1991. The Greater Yellowstone Ecosystem: Redefining America's wilderness heritage. Yale University Press, New Haven.

Keiter, R. B., and H. Locke. 1996. Law and large carnivore conservation in the Rocky Mountains of the U.S. and Canada. Conservation Biology 10(4):1003–12.

Keith, L. B. 1990. Dynamics of snowshoe hare populations. Pp. 119–95 *in* H. H. Genoways, ed., Current mammalogy, vol. 2. Plenum Publishing, New York.

Keith, L. B., S. E. M. Bloomer, and T. Willebrand. 1993. Dynamics of a snowshoe hare population in fragmented habitat. Canadian Journal of Zoology 71(7):1385–92.

Keitt, T. H., D. L. Urban, and B. T. Milne. 1997. Detecting critical scales in fragmented landscapes. Conservation Ecology [online] 1(1):4. http://www.consecol.org/vol1/iss1/art4

Kemp, G. A. 1976. The dynamics and regulation of black bear, *Ursus americanus,* populations in northern Alberta. International Conference on Bear Research and Management 3:191–95.

Kendall, B. E. 1998. Estimating the magnitude of environmental stochasticity in survivorship data. Ecological Applications 8(1):184–93.

Kernohan, B. J., J. J. Millspough, J. A. Jenks, and D. E. Naugle. 1998. Use of an adaptive kernel home-range estimator in a GIS environment to calculate habitat use. Journal of Environmental Management 53(1):83–89.

Kerr, J. T. 1997. Species richness, endemism, and the choice of areas for conservation. Conservation Biology 11(5):1094–1100.

Kerr, J. T., and L. Packer. 1997. Habitat heterogeneity as a determinant of mammal species richness in high-energy regions. Nature 385(6613):252–54.

Khanina, L. 1998. Determining keystone species. Conservation Ecology [online] 2(2):R2. http://www.consecol.org/Journal/vol2/iss2/resp2

King, C. M. 1984. Immigrant killers: Introduced predators and the conservation of birds in New Zealand. Oxford University Press, New York.

——. 1989. The natural history of weasels and stoats. Comstock Publishing Associates, Cornell University Press, Ithaca.

Kleiman, D. G. 1977. Monogamy in mammals. Quarterly Review of Biology 52:39–69.

Kleiman, D. G., M. E. Allen, K. V. Thompson, and S. Lumpkin, eds. 1996. Wild mammals in captivity: Principles and techniques. University of Chicago Press, Chicago.

Kleiman, D. G., and C. A. Brady. 1978. Coyote behavior in the context of recent canid research: Problems and perspectives. Pp. 163–88 *in* M. Bekoff, ed., Coyotes: Biology, behavior, and management. Academic Press, New York.

Kleiman, D. G., and J. F. Eisenberg. 1973. Comparison of canid and felid social systems from an evolutionary perspective. Animal Behaviour 21:637–59.

Klemola, T., M. Koivula, E. Korpimaki, and K. Norrdahl. 1997. Small mustelid predation slows population growth of Microtus voles: A predator reduction experiment. Journal of Animal Ecology 66(5):607–14.

Knowlton, F. F. 1972. Preliminary interpretations of coyote population mechanics with some management implications. Journal of Wildlife Management 36:369–82.

Koenig, W. D., and J. M. H. Knops. 1998. Testing for spatial autocorrelation in ecological studies. Ecography 21(4):423–29.

Kohn, M. H., and R. K. Wayne. 1997. Facts from feces revisited. Trends in Ecology and Evolution 12(6):223–27.

Komers, P. E. 1997. Behavioural plasticity in variable environments. Canadian Journal of Zoology 75(2):161–69.

Konishi, M., S. T. Emlen, R. E. Ricklefs, and J. C. Wingfield. 1989. Contributions of bird studies to biology. Science 246:465–72.

Korpimaki, E., and K. Norrdahl. 1998. Experimental reduction of predator reverses the crash phase of small-rodent cycles. Ecology 79(7):2448–55.

Korpimaki, E., K. Norrdahl, and J. Valkama. 1994. Reproductive investment under fluctuating predation risk: Microtine rodents and small mustelids. Evolutionary Ecology 8(4):357–68.

Krebs, C. J. 1991. The experimental paradigm and long-term population studies. Ibis 133(1 suppl.):3–8.

——. 1996. Population cycles revisited. Journal of Mammalogy 77(1):8–24.

Krebs, C. J., S. Boutin, R. Boonstra, A. R. E. Sinclair, J. N. M. Smith, M. R. T. Dale, K. Martin, and R. Turkington. 1995. Impact of food and predation on the snowshoe hare cycle. Science 269:1112–15.

Krebs, C. J., B. L. Keller, and R. H. Tamarin. 1969. Microtus population biology: Dispersal in fluctuating populations of *M. ochrogaster* and *M. pennsylvanicus* in southern Indiana. Ecology 50:587–607.

Krohne, D. T. 1997. Dynamics of metapopulations of small mammals. Journal of Mammalogy 78(4):1014–26.

Kruuk, H. 1972. The spotted hyena: A study of predation and social behavior. University of Chicago Press, Chicago.

——. 1978. Spatial organization and territorial behavior of the European badger *Meles meles*. Journal of Zoology 184:1–19.

——. 1989. The social badger: Ecology and behaviour of a group-living carnivore *(Meles meles)*. Oxford University Press, New York.

——. 1995. Wild otters: Predation and populations. Oxford University Press, New York.

Kruuk, H., M. L. Gorman, and T. Parish. 1980. The use of zinc-65 for estimating populations of carnivores. Oikos 34:206–8.

Kurten, B. 1968. Pleistocene mammals of Europe. Weidenfeld and Nicolson, London.

——. 1972. The age of mammals. Columbia University Press, New York.

Lacy, R. C. 1997. Importance of genetic variation to the viability of mammalian populations. Journal of Mammalogy 78(2):320–35.

Lambert, W. D., and C. S. Holling. 1998. Causes of ecosystem transformation at the end of the Pleistocene: Evidence from mammal body-mass distributions. Ecosystems 1(2):157–75.

Landres, P. B., J. Verner, and J. W. Thomas. 1988. Ecological uses of vertebrate indicator species: A critique. Conservation Biology 2:316–28.

Laska, M. S., and J. T. Wootton. 1998. Theoretical concepts and empirical approaches to measuring interaction strength. Ecology 79(2):461–76.

Law, B. S., and C. R. Dickman. 1998. The use of habitat mosaics by terrestrial vertebrate fauna: Implications for conservation and management. Biodiversity and Conservation 7(3):323–33.

Law, R. 1979. Optimal life histories under age-specific predation. American Naturalist 114:399–417.

Lawler, S. P., and P. J. Morin. 1995. Food web architecture and population dynamics: Theory and empirical evidence. Annual Review of Ecology and Systematics 26:505–29.

Lawton, J. H. 1995. Ecological experiments with model systems. Science 269:328–31.

Lawton, J. H., D. E. Bignell, B. Bolton, G. F. Bloemers, P. Eggleton, P. M. Hammond, M. Hodda, R. D. Holt, T. B. Larsen, N. A. Mawdsley, N. E. Stork, D. S. Srivastava, and A. D. Watt. 1998. Biodiversity inventories, indicator taxa and effects of habitat modification in tropical forest. Nature 391(6662):72–76.

Lawton, J. H., and R. M. May, eds. 1995. Extinction rates. Oxford University Press, New York.

Legendre, P. 1993. Spatial autocorrelation: Trouble or new paradigm? Ecology 74:1659–63.

Leibold, M. A. 1996. A graphical model of keystone predators in food webs: Trophic regulation of abundance, incidence, and diversity patterns in communities. American Naturalist 147(5):784–812.

Leibold, M. A., J. M. Chase, J. B. Shurin, and A. L. Downing. 1997. Species turnover and the regulation of trophic structure. Annual Review of Ecology and Systematics 28:467–94.

Lele, S., M. L. Taper, and S. Gage. 1998. Statistical analysis of population dynamics in space and time using estimating functions. Ecology 79(5):1489–1502.

Lembeck, M. 1986. Long-term behavior and population dynamics of an unharvested bobcat population in San Diego County, California. Pp. 305–10 *in* S. D. Miller and D. D. Everett, eds., Cats of the world: Biology, conservation, and management. National Wildlife Federation, Washington.

Leopold, A. S. 1933. Game management. Charles Scribner's Sons, New York.

Levin, S. A. 1992. The problem of pattern and scale in ecology. Ecology 73:1943–67.

Lewis, M. A., K. A. J. White, and J. D. Murray. 1997. Analysis of a model for wolf territories. Journal of Mathematical Biology 35(7):749–74.

Lewis, M. E. 1997. Carnivoran paleoguilds of Africa: Implications for hominid food procurement strategies. Journal of Human Evolution 32(2–3):257–88.

Leyhausen, P. 1979. Cat behavior: The predatory and social behavior of domestic and wild cats. Garland STPM Press, New York.

Lidicker, W. Z. 1962. Emigration as a possible mechanism permitting the regulation of population density below carrying capacity. American Naturalist 46:29–33.

———. 1994. Population ecology. Pp. 323–47 *in* E. C. Birney and J. R. Choate, eds., Seventy-five years of mammalogy (1919–1994). Special Publication no. 11, American Society of Mammalogists.

———, ed. 1995. Landscape approaches in mammalian ecology and conservation. University of Minnesota Press, Minneapolis.

Liebhold, A. M., R. E. Rossi, and W. P. Kemp. 1993. Geostatistics and geographic information systems in applied insect ecology. Annual Review of Entomology 38:303–27.

Lima, S. L. 1998. Stress and decision making under the risk of predation: Recent developments from behavioral, reproductive, and ecological perspectives. Advances in the Study of Behavior 27:215–90.

Lima, S. L., and L. M. Dill. 1990. Behavioral decisions made under the risk of predation: A review and prospectus. Canadian Journal of Zoology 68:619–40.

Lima, S. L., and P. A. Zollner. 1996. Towards a behavioral ecology of ecological landscapes. Trends in Ecology and Evolution 11(3):131–35.

Lindenmayer, D. B., and H. P. Possingham. 1996. Modelling the inter-relationships between habitat patchiness, dispersal capability and metapopulation persistence of the endangered species, Leadbeater's possum, in south-eastern Australia. Landscape Ecology 11(2):79–105.

Lindstedt, S. L., B. J. Miller, and S. W. Buskirk. 1986. Home range, time, and body size in mammals. Ecology 67: 413–18.

Lindström, E. R. 1989. The role of medium-sized carnivores in the Nordic Sweden boreal forest. Finnish Game Research (46):53–63.

Lindström, E. R., H. Andren, P. Angelstam, G. Cederlund, B. Hornfeldt, L. Jaderberg, P. A. Lemnell, B. Martinsson, K. Skold, and J. E. Swenson. 1994. Disease reveals the predator: Sarcoptic mange, red fox predation, and prey populations. Ecology 75(4):1042–49.

Lindström, J., H. Kokko, E. Ranta, and H. Linden. 1998. Predicting population fluctuations with artificial neural networks. Wildlife Biology 4(1):47–53.

Linhart, S. B., and F. F. Knowlton. 1975. Determining the relative abundance of coyotes by scent station lines. Wildlife Society Bulletin 3:119–24.

Link, W. A., R. J. Barker, J. R. Sauer, and S. Droege. 1994. Within-site variability in surveys of wildlife populations. Ecology 75:1097–1108.

Linnell, J. D. C., R. Aanes, and J. E. Swenson. 1997. Translocation of carnivores as a method for managing problem animals: A review. Biodiversity and Conservation 6(9):1245–57.

Lockie, J. D. 1966. Territory in small carnivores. Symposia Zoological Society London 18:143–65.

Loehle, C., and B. L. Li. 1996. Statistical properties of ecological and geologic fractals. Ecological Modelling 85(2–3):271–84.

Lomolino, M. V. 1984. Mammalian island biogeography: Effects of area, isolation and vagility. Oecologia 61:376–82.

———. 1986. Mammalian community structure on islands: The importance of immigration, extinction, and interactive effects. Biological Journal of the Linnean Society 28:1–21.

Lomolino, M. V., and R. Channel. 1995. Splendid isolation: Patterns of geographic range collapse in endangered mammals. Journal of Mammalogy 76(2):335–47.

Lott, D. F. 1991. Intraspecific variation in the social systems of wild vertebrates. Cambridge University Press, New York.

Ludeke, A., R. Maggio, and L. Reid. 1990. An analysis of anthropogenic deforestation using logistic regression and GIS. Journal of Environmental Management 31:247–59

Ludwig, D. 1994. Bad ecology leads to bad public policy. Trends in Ecology and Evolution 9(10):411.

Ludwig, D., D. D. Jones, and C. S. Holling. 1978. Qualitative analysis of insect outbreak systems: The spruce budworm and forest. Journal of Animal Ecology 44:315–32.

Ludwig, D., B. Walker, and C. S. Holling. 1997. Sustainability, stability, and resilience. Conservation Ecology [online] 1(1):7. http://www.consecol.org/vol1/iss1/art7

Macdonald, D. W. 1983. The ecology of carnivore social behaviour. Nature 301:379–84.

———. 1985. The carnivores: Order Carnivora. Pp. 619–722 in R. E. Brown and D. W. Macdonald, eds., Social odours in mammals, vol. 2. Oxford University Press, New York.

———. 1987. Running with the fox. Facts on File Publications, New York.

———. 1993. The velvet claw: A natural history of the carnivores. Parkwest Books, New York, for British Broadcasting Corporation.

Maehr, D. S. 1997. The Florida panther: Life and death of a vanishing carnivore. Island Press, Washington, D.C.

Maehr, D. S., and J. A. Cox. 1995. Landscape features and panthers in Florida. Conservation Biology 9(5):1008–19.

Mahon, P. S., P. B. Banks, and C. R. Dickman. 1998. Population indices for wild carnivores: A critical study in sand-dune habitat, south-western Queensland. Wildlife Research 25(1):11–22.

Major, M., M. K. Johnson, W. S. Davis, and T. F. Kellogg. 1980. Identifying scats by recovery of bile acids. Journal of Wildlife Management 44:290–93.

Manly, B. F. J., L. L. McDonald, and D. L. Thomas. 1993. Resource selection by animals: Statistical design and analysis for field studies. Chapman and Hall, New York.

Mares, M. A., and G. N. Cameron. 1994. Community and ecosystem ecology. Pp. 348–76 in E. C. Birney and J. R. Choate, eds., Seventy-five years of mammalogy (1919–1994). Special Publication no. 11, American Society of Mammalogists.

Marti, C. D. 1987. Predator-prey interactions: A selective review of North American research results. Revista Chilena de Historia Natural 60:203–19.

Marti, C. D., K. Steenhof, M. N. Kochert, and J. S. Marks. 1993. Community trophic structure: The roles of diet body size and activity time in vertebrate predators. Oikos 67(1):6–18.

Matlock, R. B., Jr., J. B. Welch, and F. D. Parker. 1996. Estimating population density per unit area from mark, release, recapture data. Ecological Applications 6(4):1241–53.

Mattson, D. J., S. Herrero, R. G. Wright, and C. M. Pease. 1996. Science and management of Rocky Mountain grizzly bears. Conservation Biology 10(4):1013–25.

Maurer, B. A. 1994. Geographical population analysis: Tools for the analysis of biodiversity. Blackwell Scientific Publications, Boston.

May, R. M. 1976. Simple mathematical models with very complicated dynamics. Nature 261:459–67.

———. 1981. Models for two interacting populations. Pp. 78–104 in R. M. May, ed., Theoretical ecology: Principles and applications. Blackwell Scientific Publications, Oxford.

———. 1994. Graeme Caughley and the emerging science of conservation biology. Trends in Ecology and Evolution 9(10):368–69.

McClean, S. A., M. A. Rumble, R. M. King, and W. L. Baker. 1998. Evaluation of resource selection methods with different definitions of availability. Journal of Wildlife Management 62(2):793–801.

McCullough, D. R., ed. 1996. Metapopulations and wildlife conservation. Island Press, Washington, D.C.

McCullough, D. R., and R. H. Barrett, eds. 1992. Wildlife 2001: Populations. Elsevier Applied Science, New York.

McEuen, A. 1993. The wildlife corridor controversy: A review. Endangered Species Update 10(11–12):1–6.

McKelvey, K., B. R. Noon, and R. H. Lamberson. 1993. Conservation planning for species occupying fragmented landscapes: The case of the northern spotted owl. Pp. 424–50 in P. M. Kareiva, J. G. Kingsolver, and R. B. Huey, eds., Biotic interactions and global change. Sinauer Associates, Sunderland, Massachusetts.

McKenney, D. W., R. A. Sims, M. E. Soulé, B. G. Mackey, and K. L. Campbell, eds. 1994. Towards a set of biodiversity indicators for Canadian forests. Natural Resouces Canada, Canadian Forest Service, Ontario.

McKinney, M. L. 1997. Extinction vulnerability and selectivity: Combining ecological and paleontological views. Annual Review of Ecology and Systematics 28:495–516.

McLaren, B. E., and R. O. Peterson. 1994. Wolves, moose, and tree rings on Isle Royale. Science 266(5190):1555–58.

McNamara, J. M., and A. I. Houston. 1996. State-dependent life histories. Nature 380(6571):215–21.

McShea, W. J., J. H. Rappole, and H. B. Underwood. 1997. The science of overabundance: Deer ecology and population management. Smithsonian Institution Press, Washington.

Mech, L. D. 1966. The wolves of Isle Royale. U.S. Government Printing Office, Washington.

———. 1970. The wolf: The ecology and behavior of an endangered species. Natural History Press, Doubleday, New York.

———. 1995. The challenge and opportunity of recovering wolf populations. Conservation Biology 9(2):270–78.

———. 1996. A new era for carnivore conservation. Wildlife Society Bulletin 24(3):397–401.

Meir, E., and P. M. Kareiva. 1998. Contributions of spatially explicit landscape models to conservation biology. Pp. 497–507 in P. L. Fiedler and P. M. Kareiva, eds., Conservation biology, 2d ed. Chapman and Hall, New York.

Melquist, W. E., and M. G. Hornocker. 1983. Ecology of river otters in west-central Idaho. Wildlife Monographs 83:1–60.

Menge, B. A. 1995. Indirect effects in marine rocky intertidal interaction webs: Patterns and importance. Ecological Monographs 65(1):21–74.

Meserve, P. L., J. R. Gutierrez, J. A. Yunger, L. C. Contreras, and F. M. Jaksic. 1996. Role of biotic interactions in a small mammal assemblage in semiarid Chile. Ecology 77(1):133–48.

Messick, J. P., and M. G. Hornocker. 1981. Ecology of the badger in southwestern Idaho. Wildlife Monographs 76:1–53.

Messier, F. 1995. Trophic interactions in two northern wolf-ungulate systems. Wildlife Research 22(1):131–46.

Miller, B., R. P. Reading, and S. Forrest. 1996. Prairie night: Black-footed ferrets and the recovery of endangered species. Smithsonian Institution Press, Washington.

Miller, S. D., G. C. White, R. A. Sellers, H. V. Reynolds, J. W. Schoen, K. Titus, V. G. Barnes, Jr., R. B. Smith, R. R. Nelson, W. B. Ballard, and C. C. Schwartz. 1997. Brown and black bear density estimation in Alaska using radiotelemetry and replicated mark-resight techniques. Wildlife Monographs 133:1–55.

Mills, L. S., M. E. Soulé, and D. F. Doak. 1993. The keystone-species concept in ecology and conservation. BioScience 43:219–24.

Mills, M. G. L. 1990. Kalahari hyaenas: Comparative behavioural ecology of two species. Unwin Hyman, Boston.

Milne, B. T. 1992. Spatial aggregation and neutral models in fractal landscapes. American Naturalist 139:32–57.

Minta, S. C. 1992. Tests of spatial and temporal interaction among animals. Ecological Applications 2(2):178–88.

———. 1993. Sexual differences in spatio-temporal interaction among badgers. Oecologia 96(3):402–9.

Minta, S. C., and P. M. Kareiva. 1994. A conservation science perspective: Conceptual and experimental improvements. Pp. 275–304 *in* T. W. Clark, R. P. Reading, and A. L. Clarke, eds., Endangered species recovery: Finding the lessons, improving the process. Island Press, Washington, D.C.

Mladenoff, D. J., and T. A. Sickley. 1998. Assessing potential gray wolf restoration in the northeastern United States: A spatial prediction of favorable habitat and potential population levels. Journal of Wildlife Management 62(1):1–10.

Mladenoff, D. J., T. A. Sickley, R. G. Haight, and A. P. Wydevan. 1995. A regional landscape analysis and prediction of favorable gray wolf habitat in the northern Great Lakes region. Conservation Biology 9(2):279–94.

Moa, P. F., A. Negard, and T. Kvam. 1998. Area use and movement pattern in lynx in a mid-Norwegian coniferous forest landscape. Fauna (Oslo) 51(1):10–23.

Moehlman, P. D. 1989. Intraspecific variation in canid social systems. Pp. 143–63 *in* J. L. Gittleman, ed., Carnivore behavior, ecology, and evolution. Cornell University Press, Ithaca.

Moilanen, A., and I. Hanski. 1998. Metapopulation dynamics: Effects of habitat quality and landscape structure. Ecology 79(7):2503–15.

Monkkonen, M., and J. Aspi. 1998. Sampling error in measuring temporal density variability in animal populations and communities. Annales Zoologici Fennici 35(1):47–57.

Moors, P. J. 1984. Coexistence and interspecific competition in the carnivore genus Mustela. Acta Zoologica Fennica 172:37–40.

Morris, D. W., Z. Abramsky, B. J. Fox, and M. R. Willig, eds. 1989. Patterns in the structure of mammalian communities. Special Publication of the Museum of Texas Tech University, no. 28.

Murie, A. 1940. Ecology of the coyote in the Yellowstone. Fauna of the national parks of the United States. Fauna Series no. 4. U.S. Government Printing Office, Washington.

———. 1944. The wolves of Mount McKinley. Fauna of the national parks of the United States. Fauna Series no. 5. United States Department of the Interior National Park Service.

Murphy, D. D. 1989. Conservation and confusion: Wrong species, wrong scale, wrong conclusions. Conservation Biology 3:82–85.

Murphy, D. D., and B. R. Noon. 1991. Coping with uncertainty in wildlife biology. Journal of Wildlife Management 55:773–82.

Murray, D. L., and S. Boutin. 1991. The influence of snow on lynx and coyote movements: Does morphology affect behavior? Oecologia 88(4):463–69.

Murray, D. L., S. Boutin, and M. O'Donoghue. 1994. Winter habitat selection by lynx and coyotes in relation to snowshoe hare abundance. Canadian Journal of Zoology 72(8):1444 – 51.

Murray, D. L., S. Boutin, M. O'Donoghue, and V. O. Nams. 1995. Hunting behaviour of a sympatric felid and canid in relation to vegetative cover. Animal Behaviour 50(5):1203 – 10.

Mysterud, A., and R. A. Ims. 1998. Functional responses in habitat use: Availability influences relative use in trade-off situations. Ecology 79(4):1435 – 41.

Naeem, S., J. Thompson, S. P. Lawler, J. H. Lawton, and R. M. Woodfin. 1994. Declining biodiversity can alter the performance of ecosystems. Nature 368:734 – 37.

Nams, V. O. 1997. Density-dependent predation by skunks using olfactory search images. Oecologia 110(3):440 – 48.

National Research Council. 1997. Wolves, bears, and their prey in Alaska: Biological and social challenges in wildlife management. National Academy Press, Washington.

Naumov, N. P. 1948. Sketches of the comparative ecology of mouse-like rodents. Isd-vo Akademii nauk, SSSR, Moscow.

Newmark, W. D. 1986. Species-area relationship and its determinants for mammals in western North American national parks. Biological Journal of the Linnean Society 28:83 – 98.

———. 1987. A land-bridge island perspective on mammalian extinctions in western North American parks. Nature 325:430 – 32.

———. 1995. Extinction of mammal populations in western North American national parks. Conservation Biology 9(3):512 – 26.

———. 1996. Insularization of Tanzanian parks and the local extinction of large mammals. Conservation Biology 10(6):1549 – 56.

Newsome, A. E., I. Parer, and P. C. Catling. 1989. Prolonged prey suppression by carnivores: Predator-removal experiments. Oecologia 78(4):458 – 67.

Niemi, G. J., J. M. Hanowski, A. R. Lima, T. Nicholls, and N. Weiland. 1998. A critical analysis on the use of indicator species in management. Journal of Wildlife Management 61:1240 – 52.

Norrdahl, K., and E. Korpimaki. 1995a. Effects of predator removal on vertebrate prey populations: Birds of prey and small mammals. Oecologia 103(2):241 – 48.

———. 1995b. Small carnivores and prey population dynamics in summer. Annales Zoologici Fennici 32(1):163 – 69.

Noss, R. F. 1990. Indicators for monitoring biodiversity: A hierarchical approach. Conservation Biology 4:355 – 64.

Noss, R. F., H. B. Quigley, M. G. Hornocker, T. Merrill, and P. C. Paquet. 1996. Conservation biology and carnivore conservation in the Rocky Mountains. Conservation Biology 10(4):949 – 63.

Novak, M., J. A. Baker, M. E. Obbard, and B. Malloch, eds. 1987. Wild furbearers management and conservation in North America. Ontario Trappers Association, North Bay, Ontario.

Nusser, S. M., F. J. Breidt, and W. A. Fuller. 1998. Design and estimation for investigating the dynamics of natural resources. Ecological Applications 8(2):234 – 45.

O'Donoghue, M., S. Boutin, C. J. Krebs, D. L. Murray, and E. J. Hofer. 1998a. Behavioural responses of coyotes and lynx to the snowshoe hare cycle. Oikos 82(1):169 – 83.

O'Donoghue, M., S. Boutin, C. J. Krebs, G. Zuleta, D. L. Murray, and E. J. Hofer. 1998b. Functional responses of coyotes and lynx to the snowshoe hare cycle. Ecology 79(4):1193 – 1208.

Oehler, J. D., and J. A. Litvaitis. 1996. The role of spatial scale in understanding responses of medium-sized carnivores to forest fragmentation. Canadian Journal of Zoology 74(11):2070 – 79.

Okarma, H. 1995. The trophic ecology of wolves and their predatory role in ungulate communities of forest ecosystems in Europe. Acta Theriologica 40(4):335 – 86.

Okarma, H., B. Jedrzejewska, W. Jedrzejewski, Z. A. Krasinski, and L. Milkowski. 1995. The roles of predation, snow cover, acorn crop, and man-related factors on ungulate mortality in Bialowieza Primeval Forest, Poland. Acta Theriologica 40(2):197 – 217.

Okarma, H., W. Jedrzejewski, K. Schmidt, R. Kowalczyk, and B. Jedrzejewska. 1997. Predation of Eurasian lynx on roe deer and red deer in Bialowieza Primeval Forest, Poland. Acta Theriologica 42(2):203–24.

Okarma, H., W. Jedrzejewski, K. Schmidt, S. Sniezko, A. N. Bunevich, and B. Jedrzejewska. 1998. Home ranges of wolves in Bialowieza Primeval Forest, Poland, compared with other Eurasian populations. Journal of Mammalogy 79(3):842–52.

Oksanen, L., J. Moen, P. A. Lundberg. 1992. The time-scale problem in exploiter-victim models: Does the solution lie in ratio-dependent exploitation? American Naturalist 140(6):938–60.

Oksanen, T., and H. Henttonen. 1996. Dynamics of voles and small mustelids in the taiga landscape of northern Fennoscandia in relation to habitat quality. Ecography 19(4):432–43.

Oksanen, T., L. Oksanen, and M. Gyllenberg. 1992a. Exploitation ecosystems in heterogeneous habitat complexes II: Impact of small-scale heterogeneity on predator-prey dynamics. Evolutionary Ecology 6(5):383–98.

Oksanen, T., L. Oksanen, and M. Norberg. 1992b. Habitat use of small mustelids in north Fennoscandian tundra: A test of the hypothesis of patchy exploitation ecosystems. Ecography 15(2):237–44.

Otis, D. L. 1997. Analysis of habitat selection studies with multiple patches within cover types. Journal of Wildlife Management 61:1016–22.

Otis, D. L., L. L. McDonald, and M. A. Evans. 1993. Parameter estimation in encounter sampling surveys. Journal of Wildlife Management 57:543–48.

Owen-Smith, N. 1988. Megaherbivores: The influence of very large body size on ecology. Cambridge University Press, New York.

———. 1989. Megafaunal extinctions: The conservation message from 11,000 years B.C. Conservation Biology 3:405–11.

Pagel, M. D., R. M. May, and A. R. Collie. 1991. Ecological aspects of the geographical distribution and diversity of mammalian species. American Naturalist 137(6):791–815.

Paine, R. T. 1974. Intertidal community structure: Experimental studies on the relationship between a dominant competitor and its principal predator. Oecologia 15:93–120.

———. 1992. Food-web analysis through field measurements of per capita interaction strength. Nature 355(6355):73–75.

———. 1993. A salty and salutary perspective on global change. Pp. 347–55 in P. M. Kareiva, J. G. Kingsolver, and R. B. Huey, eds., Biotic interactions and global change. Sinauer Associates, Sunderland, Massachusetts.

Palmqvist, P., B. Martinez-Navarro, and A. Arribas. 1996. Prey selection by terrestrial carnivores in a lower Pleistocene paleocommunity. Paleobiology 22(4):514–34.

Palomares, F., P. Ferreras, J. M. Fedriani, and M. Delibes. 1996. Spatial relationships between Iberian lynx and other carnivores in an area of south-western Spain. Journal of Applied Ecology 33(1):5–13.

Palomares, F., P. Gaona, P. Ferreras, and M. Delibes. 1995. Positive effects on game species of top predators by controlling smaller predator populations: An example with lynx, mongooses, and rabbits. Conservation Biology 9(2):295–305.

Paquet, P. C., and A. Hackman. 1995. Large carnivore conservation in the Rocky Mountains: A long-term strategy for maintaining free-ranging and self-sustaining populations of carnivores. World Wildlife Fund–Canada, Toronto.

Parker, P. G., A. A. Snow, M. D. Schug, G. C. Booton, and P. A. Fuerst. 1998. What molecules can tell us about populations: Choosing and using a molecular marker. Ecology 79(2):361–82.

Parsons, P. A. 1987. Features of colonizing animals: Phenotypes and genotypes. Pp. 133–54 in A. J. Gray, M. J. Crawley, and P. J. Edwards, eds., Colonization, succession, and stability. Blackwell Scientific Publications, Boston.

Pastor, J., B. Dewey, R. Moen, D. J. Mladenoff, M. White, and Y. Cohen. 1998a. Spatial patterns in the moose-forest-soil ecosystem on Isle Royale, Michigan, USA. Ecological Applications 8(2):411–24.

Pastor, J., S. Light, and L. Sovell. 1998b. Sustainability and resilience in boreal regions: Sources and consequences of variability. Conservation Ecology [online] 2(2):16. http://www.consecol.org/vol2/iss2/art16

Pastor, J., R. Moen, and Y. Cohen. 1997. Spatial heterogeneities, carrying capacity, and feedbacks in animal-landscape interactions. Journal of Mammalogy 78(4):1040–52.

Pech, R. P., A. R. E. Sinclair, and A. E. Newsome. 1995. Predation models for primary and secondary prey species. Wildlife Research 22(1):55–64.

Pech, R. P., A. R. E. Sinclair, A. E. Newsome, and P. C. Catling. 1992. Limits to predator regulation of rabbits in Australia: Evidence from predator-removal experiments. Oecologia 89(1):102–12

Pelton, M. R., and L. C. Marcum. 1977. The potential use of radioisotopes for determining densities of black bears and other carnivores. Pp. 221–36 *in* R. L. Phillips and C. Jonkel, eds., Proceedings of the 1975 Predator Symposium, University of Montana, Missoula.

Pelton, M. R., and F. T. Van Manen. 1996. Benefits and pitfalls of long-term research: A case study of black bears in Great Smoky Mountains National Park. Wildlife Society Bulletin 24(3):443–50.

Pereira, J. M. C., and R. Itami. 1991. GIS-based habitat modeling using logistic multiple regression: A study of the Mt. Graham red squirrel. Photogrammetric Engineering and Remote Sensing 57(11):1475–86.

Perry, D. A. 1998. The scientific basis of forestry. Annual Review of Ecology and Systematics 29:435–66.

Peters, R. H. 1991. A critique for ecology. Cambridge University Press, New York.

Peterson, R. O. 1988. The pit or the pendulum: Issues in large carnivore management in natural ecosystems. Pp. 105–17 *in* J. K. Agee and D. R. Johnson, eds., Ecosystem management for parks and wilderness. University of Washington Press, Seattle.

Peterson, R. O., N. J. Thomas, J. M. Thurber, J. A. Vucetich, and T. A. Waite. 1998. Population limitation and the wolves of Isle Royale. Journal of Mammalogy 79(3):828–41.

Pfister, C. A. 1998. Patterns of variance in stage-structured populations: Evolutionary predictions and ecological implications. Proceedings of the National Academy of Sciences of the United States of America 95(1):213–18.

Pickett, S. T. A., R. S. Ostfeld, M. Shachak, and G. E. Likens, eds. 1997. The ecological basis of conservation: Heterogeneity, ecosystems, and biodiversity. Chapman and Hall, New York.

Pickett, S. T. A., V. T. Parker, and P. L. Fiedler. 1992. The new paradigm in ecology: Implications for conservation biology above the species level. Pp. 65–88 *in* P. L. Fiedler and S. K. Jain, eds., Conservation biology: The theory and practice of nature conservation, preservation, and management. Chapman and Hall, New York.

Polis, G. A., W. B. Anderson, and R. D. Holt. 1997. Toward an integration of landscape and food web ecology: The dynamics of spatially subsidized food webs. Annual Review of Ecology and Systematics 28:289–316.

Polis, G. A., and R. D. Holt. 1992. Intraguild predation: The dynamics of complex trophic interactions. Trends in Ecology and Evolution 7(5):151–54.

Polis, G. A., R. D. Holt, B. A. Menge, and K. O. Winemiller. 1996. Time, space, and life history: Influences on food webs. Pp. 435–60 *in* G. A. Polis and K. O. Winemiller, eds., Food webs: Integration of patterns and dynamics. Chapman and Hall, New York.

Polis, G. A., C. A. Meyers, and R. D. Holt. 1989. The ecology and evolution of intraguild predation: Potential competitors that eat each other. Annual Review of Ecology and Systematics 20:297–330.

Poole, K. G. 1997. Dispersal patterns of lynx in the Northwest Territories. Journal of Wildlife Management 61(2):497–505.

Poole, K. G., L. A. Wakelyn, and P. Nicklen. 1996. Habitat selection by lynx in the Northwest Territories. Canadian Journal of Zoology 74(5):845–50.

Porter, W. F., and Church, K. E. 1987. Effects of environmental pattern on habitat preference analysis. Journal of Wildlife Management 51:681–85.

Possingham, H. P., and I. Davies. 1995. ALEX: A model for the viability analysis of spatially structured populations. Biological Conservation 73(2):143–50.

Post, E., and N. C. Stenseth. 1998. Large-scale climatic fluctuation and population dynamics of moose and white-tailed deer. Journal of Animal Ecology 67(4):537–43.

Powell, R. A. 1979a. Ecological energetics and foraging strategies of the fisher *(Martes pennanti)*. Journal of Animal Ecology 48:195–212.

——. 1979b. Mustelid spacing patterns: Variations on a theme by *Mustela*. Zeitschrift fur Teirpsychologie 50:153–65.

——. 1980. Stability in a one-predator–three-prey community. American Naturalist 115:567–79.

——. 1993. The fisher: Life history, ecology, and behavior. 2d ed. University of Minnesota Press, Minneapolis.

——. 1994. Effects of scale on habitat selection and foraging behavior of fishers in winter. Journal of Mammalogy 75(2):349–56.

Powell, R. A., J. W. Zimmerman, D. E. Seaman, and J. F. Gilliam. 1996. Demographic analyses of a hunted black bear population with access to a refuge. Conservation Biology 10(1):224–34.

Powell, R. A., J. W. Zimmerman, D. E. Seaman, and C. Powell. 1997. Ecology and behaviour of North American black bears: Home ranges, habitat, and social organization. Chapman and Hall, New York.

Power, M. E., D. Tilman, J. A. Estes, B. A. Menge, W. J. Bond, L. S. Mills, G. Daily, J. C. Castilla, J. Lubchenco, and R. T. Paine. 1996. Challenges in the quest for keystones. BioScience 46: 609–20.

Powers, D. A. 1989. Fish as model systems. Science 246:352–58.

Prendergast, J. R., R. M. Quinn, J. H. Lawton, B. C. Eversham, and D. W. Gibbons. 1993. Rare species: The coincidence of diversity hotspots and conservation strategies. Nature 365(6444):335–37.

Pulliainen, E., and P. Ollinmaki. 1996. A long-term study of the winter food niche of the pine marten *(Martes martes)* in northern boreal Finland. Acta Theriologica 41(4):337–52.

Pulliam, H. R., J. B. Dunning, and J. Liu. 1992. Population dynamics in complex landscapes: A case study. Ecological Applications 2:165–77.

Putman, R. J. 1984. Facts from faeces. Mammal Review 14:79–97.

——, ed. 1989. Mammals as pests. Chapman and Hall, New York.

Quade, C., R. Bigley, and L. Young. 1996. Lynx habitat management plan. Washington State Department of Natural Resources, Seattle.

Rabinowitz, A. 1986. Jaguar: Struggle and triumph in the jungles of Belize. Arbor House, New York.

Radinsky, L. B. 1988. The evolution of vertebrate design. University of Chicago Press, Chicago.

Ralls, K., and P. H. Harvey. 1985. Geographic variation in size and sexual dimorphism of North American weasels. Biological Journal of the Linnean Society 25:119–67.

Raphael, M. G., K. S. McKelvey, and B. M. Galleher. 1998. Using geographic information systems and spatially explicit population models for avian conservation: A case study. Pp. 65–74 *in* J. M. Marzluff and R. Sallabanks, eds., Avian conservation: Research and management. Island Press, Washington, D.C.

Rapport, D. J., R. Costanza, and A. J. McMichael. 1998. Assessing ecosystem health. Trends in Ecology and Evolution 13(10):397–402.

Rastetter, E. B., A. W. King, B. J. Cosby, G. M. Hornberger, R. V. O'Neill, and J. E. Hobbie. 1992. Aggregating fine-scale ecological knowledge to model coarser-scale attributes of ecosystems. Ecological Applications 2:55–70.

Real, L. A. 1980. Fitness, uncertainty, and the role of diversification in evolution and behavior. American Naturalist 115:623–38.

Reed, J. M., D. D. Murphy, and P. F. Brussard. 1998. Efficacy of population viability analysis. Wildlife Society Bulletin 26(2):244–51.

Reed, K. E. 1994. Community organization through the Plio-Pleistocene. Journal of Vertebrate Paleontology 14(3 Supplement):43A.

——. 1998. Using large mammal communities to examine ecological and taxonomic structure and predict vegetation in extant and extinct assemblages. Paleobiology 24(3):384–408.

Reynolds, J. C., and S. C. Tapper. 1996. Control of mammalian predators in game management and conservation. Mammal Review 26(2–3):127–56.

Rich, T. J., and J. L. Hurst. 1998. Scent marks as reliable signals of the competitive ability of mates. Animal Behaviour 56(3):727–35.

Ritchie, M. E. 1998. Scale-dependent foraging and patch choice in fractal environments. Evolutionary Ecology 12(3):309–30.

Rogers, C. M., and M. J. Caro. 1998. Song sparrows, top carnivores and nest predation: A test of the mesopredator release hypothesis. Oecologia 116(1–2):227–33.

Rohlf, D. J. 1991. Six biological reasons why the Endangered Species Act doesn't work—and what to do about it. Conservation Biology 5:273–82.

Rollo, C. D., ed. 1995. Phenotypes: Their epigenetics, ecology, and evolution. Chapman and Hall, New York.

Rooney, S. M., A. Wolfe, and T. J. Hayden. 1998. Autocorrelated data in telemetry studies: Time to independence and the problem of behavioural effects. Mammal Review 28(2):89–98.

Rosenzweig, M. L. 1966. Community structure in sympatric Carnivora. Journal of Mammalogy 47:602–12.

———. 1968. The strategy of body size in mammalian carnivores. American Midland Naturalist 80:299–315.

———. 1991. Habitat selection and population interactions: The search for mechanism. American Naturalist 137:S5–S28.

———. 1995. Species diversity in space and time. Cambridge University Press, New York.

Rossi, R. E., D. J. Mulla, A. G. Journel, and E. H. Franz. 1992. Geostatistical tools for modeling and interpreting ecological spatial dependence. Ecological Monographs 62:277–314.

Rossiter, M. 1996. Incidence and consequences of inherited environmental effects. Annual Review of Ecology and Systematics 27:451–76.

Ruckelshaus, M., C. Hartway, and P. Kareiva. 1997. Assessing the data requirements of spatially explicit dispersal models. Conservation Biology 11(6):1298–1306.

Ruggiero, L. F., K. B. Aubry, S. W. Buskirk, L. J. Lyon, W. J. Zielinski, eds. 1994. The scientific basis for conserving forest carnivores: American marten, fisher, lynx, and wolverine in the western United States. General Technical Report RM-254. United States Department of Agriculture Forest Service, Rocky Mountain Forest and Range Experiment Station, Fort Collins, Colorado.

Ruggiero, L. F., D. E. Pearson, and S. E. Henry. 1998. Characteristics of American marten den sites in Wyoming. Journal of Wildlife Management 62(2):663–73.

Saether, B. E. 1997. Environmental stochasticity and population dynamics of large herbivores: A search for mechanisms. Trends in Ecology and Evolution 12(4):143–49.

Sandell, M. 1990. The evolution of seasonal delayed implantation. Quarterly Review of Biology 65:23–42.

Sargeant, A. B. 1972. Red fox spatial characteristics in relation to waterfowl predation. Journal of Wildlife Management 36:225–35.

Sargeant, A. B., S. H. Allen, and R. T. Eberhardt. 1984. Red fox predation on breeding ducks in midcontinent North America. Wildlife Monographs 89:1–41.

Schaller, G. B. 1967. The deer and the tiger: A study of wildlife in India. University of Chicago Press, Chicago.

———. 1972. The Serengeti lion: A study of predator-prey relations. University of Chicago Press, Chicago.

———. 1993. The last panda. University of Chicago Press, Chicago.

Scheiner, S. M. 1998. The genetics of phenotypic plasticity VII: Evolution in a spatially-structured environment. Journal of Evolutionary Biology 11(3):303–20.

Schindler, D. W. 1998. Replication versus realism: The need for ecosystem-scale experiments. Ecosystems 1(4):323–34.

Schippers, P., J. Verboom, J. P. Knaapen, and R. Van Apeldoorn. 1996. Dispersal and habitat connectivity in complex heterogeneous landscapes: An analysis with a GIS-based random walk model. Ecography 19(2):97–106.

Schmitz, O. J., A. P. Beckerman, and K. M. O'Brien. 1997. Behaviorally mediated trophic cascades: Effects of predation risk on food web interactions. Ecology 78(5):1388–99.

Schmitz, O. J., D. S. Hik, and A. R. E. Sinclair. 1992. Plant chemical defense and twig selection by snowshoe hare: An optimal foraging perspective. Oikos 65(2):295–300.

Schoener, T. W. 1969. Models of optimal size for solitary predators. American Naturalist 103:277–313.

———. 1983. Rate of species turnover declines from lower to higher organisms: A review of the data. Oikos 41:372–77.

Schoener, T. W., and G. H. Adler. 1991. Greater resolution of distributional complementarities by controlling for habitat affinities: A study with Bahamian lizards and birds. American Naturalist 137:669–93.

Schonewald-Cox, C., R. Azari, and S. Blume. 1991. Scale, variable density, and conservation planning for mammalian carnivores. Conservation Biology 5(4):491–95.

Schulze, E. D. 1982. Plant life forms and their carbon, water, and nutrient relations. Pp. 616–76 in O. L. Lange, P. S. Nobel, C. B. Osmond, and H. Ziegler, eds., Physiological plant ecology II: Water relations and carbon assimilation. Springer-Verlag, Berlin.

Schumaker, N. H. 1996. Using landscape indices to predict habitat connectivity. Ecology 77(4):1210–25.

Seidensticker, J. C., M. G. Hornocker, W. V. Wiles, and J. P. Messick. 1973. Mountain lion social organization in the Idaho Primitive Area. Wildlife Monographs 35:1–60.

Shrader-Frechette, K. S., and E. McCoy, eds. 1993. Method in ecology: Strategies for conservation. Cambridge University Press, New York.

Sibly, R. M. 1995. Life-history evolution in spatially heterogeneous environments, with and without phenotypic plasticity. Evolutionary Ecology 9(3):242–57.

Sibly, R. M., and R. H. Smith. 1998. Identifying key factors using lambda contribution analysis. Journal of Animal Ecology 67(1):17–24.

Sih, A., P. Crowley, M. McPeek, J. Petranka, and K. Strohmeier. 1985. Predation, competition, and prey communities: A review of field experiments. Annual Review of Ecology and Systematics 16:269–311.

Sih, A., G. Englund, and D. Wooster. 1998. Emergent impacts of multiple predators on prey. Trends in Ecology and Evolution 13(9):350–55.

Simberloff, D. 1998. Flagships, umbrellas, and keystones: Is single-species management passé in the landscape era? Biological Conservation 83(3):247–57.

Simonetti, J. A. 1988. The carnivorous predatory guild of central Chile: A human-induced community trait? Revista Chilena de Historia Natural 61(1):23–26.

Simons, A. M., and M. O. Johnston. 1997. Developmental instability as a bet-hedging strategy. Oikos 80(2):401–6.

Simpson, G. G. 1940. Mammals and land bridges. Journal of the Washington Academy of Sciences 30:137–63.

———. 1944. Tempo and mode in evolution. Columbia University Press, New York.

———. 1953. The major features of evolution. Columbia University Press, New York.

Sinclair, A. R. E. 1989. Population regulation in animals. Pp. 197–242 in J. M. Cherrett, ed., Ecological concepts: The contribution of ecology to an understanding of the natural world. Blackwell Scientific Publications, Cambridge, Massachusetts.

Sinclair, A. R. E., and P. Arcese, eds. 1995. Serengeti II: Dynamics, management, and conservation of an ecosystem. University of Chicago Press, Chicago.

Sinclair, A. R. E., J. M. Gosline, G. Holdsworth, C. J. Krebs, S. Boutin, J. N. M. Smith, R. Boonstra, and M. Dale. 1993. Can the solar cycle and climate synchronize the snowshoe hare cycle in Canada? Evidence from tree rings and ice cores. American Naturalist 141(2):173–98.

Sinclair, A. R. E., and M. Norton Griffiths, eds. 1979. Serengeti: Dynamics of an ecosystem. University of Chicago Press, Chicago.

Sinclair, A. R. E., C. J. Krebs, J. N. M. Smith, and S. Boutin. 1988. Population biology of snowshoe hares III: Nutrition, plant secondary compounds and food limitation. Journal of Animal Ecology 57(3):787–806.

Sinclair, A. R. E., P. D. Olsen, and T. D. Redhead. 1990. Can predators regulate small mammal populations? Evidence from house mouse outbreaks in Australia. Oikos 59(3):382–92.

Sinclair, A. R. E., and R. P. Pech. 1996. Density dependence, stochasticity, compensation, and predator regulation. Oikos 75(2):164–73.

Sinclair, A. R. E., R. P. Pech, C. R. Dickman, D. Hik, P. Mahon, and A. E. Newsome. 1998. Predicting effects of predation on conservation of endangered prey. Conservation Biology 12(3):564–75.

Sit, V., and B. Taylor, eds. 1998. Statistical methods for adaptive management studies. Land

Management Handbook, no. 42. Research Branch, British Columbia Ministry of Forestry, Victoria, Canada.

Skalski, J. R. 1991. Using sign counts to quantify animal abundance. Journal of Wildlife Management 55:705–15.

———. 1994. Estimating wildlife populations based on incomplete area surveys. Wildlife Society Bulletin 22:192–203.

Skalski, J. R., and D. S. Robson. 1992. Techniques for wildlife investigations: Design and analysis of capture data. Academic Press, New York.

Skogland, T. 1991. What are the effects of predators on large ungulate populations? Oikos 61(3):401–11.

Slobodkin, L. B. 1972. On the inconstancy of ecological efficiency and the form of ecological theories. Pp. 291–306 *in* E. Deevey, ed., Growth by intussusception: Ecological essays in honor of G. Evelyn Hutchinson. Transactions of the Connecticut Academy of Sciences no. 44.

———. 1992. A summary of the special feature, ratio-dependent predator-prey theory, and comments on its theoretical context and importance. Ecology 73:1564–66.

Slough, B. G., and G. Mowat. 1996. Lynx population dynamics in an untrapped refugium. Journal of Wildlife Management 60(4):946–61.

Smale, L., S. Nunes, and K. E. Holekamp. 1997. Sexually dimorphic dispersal in mammals: Patterns, causes, and consequences. Advances in the Study of Behavior 26:181–250.

Smallwood, K. S., and C. Schonewald. 1996. Scaling population density and spatial pattern for terrestrial, mammalian carnivores. Oecologia 105(3):329–35.

———. 1998. Study design and interpretation of mammalian carnivore density estimates. Oecologia 113(4):474–91.

Smith, J. L. D., C. McDougal, and D. Miquelle. 1989. Scent marking in free-ranging tigers, *Panthera tigris*. Animal Behaviour 37:1–10.

Solbrig, O. T. 1994, Plant traits and adaptive strategies: Their role in ecosystem function. Pp. 97–116 *in* E. D. Schulze and H. A. Mooney, eds., Biodiversity and ecosystem function. Springer-Verlag, Berlin.

Solomon, N. G., and J. A. French, eds. 1997. Cooperative breeding in mammals. Cambridge University Press, New York.

Songer, M. A., M. V. Lomolino, and D. R. Perault. 1997. Niche dynamics of deer mice in a fragmented, old-growth-forest landscape. Journal of Mammalogy 78(4):1027–39.

Soulé, M. E., ed. 1986. Conservation biology: The science of scarcity and diversity. Sinauer Associates, Sunderland, Massachusetts.

———. 1987. Viable populations for conservation. Cambridge University Press, New York.

———. 1996. An unflinching vision: Networks of people defending networks of lands. Pp. 1–8 in D. A. Saunders, J. L. Craig, and E. M. Mattiske, eds., Nature Conservation IV: The role of networks. Surrey Beatty and Sons, Chipping Norton, Australia.

———. 1999. The policy and science of regional conservation. Pp. 1–27 in M. E. Soulé and J. Terborgh, eds. Continental conservation: Scientific foundations of regional reserve networks. Island Press, Washington, D.C.

Soulé, M. E., D. T. Bolger, A. C. Alberts, J. Wright, M. Sorice, and S. Hill. 1988. Reconstructed dynamics of rapid extinctions of chaparral-requiring birds in urban habitat islands. Conservation Biology 2(1):75–92.

Soulé, M. E., and K. A. Kohm, eds. 1989. Research priorities for conservation biology. Society for Conservation Biology and Island Press, Washington, D.C.

Soulé, M. E., and R. Noss. 1998. Rewilding and biodiversity: Complementary goals for continental conservation. Wild Earth 8(3):18–29.

Soulé, M. E., and J. Terborgh, eds. 1999. Continental conservation: Scientific foundations of regional reserve networks. Island Press, Washington, D.C.

Southwood, T. R. E. 1977. Habitat: The templet for ecological strategies? Journal of Animal Ecology 46:337–65.

———. 1988. Tactics, strategies and templets. Oikos 52(1):3–18.

———. 1996. Natural communities: Structure and dynamics. Philosophical Transactions of the Royal Society of London B Biological Sciences 351(1344):1113–29.

Sovada, M. A., A. B. Sargeant, and J. W. Grier. 1995. Differential effects of coyotes and red foxes on duck nest success. Journal of Wildlife Management 59(1):1–9.

Stamps, J. A., and M. Buechner. 1985. The territorial defense hypothesis and the ecology of insular vertebrates. Quarterly Review of Biology 60:155–81.

Stander, P. E. 1998. Spoor counts as indices of large carnivore populations: The relationship between spoor frequency, sampling effort, and true density. Journal of Applied Ecology 35(3):378–85.

Stearns, S. C. 1989. The evolutionary significance of phenotypic plasticity. Bioscience 39(7):436–45.

———. 1992. The evolution of life histories. Oxford University Press, New York.

Stenseth, N. C. 1995. Snowshoe hare populations: Squeezed from below and above. Science 269:1061–62.

Stenseth, N. C., W. Falck, O. N. Bjornstad, and C. J. Krebs. 1997. Population regulation in snowshoe hare and Canadian lynx: Asymmetric food web configurations between hare and lynx. Proceedings of the National Academy of Sciences (USA) 94(10):5147–52.

Stenseth, N. C., and W. Z. Lidicker, eds. 1992. Animal dispersal: Small mammals as a model. Chapman and Hall, New York.

Stoddart, D. M. 1991. The scented ape: The biology and culture of human odour. Cambridge University Press, Cambridge.

Stoms, D. M., F. W. Davis, and C. B. Cogan. 1992. Sensitivity of wildlife habitat models to uncertainties in GIS data. Photogrammetric Engineering and Remote Sensing 58:843–50.

Storm, G. L., and W. M. Tzilkowski. 1982. Furbearer population dynamics: A local and regional management perspective. Pp. 69–90 *in* G. C. Sanderson, ed., Midwest furbearer management. Kansas Chapter Wildlife Society, Bloomington, Indiana.

Stuart, A. J. 1991. Mammalian extinctions in the late Pleistocene of northern Eurasia and North America. Biological Review 66:453–62.

Sunde, P., and T. Kvam. 1997. Diet patterns of Eurasian lynx *(Lynx lynx):* What causes sexually determined prey size segregation? Acta Theriologica 42(2):189–201.

Sunde, P., K. Overskaug, and T. Kvam. 1998a. Culling of lynxes *(Lynx lynx)* related to livestock predation in a heterogeneous landscape. Wildlife Biology 4(3):169–75.

Sunde, P., S. O. Stener, and T. Kvam. 1998b. Tolerance to humans of resting lynxes *(Lynx lynx)* in a hunted population. Wildlife Biology 4(3):177–83.

Sunquist, F., and M. E. Sunquist. 1988. Tiger moon. University of Chicago Press, Chicago.

Swanson, B. J. 1998. Autocorrelated rates of change in animal populations and their relationship to precipitation. Conservation Biology 12(4):801–8.

Swetnam, R. D., P. Ragou, L. G. Firbank, S. A. Hinsley, and P. E. Bellamy. 1998. Applying ecological models to altered landscapes: Scenario-testing with GIS. Landscape and Urban Planning 41(1):3–18.

Tanner, J. E., T. P. Hughes, and J. H. Connell. 1994. Species coexistence, keystone species, and succession: A sensitivity analysis. Ecology 75(8):2204–19.

Taylor, A. D. 1990. Metapopulations, dispersal, and predator-prey dynamics: An overview. Ecology 71:429–33.

Taylor, R. J. 1984. Predation. Chapman and Hall, New York.

Taylor, V. J., and N. Dunstone, eds. 1996. The exploitation of mammal populations. Chapman and Hall, New York.

Terborgh, J. 1988. The big things that run the world: A sequel to E. O. Wilson. Conservation Biology 2:402–3.

Terborgh, J., J. A. Estes, P. C. Paquet, K. Ralls, D. Boyd-Heger, B. Miller, and R. Noss. 1999. The role of top carnivores in regulating terrestrial ecosystems. Pp. 60–103 *in* M. E. Soulé and J. Terborgh, eds. Continental conservation: Scientific foundations of regional reserve networks. Island Press, Washington, D.C.

Thomas, C. D. 1994. Extinction, colonization, and metapopulations: Environmental tracking by rare species. Conservation Biology 8:373–78.

Thomas, D. L., and E. J. Taylor. 1990. Study designs and tests for comparing resource use and availability. Journal of Wildlife Management 54:322–30.

Thompson, I. D. 1991. Could marten become the spotted owl of eastern Canada? Forestry Chronicle 67(2):136–40.

Thompson, I. D., I. J. Davidson, S. O'Donnell, and F. Brazeau. 1989. Use of track transects to measure the relative occurrence of some boreal mammals in uncut forest and regeneration stands. Canadian Journal of Zoology 67(7):1816–23.

Thompson, S. K. 1992. Sampling. John Wiley and Sons, New York.

Tienderen, P. H. v. 1997. Generalists, specialists, and the evolution of phenotypic plasticity in sympatric populations of distinct species. Evolution 51(5):1372–80.

Tilman, D., and J. A. Downing. 1994. Biodiversity and stability in grassland. Nature 367:363–65.

Tilman, D., and P. Kareiva, eds. 1997. Spatial ecology: The role of space in population dynamics and interspecific interactions. Princeton University Press, Princeton.

Tilman, D., C. L. Lehman, and C. Yin. 1997. Habitat destruction, dispersal, and deterministic extinction in competitive communities. American Naturalist 149(3):407–35.

Todd, C. R., and M. A. Burgman. 1998. Assessment of threat and conservation priorities under realistic levels of uncertainty and reliability. Conservation Biology 12(5):966–74.

Travaini, A., M. Delibes, P. Ferreras, and F. Palomares. 1997. Diversity, abundance of rare species as a target for the conservation of mammalian carnivores: A case study in southern Spain. Biodiversity and Conservation 6(4):529–35.

Trexler, J. C., and J. Travis. 1993. Nontraditional regression analyses. Ecology 74:1629–37.

Turchin, P. 1991. Translating foraging movements in heterogeneous environments into the spatial distribution of foragers. Ecology 72:1253–66.

——. 1996. Fractal analyses of animal movement: A critique. Ecology 77(7):2086–90.

——. 1998. Quantitative analysis of movement: Measuring and modeling population redistribution in animals and plants. Sinauer Associates, Sunderland, Massachusetts.

Turner, A., and M. Anton. 1997. The big cats and their fossil relatives: An illustrated guide to their evolution and natural history. Columbia University Press, New York.

van de Koppel, J., M. Rietkerk, and F. J. Weissing. 1997. Catastrophic vegetation shifts and soil degradation in terrestrial grazing systems. Trends in Ecology and Evolution 12(9):352–56.

Van Sickle, W. D., and F. G. Lindzey. 1991. Evaluation of a cougar population estimator based on probability sampling. Journal of Wildlife Management 55:738–43.

Van Tienderen, P. H. 1997. Generalists, specialists, and the evolution of phenotypic plasticity in sympatric populations of distinct species. Evolution 51(5):1372–80.

Van Valkenburgh, B. 1988. Trophic diversity in past and present guilds of large predatory mammals. Paleobiology 14(2):155–73.

——. 1989. Carnivore dental adaptations and diet: A study of trophic diversity within guilds. Pp. 410–36 *in* J. L. Gittleman, ed., Carnivore behavior, ecology, and evolution. Cornell University Press, Ithaca.

——. 1995. Tracking ecology over geological time: Evolution within guilds of vertebrates. Trends in Ecology and Evolution 10(2):71–76.

Van Valkenburgh, B., and C. M. Janis. 1993. Historical diversity patterns in North American large herbivores and carnivores. Pp. 330–40 *in* R. E. Ricklefs, and D. Schluter, eds., Species diversity in ecological communities: Historical and geographical perspectives. University of Chicago Press, Chicago.

Varley, J. D., and W. G. Brewster, eds. 1990. Wolves for Yellowstone? A report to the United States Congress, volumes 1–4. Yellowstone National Park, University of Wyoming, University of Idaho, Interagency Grizzly Bear Study Team, and University of Minnesota Cooperative Park Studies Unit.

Verbyla, D. L., and J. A. Litvaitis. 1989. Resampling methods for evaluating classification accuracy of wildlife habitat models. Environmental Management 13:783–87.

Ver Hoef, J. M., and N. Cressie. 1993. Spatial statistics: Analysis of field experiments. Pp. 319–41 *in* S. M. Scheiner and J. Gurevitch, eds., The design and analysis of ecological experiments. Chapman and Hall, New York.

Via, S., R. Gomulkiewicz, G. De Jong, S. M. Scheiner, C. D. Schlichting, and P. H. Van Tienderen. 1995. Adaptive phenotypic plasticity: Consensus and controversy. Trends in Ecology and Evolution 10(5):212–17.

Via, S., and R. Lande. 1985. Genotype-environment interactions and the evolution of phenotypic plasticity. Evolution 39:505–22.

Vickery, P. D., M. L. Hunter, Jr., and J. V. Wells. 1992. Evidence of incidental nest predation and its effects on nests of threatened grassland birds. Oikos 63:281–88.

Voigt, D. R., and W. E. Berg. 1987. Coyote. Pp. 345–57 *in* M. Novak, J. A. Baker, M. E. Obbard, and B. Malloch, eds., Wild furbearers management and conservation in North America. Ontario Trappers Association, North Bay, Ontario.

Wagner, F. H. 1988. Predator control and the sheep industry: The role of science in policy formation. Regina Books, Claremont, California.

Webb, S. D., R. C. Hulbert, Jr., and W. D. Lambert. 1995. Climatic implications of large-herbivore distributions in the miocene of North America. Pp. 91–108 *in* E. S. Vrba, G. H. Denton, T. C. Partridge, and L. H. Burckle, eds., Paleoclimate and evolution, with emphasis on human origins. Yale University Press, New Haven.

Weber, W., and A. Rabinowitz. 1996. A global perspective on large carnivore conservation. Conservation Biology 10(4):1046–54.

Wennergren, U., M. Ruckelshaus, and P. Kareiva. 1995. The promise and limitations of spatial models in conservation biology. Oikos 74(3):349–56.

Werner, E. E. 1992. Individual behavior and higher-order species interactions. American Naturalist 140:S5–S32.

Werner, E. E., and B. R. Anholt. 1996. Predator-induced behavioral indirect effects: Consequences to competitive interactions in anuran larvae. Ecology 77(1):157–69.

White, G. C., and R. A. Garrott. 1990. Analysis of wildlife radio-tracking data. Academic Press, San Diego.

Whitlock, M. C., P. C. Phillips, F. B.-G. Moore, and S. J. Tonsor. 1995. Multiple fitness peaks and epistasis. Annual Review of Ecology and Systematics 26:601–29.

Wiens, J. A., N. C. Stenseth, B. Van Horne, and R. A. Ims. 1993. Ecological mechanisms and landscape ecology. Oikos 66:369–80.

Wikramanayake, E. D., E. Dinerstein, J. G. Robinson, U. Karanth, A. Rabinowitz, D. Olson, T. Mathew, P. Hedao, M. Conner, G. Hemley, and D. Bolze. 1998. An ecology-based method for defining priorities for large mammal conservation: The tiger as case study. Conservation Biology 12(4):865–78.

Wileyto, E. P., W. J. Ewens, and M. A. Mullen. 1994. Markov-recapture population estimates: A tool for improving interpretation of trapping experiments. Ecology 75:1109–17.

Williamson, M. 1981. Island populations. Oxford University Press, Oxford.

Willis, K. J. 1993. How old is ancient woodland? Trends in Ecology and Evolution 8(12):427–28.

Wilson, D. S. 1998. Adaptive individual differences within single populations. Philosophical Transactions of the Royal Society of London B Biological Sciences 353:199–205.

Wilson, E. O. 1975. Sociobiology, the new synthesis. Harvard University Press, Cambridge, Massachusetts.

With, K. A. 1997. The application of neutral landscape models in conservation biology. Conservation Biology 11(5):1069–80.

With, K. A., and T. O. Crist. 1995. Critical thresholds in species' responses to landscape structure. Ecology 76(8):2446–59.

With, K. A., R. H. Gardner, and M. G. Turner. 1997. Landscape connectivity and population distributions in heterogeneous environments. Oikos 78(1):151–69.

Witmer, G. W., S. K. Martin, R. D. Sayler, and T. M. Quickley. 1998. Forest carnivore conservation and management in the interior Columbia Basin: Issues and environmental correlates. General Technical Report PNW-GTR-420. Pacific Northwest Research Station, Forest Service, U.S. Department of Agriculture, Portland, Oregon.

Wolff, J. O. 1997. Population regulation in mammals: An evolutionary perspective. Journal of Animal Ecology 66(1):1–13.

Woodroffe, R. 1997. Review of "Carnivore behavior, ecology, and evolution, v. 2." Trends in Ecology and Evolution 12(6):240.

Woodroffe, R., and J. R. Ginsberg. 1998. Edge effects and the extinction of populations inside protected areas. Science 280:2126–28.

Wootton, J. T. 1994. The nature and consequences of indirect effects in ecological communities. Annual Review of Ecology and Systematics 25:443–66.

Wright, S. J., M. E. Gompper, and B. Deleon. 1994. Are large predators keystone species in neotropical forests? The evidence from Barro Colorado Island. Oikos 71(2):279–94.

Yahner, T. G., N. Korostoff, T. P. Johnson, A. M. Battaglia, and D. R. Jones. 1995. Cultural landscapes and landscape ecology in contemporary greenway planning, design and management: A case study. Landscape and Urban Planning 33(1–3):295–316.

Yoneda, M., and K. Maekawa. 1982. Effect of hunting on age structure and survival rates of red fox in eastern Hokkaido. Journal of Wildlife Management 46:781–86.

Yoshimura, J., and C. W. Clark. 1991. Individual adaptations in stochastic environments. Evolutionary Ecology 5(2):173–92.

Young, S. P. 1958. The bobcat of North America: Its history, life habits, economic status and control, with list of currently recognized subspecies. Stackpole, Harrisburg, Pennsylvania.

Young, S. P., and E. A. Goldman. 1944. The wolves of North America. American Wildlife Institute, Washington.

——. 1946. The puma, mysterious American cat. American Wildlife Institute, Washington.

Young, S. P., and H. H. T. Jackson. 1951. The clever coyote. Stackpole, Harrisburg, Pennsylvania, and Wildlife Management Institute, Washington.

Young, T. P. 1994. Natural die-offs of large mammals: Implications for conservation. Conservation Biology 8(2):410–18.

Zakrzewski, R. J., and J. A. Lillegraven. 1994. Paleomammalogy. Pp. 200–214 *in* E. C. Birney and J. R. Choate, eds., Seventy-five years of mammalogy (1919–1994). Special Publication no. 11 of Mammalogists.

Zezulak, D. S. 1998. Spatial, temporal, and population characteristics of two bobcat, *Lynx rufus* (Carnivora: Felidae), populations in California. Ph.D. diss., University of California, Davis.

Zezulak, D. S., and R. G. Schwab. 1979. A comparison of density, home range, and habitat utilization of bobcat populations at Lava Beds and Joshua Tree National Monuments, California. National Wildlife Scientific Technical Series 6:74–79.

Zielinski, W. J., and T. E. Kucera, eds. 1995. American marten, fisher, lynx, and wolverine: Survey methods for their detection. General Technical Report PSW-GTR-157. Pacific Southwest Research Station, Forest Service, U.S. Department of Agriculture, Albany, California.

Zielinski, W. J., and H. B. Stauffer. 1996. Monitoring *Martes* populations in California: Survey design and power analysis. Ecological Applications 6(4):1254–67.

Zimov, S. A., V. I. Chuprynin, A. P. Oreshko, F. S. Chapin III, M. C. Chapin, and J. F. Reynolds. 1995a. Effects of mammals on ecosystem change at the Pleistocene-Holocene boundary. Pp. 127–35 in F. S. Chapin III and C. Koerner, eds., Arctic and alpine biodiversity: Patterns, causes and ecosystem consequences. Springer-Verlag, New York.

Zimov, S. A., V. I. Chuprynin, A. P. Oreshko, F. S. Chapin III, J. F. Reynolds, and M. C. Chapin. 1995b. Steppe-tundra transition: A herbivore-driven biome shift at the end of the pleistocene. American Naturalist 146(5):765–94.

Contributors

ERIC M. ANDERSON is an associate professor in the wildlife department at the University of Wisconsin, Stevens Point, where he teaches courses in conservation biology, ecology, and biostatistics. He earned his M.S. and Ph.D. from Colorado State University in wildlife biology and his B.S. from Michigan State University in general science and environmental education. Eric's research has focused on resource selection by North American carnivores and the impacts of human development. In 1996, for instance, he published two articles on bobcats, "Bobcat movements and home ranges relative to roads in Wisconsin" and "Bobcat home range size and habitat use in northwestern Wisconsin." He is also coauthor of a chapter on "Measuring vertebrate use of terrestrial habitats and foods" in the fifth edition of the Wildlife Society's *Wildlife techniques manual*. Eric and his colleagues have been investigating the impact of roads on the movement and behavior of wolves in the Great Lakes region.

EDWARD E. BANGS is the wolf recovery coordinator for the U.S. Fish and Wildlife Service. He received a B.S. from Utah State University and an M.S. from the University of Nevada, Reno. Since 1988 he has been in Helena, Montana. Initially, he led wolf recovery efforts in Montana, followed by preparation of the congressionally mandated EIS on wolf reintroduction into central Idaho and Yellowstone National Park and actual reintroduction of wolves. Ed has published many popular and scientific articles, including "Reintroducing the gray wolf to central Idaho and Yellowstone National Park" (1996 with S. H. Fritts) and "Wolf hysteria: Reintroducing wolves to the West" in *The war against the wolf* (R. McIntyre 1995, Voyageur Press). He has received numerous awards, including the Meritorious Service Award from the secretary of the interior in 1994 and the Distinguished Service Award of the Montana chapter of the Wildlife Society in 1996.

BONNIE M. BLANCHARD has conducted wildlife research for twenty-three years, the last nineteen as a permanent wildlife biologist with the Interagency Grizzly Bear Study Team. She earned both her B.S. and M.S. degrees from Montana State University in Bozeman. Over the years Bonnie has been involved in nearly all aspects of the study team's grizzly bear research. She has published more than twenty papers in professional journals and proceedings and more than twenty technical reports, including the study team annual reports since 1978. Among her major papers is one in *Biological Conservation* on the movements of Yellowstone's grizzlies. She has given approximately twenty presentations concerning the study team's research results to private and agency groups.

MARK S. BOYCE received his B.S. from Iowa State University, his M.S. from University of Alaska, Fairbanks, and his M.Phil. and Ph.D. from Yale University. He has conducted field research and theoretical studies of numerous birds and mammals, including whooping cranes, spotted owls, elk, swift foxes, wolves, and bighorn sheep. Mark has done research in the Greater Yellowstone Ecosystem for more than two decades and is a leading expert on simulation modeling and ecosystem management. Mark is the Vallier Chair of Ecology and a Wisconsin Distinguished Professor at the University of Wisconsin, Stevens Point. He is also editor in chief of the *Journal of Wildlife Management*. His many publications include *Ecosystem management: Applications for sustainable forest and wildlife resources* (1997, Yale University Press), *The Greater Yellowstone Ecosystem: Redefining America's wilderness heritage* (1994, Yale University Press), and *The Jackson elk herd: Intensive wildlife management in North America* (1989, Cambridge University Press).

WAYNE G. BREWSTER received his B.S. in wildlife management and his M.S. in wildlife and fisheries science from South Dakota State University. After serving four years as a wildlife biologist for the U.S. Fish and Wildlife Service, Wayne supervised the agency's endangered species program in Montana and Wyoming from 1979 to 1988. During this time he implemented recovery programs for bald eagles, peregrine falcons, grizzly bears, black-footed ferrets, and wolves. From 1988 to 1991 Wayne was the wolf management coordinator for Glacier National Park, and in 1991 he moved to Yellowstone National Park to serve as the technical chairman for the Wolf Management Committee. He oversaw preparation of the *Wolves for Yellowstone?* reports (1992, Yellowstone National Park) and represented the Park Service on the planning team that prepared the EIS for wolf reintroduction to Yellowstone and central Idaho. Wayne is deputy director for the Yellowstone Center for Resources and program supervisor for the wolf reintroduction program in Yellowstone.

STEVEN W. BUSKIRK earned his Ph.D. from the University of Alaska for his work with American martens. He is head of the Department of Zoology and Physiology at the University of Wyoming. Steve has studied a wide range of theoretical and practical problems involving mammals, habitat ecology, physiological ecology, community ecology, and population genetics. He has conducted field research on mammalian carnivores in Wyoming, Alaska, and China. Steve is best known for his work with mustelids and such publications as *Martens, sables, and fishers: Biology and conservation* (1994, Cornell University Press) and *Conserving circumboreal forests for martens and fishers* (1992, Conservation Biology).

TIM W. CLARK is profesor adjunct in the School of Forestry and Environmental Studies at Yale University, fellow in the Institution for Social and Pol-

icy Studies at Yale, and president of the Northern Rockies Conservation Cooperative in Jackson, Wyoming. He received his Ph.D. from the
University of Wisconsin, Madison. His research interests include conservation biology, policy sciences, and organization theory and management. He has worked for more than twenty-five years on endangered species conservation in the United States and overseas. Tim has
published more than two hundred papers and several books and
monographs, including *Endangered species recovery: Finding the lessons,
improving the process* (coedited with R. P. Reading and A. L. Clarke, 1994,
Island Press), *People and nature: Perspectives on private land use and endangered species recovery* (coedited with A. Bennett and G. Backhouse,
1995, Transactions of the Royal Zoological Society of New South
Wales), and *Averting extinction: Reconstructing endangered species recovery*
(1997, Yale University Press). In 1996 he coedited a 120-page special section of the journal *Conservation Biology* on "Large carnivore conservation in the Rocky Mountains of the United States and Canada," with
contributions from twenty-five authors.

ROBERT L. CRABTREE received his B.S. at the University of Idaho and his M.S. at
Utah State University, and he returned to Idaho for his Ph.D. in
forestry, wildlife, and range sciences. Bob accepted a visiting scholar–
postdoctoral appointment at University of California, Berkeley, in
1989 to work on survey techniques and landscape modeling. He was
awarded numerous grants to conduct research in Yellowstone National Park to examine canid ecology and the effects of wolf reintroduction on ecosystems. In 1993 he founded a nonprofit organization
dedicated to large-scale and long-term ecological research, Yellowstone Ecosystem Studies. Bob is also affiliate faculty of biology at
Montana State University and research ecologist with the Yellowstone
Cooperative Park Studies Unit. He is primarily interested in applied
and conservation ecology (especially predator-prey relations), predator behavior, population dynamics, and survey methods for carnivores and large herbivores. Among his recent publications is "Social
and nutritional factors influencing the dispersal of resident coyotes"
in *Animal Behavior*.

F. LANCE CRAIGHEAD received his B.S. from Carleton College, his M.S. from the
University of Wisconsin, Madison, and his Ph.D. from Montana State
University. After working as a Peace Corps volunteer in the South Pacific, he studied Pribilof Island seabirds in Alaska for his master's
thesis and spent the next ten years on numerous wildlife research
projects for state and federal agencies and consulting firms, primarily in Alaska. Lance's Ph.D. dissertation focused on the conservation
genetics of grizzly bears based on DNA microsatellite analysis of a population in the western Brooks Range of Alaska. He is an affiliate asso-

ciate professor of biology at Montana State University, executive director of the Craighead Environmental Research Institute, and a consultant for the Montana Department of Fish, Wildlife, and Parks, American Wildlands, and some private firms. His research focus is analysis and mapping of wildlife corridors for large carnivores in the Northern Rocky Mountains of the United States and Canada.

A. PEYTON CURLEE earned her M.S. in wildlife ecology with a concentration in conservation biology from Yale University's School of Forestry and Environmental Studies. Her thesis work focused on a small parrot species in the central grasslands of Venezuela. Previous work as a field assistant included seabird studies off the Pribilof Islands and the eastern United States. She has worked as a freelance editor, writer, and graphic designer and served as associate editor of *Earth Island Journal* and assistant editor of Friends of the Earth's *Not Man Apart* journal. Peyton is executive director of the Nothern Rockies Conservation Cooperative (NRCC) in Jackson, Wyoming. Carnivore conservation in the northern Rockies of the United States and Canada is one of her major research areas. Recent publications she has cowritten or coedited include a special section of *Conservation Biology*, "Large carnivore conservation in the Rocky Mountains of the United States and Canada" (1996).

MICHAEL E. GILPIN is a professor of conservation biology at the University of California, San Diego. He earned an M.S. in theoretical physics before joining the Peace Corps in 1967 and teaching computer science at the Middle East Technical University in Ankara, Turkey. After returning to the United States, he earned a Ph.D. in theoretical ecology at the University of California, Irvine. He joined with Michael Soulé, initially a colleague at U.C., San Diego, to investigate genetic, demographic, and metapopulational aspects of minimum viable populations, and they went on to develop the approach of Population Viability Analysis (PVA). Since 1986 Mike has been involved in modeling the viability of a number of endangered species, including condors, tortoises, kangaroo rats, squawfish, bull trout, grizzly bears, and bighorn sheep. He has written and edited six books, the most recent of which is *Metapopulation biology: Genetics, evolution, and ecology* (1996, Academic Press).

MAURICE G. HORNOCKER earned his B.S. and M.S. degrees at the University of Montana and a Ph.D. at the University of British Columbia. Maurice is best known for his pioneering studies of cougars. He was the first to mark individuals and use radiotelemetry in the study of cougars, bobcats, lynx, wolverines, badgers, and river otters in North America and leopards in Africa. Maurice has also studied jaguars in Central America and South America, tigers in Asia, and ocelots in south Texas. After seventeen years as leader of the Idaho Cooperative Wildlife Research Unit, in 1985 he founded the Hornocker Wildlife Institute, which he directs. He has written fifty scientific publications and more than fifteen popular articles, and he participated in the production of four documentary films. He and his colleagues are pursuing long-term studies of cougars in Yellowstone and Glacier national parks, black bears in New Mexico, and tigers in Siberia.

KURT A. JOHNSON is a zoologist with the Office of Scientific Authority, U. S. Fish and Wildlife Services, in Arlington Virginia. He earned his B.S. in wildlife science and his Ph.D. in animal ecology at Utah State University, as well as an M.S. in wildlife ecology from the University of Wisconsin, Madison. Kurt has more than twenty years' experience in the fields of wildlife biology and international conservation. He has worked on such varied problems as international wildlife trade, conservation of caprids in Pakistan, carnivore-prey relations in southern Chile, small-mammal survival and recolonization following the eruption of Mount St. Helens, protected area planning on the Tibetan Plateau, and medicinal plant conservation. Kurt's work centers on implementation of c.i.t.e.s. and e.s.a. listings of foreign species.

PETER M. KAREIVA received his B.A. in zoology from Duke University, his M.S. in ecology from the University of California, Irvine, and his Ph.D. in ecology and evolutionary biology from Cornell University. His research has included such topics as insect herbivory, dispersion, and conservation; desert tortoise population viability analysis; and the effects of nonnative and genetically engineered organisms. Peter has written many articles, essays, and book chapters in addition to editing such books as *Conservation biology for the next decade* (1997, Chapman and Hall) and *Biotic interactions and global change* (1993, Sinauer). As professor of zoology at the University of Washington, he teaches conservation biology and conducts research in theoretical ecology, conservation, and agricultural ecology. Peter has served on the editorial boards of more than ten journals and on the board of governors for the Nature Conservancy and the science advisory boards of the Environmental Protection Agency and the National Center for Ecological Analysis and Synthesis.

RICHARD R. KNIGHT has been engaged in wildlife research for more than thirty years. He earned his Ph.D. at the University of Minnesota. He served as a research biologist for the Montana Department of Fish, Wildlife, and Parks and as a faculty member at the University of Idaho. He is best known for his work as the founding leader of the Interagency Grizzly Bear Study Team, which he headed for twenty-four years. Dick has written more than thirty-five technical publications and more than forty technical reports and white papers on large mammals such as bears and elk, among them *The Sun River elk herd* (1970, Wildlife Monographs no. 23). He serves as an adjunct associate professor of biology at Montana State University.

JOHN A. MACK is a wildlife biologist with the Yellowstone Center for Resources. He earned his B.S. in biological sciences and his M.S. in fish and wildlife management from Montana State University. John conducted research on black and grizzly bears in Montana and collected data for a habitat enhancement project for bighorn sheep in northwestern Montana. From 1989 to 1995 he worked on wildlife resources, ungulate monitoring, and gray wolf research in Yellowstone. Since then he has been responsible for ungulate inventory and monitoring and other research efforts in the park, including bison planning, management, and policy issues. John has published papers on black bears in the International Conference on Bear Research and Management and on wolves and related topics in *Wolves for Yellowstone?* as well as a monograph, *Ecological issues on reintroducing wolves into Yellowstone National Park.* A cowritten paper on Northern Range elk will soon be published in the *Intermountain Journal.*

STEVEN C. MINTA is on the research faculty in wildlife ecology Environmental Studies Board, University of California, Santa Cruz. He began working in the Yellowstone region in 1982 on carnivore ecology and research methods for his dissertation from the University of California, Davis. From 1988 to 1993 he consulted with nongovernmental organizations and government agencies, bringing the tools of academic research to bear on practical problems of land use and species management. He has collaborated with groups developing GIS as a platform for spatial modeling and environmental decision making and as an aid to interdisciplinary efforts in the emerging fields of landscape ecology and ecosystem management. He is collaborating on a long-term project to develop both large-scale monitoring methods for carnivores and a predictive approach for relating landscape attributes to carnivore distribution in the western Yellowstone. Steve has contributed chapters to edited volumes on conservation science and published in *Ecology, Ecological Applications,* and *Oecologia.*

KERRY M. MURPHY completed his B.S. and M.S. in wildlife biology at the University of Montana and his Ph.D. in forestry, range, and wildlife science at the University of Idaho. His Ph.D. work included a nine-year study of cougar population dynamics, social ecology, and predator-prey relations in Yellowstone National Park and vicinity. This comprehensive research has provided important baseline information for resource managers in Yellowstone and surrounding states. Kerry has published several professional and popular articles on aspects of cougar ecology and biology. He has worked as a research ecologist and wildlife technician for the Montana Department of Fish, Wildlife and Parks, Idaho Department of Fish and Game, Washington Department of Wildlife, and the Bureau of Indian Affairs. Kerry is employed as a research ecologist with the Hornocker Institute and Yellowstone National Park's Wolf Project.

P. IAN ROSS earned his honors B.S. in wildlife biology from the University of Guelph. He is senior wildlife biologist with Arc Wildlife Services Ltd. in Calgary, a consulting firm specializing in wildlife research and management. Ian's particular interest is the ecology, management, and conservation of large carnivores and ungulates and their habitats in western and northern Canada. With his partners, he conducted the most intensive study of cougar ecology in Canada, investigating population characteristics, food habits, and habitat use from 1981 to 1994. Ian has conducted long-term population and habitat studies of grizzly bears in Alberta and British Columbia, and population genetics and habitat research on bighorn sheep in the Rocky Mountains. He has also conducted surveys of caribou, musk oxen, and raptors on the Canadian central Arctic coast.

PAUL SCHULLERY has worked for about twelve of the past twenty-five years in Yellowstone National Park as a ranger-naturalist, park historian, resource naturalist, chief of cultural resources, and senior editor of the Yellowstone Center for Resources. He is an affiliate professor of history at Montana State University and an adjunct professor of American studies at the University of Wyoming. Paul earned his M.A. in American history from Ohio University and was awarded an honorary doctorate of letters by Montana State University in 1997. His research emphasis has been on conservation history, especially resource management and science in the national parks. Paul is the author, coauthor, or editor of twenty-seven books about conservation, natural history, and outdoor sports, including six books about bears. He edited *The Yellowstone wolf: A guide and sourcebook* (1996, High Plains Publishing), the most complete reference work on the subject. His environmental history, *Searching for Yellowstone: Ecology and wonder in the last wilderness,* was published in 1997 (Houghton Mifflin).

JENNIFER W. SHELDON works as a research biologist on the canid ecology project with Yellowstone Ecosystem Studies in Bozeman, Montana. She is also a Ph.D. candidate in the biology department at Montana State University. After receiving a B.A. in natural sciences and mathematics from Bennington College, she graduated with an M.S. in zoology from Colorado State University. Her primary area of interest is the demographics and behavioral ecology of canids. She is the author of the book *Wild dogs: The natural history of the non-domestic Canidae* (Academic Press, 1992).

FRANCIS J. SINGER is a research ecologist with the biological resources division of the U.S. Geological Survey and senior scientist with the Natural Resource Ecology Lab at Colorado State University. He directs a large research program on the conservation biology and restoration ecology of bighorn sheep in sixteen national parks and monuments, plant-ungulate interactions and management of ungulates in six national parks, and conservation biology and population dynamics of wild horses on Department of Interior lands. From 1985 to 1991 he was a research ecologist at Yellowstone National Park, studying the ecology of elk, the effects of natural regulation policy on the northern Yellowstone winter range, and the predicted effects of wolf restoration on the ungulates of the Yellowstone ecosystem. For his work in Yellowstone, Frank received the 1990 Regional Directors and also the Directors of the National Park Service Award for Science. Frank has published more than fifty papers on the ecology and management of eight species of ungulates.

DOUGLAS W. SMITH received his B.S. in wildlife biology from the University of Idaho, his M.S. in biology from Michigan Technological University, and his Ph.D. from the University of Nevada, Reno, in the program of ecology, evolution, and conservation biology. His dissertation research was on the dispersal and cooperative breeding behavior of beavers. Doug is project leader and biologist for the Yellowstone Gray Wolf Restoration Project in Yellowstone National Park. He previously worked with wolves in Isle Royale, Michigan, and in Minnesota. Doug has published numerous articles about beavers and wolves, including "Denning behavior of non-gravid wolves" (with L. D. Mech, T. J. Kreeger, and M. K. Phillips, 1996), and he cowrote *The wolves of Yellowstone* (Voyageur Press, 1996) with Michael Phillips.

ERNEST R. VYSE is professor of biology and interim department head at Montana State University, where he teaches courses in cell and molecular biology and genetics. He earned his B.S. at the University of Idaho, his M.S. at the University of Oregon, and his Ph.D. at the University of Alberta in Edmonton. Ernie is interested in genetic diversity within and between wildlife populations and in the conservation biology of small populations and endangered species. His research efforts include the study of mating systems, paternity, dispersal, and migration in wildlife populations through the application of molecular genetics. He has lived, hunted, and done research in the Greater Yellowstone Ecosystem for more than twenty years. In 1994 Ernie received the Wildlife Society award for the outstanding publication in wildlife ecology and management.

LEE H. WHITTLESEY is archivist for history for the National Park Service at Yellowstone National Park, where he manages the park archives and research library and comanages the park historic photographic collection. The Yellowstone archives is one of only five national satellites of the National Archives in Washington, D.C. Lee has been researching the history of the Yellowstone region for twenty-five years and has produced six books and numerous journal articles, including *Death in Yellowstone: Accidents and foolhardiness in the first national park* (1995, Roberts Rinehart) and *A history of large mammals on the Yellowstone Plateau, 1806–1885* (in preparation with Paul Schullery). He earned his J.D. from the University of Oklahoma and is an affiliate professor in the department of history and philosophy at Montana State University.

Index

Note: **Boldface** page numbers signify figures and tables.

Abundance. *See* Population estimates/ densities
Acclimation, 111
Adaptations, physiological, 167, 172, 176– 78
Adaptive management, 281
Aerial gunning, 120
Africa, 228–229, 358
African wild dog, 151
Age structure, grizzlies, 55, **56**
Alaska, 118, 119, 359
Alberta, Canada, 91, 129
Albright, Horace, 28–29, 53
Algonquin Provincial Park, 106–7, 119
Allendorf, F. W., 289, 296, 305
Allen, S. H., 342
Alley, J. J., 252
Althoff, D. P., 134
American Association for the Advancement of Science, 28
American Society of Mammalogists, 28
Amphibians, 260–61
Amstrup, S. C., 307
Anderson, A. E., 88, 91

Anderson, E. M., 337, 339
Anderson, Jack, 34
Animal Damage Control. *See* Wildlife Services
Antipredatory behavior, 218
Arcese, P., 228–29, 334
Archaeological studies, 14–15
Aspen, 207, 225, 259
Attitudes: and bears in Yellowstone, 32, 69, 71; and carnivore conservation, 324, 374–75; environmentalism and, 36–38; of Euro-American settlers, 3, 21; and maintaining genetic diversity, 313; and mountain lions, 94, 96, 312; and predator management policy in NPS, 11–12, 27–28, 33, 42–43; and wolf restoration, 39, 41–42, 109, 118–20
Autocorrelation function (ACF), 271–73, **272, 274,** 275–76, **275, 276, 277**

Bader, M., 297
Badger *(Taxidea taxus):* abundance in GYE, 165, 168, 170; and coyote, 175; food sources, 193; as habitat generalist/ feeding specialist, 256–57; history in Yellowstone, 13, 22; range, 311; and wolf restoration, 180, 259

Bailey, V., 170, 179
Balharry, D., 354
Ballard, W. B., 220
Bannock Indians, 15–16
Barbee, Robert, 64–65
Barnhurst, D., 87
Barrette, C., 136, 137, 138
Barrett, R. H., 83, 88
Barrowclough, G. F., 289
Bartholow, J., 219
Bats, 243
Bear *(Ursidae):* body sizes/shapes and, 176;
    and carrion, 77, 214; controversy on
    managing Yellowstone, 33–34; and
    cougars, 86; defining the GYE and
    research on, 3, 4; genetic diversity and,
    305–7; human impact and, 342;
    symbolism and, 1–3; as tourist
    attraction in Yellowstone, 25–26, 31–
    32, 51; ungulates and, 193, 197. *See also*
    Bear, black; Bear, grizzly
Bear, black *(Ursus americanus):* and
    domestic sheep, 61; food sources, 240;
    and genetic diversity, 305–6; as
    habitat/feeding generalist, 256; history
    in Yellowstone, 13, 18; population
    estimates, 67–68, 307; as tourist
    attractions, 51; ungulates and, 192,
    215; wolves and, 114, 225
Bear, grizzly *(Ursus arctos):* effective
    population size, 289, 297; estimating
    population structure/condition of, 54–
    56, **56**, 57–59, **58, 59**, 63–64, **64**, 306;
    food sources, 240, 241; genetic diversity
    and, 293, 305, 307, 315; as habitat/
    feeding generalist, 256; habitat
    requirements, 62, 70, 171, 312; history
    and early park policy, 13, 14, 18, 52–54;
    infanticide and male, 90; predation/
    foraging behavior, 59–63, 148; recovery
    effort, 41, 56–57, 68–72; research on,
    33–34, 51; territory/range, 301–2, **301**,
    306–7; and ungulates, 192–93, 214,
    215; and wolves, 112–14, 225
Bear, polar, 305
Beauvais, G. P., 251
Beaver *(Castor canadensis):* and Algonquin
    Park study, 107; and early YNP manage-
    ment policy, 23; and Isle Royale study,
    106; population estimates, 205–7, 228,
    246; and wolves, 225, 258–59, 260

Beecham, J., 307
Before-after-control-impact (BACI)
    analysis, 340
Behavioral ecology, 328, 329;
    antipredatory behavior, 218; and
    research on coyotes, 128–29. *See also*
    Behavioral instability; Behavioral
    phenotypes
Behavioral instability, 342–43
Behavioral phenotypes, 364–69, 371
Beier, P., 81, 83, 84–85, 88
Beissinger, S. R., 355
Bekoff, M., 128, 134, 136, 137, 139, 331
Belden, R. C., 84
Berger, J., 86
Berg, W. E., 142
Bialowieza Primeval Forest, Poland, 358
Big Bend National Park, 85
Big Cypress National Preserve, 84
Biggs, T., 217
Bighorn sheep *(Ovis canadensis):* and
    antipredatory behavior: 218;
    competition with elk, 209; and
    cougars, 91; history in GYE, 13, 14, 18,
    194; population estimates, 204, 207,
    228; and wolf restoration, 225
Biochemical analysis, 329
Biogeography, 328, 332
Biopolitics, 34
Birds, 241, 260–61
Bison *(Bison bison):* and antipredatory
    behavior, 218; causes of mortality, 214;
    effect of 1988 fires on, 212; effect on
    vegetation of, 193; and grizzly bear,
    54–55; history in GYE, 13, 14, 18, 21,
    194; jump sites, 17; management policy
    and, 195, 201–2; population, 201–2,
    207, 228; range, 207; wolves and, 112
Bissonette, J. A., 245
Bobcat *(Lynx rufus):* and cougar, 86;
    dispersal, 308; history in Yellowstone,
    14; population estimates/range, 165,
    168, 170, 309; research on, 168
Body-mass distribution, 172–75, **174**
Body size/shape, 172–75, 176–78, **177**
Boone and Crockett Club, 28
Boutin, S., 197, 279
Bowen, W. D., 136, 137
Bowers, M. A., 369
Bowers, W. Z., 369
Boyce, M. S., 201, 219, 274, 337, 339, 351

Boyd, D. K., 302
Brady, C. A., 341
Brainerd, S. M., 309
Branch of Research and Education, 30
Bronson, F. H., 367
Brussard, P. F., 351
Bunnell, F. L., 342
Buskirk, S. W., 171, 241, 248, 256, 257

Cahalane, V. H., 21, 28
Camenzind, F. J., 128, 139, 143
Campbell, T. M., 251
Canada, 91, 119, 359
Canids *(Canidae)*: body sizes/shapes and,
    176; coexistence and, 150, 154–55, 156,
    177, 224; genetic status in GYE, 302–5;
    hybridization of, 156; interspecific
    competition and, 145–47, 150–54; and
    research on carnivores, 127–28; three-
    species pattern for, 151–52. *See also*
    Coyote; Fox, red; Wolf, gray
Cannon, K. P., 14
Carbyn, L. N., 220
Caribou, **199**, 218
Carnivore Behavior, Ecology, and Evolution
    (Gittleman), 328, 333–34
Carnivore management: adaptive, 339–
    40; environmentalism and, 37–38;
    and identifying threats, 354, 358;
    integration with carnivore research,
    334–35, 373–74; and mesocarnivores,
    181–82; and predator-prey theory, 337–
    40; recovery efforts, 39–42; role of
    genetics in, 285, 287, 293–94, 295–99,
    315–16; Yellowstone and, 11–12, 18,
    21–27, **26**, 27–36, 42–43. *See also*
    Carnivores; Wildlife management;
    Wolf restoration
Carnivores: body-mass distribution in
    GYE, **174**; and community interaction,
    345–53; ecological/conservation
    research and, 327–45, 353–56, 366–72;
    effect on ecosystem of, 265–66; as
    experimental subjects, 325–27, 372–
    74; genetic diversity and, 285, 315–16;
    as indicators of ecosystem health, 323–
    24, 353; paleohistory of, 356–59;
    population genetics of, 299–312, **302**;
    populations in GYE of, 286–87, **286**;
    traits, 330. *See also* Carnivore
    management; Carnivores, large;

Carnivores, small; Mesocarnivores;
    specific carnivore species
Carnivores, large, 253–56; and
    controlling prey populations, 30–31,
    32, 35; as habitat and feeding
    generalists, 256; history in Yellowstone,
    13, 14, 18, **20**; widespread destruction
    of, 22, 23, 24–25, **26**; and Yellowstone
    management policy, 43. *See also specific
    species*
Carnivores, small: 253; as habitat and
    feeding generalists, 256; history in
    Yellowstone, 13, 14, 22. *See also*
    Mesocarnivores; *specific species*
Carnivores, The (Ewer), 328
Carpenter, S. R., 349
Carrion: and coyotes, 129, 150; as food
    source for predators, 77, 153–54, 155,
    156, 192, 193, 213–14; and
    mesocarnivores, 175–76
Cattle, 61, 115
Caughley, G., 269, 279, 290, 355
Cavallini, P., 354
Character displacement, 153
Chesness, R. A., 142
Chile, 359
Chilla *(Dusicyon griseus)*, 153
Chipmunk *(Tamias)* family, 244, 249, 250,
    252
Chipmunk, least *(Tamias minimus)*, 244
Chipmunk, Uinta *(Tamias umbrinus)*, 244–
    45
Chipmunk, yellow-pine *(Tamias amoenus)*,
    244
Clark, T. W., 243, 246, 248, 251
Climate: changes and extinctions, 287,
    356, 357; effect on ecosystem of, 227–
    28; global warming, 70; as habitat
    limit, 196; and population dynamics,
    106, 279; and small mammals, 250,
    251; and ungulates, 207; of YNP
    northern winter range, 194
Coexistence, canid, 150, 154–55, 156, 177,
    224
Cole, G. F., 68
Communication, 78–79
Community ecology: community
    interactions and carnivore studies, 327,
    345–53; as a new field, 332
Compensation: and carnivore life
    histories, 367; and predation, 215;

Compensation (*continued*)
and wolf restoration, 197–98, 221–24, **222, 223**, 225, 229
Competition, interspecific: canid, 145–47, 150–54; carnivore research and, 348; cougars and, 85, 86–87; mesocarnivores and, 172–76, **173**; and moose, 205; and pronghorn, 202; and ungulates, 207
Competition, intraspecific: coyotes and, 142, 144; and ungulates, 207, 229
Congress, U.S., 109–10
Conservation biology: carnivores and, 331, 353–56, 373–74; and mesocarnivores, 178–82; as a new field, 334–35, 369; public attitudes and carnivore, 324, 374–75; research scale and, 343–45; wolf ecological studies and, 121. *See also* Carnivore management
Cooper Ornithological Club, 28
Copeland, J., 310
Corridors, travel, 298, 315–16; and cougars, 82–83, 95, 96; and grizzlies, 71
Cottingham, K. L., 349
Cottontail, desert (*Sylvilagus audubonii*), 243
Cottontail, mountain (*Sylvilagus nuttallii*), 243
Cougar. *See* Mountain lion
Coughenour, M. B., 219
Cover: 1988 fires and, 209; and cougars, 81, 85, 86; and elk, 218; and lynx, 351, 352; and mesocarnivores, 171, 172
Coyote (*Canis latrans*): and badgers, 175; and canid interspecific competition/coexistence, 145–47, 152, 154–55; carrion as food source for, 77, 129, 150, 214; and cougars, 86; dispersal and, 303–4; flexibility and, 372; genetic diversity and, 303; habitat, 146, 147, 351; as habitat/feeding generalist, 256; history in Yellowstone, 13, 18, 22, 23, 25, 26, 29, 30; human impact on, 143–44, 342–43; as a keystone species, 350, 351; and lynx, 351; management of, 155–56; population estimates/demographics, 139–45, 165, 168, 169, 304; predation/food sources, 129–34, 148–50, **148, 149**, 241; and pronghorn,

202, 213, 216; reproductive patterns, 341; research on, 32, 127–29, **130–33**; role in ecosystem of, 147–50; and snowshoe hare, 352; social ecology, 134–39, 143; territory/range, 136, 137, 144, 304; ungulates and, 192, 193, 215; and wolves, 104, 112, 155, 156, 224, 259
Crabtree, R. L., 136–37, 142, 144–45, 179, 240, 258, 354
Craighead, Frank, 3, 33–34, 324, 329
Craighead, John, 3, 16–17, 33–34, 35, 54, 65, 324, 329
Crawley, M. J., 338
Crayfish, 240
Cricetidae, 246
Cronin, M. A., 295, 305
Culpeo (*Dusicyon culpaeus*), 153
Cumulative effects analysis (CEA), 62, 70
Curio, E., 331
Curnow, E., 27

Davison, R. P., 142
Declining-population paradigm, 355
Deer (*Odocoileus*): history in Yellowstone, 13, 14, 18; population estimates, 205
Deer, mule (*Odocoileus hemionus*): and cougar, 86, 91–92; effect of 1988 fires on, 209; history in GYE, 194; management policy and, 195; population estimates, 203, 207; and wolf restoration, 193, 221, 225–26
Deer, white-tailed (*Odocoileus virginianus*): and Algonquin Park study, 107; and cougars, 85–86; management policy and, 195; population estimates, 204–5
Demographic research, 355–56. *See also* Population estimates/densities
Denali National Park, 119, 217
Density-dependent factors, 196, 197, 214; and elk, 201, 228–29; and population regulation, 266–68, 280
Despain, D. G., 351
Dhole, 151
Diet: flexibility in, 366; grizzly bear, 52–53, 54, 59–63, 70, 214; and mesocarnivores, 167; and sympatric canids, 151–52; and ungulate niches, **208**. *See also* Predation
Disease: brucellosis and bison, 202, 228; and coyote mortality, 141; Euro-American introduction of, 15, 16;

genetic fitness and vulnerability to, 287, 292

Dispersal/migration: bear, 306; and carnivores, 299–300, **300**; cougar, 79–80, 89, 95, 308; coyote, 138–39, 142–43, 144, 303–4; elk, **190**, 200, 228; felids and, 308; fisher, 310; genetic diversity and, 291, 294, 296; modeling for, 363; and ungulates, 194, 198, 201, 202–3; wolverine, 310; and wolves, 104, 217, 303–4; and Yellowstone 1988 fires, 210

DNA analysis, 295, 362

Doak, D. F., 316

Dobyns, H. F., 16

Documentary record, 17–18

Dunstone, N., 334

Eagle, bald, 214

Eagle, golden, 202, 214

East Africa, 153

Ecocenters, 17

Ecological carrying capacity (KCC): and grizzlies, 63–64

Ecological modeling, 325–35, 339, 345–53, 360, 362–64, 369–71; elk population models with wolf restoration, 219–24, **222**, **223**; population modeling, 64–67; three-trophic-level, **268**, 269–81, **269**, **270–77**, **278**

Ecological Society of America, 28

Ecology of the Coyote in Yellowstone, 32

Economic carrying capacity (ECC), 195, 196, 211

Economic impact of carnivore conservation, 41–42, 61, 86, 92, 115, 116, 117–18, 134

Ecosystem: effect of small mammals on, 252–53; effect of wolves on, 104–8; management, 4–7, 324, 334–35, 336; role of carnivores in, 265–66, 323–24; role of coyote in, 147–50; role of mesocarnivores in, 166–68; science, 128, 335–43, 369, 372–73; three-trophic-level model, **268**, 269–78, **269**, **270–77, 278**; ungulate effect on, 198–200; YNP as an, 226–29, **227**. *See also* Ecological modeling; Greater Yellowstone Ecosystem; Landscape ecology

Ehrlich, P. R., 352

Eisenberg, J. F., 331–32, 333–34

Elk: and antipredatory behavior, 218; causes of mortality, **199**, 214–15; and cougars, 85, 86, 91, 148; and coyotes, 134, 148, 150; as dominant prey species, 193–94, 214–16; effect of 1988 fires on, 209, 210, 211, **211**, 212; grizzly bear and, 54–55, 61, 148; history in GYE, 13, 14, 18, 21, 22, 26, 194–95; management, 267; perceived overpopulation of, 30, 31, 35, 38; population estimates, 200–201, 205, 207, 228; population models with wolf restoration, 219–24, **222**, **223**; ranges/migration, **190**, 200; three-trophic-level model and, 269–78; wolves and, 105, 112

Endangered Species Act (ESA): as commitment to conservation, 3, 4; and grizzly bear, 52, 70–71; and mesocarnivores, 168; public education and, 37, 39; wolves and, 108–9, 114, 118–19

Environmental impact statement (EIS), 110, 114–17

Environmentalism, 36–37

Equilibrium model, 332

Ermine *(Mustela erminea)*: food sources, 240, 241; as habitat/feeding generalist, 256; history in Yellowstone, 13, 22; and wolf restoration, 259

Estes, J. A., 349

Ethology. *See* Behavioral phenotypes

Euro-Americans, 15–17, 108

Evolutionary perspectives, 327, 328; and phenotypic design, 367–68

Ewer, R. F., 328

Exotic species, 357

Extinction: and mesocarnivores, 180; probability of, 66; role of genetics in, 287, 288–90, 293, 296; studying past, 356–59

Fain, Steven, 310

Feeding generalists/specialists, 256–58

Felids *(Felidae)*, 307–9. *See also* Bobcat; Lynx; Mountain lion

Ferret, black-footed *(Mustela nigripes)*, 165

Findley, J. S., 169

Fir, balsam, 106

Fire: effect on cougars of, 83–84; effect on ecosystem of, 227–28; effects on

Fire (*continued*)
small mammals of, 251–52; and
paleohistory, 357. *See also* Fires, 1988
Yellowstone
Fires, 1988 Yellowstone: effect on
grizzlies, 62–63; effects on marten,
257; effects on ungulates, 193, 209–14,
211; habitat disturbance and, 240; and
small mammals, 252, 259
Fish, 241. *See also specific species*
Fisher *(Martes pennanti)*: dispersal, 310;
effect of logging on, 181; food sources,
193, 240–41; genetic diversity and, 309,
315; habitat, 171, 172, 175, 311, 313; as
habitat specialist/feeding generalist,
257; as a keystone species, 350;
population estimates, 165, 168, 169,
179, 311; research on, 168
Flowers, C., 374–75
Follman, E. H., 153
Food. *See* Diet; Predation
Foraging/scavenging: efficiency, 137; and
interspecific competition, 153–54
Forbes, S. H., 302
Forest: and 1988 fires, 210–11, 252;
regeneration, 107; and small
mammals, 249. *See also* Logging
Fox, B. J., 173
Fox, red *(Vulpes vulpes)*, 165; and canid
interspecific competition/coexistence,
145–46, 154–55; competition between
red and gray fox, 153; flexibility and,
372; food sources, 175, 193, 240;
genetic diversity and, 303; habitat, 146;
history in GYE, 13, 18, 22, 179; human
impact and reproduction changes in,
341–42; population estimates, 169,
304–5; territory/range, 153, 304–5;
and wolves, 104, 180, 225, 259
Frame, G. W., 152
Frank, D. A., 200
Frankel, O. H., 36, 297
Frankham, R., 289, 293
Frank, L., 143
Franklin, I. A., 295
Franzmann, A. W., 220
French, M. G., 216
French, S. P., 216
Frison, G. C., 14
Frogs, 241
Fruit, 129

Fuller, T. K., 259
Fulmer, K. F., 137
Functional groups, 346
Fur harvest: effects of fur trade, 16;
research based on, 326. *See also*
Trapping

Garton, E. O., 156, 170, 267
Gender ratios: and grizzly bears, 55
Gene flow, 291, 294, 299–300, 313, 316.
See also Dispersal/Migration
Genetic diversity: carnivore management
and, 287, 293–94, 295–99, 315–16;
carnivores and, 285, 302–15, **302**; and
effective population size, 288–90; and
island populations, 6, 71, 108, 287, 293,
297, 313; processes in, 290–93
Genetic drift, 292
Genetic isolation. *See* Island populations
Genetic tools/markers, 295
Geographic Information Systems (GIS),
359, 363
Geostatistical approaches, 362
Gese, E. M., 138, 143
Gilpin, M. E., 296
Gipson, P. S., 134
Gittleman, J. L., 328, 334
Glacier National Park, 302
Global Positioning Systems (GPS), 359, 362
Global warming, 70. *See also* Climate
Goldman, D. P., 295
Gopher, northern pocket *(Thomomys
talpoides)*: and coyotes, 150; effect on
ecosystem of, 253; effects of habitat
disturbance on, 251, 252; in the GYE,
245–46, 249, 250; as habitat generalist/
feeding specialist, 257
Gorman, M. L., 334
Grand Teton National Park (GTNP), 250;
management policy, 195, 196. *See also*
Greater Yellowstone Ecosystem
Grassland: and 1988 fires, 210; and small
mammals, 248–49, 250
Greater Yellowstone Ecosystem (GYE), **2**,
297, **298**, 299; and canid ranges, 146;
carnivore populations in, 286–87, **286**;
as a concept, 35, 37; and cougars, 95;
coyotes in, 147; as ecosystem study
area, 107–8, 194, 261, 324, 343, 350;
genetic status of carnivores in, **302**,
312–16; mammal body-mass

distributions in, **174**; mesocarnivores in, 165–66, 168–70, 178–82; and small mammals, 239–40; species of the, 1, 3–4, 6; ungulates in, 189; wolf restoration/management in, 119, 190–92. *See also* Yellowstone National Park
Great Yellowstone Grizzly Bear Recovery Zone, 297
Grinnell, George Bird, 37
Gruell, G. E., 86
Gunther, K. A., 68

Habitat: change and indicator species, 352–53; cougar, 78, 80–82, 82–85, 96, 97, 171; coyote, 146, 147, 351; fisher, 171, 172, 175, 311, 313; generalists, 256–57; genetic diversity and changing, 314–15; and grizzlies, 62, 70, 171, 312; in the GYE, 194; human disturbance and, 250–51, 314–15; as limiting factor, 196, 287, 312, 313; lynx, 313, 351; marten, 171, 172, 175; and mesocarnivores, 167, 171–72, 175, 256–57; needed research on, 260; red fox, 146; and river otters, 311–12, 313; and small mammals, 239–40, 243–48, 250–51; specialists, 257–58; and species diversity, 250, 251; ungulate niches, **208**; wolf, 171–72; wolverine, 312–13. *See also* Cover
Habituation, 93
Hadly, E. A., 14, 245, 251
Halfpenny, J. C., 85
Hanski, I., 364
Hare, snowshoe (*Lepus americanus*): and coyotes, 129, 150; in the GYE, 243–44; and Isle Royale study, 106; and lynx, 257, 260; population cycle of, 280, 350–51, 352; and wolves, 258–59
Hargis, C. D., 251, 257
Harris, Moses, 24–25
Harris, R. B., 289
Harter, M. K., 240
Hash, H. S., 310
Hatier, K. G., 139
Heath, M. L., 251
Hedrick, P. W., 295
Herptiles, 241
Heterozygosity: and genetic fitness/ diversity, 292, 294; and grizzly bears, 293, 305; measuring, 288–89

Hoak, J. H., 170, 311
Hodgson, J. R., 246
Hornocker, Maurice, 86, 91, 309, 310, 330–31, 341, 342
Horses, 15
Houston, D. B., 17, 194, 219, 311
Human impacts: and bear migration, 306; changing habitats due to, 250–51, 314–15; and coexistence with wildlife, 6; and cougars, 78, 82–93, 96, 97; and coyotes, 129, 143–44; and mesocarnivores, 166–67, 168, 180–81; paleolithic hunters, 357; research and evaluating, 39–40, 341–43, 373–74; small mammals and, 250–51; and wolf mortality, 217
Hunting: and bear mortality, 342; and carnivore mortality rates, 367; and cougars, 85, 87, 89–93, 97; and elk mortality, 195, 216; as factor in population models, 219, 220, 221, 223–24, 229; and grizzly bears, 52; market, 21–22, 23–24; and mesocarnivores, 167, 178; and pronghorns, 203; research based on, 326; and ungulate management, 195, 197–98, 201–2, 203; Yellowstone 1988 fires and, 209
Huntley, N. J., 253
Hybridization: wolf/coyote, 303

Idaho: and grizzly bears, 52, 68; river otter in, 168; and wolf restoration project, 103, 109, 111, 116, 118
Inbreeding, 292; depression, 292–93, 313
India, 359
Indicator/keystone species, 345–53
Individual variation, 365–66
Infanticide, 90
Injury, human, 32
Insects, 240–41, 243
Interagency Grizzly Bear Committee (IGBC), 56
Interagency Grizzly Bear Study Team (IGBST), 55
Interference competition, 152–53; and mesocarnivores, 172–75, **173**; size-structuring effects of, **174**, 176–77
International Biological Program (IBP), 331
Interspecies interaction, 373–74
Invertebrates, 240–41

Island of Rhum, 215
Island populations, 6, 71; equilibrium theory of, 297; and genetic variation, 293, 313; and the GYE, 287; and wolf restoration in GYE, 108
Isle Royale National Park, 105–6, 152, 265, 275, 276, **276**, 279, 338, 348–49

Jackal species, 153
Jackrabbits, black-tailed *(Lepus californicus)*, 129
Jackrabbit, whitetail *(Lepus townsendii)*, 243, 244
Jackson Hole, Wyoming, 128, 129, 175
Johnson, A., 14
Johnson, W. E., 150, 151, 156
Jones, J. K., 165
Jordan, P. A., 338

Kay, C. E., 14, 16, 17, 21, 195, 205
Keddy, P. A., 172
Keiter, R. B., 351
Keith, L. B., 331
Kleiman, D. G., 341
Knowlton, Fred F., 134, 139, 140, 141, 142, 143, 145, 330
Knudsen, J. J., 140, 305
Koehler, G. M., 86, 309
Kohn, M., 295
Kolmogorov, A. N., 269

Lacy, R. C., 292, 293
Lagomorphs, 243
Laing, S. P., 81
Lamar Valley cave site, 13, 245, 251
Lande, R., 289
Landscape ecology, 324, 364–69, 369–72; and study scale, 344
Land-use restrictions, 115–16
Lawton, J. H., 346
Learning, 368
Legislation: and conservation, 331; and wolf restoration, 109–10. *See also* Policy
Leopold, Aldo, 109
Leopold Report, 35–36
Levin, S. A., 343
Leyhausen, P., 331
Lidicker, W. Z., 327
Lime, D., 217
Limiting factors, 280, 312–13; coyote, 139–43, 144–45; defined, 196; density-dependent, 196, 197, 201, 228–29, 266–67; and elk, 215, 228–29; habitat as, 196, 287, 312, 313; and mountain lion, 308, 312; and pronghorn, 216; and small mammals, 245; and wolves, 258, 312
Lindzey, F. G., 81, 91, 92
Lions, African *(Panthera leo)*, 90
Litter size: and coyotes, 140, 150
Litvaitis, J. A., 180
Livestock: and cougars, 86, 92; and wolves, 115, 116, 117–18
Lizards, 241
Locomotor adaptations, 171
Logan, K. A., 90, 91, 92
Logging: ecosystem studies and, 107; effect on mesocarnivores of, 179, 181; effects on cougar, 83, 84, 86
Long, C. A., 170
Ludwig, D., 356
Lynx *(Lynx canadensis)*: dispersal, 308; effect of logging on, 181; food sources, 241; genetic diversity and, 308, 315; habitat, 313, 351; as habitat/feeding specialist, 257; as a keystone species, 350–52; population estimates, 165, 168, 170, 309; research on, 168; and snowshoe hare, 259; territory/range, 4, 167, 309

Mack, John A., 219, 259, 275
Macroecology, 328
Maehr, D. S., 81, 82, 84, 88
Magpie, black-billed, 214
Maj, M., 170
Malecki, R. A., 82
Mammals: body-mass distribution in GYE, **174**; history in Yellowstone, 12–18, **19**, **20**, 32–33. *See also* Carnivores; Mammals, small; Ungulates
Mammals as Predators (Dunstone and Gorman), 334
Mammals, small: abundance/habitat in the GYE, 239–40, 242–48, **242**, **247**; communities, 248–50; and coyotes, 129, 148–50; defined, 242; effect on ecosystem of, 252–53; effects of disturbance on, 250–52; history in Yellowstone, 18; as prey, 239, 253, **254–55**; wolf restoration and, 258–59. *See also* Mesocarnivores

Marmot *(Marmota):* and wolves, 258–59
Marmot, yellow-bellied *(Marmota flaviventris),* 244
Marten, American *(Martes americana):* abundance/range, 165, 168, 169, 311; as an indicator species, 349, 353; effect of logging on, 181, 251; effect of wolf restoration on, 180; food sources, 176, 240–41; and genetic diversity, 309, 310; habitat, 171, 172, 175; as habitat specialist/feeding generalist, 257; history in Yellowstone, 13, 22; and red squirrel middens, 245; research on, 168, **361**
Marten, Humboldt, 168
Mattson, D. J., 245
McLaren, B. E., 106, 276, 279, 280
Meagher, M. M., 17, 170, 212
Mech, L. D., 220, 331
Merrill, E. H., 271, 274
Mesocarnivores: abundance in the GYE, 165–66, 168–70; conservation of, 178–82; habitat/feeding behavior, 171–72, 256–58; and interspecific competition/coexistence, 172–78; research on, **166**; role in ecosystems of, 166–68, 193. *See also specific mesocarnivore species*
Mesopredator release, 348
Messier, F., 136, 137, 138, 217, 228, 280
Metzgar, L. H., 297
Microtines: abundance in the GYE, 246; and coyotes, 148–50; effect on ecosystem of, 253
Migration. *See* Dispersal/migration
Mills, L. S., 142, 296, 316
Minimal management, 201
Mining: water pollution and, 178, 181
Mink *(Mustela vison):* abundance of, 165, 168, 169; food sources, 240, 241; history in Yellowstone, 14, 22; and introduction of lake trout, 180; research on, 168; and wolf restoration, 259
Minnesota, 118, 359
Minta, S. C., 175
Mitton, J. B., 310
Moa, P. F., 352
Models. *See* Ecological modeling
Moehlman, P., 136
Moilanen, A., 364
Mollusks, 240

Montana: and cougars, 92; and deer, 85–86; and grizzly bears, 52, 68; lynx and river otter in, 168; and wolf restoration project, 116, 118
Montana Fish, Wildlife, and Parks, 201
Moore, R. E., 252
Moose *(Alces alces):* and Algonquin Park study, 107; antipredatory behavior, 218; and cougars, 91; effect of predation on, 105; effect on vegetation of, 193; history in GYE, 14, 18, 194; and Isle Royale study, 105–6, 265, 275, 276, 279, 338; population estimates, 203–4, 205, 228; wolves and, 112, 221, 225; and Yellowstone 1988 fires, 210, 211–12
Mortality: causes of coyote, 112, 134, 139, 141, 143–44; cougars and, 87–90, **88**, 97; human-caused, 342, 367 *(see also Hunting; Poison; Trapping);* interspecific carnivore kills, 152; and lynx, 352; and ungulates, 197–98, **199**, 214–15; wolf, 217; and wolf restoration, 112; and Yellowstone 1988 fires, 209–10
Moths, cutworm, 62, 70
Mountain goat *(Oreamnos americanus):* history in Yellowstone, 18; population estimates, 205
Mountain lion *(Felis concolor):* behavior/reproduction, 78–79, **80**; dispersal, 79–80, 308; effective population size, 297; genetic diversity/fitness, 294, 307–8; habitat, 78–79, 80–82, 171; as habitat/feeding generalist, 256; history in Yellowstone, 18, 22, 23, 26, 147; human impact on, 82–93, 342; management and, 94–97; population estimates, 77–78, 308; population limiting factors, 308, 312; predation/food sources, 81–82, 148, 192, 193, 215, 216; researching, 93–94; role in ecosystem, 77; territory/range, 78–79, 300, 308; and wolf restoration, 225
Mouse, deer *(Peromyscus maniculatus):* abundance in the GYE, 246, 249; effect on ecosystem of, 253; habitat disturbance and, 251, 252
Mouse, western jumping *(Zapus princeps),* 248, 249, 250
Murie, Adolph, 17, 26, 32–33, 127, 148, 150, 169, 170
Murie, Olaus, 33

Murphy, K. A., 308
Murphy, K. M., 87, 91
Muskrat *(Ondatra zibethicus)*, 248
Mustelids *(Mustelidae)*: body sizes/shapes
 and, 176; and genetic diversity, 309–12
Mutation, 291

National Elk Refuge, 214
National Environmental Policy Act
 (NEPA), 4, 37, 331
National Park Service (NPS): wildlife
 management policies, 11–12, 27–36;
 and wolf restoration to YNP, 111, 116
National Park Service Act, 27
Native Americans: carnivores,
 symbolism, and, 3; and ecology of
 Yellowstone, 16–17, 18; Euro-American
 influences on, 15–16; and wolves, 108
Natural process management, 201
Natural regulation, 36, 229, 267; effect on
 elk of, 201; as management policy, 195,
 196–97
Natural resource economics, 42
Natural selection, 291–92, 314
Negus, N. C., 169, 246
Newmark, W. D., 180
New York Zoological Society, 28
Nez Perce tribe, 116
Niemi, G. J., 346
Nonmammalian prey, 240–41, 260–61
Nordyke, K. A., 248
Norris, Philetus, 22–23
Northern Continental Divide Ecosystem
 (NCDE), 300
Northern Range Small Mammal Study,
 239, 240, 248–49, 252, 260

O'Donoghue, M., 351
O'Gara, B. W., 202
Optimal foraging theory, 134
Osprey, 37
Otter, river *(Lutra canadensis)*: effect of
 logging on, 181; food sources, 241;
 genetic diversity, 309, 315; habitat,
 311–12, 313; as habitat/feeding
 specialist, 257–58; history in
 Yellowstone, 14, 22; and introduction
 of lake trout, 180; physiological
 adaptations, 172; population estimates,
 165, 168, 170, 179, 312; research on,
 168

Otter, sea, 349
Owen-Smith, N., 357

Paetkau, D., 293, 295, 305–6
Paleoecology, 328
Paleohistory, 356–59
Paleontology, 13, 328
Palomares, F., 180, 348
Panther, Florida, 83, 84
Paquet, P., 154, 155, 224, 297
Pastor, J., 333
Pattie, D. L., 248
Peek, J. M., 219
Peterson, Rolf O., 106, 258, 275, 276, 279,
 280
Peters, R. H., 167
Picton, H., 301
Pika *(Ochotona princeps)*, 243
Pimlott, Douglas H., 106, 220
Pine, whitebark, 60–61, 69–70, 176, 245
Pinter, A. J., 250
Pitcher, John, 25
Poaching, 24
Poison, 22, 25, 26; banning of, 38; and
 grizzly bears, 52; and mesocarnivores,
 169, 178
Policy: grizzly bear, 56–57, 62;
 interdisciplinary approach to
 formulating, 7; and national park
 system, 11–12, 27–36; and population
 modeling, 64–67; and public attitudes,
 324 *(see also* Attitudes); Yellowstone
 predator, before NPS creation, 18, 21–
 27. *See also* Carnivore management
Polis, G. A., 344
Politics: and wolf management, 120. *See
 also* Policy
Polymorphism, 305
Population estimates/densities: black
 bear, 67–68, 307; bobcat, 309; and
 body-mass distribution, **174**; cougar,
 77–78, 89–90, 96–97, 308; coyote,
 139–43, 144–45, 147, 155–56, 304;
 determining, 361–62; effective, 288–
 90, 295–99, 360–61; fishers, 311;
 grizzly bear, 53–54, 55–56, 63–64, 65–
 66, 71–72, 306; GYE carnivore, 286–87,
 286; GYE ungulate, **191**, 200–209, **206**,
 228–29; lynx, 309; marten, 311;
 mesocarnivores, 167, 168–70; red fox,
 304–5; river otter, 312; small mammal,

248–50; weasel, 311; wolf, 120, 217, 304; Yellowstone 1988 fires and ungulate, 211–13. *See also* Limiting factors; Reproduction; Top-down *versus* bottom-up influences; *specific species*
Population genetics, 299–312
Population models, 64–67, **268**; for elk given wolf restoration, 219–24; spatially explicit, 362–64; three-trophic-level, **268**, 269–78, **269, 270, 271, 272, 273, 274, 275, 276, 277, 278**. *See also* Ecological modeling
Population viability analysis (PVAS), 360–61
Porcupine *(Erethizon dorsatum)*, 248
Powell, R. A., 171, 311, 331
Power, M. E., 349
Prairie dog, white-tailed *(Cynomys leucurus)*, 244, 253
Predation: and cougars, 81–82, 85–86, 91–92; and coyotes, 129–34, 148–50, **148, 149**; density dependence and, 182, 228–29; food specialists *versus* generalists, 172, 256–58; and pronghorn, 202, 213, 216; and small mammals, **254–55**; and top-down *versus* bottom-up influences, 106–7, 266–67, 276, 278–81, 345; and ungulates, 192–93, 213–16; wolves and, 104–5, 112, 115, 116, 218–24, **220, 222, 223**; in Yellowstone's Northern Range, **135**. *See also specific predator species*
Predator-prey theory, 326, 332, 336–40
Prey: effect of predation on, 104–5; nonmammalian, 240–41, 260–61. *See also* Predation; *specific prey species*
Prey specialization: coyotes and, 134
Private land: and cougars, 82; and grizzlies, 69; and wolf restoration controversy, 109, 115
Pronghorn antelope *(Antilocapra americana)*: antipredatory behavior, 218; and coyotes, 134; effect of 1988 fires on, 209, 212–13; history in GYE, 18, 194; limiting factors for, 216; population estimates, 202–3, 207, 228; and wolf restoration, 225
Protein electrophoresis, 295
Public education/involvement: and carnivore conservation, 374–75; endangered species and, 37, 39; and

maintaining carnivore genetic diversity, 316; and restoration initiatives, 43, 110, 118–19. *See also* Attitudes
Pup protection hypothesis, 139

Quammen, D., 313

Rabbits/hares *(Leporidae)*, 253. *See also specific species of rabbits/hares*
Raccoon *(Procyon lotor)*: abundance in GYE, 165, 168, 169; food sources, 240; habitat structure, 171; research on, 168
Ramenofsky, A. F., 16
Raphael, M. G., 310
Rarity, 353–55
Ravens, 214
Reeve, A., 170
Regulating factors, 280
Reinhart, D. P., 245
Reintroduction: as restoration strategy, 109
Reproduction: bear breeding behavior, 306; and cougars, 79, **80**, 90; and coyotes, 134, 138, 139–42, 144; environmental factors affecting, 367; fox breeding behavior, 303; grizzly bear rate of, 55–56, 58–59, **59**, 63; and ground squirrels, 244; human impacts on carnivore, 341–42; and wolves, 104
Reptiles, 260–61
Research: behavioral phenotypes and, 365–69; carnivore, 32–33, 325–27, 327–35, 356; on community interactions, 345–53; on cougars, 93–94, 95–96; coyotes and, 127–29, **130–33**; ecosystem, 105–8, 335–43, 372–73, 374; on grizzlies, 51; improved methodological tools, 359–65; and landscape ecology, 369–72; on mesocarnivores, **166**, 168; paleohistory and, 356–59; role in wildlife management of, 29, 33, 34–35; and small mammals, 248, 259–61; spatial and temporal scales, 343–45; and wolf recovery, 41–42, 114–17, 120–21, 281. *See also* Carnivore management; Ecosystem
Resource partitioning, 151–52, 154
Restoration: policy and attitudes toward, 38–42, 43. *See also* Wolf restoration

Restriction-Fragment Length Polymorphism, 305
Rewilding, 374
Riding Mountain National Park, Manitoba, 217, 224
Rieger, J. F., 244
Riley, S. J., 82
Roads: effect on cougars, 87
Rodents *(Rodentia)*, 244–47, 332–33
Romme, W. H., 194, 251–52
Rosenzweig, M. L., 173, 328
Royal Teton Ranch, 203, 213
Ruff, R. L., 307
Ruggiero, L. F., 171, 173, 241
Russell, O., 169
Ruth, T. K., 85

Sable, 22
Sagebrush steppe, 213, 249, 252
Salamanders, 241
Salmon-Selway Ecosystem (sse), 300
Salwasser, H., 297
Sargeant, Al, 330
Satiation, 266–67
Scandinavia, 359
Scat marking, 329
Scavenging. *See* Foraging/Scavenging
Schaller, G. B., 334
Schoener, T. W., 328
Schoenwald-Cox, C. M., 297
Schullery, P., 17, 21, 28, 52, 146, 205, 224
Scotland, 359
Scott, D., 202
Seals, elephant, 293
Seals, Hawaiian monk, 293
Seidensticker, J. C., 308
Serengeti ecosystem, 324, 328–29, 358
Seton, Ernest Thompson, 168, 169, 170
Shaffer, M. L., 66, 296
Shaw, H. G., 91
Sheep, domestic, 61, 92, 115, 134
Sheldon, J. W., 137–38, 354
Sherburne, S. S., 245
Shields, G. F., 295
Shrew *(Soricidae)* family, 13, 243, 249; effects of habitat disturbance on, 251, 252
Shrew, Merriam's *(Sorex merriami)*, 13
Shrew, pygmy *(Sorex hoyi)*, 13
Shrew, water *(Sorex palustris)*, 13

Simberloff, D., 346
Sinclair, A. R. E., 197, 228–29, 324, 334
Singer, F. J., 216, 219, 259
Skinner, Milton P., 17, 29, 53, 147, 149, 170
Skunk species: history in Yellowstone, 13, 14, 22; widespread destruction of, 24–25
Skunk, striped *(Mephitis mephitis)*: abundance, 165, 168, 170; as habitat/feeding generalist, 256
Skunk, western spotted *(Spilogale putorius)*, 240, 241
Slobodkin, L. B., 335
Small-population paradigm, 355
Snakes, 241
Snowtracking, 359, 362
Social structure: coyote, 134–39, 143; human impacts on carnivore, 341, 342–43, 354; and mesocarnivores, 167; wolf, 104. *See also* Behavioral ecology
Soulé, M. E., 36, 165, 295, 296, 297, 334, 348, 374
Spatial/temporal partitioning, 152, 178
Squirrel *(Sciuridae)* family, 244; ground, 150, 253
Squirrel, golden-mantled ground *(Spermophilus lateralis)*, 244
Squirrel, northern flying *(Glaucomys sabrinus)*, 244, 245
Squirrel, pine *(Tamiasciurus spp.)*, 176
Squirrel, red *(Tamiasciurus hudsonicus)*, 244, 245; and lynx, 351
Squirrel, Uinta ground *(Spermophilus armatus)*, 244, 249, 250; effects of habitat disturbance on, 251, 252; as habitat generalist/feeding specialist, 257; as prey, 260
Statistical/simulation tools, 359–60
Stelfox, J., 218
Stenseth, N. C., 351
Stochastic variance, 270, **272, 273,** 276–78, 290, 313
Stoddart, L. C., 139, 142, 143, 145
Storm, G. L., 341
Strobeck, C., 295, 306
Sucker, mountain *(Catastomus platyrhynchus)*, 241
Sunde, P., 352

Tait, D. E. N., 342
Talbot, S. L., 295

Telemetry: and coyote research, 141, 143; and grizzly bears, 55, 62, 70, 72; as a research tool, 128, 329–30, 359, 362

Teller, H. M., 24

Terborgh, J., 180, 346–47, 349

Territorialism, 342

Territory/range: badger, 311; and bison, 207; black bear, 307; bobcat, 309; and carnivores, **300;** cougar, 78–79, 300, 308; coyote, 136, 137, 144, 304; elk, **190,** 200, 228; grizzly bear, 301–2, **301,** 306–7; home range interspersion, 153; lynx, 4, 167, 309; marten, 311; and mesocarnivores, 167; and North American canids, 146, 154; red fox, 303–4, 304–5; wolf, 104, 112, **113,** 217, 304; wolverine, 310

Theberge, John and Mary, 106

Thomas, E. M., 169, 170

Tick outbreaks, 106

Till, J. A., 134

Toads, 241

Top-down *versus* bottom-up influences: ecosystem studies on, 106–7, 345; wolf restoration and, 265–67, 276, 278–81

Translocation: and cougars, 92, 95

Trapping: and accidental cougar deaths, 87–88; and early YNP management policy, 23; and lynx, 352; and mesocarnivores, 167, 178, 179

Trout, cutthroat *(Oncorhynchus clarki bouvieri),* 37; lake trout threat to, 70, 180; as prey, 54–55, 61–62, 241

Trout, lake *(Salvelinus namaycush),* 70, 180, 257

Tyers, D. B., 205

Tzilkowski, W. M., 128, 341

Umbrella species, 346

Ungulates: biomass in YNP, **192,** 213–16; coyotes and, 134, 148; effect of 1988 fire on, 209–16; effect on ecosystem of, 198–200; history in Yellowstone, 13–18, **19,** 26; management of, 194–98; niche/diet overlap, **208;** perceived overpopulation of, 30–31, 35, 38; population estimates, **191,** 200–209, **206,** 228–29; as prey, 192–94; ranges/migration, 194, 198; wolf restoration and, 218–24, **220,** 225–26, **226,** 265–67. *See also specific ungulate species*

U.S. Fish and Wildlife Service, 110, 116

Utah: research in, 129

Vales, D. J., 219

Van Dyke, F. G., 83, 84

Van Valkenburgh, B., 153, 171

Varley, J. D., 142

Vegetation: effect of wolf restoration on, 225, 259, 265–66; in the GYE, 239; small mammal effects on, 253; three-trophic-level model and, 269–78; ungulate effects on, 193, 197, 198–200, 267; and Yellowstone 1988 fires, 210–11, 213. *See also* Forest; Grassland

Verbeek, N. A. M., 248

Villafuerte, R., 180

Viverridae: body sizes/shapes and, 176

Vole family: and coyote, 351

Vole, heather *(Phenacomys intermedius),* 246

Vole, long-tailed *(Microtus longicaudus),* 246, 249, 250

Vole, meadow *(Microtus pennsylvanicus),* 246, 248–49, 250; effects of habitat disturbance on, 252

Vole, montane *(Microtus montanus),* 246, 248–49, 250; effects of habitat disturbance on, 252

Vole, red-backed *(Clethrionomys gapperi),* 246, 248, 249, 250; effect on ecosystem of, 253; effects of habitat disturbance on, 251, 252

Vole, Richardson's/water *(Microtus richardsoni),* 246, 249, 250

Vole, sagebrush *(Lemmiscus curtatus),* 248

Vulnerability, 220, 366

Wagner, F. H., 205

Wardell site, Wyoming, 17

Washington: research in, 142

Water pollution, 178, 181

Wayne, R. K., 153

Weaponry/iron tools, 16

Weasel family, 311

Weasel, long-tailed *(Mustela frenata),* 13; food sources, 240, 241; as habitat/feeding generalist, 256; and wolf restoration, 259

Weasel, short-tailed *(Mustela erminea),* 172

Weaver, J. L., 26, 108, 128, 299

Wells, M. C., 134, 136, 137, 139

Wennergren, U., 364

Westphal, M. I., 355
Whitefish, mountain *(Prosopium williamsoni)*, 241
Whittlesey, L., 17, 21, 146, 205, 224
Wilcox, B. A., 289
Wildlife management: global environmentalism and, 37–43; natural regulation and, 36, 195, 196–97, 201, 229, 267; NPS policies, 11–12, 27–36; population models and, 278–81; role of research in, 29, 33, 34–35; and ungulates, 194–200, 201–2, 229; using ecological studies in, 326–27 *(see also* Ecological modeling); Yellowstone prior to NPS creation, 18, 21–27, **26**. *See also* Carnivore management
Wildlife Services (formerly Animal Damage Control), 116
Williams, J. S., 84, 308
Willow, 207, 225, 259
Wilson Ornithological Club, 28
Winter, B., 180
Winter kill: and ungulates, 214; wolf restoration, compensation, and, 197–98; and Yellowstone 1988 fires, 209, 210. *See also* Climate
Wolda, H., 229
Wolf, gray *(Canis lupus):* and canid interspecific competition/coexistence, 145–47, 154–55, 156; and cougar competition, 86–87; dispersal, 303–4; effective population size, 297; genetic diversity and, 302–3, 315; as habitat/feeding generalist, 256; habitat structure of, 171–72; history in Yellowstone, 13, 18; and Isle Royale study, 338; as a keystone species, 350; management, 117–21; population estimates/density, 120, 304; population limiting factors, 258, 312; predation/food sources, 150, 258–59; role in ecosystem of, 104–8; social structure/reproduction, 104; symbolism and, 3; territory/range, 104, 112, **113**, 217, 304; and ungulates, 192, 193, 197, 215, 218–24, **220**, 267; widespread destruction of, 22, 23, 26, 27–28, 108, 151, 304. *See also* Wolf restoration
Wolf Management Committee, 109

Wolf restoration, 108–14; compensatory mechanisms and, 197–98; controversy surrounding, 103–4, 374; effect on ecosystem of, 190–92, 227–28, **227**, 265–66, 279; effect on mesocarnivores of, 180; effect on small mammals of, 258–59; effect on ungulates of, 216–26, **220**, **226**, 229, 267; and genetic diversity, 304; and GYE as an ecosystem study area, 107–8, 156; and human attitudes/interactions, 39, 41–42, 373; predictions and results, 114–17, **115**; three-trophic-level model for, 269–78
Wolverine *(Gulo gulo):* dispersal, 310; food sources, 176, 193, 241; genetic diversity/fitness, 294, 315; habitat, 312–13; history in Yellowstone, 14, 18, 22; population estimates/densities, 165, 167, 168, 169–70, 179, 310–11; range, 4, 310; research on, 168; and wolf restoration, 225
Wolves for Yellowstone?, 41
Wood, M. A., 252
Woodrat, bushy-tailed *(Neotoma cinerea),* 246, 251
Wright, G., 14
Wright, George, 30, 33
Wright, R. G., 28
Wright, William, 53
Wyoming: and grizzly bears, 52, 68; lynx and river otter in, 168; and wolf restoration project, 116, 118

Yellowstone Ecosystem Studies (YES), 239
Yellowstone Lake: and mesocarnivores, 180
Yellowstone National Park (YNP): and canids, 146–47, 154–55; coyotes and, 127, 142, 147–50; ecology before 1872, 12–18; as ecosystem, 226–29, **227**; global environmentalism and wildlife management policy, 37–43; and mesocarnivores, 179; natural regulation as management policy in, 196; and the Northern Range Small Mammal Study, 240; northern winter range, **135**, 194; and NPS predator management policy, 11–12, 27–36; predator policy prior to NPS creation,

18–27, **26**; wolf restoration project, 103, 109–14, 117 (*see also* Wolf restoration). *See also* Greater Yellowstone Ecosystem
Yellowstone Park Act, 18
Youmans, C. C., 245

Young, B. F., 307
Young, S. B. M., 25

Zapodidae family, 248
Zimmerman, E. G., 305